Birkhäuser

Frontiers in Mathematics

Advisory Editorial Board

Leonid Bunimovich (Georgia Institute of Technology, Atlanta)
Benoît Perthame (Université Pierre et Marie Curie, Paris)
Laurent Saloff-Coste (Cornell University, Ithaca)
Igor Shparlinski (Macquarie University, New South Wales)
Wolfgang Sprössig (TU Bergakademie Freiberg)
Cédric Villani (Institut Henri Poincaré, Paris)

Yuan-Jen Chiang

Developments of Harmonic Maps, Wave Maps and Yang-Mills Fields into Biharmonic Maps, Biwave Maps and Bi-Yang-Mills Fields

Birkhäuser

Yuan-Jen Chiang
Department of Mathematics
University of Mary Washington
Fredericksburg, VA
USA

ISSN 1660-8046 ISSN 1660-8054 (electronic)
ISBN 978-3-0348-0533-9 ISBN 978-3-0348-0534-6 (eBook)
DOI 10.1007/978-3-0348-0534-6
Springer Basel Heidelberg New York Dordrecht London

Library of Congress Control Number: 2013940741

Mathematics Subject Classification (2010): 58E20, 58E15, 58E12, 81T13, 53A10, 53C07, 53C12, 53C43, 49Q05, 35J47, 35J48, 35K05, 35L70, 35J10, 32Q15

Printed on acid-free paper

Springer Basel is part of Springer Science+Business Media (www.birkhauser-science.com)

In Memory of Professor James Eells and Professor Joseph H. Sampson

The names of these two pioneers of the theory of harmonic maps will be engraved in the minds of all mathematicians who work on harmonic maps, wave maps, and Yang-Mills fields, for their great and everlasting contributions.

Introduction

We present an overview of the developments of harmonic maps, wave maps, and Yang-Mills fields into biharmonic maps, biwave maps, and bi-Yang-Mills fields. The theory of harmonic maps between Riemannian manifolds was first established by Eells and Sampson (Chiang's Ph.D. adviser) [129] in 1964. Wave maps are harmonic maps on Minkowski spaces and were studied in the early 1990s. In the last two decades, there were many new developments in wave maps achieved by a number of mathematicians. Yang-Mills fields are the critical points of the Yang-Mills functionals of connections whose curvature tensors are harmonic. They were first explored by a number of physicists in the 1950s, and since then there were many new developments in this subject. Biharmonic maps, which generalize harmonic maps, were first studied by Jiang [196–198] in 1986. In recent years, there has been progress in biharmonic maps, accomplished by quite a few mathematicians. Biwave maps are biharmonic maps on Minkowski spaces which generalize wave maps, and they were first studied by Chiang [75, 76] in 2009 and with Wolak [84] later. Bi-Yang-Mills fields, which generalize Yang-Mills fields, were first investigated by Ichiyama, Inoguchi, and Urakawa [191, 192] in 2009. Moreover, exponentially harmonic maps were first introduced by Eells and Lemaire [125] in 1990. Exponential wave maps are exponentially harmonic maps on Minkowski spaces, which were first studied by Chiang and Yang [88] in 2007. Exponential Yang-Mills connections were first explored by Matsuura and Urakawa [260] in 1995. Since this book covers broad topics intervening harmonic maps, wave maps, Yang-Mills fields, biharmonic maps, biwave maps, bi-Yang-Mills fields, exponentially harmonic maps, exponential wave maps, and exponential Yang-Mills connections, it is impossible to describe details completely and extensively. However, we try to present the most crucial ingredients of the recent developments of the topics.

Harmonic maps were first introduced by Sampson in the hope of obtaining a homotopy version of the highly successful Hodge theory of cohomology in 1952. Not long after that his then colleague, John Nash (one of the three Nobel laureates in Economics in 1994) proposed a quite different but equivalent definition – both of them were Moore Instructors at MIT at the time. Fuller [150] also came upon harmonic maps in 1954. The definition, whether in terms of the energy functional or the Euler-Lagrange equations, seems very natural to us today, but it was not so obvious half a century ago.

Eells and Sampson [129] collaborated on the first paper on harmonic maps of Riemannian manifolds at the Institute for Advanced Study at Princeton in 1964. This paper is usually considered as the pioneering work in harmonic maps. They also published a second and third joint paper [130, 131] afterwards.

With an eye toward the physical concept of kinetic energy ($\frac{mv^2}{2}$), a harmonic map $f : (M^m, g_{ij}) \to (N^n, h_{\alpha\beta})$ from an m-dimensional Riemannian manifold into an n-dimensional Riemannian manifold is defined as a critical point of the energy functional

$$E(f) = \frac{1}{2}\int_M |df|^2 dv = \frac{1}{2}\int_M h_{\alpha\beta} f_i^\alpha f_j^\beta g^{ij} dv, \tag{1}$$

where dv is the volume form of M determined by the metric g. In order to derive the associated Euler-Lagrange equations, we consider a one-parameter family of maps $\{f_t\} \in C^\infty(M \times [0, 1], N)$ from a compact manifold M (without boundary) into a Riemannian manifold N such that f_t is the endpoint of a segment starting at $f(x)(= f_0(x))$ determined in length and direction by the vector field $\dot{f}(x)$. If M is a non-closed manifold, we assume that $\dot{f}(x)$ has compact support, which is contained in the interior of M. We now compute the first variation of the energy functional:

$$\begin{aligned}\dot{E}(f) = \frac{d}{dt}E(f_t)\Big|_{t=0} &= \int_M (df_t, D_t df_t)\Big|_{t=0} dv = \int_M (df, D\dot{f})\, dv \\ &= \int_M div(w)\, dv - \int_M (\tau f, \dot{f})\, dv = -\int_M (\tau f, \dot{f})\, dv, \quad \forall \dot{f},\end{aligned} \tag{2}$$

by the divergence theorem, where $\tau^\alpha(f) = trace_g(Ddf)$, D is the connection on $T^*M \otimes f^{-1}TN$ induced by the Levi-Civita connections on M and N, $div(w) = w^j_{|j}$, and $w^j = h_{\alpha\beta} f_i^\alpha \dot{f}^\beta g^{ij}$ is a vector field on M. The map $f : M \to N$ is *harmonic* if the tension field

$$\begin{aligned}\tau^\alpha(f) = trace_g(Ddf) &= g^{ij} f^\alpha_{i|j} = g^{ij}(f^\alpha_{i,j} + \Gamma'^\alpha_{\beta\gamma} f_i^\beta f_j^\gamma) \\ &= g^{ij}(f^\alpha_{ij} - \Gamma^k_{ij} f^\alpha_k + \Gamma'^\alpha_{\beta\gamma} f_i^\beta f_j^\gamma)\end{aligned} \tag{3}$$

vanishes identically, where $f_i^\alpha = \frac{\partial f^\alpha}{\partial x_i}$, $f_{ij}^\alpha = \frac{\partial^2 f^\alpha}{\partial x_i \partial x_j}$, $f^\alpha_{i,j} = f^\alpha_{ij} - \Gamma^k_{ij} f^\alpha_k$, and Γ^k_{ij} and $\Gamma'^\alpha_{\beta\gamma}$ are the Christoffel symbols of the Levi-Civita connections on M and N, respectively.

Assume that $f(x) = f_0(x)$ is harmonic and that $\xi = \frac{\partial f_t}{\partial t}$. We next compute the second variation of the energy from (2):

$$\ddot{E}(f) = \frac{d^2}{dt^2}E(f_t)\Big|_{t=0} = -\int_M \frac{d}{dt}(\tau f_t, \xi)\,dv\Big|_{t=0}$$

$$= -\int_M [(D_t\tau f_t, \xi) + (\tau f_t, D_t\xi)]\Big|_{t=0} dv. \tag{4}$$

At $t = 0$, $f(x)$ is harmonic and the above second variation becomes

$$\ddot{E}(f) = -\int_M (D_t\tau f, \xi)\Big|_{t=0} dv.$$

The components of $D_t\tau f$ are $f^\alpha_{i|j|t} = \frac{\partial f^\alpha_{i|j}}{\partial t} + \Gamma'^\alpha_{\mu\nu} f^\mu_{i|j}\xi^\nu$. By Eisenhart [137], $f^\alpha_{i|j|t} - f^\alpha_{i|t|j} = -R^k_{ijt} f^\alpha_k + R'^\alpha_{\beta\gamma\mu} f^\beta_i f^\gamma_j \xi^\mu$ and using the curvature formula on $M \times [0, 1] \to N$, the first curvature term vanishes and $f^\alpha_{i|j|t} = f^\alpha_{i|t|j} + R'^\alpha_{\beta\gamma\mu} f^\beta_i f^\gamma_j \xi^\mu$, where R' is the Riemannian curvature of N. But $f^\alpha_{i|t} = f^\alpha_{t|i} = \xi^\alpha_{|i}$, and so $f^\alpha_{i|j|t} = \xi^\alpha_{|i|j} + R'^\alpha_{\beta\gamma\mu} f^\beta_i f^\gamma_j \xi^\mu$. Therefore, $D_t\tau f$ has components

$$g^{ij}\xi^\alpha_{|i|j} + g^{ij}R'^\alpha_{\beta\gamma\mu} f^\beta_i f^\gamma_j \xi^\mu. \tag{5}$$

Denote the first term by $(\Delta\xi)^\alpha$. Thus we arrive at

$$\ddot{E}(f) = -\int_M [(\Delta\xi, \xi) + g^{ij}R'_{\alpha\beta\gamma\mu}\xi^\alpha f^\beta_i f^\gamma_j \xi^\mu]\,dv. \tag{6}$$

Using the integration by parts gives $d(D\xi, \xi) = (\Delta\xi, \xi) + (D\xi, D\xi)$, and then the divergence theorem recasts (6) into

$$\ddot{E}(f) = \int_M \left[|D\xi|^2 - g^{ij}R'_{\alpha\beta\gamma\mu}\xi^\alpha f^\beta_i f^\gamma_j \xi^\mu\right] dv. \tag{7}$$

(Recall that ξ is a section of $f^{-1}(TN)$, i.e., a vector field along f. For given ξ, we obtain a suitable variation of f by setting $f_t(x) = exp_{f(a)}t\xi(x)$, $a \in M$.) If N has negative sectional curvature, i.e., $R'_{\alpha\beta\gamma\mu}\lambda^\alpha\eta^\beta\lambda^\gamma\eta^\mu \leq 0$ for arbitrary vector fields λ and η, then it follows from (7) that $\ddot{E}(f) \geq 0$, so every harmonic map is a local minimum for the energy E.

Observe that if all the f_t are harmonic for t near 0, then $\tau f_t = 0$, and also $D_t\tau f_t = 0$. By (4)–(6), this is the Jacobi equation (at $t = 0$), namely,

$$J_f(\xi) = \Delta\xi + R'(df, df)\xi = g^{ij}\xi^\alpha_{|i|j} + g^{ij}R'^\alpha_{\beta\gamma\mu} f^\beta_i f^\gamma_j \xi^\mu = 0. \tag{8}$$

It is a linear equation for ξ. Solutions of (8) are called *Jacobi fields* along f.

Part of the main results of Eells and Sampson [129] are described as follows:

Theorem 1. *Suppose that the Ricci curvature of M is nonnegative and that the Riemannian curvature of N is nonpositive. Then a map $f : M \to N$ is harmonic if and only if it is totally geodesic. Moreover,*

(1) *If there is at least one point of M at which its Ricci curvature is positive, then every harmonic map $f : M \to N$ is constant.*
(2) *If the Riemannian curvature of N is everywhere negative, then every harmonic map $f : M \to N$ is either constant or maps onto a closed geodesic of N.*

They used the heat flow method to obtain the following theorem:

Theorem 2. *Let N be a Riemannian manifold of nonpositive Riemannian curvature and let $f_t : M \to N$ be a bounded solution of $\tau(f_t) = \frac{\partial f}{\partial t}$, $0 < t < \infty$. Then there is a sequence $t_1, t_2, t_3, \cdots$ of t-values such that the maps f_{t_i} converge uniformly, along with their first-order space derivatives, to a harmonic map f.*

Corollary 3. *Let N have non-positive Riemannian curvature and let $f : M \to N$ be a continuously differentiable map. Let f_t be the solution of $\tau(f_t) = \frac{\partial f}{\partial t}$ which reduces to f at $t = 0$. If f_t is bounded as $t \to \infty$, then f is homotopic to a harmonic map f' for which $E(f') \le E(f)$. In particular, if N is compact, then every continuous map $M \to N$ is homotopic to a harmonic map.*

Hartman [173] showed that if $f : M \to N$ is harmonic with M compact and $Riem^N \le 0$ at every point of $f(M)$ such that there is a point of $f(M)$ at which $Riem^N < 0$, then f is unique in its homotopy class (unless $f(M)$ is a closed geodesic γ of N, and in that case we have uniqueness up to rotations of γ). The above results of Eells and Sampson were extended to harmonic maps with boundary by Hamilton [170]. The curvature condition $Riem^N \le 0$ was generalized by Sacks and Uhlenbeck [310, 311] as follows. If $m = 2$ and $\pi_2(N) = 0$, then given a map $f_0 : M \to N$ there is a harmonic map f homotopic to f_0. The condition $Riem^N \le 0$ also can be replaced by the condition that the image of f_0 (and hence of f) supports a uniformly strictly convex function (cf. Jost [201, 202, 209] and von Wahl [396]). Furthermore, Schoen and Uhlenbeck [320, 321] showed a partial regularity theorem which asserts that a bounded energy minimizing map $f : M \to N$ between two Riemannian manifolds is regular (in the interior) except for a closed singular set S of Hausdorff dimension at most $m - 3$. In the mean time, Giaquinta and Giusti [153, 154] also proved a partial regularity theorem in the case where $f(M)$ is contained in a single chart of N. This theorem can be checked in some interesting cases, for instance, if N has nonpositive curvature or if the image of the map lies in a convex ball of N, then $S = \emptyset$ and any minimizing harmonic maps into such manifolds are smooth, which is exactly the Eells-Sampson's [129] case or the Hildebrandt-Kaul-Widman's [183, 184] case.

In the case when the curvatures of the target manifolds are positive, Eells and Wood [134–136], Wood [414–416], Chern and Wolfson [67–69, 71, 72], Wolfson [410–412, 414, 415, 418], Burstall and Wood [52], Burstall and Salamon [51], etc., constructed harmonic maps from Riemann surfaces into projective spaces or complex Grassmannians in the 1970s and 1980s.

Eells and Lemaire [119–125] collaborated on quite a few papers, including two well-known reports and a survey paper describing the developments of harmonic maps up to 1988. In the 1980s, Siu [342–346] and Sampson [313–316] made breakthroughs on harmonic maps of Kähler manifolds. Siu proved the well-known strong rigidity theorem and obtained related results. In 1989, Uhlenbeck [387] explored harmonic maps into Lie groups and obtained elegant results.

In the 1990s, Eells and Ratto collaborated on a paper [127] and a book [128] about harmonic maps and minimal immersions. Chiang [73, 74] studied harmonic maps of V-manifolds in 1990, and Chiang and Ratto [79] investigated harmonic maps on spaces with conical singularities in 1992. Afterwards, three books on harmonic maps, loop groups, integrable systems, conservation laws, and moving frames were published by Guest [166] in 1997 and by Helein [178, 179] in 2001 and 2002. Moreover, Eells' final monograph, *Harmonic Maps between Riemannian Polyhedra* [118], coauthored with the Danish mathematician Fuglede, was published in 2001. Later, Konderak and Wolak [230,231] studied transversally harmonic maps between manifolds with Riemannian foliations in 2003 and in 2008.

Harmonic morphisms were independently studied by Fuglede [148, 149] in 1978 and Ishihara [194] in 1979. Harmonic morphisms between Riemannian manifolds are harmonic maps which are horizontally (weakly) conformal. In the last three decades, there were many new developments in harmonic morphisms. For more details, we refer the reader to the book *Harmonic Morphisms between Riemannian Manifolds* by Baird and Wood [23], published in 2003.

The f-harmonic maps, which generalize the harmonic maps, were first introduced by Lichnerowicz [252] in 1970. They were recently studied by Course [94, 95], Ouakkas, Nasri, and Djaa [292], Chiang and Wolak [85, 86], and others. Moreover, F-harmonic maps between Riemannian manifolds were first introduced by Ara [8, 9] in 1999; they can be considered as special cases of f-harmonic maps.

Wave maps are harmonic maps on Minkowski spaces. In the 1990s, Klainerman and Machedon [222–224] and Klainerman and Selberg [227] investigated the general Cauchy problem in any dimension greater or equal than two for regular data and obtained the almost optimal local well-posedness. In the difficult case of dimension 2, Christodoulou and Tahvildar-Zadeh [90] studied the regularity of spherically symmetric wave maps by imposing a convexity condition for the target manifold. Shatah and Tahvildar-Zadeh [334, 335] also studied the optimal regularity of equivariant wave maps into two-dimensional rotationally symmetric and geodesically convex Riemannian manifolds. The study of the general wave maps problem incorporated methods that exploited the null-form structure of the wave map system, as in work of Grillakis [160], as well as the geometric structure of the equations as done by Struwe [352, 353]. Keel and Tao [216] studied the one-(spatial) dimensional case.

Tataru [365, 366], following Tao [360, 361], has used new techniques which allow one to treat the Cauchy problem with critical data. Their methods rely on harmonic analysis, such as adapted frequency and gauge theoretic geometric techniques.

Tao [360, 361] established the global regularity for wave maps from $\mathbf{R} \times \mathbf{R}^m$ into the sphere S^n for low and high dimensions m. Similar results were obtained by Klainerman and Rodnianski [226] for target manifolds that admit a bounded parallelizable structure.

Nahmod, Stefanov, and Uhlenbeck [275] studied the Cauchy problem for wave maps from $\mathbf{R} \times \mathbf{R}^m$ into a (compact) Lie group (or Riemannian symmetric spaces) when $m \geq 4$ and established global existence and uniqueness, provided the Cauchy initial data are small in the critical norm. Shatah and Struwe [332, 333] obtained similar results simultaneously, also in the case when the target is any complete Riemannian manifold with bounded curvature.

Recently, Kenig, Merle, and Duyckaerts [109, 218, 219] have investigated global well-posedness, scattering, and finite time blowup. Kenig and Merle have developed a method called the *concentration compactness/rigidity theorem* method. The ideas they used here are natural extensions of many authors to study critical nonlinear elliptic problems (e.g., Yamabe problems and harmonic maps). Moreover, Chiang and Wolak have studied transversal wave maps in [87]. For more detailed developments of wave maps, refer to Shatah and Struwe [332] and Tataru [366]. Nonlinear dispersive equations in general and nonlinear wave equations in particular are important objects of study in the current research in partial differential equations.

Let M be a compact Riemannian manifold and P be a principal G-bundle over M, where G is a compact Lie group. On the space C_P of connections on G, we consider the Yang-Mills fields functional $\mathcal{YM}(D) = \frac{1}{2}\int_M ||R^D||^2 dv_M$, where R^D is the curvature of the connection D in C_P. The critical points of the smooth function $\mathcal{YM} : C_P \to \mathbf{R}$ are those connections whose curvature tensors are *harmonic*. These critical points are called *Yang-Mills connections*, and their associated curvature tensors are called *Yang-Mills fields*.

Yang-Mills theory had a profound impact on the developments of differential and algebraic geometry in the last five decades. The main lines of work were as follows: (1) in the calculus of variations associated to the Yang-Mills functional, the emphasis was on differential geometric aspects, as in the well-known results of Bourguignon and Lawson [38, 39], and on analytic aspects, as in the famous results of Uhlenbeck [382, 383], and (2) algebraic-geometric aspects, involving Ward's description of the Yang-Mills instantons in terms of holomorphic bundles over the Penrose twistor space, leading to the description of solutions via the ADHM construction [15]. For an overview of Yang-Mills theory, see the works by Donaldson and Kronheimer [102] and Donaldson [101, 103].

In the last two decades, on one hand, Taubes [370] introduced deep techniques to attack questions in the calculus of variations. On the other hand, he [369] took the critical step of studying Yang-Mills instantons over general four-dimensional Riemannian manifolds [368] (which was different than the previous work focused on special classes of Riemannian manifolds, such as symmetric spaces or "self-dual" manifolds [16]). In both cases, one can have small, highly concentrated "bubble-like" instantons related to the conformal invariance of Yang-Mills theory in four dimensions.

In this book, we concentrate on the developments of the differential geometric and analytic aspects of Yang-Mills fields by Bourguignon and Lawson [38, 39], Uhlenbeck [382, 383], Price [304], Nakajima [276], Tian [374], and others. We discuss the interaction between Yang-Mills connections, which are critical points of Yang-Mills functionals associated to a vector bundle, and minimal submanifolds, which have been investigated for many years. We also make a brief overview of Taubes' work [368–373].

Biharmonic maps were first introduced by Jiang [196–198] in 1986. In the decades that followed, there have been growing interest and progress in biharmonic maps by Balmuç, Caddeo, Montaldo, Loubeau, Oniciuc, and Piu [24–26, 54–57]; Ou, Lu, Tang, and Wang [254, 256, 284, 285, 289, 290, 400]; Chiang, Sun, and Wolak [80–83]; Fetcu [143, 144]; and others. The regularity of biharmonic maps was studied by Chang, Wang, and Yang [63], C. Wang [398, 399], and others, who generalized the regularity of harmonic maps established by Schoen and Uhlenbeck [320, 321], Chang, Wang, and Yang [62], and others.

In 2011, Nakauchi and Urakawa [277] investigated the removable singularities, bubbling, and integrable systems of biharmonic maps, extending results on the removable singularities, bubbling, and integrable systems of harmonic maps by Sacks and Uhlenbeck [310, 311], Helein [178, 179], and others. Moreover, f-biharmonic maps between Riemannian manifolds were first explored by Ouakkas, Nasri, and Djaa [292] in 2010, generalizing the biharmonic maps of Jiang [196, 197]. Chiang and Wolak [77, 85] also studied f-biharmonic maps recently.

Bi-Yang-Mills fields were introduced by Ichiyama, Inoguchi, and Urakawa [191, 192] in 2009. They investigated the relationships between Yang-Mills fields and bi-Yang-Mills fields and the isolation phenomena of bi-Yang-Mills fields. The following relationship between bi-Yang-Mills fields and biwave equations motivates one to study biwave maps.

Let P be a principal fiber bundle over a manifold M with structure group G and canonical projection π and $\mathcal{G}$ be the Lie algebra of G. A connection A can be considered locally as a $\mathcal{G}$-valued 1-form $A = A_\mu(x)dx^\mu$. The curvature of the connection A is given by the 2-form $F = F_{\mu\nu}dx^\mu dx^\nu$ with $F_{\mu\nu} = \partial_\mu A_\nu - \partial_\nu A_\mu + [A_\mu, A_\nu]$. The bi-Yang-Mills Lagrangian is defined by

$$L_2(A) = \frac{1}{2}\int_M ||\delta F||^2 dv_M, \tag{9}$$

where δ is the adjoint of the operator of exterior differentiation d on the space of E-valued smooth forms on M ($E = End(P)$, the endomorphisms of P). Then the Euler-Lagrange equation describing the critical points of (9) takes the form

$$(\delta d + F)\delta F = 0, \tag{10}$$

which is the bi-Yang-Mills system. In particular, letting $M = \mathbf{R} \times \mathbf{R}^2$ and $G = SO(2)$, the group of orthogonal transformations on $\mathbf{R}^2$, we have that $A_\mu(x)$ is a 2×2

skew symmetric matrix A^{ij}_{μ}. The appropriate equivariant ansatz is $A^{ij}_{\mu}(x) = (\delta^i_{\mu} x^j - \delta^j_{\mu} x^i) h(t, |x|)$, where $h : M \to \mathbf{R}$ is a spatially radial function. Setting $u = r^2 h$ and $r = |x|$, the bi-Yang-Mills system (10) becomes the following equation for $u(t, r)$:

$$u_{tttt} - u_{rrrr} - \frac{3}{r} u_{rrr} + \frac{2}{r^2} u_{rr} - \frac{2}{r^3} u_r = k(t, r),$$

which is a linear nonhomogeneous biwave equation, where $k(t, r)$ is a function of t and r.

Biwave maps are biharmonic maps on Minkowski spaces, and their equations are a fourth-order hyperbolic system of PDEs, which generalizes the system for wave maps. Chiang [75, 76] has recently made a first attempt to study biwave maps and their relationships with wave maps. There are interesting and difficult problems involving local and global well-posedness, and global regularity of biwave maps into Riemannian manifolds or Lie groups (or symmetric spaces), for future exploration.

Exponentially harmonic maps were first introduced by Eells and Lemaire [125] in 1990. Hong and Yang [188] also studied exponentially harmonic maps in 1993. The regularity of exponentially harmonic functions was explored by Eells and Duc [106] in 1991. Exponential wave maps are exponentially harmonic maps on Minkowski spaces, which were first investigated by Chiang and Yang [88] in 2007. It is also interesting to study the regularity of exponentially harmonic maps and well-posedness for the equations of exponential wave maps. Furthermore, exponential Yang-Mills connections were first explored by Matsuura and Urakawa [260] in 1995. They studied the relationships between Yang-Mills connections and exponential Yang-Mills connections.

This book contains nine chapters. In the first chapter, we present an overview of the developments of harmonic maps, concentrating on crucial topics including regularity, maps of surfaces, maps into Kähler manifolds, loop groups and integrable systems, harmonic morphisms, and maps of singular spaces. In the second chapter, we discuss the recent developments in the theory of wave maps such as local and global well-posedness, global regularity, equivariant wave maps, stability, and singularities. In the third chapter, we present the developments on Yang-Mills fields focusing on the aspects of differential geometry and analysis and make a brief overview of Taubes' work. In the fourth, fifth, and sixth chapters, we present the recent developments on biharmonic maps, biwave maps, and bi-Yang-Mills fields. In the seventh, eighth, and ninth chapters, we describe exponentially harmonic maps, exponential wave maps, and exponential Yang-Mills connections.

Remarks. Professor James Eells was born in Cleveland, Ohio, in 1926, and passed away in Cambridge, England, in February 2007. He received his Ph.D. at Harvard University under the topologist and analyst Hassler Whitney in 1954. He started working at the Institute for Advanced Study at Princeton, went on to the University of California at Berkeley, and then returned to the East Coast for a position at Columbia University. He also taught at Churchill College, Cambridge, and Cornell University. Later, he became excited by the freedom and potential of the University of Warwick, joined the mathematics department there, and became a professor of analysis and differential geometry in 1969. He organized the highly successful, yearlong Warwick symposia "Global Analysis" in 1971–1972 and "Geometry of the Laplace Operator" in 1976–1977. He was the first director of the International Centre for Theoretical Physics at Trieste from 1986 to 1992. He was a prominent professor and inventive mathematician, and his mathematical influence in the field of harmonic maps was widespread internationally. He advised 38 graduate students, most of whom became academics, among them are Luc Lemaire, Domingo Toledo, John C. Wood, R. T. Smith, F. E. Burstall, James F. Glazebrook, and Andrea Ratto. There are many more throughout the world who owe their careers to him. He retired in 1992, moved to Cambridge, and continued to work on harmonic maps until he passed away. The joint paper of Chiang and Ratto [79], published in the *Bulletin of French Mathematical Society* in 1992, was dedicated to Professors J. Eells and J. H. Sampson. On July 20, 1992, I received a letter from Eells: *Dear Dr. Chiang, I have just received my copy of Bull. SMF 120 and was most surprised and pleased to read a dedication by you and our excellent friend Andrew. What a nice idea! Thank you very much, and all my best. Yours cordially, James Eells.*

Professor Joseph H. Sampson was born in Philadelphia, USA, in 1926, and passed away in the south of France in August 2003. He received his Ph.D. from Princeton University under Professor S. Bochner in 1951 (we often use Bochner's techniques in harmonic maps). He worked as a Moore Instructor at MIT afterwards. He was appointed as Visiting Assistant Professor at the Johns Hopkins University in 1955–1958, then as Assistant Professor in 1958–1963, then promoted to Associate Professor in 1963–1965, and finally promoted to Full Professor in 1965. He was an editor of the *American Journal of Mathematics* from 1978 to 1992 and the chair of the Mathematics Department at Johns Hopkins from 1969 to 1979. He retired from Johns Hopkins in 1990. He was my adviser at John Hopkins where I received my Ph.D. in 1989; the major part of my dissertation was published in [73, 74]. I admired his penetrating insight and impeccable taste which characterized the precious guidance he provided over those years. Eells and Sampson were good friends for over 40 years, and Sampson invited Eells to speak on harmonic maps at a Mathematics Department colloquium at Johns Hopkins in 1985. They both were brilliant mathematicians, teachers, and great experts on harmonic maps.

Professor Richard Hamilton of Columbia University was the originator of the idea of using Ricci flow to attack the Poincaré and Thurston conjectures about 30 years ago. At the 2006 International Congress of Mathematicians in Madrid, he said that his initial inspiration came about 40 years ago, when he attended the seminars of James Eells and Joseph H. Sampson on harmonic maps, who suggested that one might be able to utilize evolution to attack the Poincaré conjecture. About 10 years later, Hamilton began to think seriously about the possibility and hit upon the idea of using the evolution equation called the Ricci flow. Hamilton investigated the Poincaré problem for over 25 years and developed the theory of the Ricci flow and laid the foundation for Grigori Perelman's work for solving the Poincaré conjecture. The names of two pioneers of the theory of harmonic maps,

Professors James Eells and Joseph H. Sampson, will be engraved in the minds of all mathematicians who work on harmonic maps, wave maps, and Yang-Mills fields for their great and everlasting contributions.

Acknowledgements The author deeply appreciates the referee and the language editor for their many valuable comments.

Contents

Chapter 1
Harmonic Maps

The last five decades have witnessed many developments in the theory of harmonic maps. To become acquainted to some of these, the reader is referred to two reports and a survey paper by Eells and Lemaire [119, 122, 124] about the developments of harmonic maps up to 1988 for details. Several books on harmonic maps [203, 205, 206, 389, 425] are also available. In this chapter, we follow the notions and notations of harmonic maps between Riemannian manifolds by Eells-Sampson [129] in the introduction. We discuss the crucial topics in harmonic maps including fundamentals, regularity, maps of surfaces, maps of Kähler manifolds, maps into groups and Grassmannians, harmonic maps, loop groups, and integrable systems, harmonic morphisms, maps of singular spaces, and transversally harmonic maps. Since the theory of harmonic maps has been developed over half a century, it is impossible to provide full details. However, we try to present the most important components of the topics.

1.1 Fundamentals

Following the definition of a harmonic map between Riemannian manifolds given in the introduction of the book, we would like to provide some examples of harmonic maps for readers. To demonstrate the importance of harmonic maps, we will prove the theorem of Ruh and Vilms which asserts that the Gauss map of an isometric immersion of an m-dimensional Riemannian manifold M into a Euclidean space $\mathbf{R}^k$ is harmonic if and only if the mean curvature is constant. We then briefly discuss variations and Jacobi fields.

Y.-J. Chiang, *Developments of Harmonic Maps, Wave Maps and Yang-Mills Fields into Biharmonic Maps, Biwave Maps and Bi-Yang-Mills Fields*, Frontiers in Mathematics, DOI 10.1007/978-3-0348-0534-6_1, © Springer Basel 2013

1.1.1 Examples

Let us give a few examples of harmonic maps involving harmonic functions, geodesics, minimal isometric immersions, Riemannian submersions with minimal fibers, and holomorphic map between Kähler manifolds. Recall that if $f : (M, g) \to (N, h)$ is a harmonic map between Riemannian manifolds, then (3) is equivalent to

$$\tau(f)(x) = trace_g(Ddf) = \sum_{i=1}^{m}(\tilde{D}_{e_i} f_* e_i - f_* D_{e_i} e_i)(x) = 0, \;\; x \in M,$$

where $\tilde{D}$ and D are the connections on $f^{-1}TN$ and M, and $\{e_1, \cdots, e_m\}$ is a local orthonormal frame at a point $x \in M$.

Example 1 (Harmonic functions). Let $(N, h) = (\mathbf{R}^n, h_0)$ be an Euclidean n-space with $\Gamma'^{\gamma}_{\alpha\beta} = 0$ and $f = (f^1, \cdots, f^n) \in C^\infty(M, \mathbf{R}^n)$. A map $f : (M, g) \to (\mathbf{R}^n, h_0)$ is harmonic if and only if f^γ is harmonic, i.e., $\triangle f^\gamma = 0,\ 1 \le \gamma \le n$. If M is compact, then f must be a constant.

Example 2 (Geodesics). Suppose that $dim\, M = 1$, and N is a Riemannian manifold. Let us take for M the unit circle S^1. Let $f : S^1 \to N$. Then the tension field of f is given by

$$\tau(f)^\gamma = \frac{d^2 f^\gamma}{dt^2} + \Gamma'^{\gamma}_{\alpha\beta} \frac{df^\alpha}{dt}\frac{df^\beta}{dt},$$

which is exactly the equation of geodesics, and t is the arc length parameter. Thus f is harmonic if and only if f defines a closed geodesic of N. It is well known that if N is compact, then in every homotopy class of maps $S^1 \to N$ there is a harmonic map.

Example 3 (Minimal isometric immersions). A smooth map $f : (M, g) \to (N, h)$ is an *isometric immersion* if (a) for each $x \in M$ the differential $f_* : T_x(M) \to T_{f(x)}N$ is injective; (b) $f^*h = g$. In this situation, we identify $x \in M$ with $f(x) \in N$, and identify $X \in \Gamma(TM)$ with f_*X. For each $x \in M$, we decompose $T_xN = T_xM \oplus T_xM^\perp$ with respect to g_x. Based on this, we decompose $D^N_X Y = D_X Y + A(X, Y)$ for $X, Y \in \mathcal{X}(M) =$ the set of all vector fields on M, where $A : T_xM \times T_xM \to T_xN^\perp$ is called the *second fundamental form* of the isometric immersion f. If $trace(A) = \sum_{i=1}^m A(e_i, e_i) = 0$, then f is called a *minimal isometric immersion*; where $\{e_i\}_{i=1}^m$ is a local orthonormal frame at a point in M. It is well known [397] that the above condition is equivalent to

$$\frac{d}{dt} Vol(M, f_t^* h)\Big|_{t=0} = 0,$$

for all smooth variations of immersions f_t whose variation vector field V satisfies $V(x) \in T_xM^{\perp}$, $x \in M$. Therefore,

$$\sum_{i=1}^{m} A(e_i, e_i) = \sum_{i=1}^{m} (D^N_{e_i} e_i - D_{e_i} e_i) = \tau(f).$$

Hence, the minimality of the isometric immersion $f : (M, g) \to (N, h)$ is equivalent to the harmonicity of f.

Example 4 (Riemannian submersions with minimal fibers). An onto map $f : (M, g) \to (N, h)$ is a *Riemannian submersion* if (a) for each $x \in M$, the differential $f_* : T_xM \to T_{f(x)}N$ is surjective; (b) for each $x \in M$ there is a unique orthogonal decomposition $T_xM = V_x \oplus H_x$ with respect to g_x, where $V_x = Ker(f_*) = \{\xi \in T_xM : f_*(\xi) = 0\}$. The restriction of f_* to H_x, $f_*|_{H_x}$ is an isometry of (H_x, g_x) onto $(T_{f(x)}N, h_{f(x)})$. The subspaces V_x and H_x are called the *vertical* and *horizontal* spaces. For each $x \in M$, $f^{-1}(f(x))$ is called the *fiber* through x. It is known that if $f : (M^m, g) \to (N^n, h)$ is a Riemannian submersion ($m \geq n$), then for each $y \in N$, $f^{-1}(y)$ is an $(m-n)$-dimensional closed submanifold of M.

Proposition 1.1.1. *A Riemannian submersion $f : (M^m, g) \to (N^n, h)$ is harmonic if and only if for each $x \in M$ the inclusion $i : (f^{-1}(f(x)), i^*g) \subset M$ is a minimal submanifold of (M, g).*

Proof. Let $\{e'_1, \cdots, e'_n\}$ be a local orthonormal frame in a neighborhood V in (N, h), and let $\{e_1, \cdots, e_n\}$ be the horizontal lift of $\{e'_1, \cdots, e'_n\}$ to $f^{-1}V$, and $\{e_1, \cdots, e_n, e_{n+1}, \cdots, e_m\}$ be an orthonormal frame defined in a neighborhood U in $f^{-1}(V)$. Thus at each point $x \in U$, $\{e_{n+1}, \cdots, e_m\}$ spans the vertical subspace V_x of T_xM. Therefore, we can decompose the tension field as

$$\tau(f) = \sum_{i=1}^{m} \{\tilde{D}_{e_i} f_* e_i - f_* D_{e_i} e_i\} = \sum_{i=1}^{n} \{\cdots\} + \sum_{i=n+1}^{m} \{\cdots\},$$

where $\tilde{D}$ and D are the connections on $f^{-1}TN$ and M such that $\tilde{D}_{e_i} f_* e_i = D^N_{e'_i} e'_i = f_* D_{e_i} e_i$, $1 \leq i \leq m$. By O'Neill formula, $f_*(D_{X'}Y') = D^N_X Y$ (X' on M is a horizontal lift of X on N if $f_* X'_x = X_{f(x)}$ for $X'_x \in H_x$, $x \in M$). Then we have $\tilde{D}_{e_i} f_* e_i = 0$, $n+1 \leq i \leq m$ and get

$$\tau(f) = -\sum_{i=n+1}^{m} f_* D_{e_i} e_i = -f_*(\sum_{i=n+1}^{m} D_{e_i} e_i).$$

It follows that $f : (M, g) \to (N, h)$ is harmonic iff $\sum_{i=n+1}^{m} D_{e_i} e_i \in V_x$, $x \in M$. On the other hand, $tr(A)$ = the trace of the second fundamental form A of the inclusion $i : (f^{-1}(f(x)), i^*g) \to (M, g)$ is the H_x-component of $\sum_{i=n+1}^{m} D_{e_i} e_i$. Hence, we can conclude the result. □

Example 5 (Holomorphic maps between Kähler manifolds). Let M be an even (say $2m$-) dimensional smooth manifold. Complexify the tangent space T_pM at each point $p \in M$, i.e. set $T_p^{\mathbf{C}}(M) = T_pM \otimes \mathbf{C}$ and let $T^{\mathbf{C}}M = \bigcup_{p\in M} T_p^{\mathbf{C}}M$, which is a complex vector bundle over M. At each point $p \in M$ the $2m$ vector fields $\{\frac{\partial}{\partial x_1^i}, \frac{\partial}{\partial y_1^i}, \cdots, \frac{\partial}{\partial x_m^i}, \cdots, \frac{\partial}{\partial y_m^i}\}$ (in a neighborhood U_i of p) span T_pM over $\mathbf{R}$, and $T_p^{\mathbf{C}}M$ over $\mathbf{C}$. A linear map $J : T_pM \to T_pM$ over $\mathbf{R}$ and its complexification $J : T_p^{\mathbf{C}}M \to T_p^{\mathbf{C}}M$ over $\mathbf{C}$ are defined by $J(\frac{\partial}{\partial x_j^i}) = (\frac{\partial}{\partial y_j^i})$, $J(\frac{\partial}{\partial y_j^i}) = -(\frac{\partial}{\partial x_j^i})$ $(1 \le j \le m)$, and they satisfy $J^2 = J \circ J = -id$ (the identity map). J is called the *almost complex structure*, and is a (1, 1) tensor field. A smooth map $f : M \to N$ between two complex manifolds is *holomorphic* if the differential $f_* : T_pM \to T_{f(p)}N$ satisfies $J \circ f_* = f_* \circ J$ for $p \in M$. Let $(z_1, \cdots, z_m)$, $(w_1, \cdots, w_n)$ be local coordinates at $p \in M$ and $f(p) \in N$, and $z_j = x_j + \sqrt{-1}y_j$, $w_k = u_k + \sqrt{-1}v_k$, $1 \le j \le m$, $1 \le k \le n$. The map f is *holomorphic* in a neighborhood of p iff each $w_k \circ f$ is a holomorphic function in $(z_1, \cdots, z_m)$, $1 \le k \le n$, iff each $u_k \circ f$ and $v_k \circ f$ satisfy the Cauchy-Riemann equations

$$\frac{\partial(u_k \circ f)}{\partial x_j} = \frac{\partial(v_k \circ f)}{\partial y_j}, \quad \frac{\partial(u_k \circ f)}{\partial y_j} = -\frac{\partial(v_k \circ f)}{\partial x_j}, \qquad 1 \le j \le m.$$

A Riemann metric g on a complex manifold M is a *Hermitian metric* if $g(JX, JY) = g(X, Y)$, $X, Y \in \mathcal{X}(M)$. If the 2-form w given by $w(X, Y) = g(X, JY)$ is a closed form, i.e., $dw = 0$, then g is called a *Kähler metric* and (M, g) is called a Kähler manifold. It is known [229] that a sufficient and necessary condition for a hermitian metric g on a complex manifold M to be a Kähler metric is $D_X(JY) = J(D_XY)$, $X, Y \in \mathcal{X}(M)$.

Proposition 1.1.2. *Any holomorphic map $f : (M, g) \to (N, h)$ between two Kähler manifolds is harmonic.*

Proof. Let $\{e_1, \cdots, e_m, l_1, \cdots, l_m\}$ be a local orthonormal frame such that $Je_i = l_i$, $Jl_i = -e_i$, $1 \le i \le m$. Then we have

$$\begin{aligned} \tilde{D}_{l_i} f_* l_i &= \tilde{D}_{l_i} J f_* e_i = J \tilde{D}_{l_i} f_* e_i = J(\tilde{D}_{e_i} f_* l_i + f_*[l_i, e_i]) \\ &= -\tilde{D}_{e_i} f_* e_i + J f_*[l_i, e_i], \end{aligned}$$

since f is holomorphic, (N, h) is Kähler, and $\tilde{D}_{l_i}(f_* e_i) - \tilde{D}_{e_i}(f_* l_i) = f_*([l_i, e_i])$. Similarly,

$$\begin{aligned} f_* D_{l_i} l_i &= f_* J D_{l_i} e_i = f_* J(D_{e_i} l_i + [l_i, e_i]) \\ &= -f_* D_{e_i} e_i + J f_*[l_i, e_i], \end{aligned}$$

since f is holomorphic and (M, g) is Kähler. Hence, we obtain

$$\sum_{i=1}^{m}(\tilde{D}_{l_i} f_* l_i - f_* D_{l_i} l_i) = -\sum_{i=1}^{m}(\tilde{D}_{e_i} f_* e_i - f_* D_{e_i} e_i)$$

which implies that $\tau(f) = 0$ at any point in M. □

1.1.2 Harmonicity of Gauss Maps

We examine the relationship between a Riemannian immersion $f : M^m \to \mathbf{R}^k$ and its Gauss map $\gamma : (M^m, g) \to G(k, m)$, and prove a well-known result of Ruh and Vilms [309] in Theorem 1.1.3. Recall that the mean curvature of f is $(1/m)trace(Ddf) = (1/m)\tau(f)$. It is a section of $f^{-1}T\mathbf{R}^k$ which is normal to the image of N, so that it can be viewed as a section of the normal bundle $V(N, M)$. Its covariant derivative in that bundle is defined as the projection in $V(N, M)$ of its derivative in $f^{-1}T\mathbf{R}^k$, and we denote it by $D^{\perp}((1/m)trace(Ddf))$. The map f has *constant mean curvature* if $D^{\perp}((1/m)trace(Ddf)) = 0$. This condition implies that $|(1/m)trace(Ddf)|$ is constant.

Let $G(k, m)$ be the Grassmannian manifold of m-spaces through the origin in $\mathbf{R}^k$. The Gauss map $\gamma : M \to G(k, m)$ associated with the immersion $f : M \to \mathbf{R}^k$ assigns to each point $x \in M$ to the m-space tangent to $f(M)$ at $f(x)$, translated to the origin of $\mathbf{R}^k$. It follow from [116, 229] that if L is an m-space through the origin of $\mathbf{R}^k$ and $\bar{L}$ is the corresponding point in $G(k, m)$, the tangent space of $G(k, m)$ at $\bar{L}$ can be viewed as the space of linear maps from L to its orthogonal complement. If K is the bundle in $G(k, m)$ whose fibre at $\bar{L}$ is L, then we have $TG(k, m) = K^* \times K^{\perp}$. For a point $\bar{L} \in G(k, m)$ we can choose an orthonormal frame $\{e_1, \cdots, e_k\}$ of $\mathbf{R}^k$ such that $\{e_1, \cdots, e_m\}$ is a basis of L and $\{e_{m+1}, \cdots, e_k\}$ is a basis of $\bar{L}$. Then the canonical Riemannian structure on $G(k, m)$ is defined by $e_i^* \otimes e_l$ $(i = 1, \cdots, m, l = m + 1, \cdots, k)$, an orthonormal basis of $T_{\bar{L}}G(k, m)$. Another interpretation of the tangent space is as follows. For the basis chosen above, represent L by the m-vector $e_1 \wedge \cdots \wedge e_m$. Then $e_i^* \otimes e_l$ can be identified with the plane $E_i^l = e_1 \wedge \cdots \wedge e_{i-1} \wedge e_l \wedge e_{i+1} \wedge \cdots \wedge e_m$.

Theorem 1.1.3 ([309]). *Let $f : (M, g) \to \mathbf{R}^k$ be an isometric immersion. Then the tension field of the Gauss map γ can be identified with the covariant derivative in the normal bundle of m times the mean curvature of f : $\tau(\gamma) = D^{\perp}\tau(f)$. Hence, f has constant mean curvature if and only if γ is harmonic.*

Proof. Let (x_i) be normal coordinates at a point $p \in M$. In these coordinates, γ maps a point x into $\frac{\partial f}{\partial x_1} \wedge \cdots \wedge \frac{\partial f}{\partial x_m}$, and the differential of γ is

$$d\gamma(\frac{\partial}{\partial x_i}) = \sum_{j=1}^{m} \frac{\partial f}{\partial x_1} \wedge \cdots \wedge \frac{\partial f}{\partial x_{j-1}} \wedge \frac{\partial^2 f}{\partial x_i \partial x_j} \wedge \frac{\partial f}{\partial x_{j+1}} \wedge \cdots \wedge \frac{\partial f}{\partial x_m}.$$

Choosing an orthonormal basis $e_1, \cdots, e_k$ of $\mathbf{R}^k$ such that for $i = 1, \cdots, m$, we get $e_i = df \cdot \frac{\partial}{\partial x_i}\Big|_p$, and so at p: $\frac{\partial^2 f}{\partial x_i \partial x_j} = Ddf(\frac{\partial}{\partial x_i}, \frac{\partial}{\partial x_j})$. Since the coordinates x_i are normal and $\mathbf{R}^k$ is flat, the second fundamental form of an immersion has only normal components, denoted by $h^l_{ij} = [Ddf(\partial/\partial x_i, \partial/\partial x^j)]^l$, such that

$$d\gamma(\frac{\partial}{\partial x_i})\Big|_p = \sum_{l=m+1}^{k} \sum_{j=1}^{m} h^l_{ij} e_1 \wedge \cdots \wedge e_{j-1} \wedge e_l \wedge e_{j+1} \wedge \cdots \wedge e_m = \sum_l \sum_j h^l_i E^l_j.$$

On the other hand, we obtain

$$Ddf(X, \frac{\partial}{\partial x_i}) = (D_{Xdf})(\frac{\partial}{\partial x_i}) = e^*_j(X) D_{e_j} df(\frac{\partial}{\partial x_i}) \text{ or } Ddf(\frac{\partial}{\partial x_i})\Big|_p = h^l_{ij} e^*_j \otimes e_l.$$

Using the above identification, we observe that $d\gamma = Ddf$, where $d\gamma \in \mathcal{C}(T^*M \otimes \gamma^{-1}TG(k,m))$ and $Ddf \in \mathcal{C}(\otimes^2 T^*M \otimes V(\mathbf{R}^k, M))$. (Let $\xi : V \to M$ be a smooth vector bundle over M of finite rank. We denote by $\mathcal{C}(V)$ the vector space of smooth sections of V, i.e., of smooth maps $\sigma : M \to V$ such that $\xi \circ \sigma = id_M$, the identity map on M). Through this identification the two bundles are isometric and they have the same connection, so that $Dd\gamma = D^\perp Ddf$, where $D^\perp$ is the connection in $\otimes^2 T^*M \otimes V(\mathbf{R}^k, M)$. Regarding Ddf as a section of $\otimes^2 T^*M \otimes f^{-1}T\mathbf{R}^k$, we then get $D^\perp Ddf = (DDdf)^\perp$, the projection of $DDdf$ on $V(N, M)$. Taking the trace of the above equality and using the notation *trace* $D_- D_-$ to indicate that the trace is taken on the two marked vectors, we obtain

$$\begin{aligned}\tau(\gamma) &= trace\ (Dd\gamma) = (trace\ (D_- D_- df))^\perp = \Big(trace(D_- Ddf(-))\Big)^\perp \\ &= \Big(trace(DD_- df(-) + R(\,,-)df(-))\Big)^\perp,\end{aligned}$$

the curvature being that of $T^*M \otimes f^{-1}T\mathbf{R}^k$, i.e., minus that of TM. Hence,

$$\tau(\gamma) = D^\perp trace(Ddf) + \Big(df \cdot R^M(-,\,)-\Big)^\perp = D^\perp \tau(f) + 0. \qquad \square$$

Corollary 1.1.4. *A Riemannian immersion* $(M, g) \to \mathbf{R}^k$ *has parallel second fundamental form if and only if its Gauss map is totally geodesic.*

1.1.3 Variations

A vector field v along a map $f : M \to N$ of Riemannian manifolds is a section of $f^{-1}T(N) \to M$. It defines a variation of f by $f_t(x) = exp_{f(x)}(tv(x))$, which is well defined as a smooth map $M \times \mathbf{R} \to N$ with $f_0 = f$, provided that N is complete. Recall from the introduction of the book that if v has compact support, then

$$D_v E(f) = \frac{dE(f_t)}{dt}\Big|_{t=0} = -\int_M (\tau f(x), v(x))dx.$$

In order to determine the behavior of E near a harmonic map, we consider its second variation. Given two vector fields v and w along f, we choose a 2-parameter variation $f_{s,t}$ such that

$$v = \frac{\partial f_{s,t}}{\partial s}\Big|_{(s,t)=0}, \quad w = \frac{\partial f_{s,t}}{\partial t}\Big|_{(s,t)=0}.$$

Then the *Hessian* of f is the symmetric bilinear form on $C(f^{-1}T(N))$ given by

$$H_f(v, w) = \frac{\partial^2 E(f_{s,t})}{\partial s \partial t}\Big|_{(s,t)=0}.$$

If D_f is the induced connection on the bundle $f^{-1}T(N)$ and R^N is the curvature tensor of (N, h), then

$$\begin{aligned} H_f(v, w) &= \int_M \left\{(D_f v,\ D_f w) - (trace(R^N(df,\ v)df), w)\right\} dx \\ &= -\int_M \left(trace(D_f^2 v) + trace(R^N(df, v)df), w\right) dx = \int_M (J_f(v),\ w)dx, \end{aligned}$$

for compactly supported variations. The solutions v of $J_f(v) = 0$ are called the *Jacobi fields* along f. Notice that if (f_t) is a variation of f through harmonic maps, then $v = \frac{\partial f_t}{\partial t}\Big|_{t=0}$ is a Jacobi field. The *nullity* of a harmonic map f is defined as $dim\, Ker(J_f)$; it is finite if M is compact, since J_f is an elliptic operator. The *index* of f is the dimension of the largest subspace of $C(f^{-1}T(N))$ on which H_f is negative definite.

Example 1. Assume that $dim\, M = 1$. Then $Ddf = \tau(f) = \frac{D^2 f}{dt^2}$ is the curvature (or acceleration) vector of the path f. Therefore, f is a harmonic map if and only if f is a geodesic of N. The condition $\tau(f) = 0$ requires that the velocity vector f' be parallel, and its length be constant along f. Let M be the circle S^1 and $f : S^1 \to N$ be a closed geodesic. Then its Jacobi fields are the solutions of

$$\frac{D^2 f}{dt^2} + R^N(f', f')f' = 0.$$

Example 2. Smith [349] computed the index and nullity of the identity map id_M of compact oriented Einstein manifolds. For instance, if M is the Euclidean sphere S^m, then

$$index(id_{S^m}) = \begin{cases} 0, & \text{if } m = 1, 2, \\ m+1, & \text{if } m \geq 3. \end{cases}$$

All harmonic maps $S^2 \to S^2$ have index 0. If M is a Grassmannian of real or complex subspaces, then $index(id_M) = 0$; if it is a Grassmannian of quaternion subspaces, then $index(id_M) \geq 1$.

Moreover, Xin [423,424] showed that any non-constant harmonic map $f : S^n \to (N, h)$ has $index(f) > 0$ for $n \geq 3$. Eells and Lemaire [122] proved that $index(f) \geq k+1$, where k is the maximal rank of f. Leung [250] showed that any non-constant harmonic map $(M, g) \to S^n$ $(n \geq 3)$ of a compact manifold has $index(f) > 0$. The totally geodesic embedding $f : S^m \to S^n$ for $n > m \geq 3$ has $index(f) = n+1$. In the case where (N, h) is compact and $Ricc^{(N,h)} \leq 0$, Smith [350] and Urakawa [386] proved that $index(id_N) = 0,\ nullity(id_N) \leq n$.

1.2 Regularity

In this section, we briefly review some significant results concerning spaces of maps, smoothness and partial regularity, existence, and removable singularities. We concentrate on the well-known regularity theory for harmonic maps constructed by Scheon and Uhlenbeck [320, 321].

1.2.1 *Spaces of Maps*

Let (M^m, g_{ij}) be a compact m-dimensional Riemannian manifold and $(N^n, h_{\alpha\beta})$ be an n-dimensional Riemannian manifold. The natural space of maps from (M, g) to (N, h) in which to study the existence problem for harmonic maps is that of maps bounded a.e., whose first derivatives are square integrable. By Nash's embedding theorem, we assume that (N, h) is embedded into $\mathbf{R}^k$ for some k. Let $L_1^2(M, \mathbf{R}^k)$ be the Hilbert space of square integrable maps $(M, g) \to \mathbf{R}^k$ whose first derivatives in the charts of M are square integrable. Define

$$L_1^2(M, N) = \{f \in L_1^2(M, \mathbf{R}^k) | f(x) \in N \ a.e.\}.$$

An element of $L_1^2(M, N)$ is a class of equivalent maps, i.e., two maps are equivalent if they agree almost everywhere.

When $dim\, M = m = 1$, each such class f contains a continuous map, so we say that the map f is continuous. For $m \geq 2$ not every class f contains a continuous map; but each L_1^2 class can be represented by a map that is absolutely continuous along almost every coordinate line. We say such a map an L_1^2 map. The map f is called continuous if its class contains a continuous representative.

The set $L_1^2(M, N)$ inherits a strong and a weak topology from those of the Hilbert space $L_1^2(M, \mathbf{R}^k)$; it is both strongly and weakly closed. Actually, $C^i(M, \mathbf{R}^k)$ is dense in $L_1^2(M, \mathbf{R}^k)$ for all $i \geq 0$. Similarly, $C^\infty(M, N)$ is dense in $C^0 \cap L_1^2(M, N)$.

Schoen and Uhlenbeck [320] showed that *if* $\dim M = 2$, *then* $C^i(M, N)$ *is dense in* $L_1^2(M, N)$ *for all* $i \geq 0$. However, *for* $m \geq 3$, *the space* $C^i(M, N)$ *is not dense in* $L_1^2(M, N)$. For instance, a map $f : S^3 \to S^2$ defined on $S^3 - \{\text{poles}\}$ by projection along the meridians of the hemispheres onto the equator is in $L_1^2(S^3, S^2)$, but can not be L_1^2-approximated by C^i maps.

Bethuel and Zheng [32, 33] proved that $C^\infty(M, S^n)$ *is dense in* $L_1^p(M, S^n)$ *for* $1 \leq p < n$; *but* $C^\infty(B^m, N)$ *is not dense in* $L_1^p(B^m, N)$ *if* $\pi_i(N) \neq 0$ *and* $i \leq p < i+1 \leq m$, *where* B^m *is the closed Euclidean unit m-ball.*

Burstall [47] showed that $C^0 \cap L_1^2(M, N)$ *is a closed separable submanifold of the Banach manifold* $L^\infty \cap L_1^2(M, N)$; *the latter is non-separable for* $m \geq 2$.

In terms of partial derivatives in the charts, the energy of an L_1^2-map is well defined. (Strictly speaking, it is not $\frac{1}{2}\int |df|^2 dx$ since there are L_1^2-maps whose differentials do not exist a.e.; but the partial derivatives do exist.) It defines a smooth functional $E : L^\infty \cap L_1^2(M, N) \to \mathbf{R}$, being the restriction of the quadratic form E on $L_1^2(M, \mathbf{R}^k)$. We are more interested in its critical points (which form a closed set) and those of the restriction $E : C^0 \cap L_1^2(M, N) \to \mathbf{R}$, which are precisely the harmonic maps $f : (M, g) \to (N, h)$.

(1.2.A) A map $f \in L_1^2(M, N)$ is *energy minimizing* if each point of M has a neighborhood U such that $E(f) \leq E(\phi)$ for every $\phi \in L_1^2(M, N)$ for which $f = \phi$ on $M \backslash U$.

Let $\xi \in C^\infty(M, \mathbf{R}^k)$ and put $f_t(x) = \pi \circ (f(x) + t\xi(x))$, where π is the orthogonal projection of $\mathbf{R}^k$ onto N (well-defined for sufficiently small t). Thus $f_0 = f$ and $f_t \in L_1^2(M, N)$. The map f is *weakly harmonic* if

$$\frac{d}{dt}E(f_t)\Big|_{t=0} = 0 \text{ for all } \xi \in C^\infty(M, \mathbf{R}^k).$$

The map f is weakly harmonic if and only if it satisfies the tension field equations weakly; i.e.,

$$\int_M \sum_\alpha g^{ij}\left[f_i^\alpha \xi_j^\alpha + A(f(x))(f_i, f_j)\xi^\alpha\right] dv = 0,$$

for all $\xi \in C^\infty(M, \mathbf{R}^k)$, where $dv = \sqrt{\det(g_{ij})}dx$ is the volume form of M, f^α, ξ^α are the components of f, ξ in $\mathbf{R}^k$, and A is the second fundamental form of the embedding of N in $\mathbf{R}^k$ (cf. [340]).

Burstall [48] proved that *the space of harmonic maps is locally compact in* $L^\infty \cap L_1^2(M, N)$, *and is open in the space of weakly harmonic maps.*

In order to find a (smooth) harmonic map in a prescribed class $\mathcal{F}$ of maps from M to N, we have direct methods of variational theory as follows: Choose a minimizing sequence $(f_k) \subset \mathcal{F}$ such that

$$\lim_{k\to\infty} E(f_k) = inf\{E(\phi) : \phi \in \mathcal{F}\}.$$

Since (f_k) is bounded in $L_1^2(M,N)$, it contains a subsequence which converges weakly to some $f \in L_1^2(M,N)$, and $E(f) \le lim(f_k)$.

1.2.2 Smoothness and Partial Regularity

An important regularity theorem states that *any continuous weakly harmonic map is harmonic (and hence smooth). In particular, any continuous energy minimizing map is harmonic.*

The proof of the above theorem is based on Morrey [271, 272], Ladyzhenskaya and Ural'tseva [238], and Hildebrandt [182]. It requires the special form (0.3) of the tension field equation $\tau(f) = 0$; it is a second order semi-linear elliptic system of divergence type, whose second order terms form a diagonal matrix with the same Laplacian in each entry, and whose first derivatives have quadratic growth.

When $1 < p < \infty$, we define the p-energy functional $E_p : L_1^p(M,N) \to \mathbf{R}$ by $E_p(f) = \frac{1}{p}\int_M |df|^p dv$. For $p \ne 2$ there are extrema of E_p which are $C^{1+\alpha}$ for some $0 < \alpha < 1$ by Uhlenbeck [205, 380], but not C^2 by Bojarski and Iwaniec [35] and Tolkdorf [375]. On the other hand, there is a partial regularity theory for E_p-minimizers due to Hardt and Lin [171].

Suppose that $f \in L_1^2(M,N)$ is an energy minimizing map as in (1.2.A). If $m = 1$, then every L_1^2-map is continuous; hence f is harmonic and smooth. If $m = 2$, then a theorem of Morrey [270, 272] asserts that f is harmonic and smooth, provided N is compact or satisfies the following uniformity condition: There are two positive constants c_1 and c_2 such that any point of N is in the domain of a coordinate chart $\phi : V \to \mathbf{R}^k$ whose image is the unit ball and

$$c_1|d\phi(y)Y|^2_{\mathbf{R}^k} \le h_y(Y,Y) \le c_2|d\phi(y)Y|^2_{\mathbf{R}^k},$$

for any $y \in V$ and $Y \in T_y(N)$.

If $m \ge 3$, f is not necessarily continuous and the size of its singular set can be estimated as follows. Schoen and Uhlenbeck [320–322] obtained the main partial regularity theorem for harmonic maps in the general case. Giaquinta and Giusti [153, 154] studied the theorem in the case where $f(M)$ is contained in a single chart of N. From now on, we focus on the work of Schoen and Uhlenbeck and try to outline their main results.

Let (M^m, g) and (N^n, h) be Riemannian manifolds of dimension m and n. By the Nash imbedding theorem, we can assume that $N \subset \mathbf{R}^k$ is isometrically embedded in the Euclidean space. Suppose that M is compact, possibly with boundary, and that

N is an open manifold. Let $f \in L_1^2(M,N)$ be an energy minimizing map whose image lies in a compact subset of N. Recall that

$$L_1^2(M,N) = \{f \in L_1^2(M,\mathbf{R}^k) | f(x) \in N \text{ a.e.}\}.$$

For $f \in L_1^2(M,\mathbf{R}^k)$, the norm is given by

$$||f||_{1,2}^2 = E(f) + \int_M \sum_\alpha (f^\alpha(x))^2 dv$$

where $E(f) = \int_M (df, df) dv$ and dv is the volume form of M.

Let

$$\tilde{E}(f) = E(f) + V(f)$$

by adding additional lower terms to $E(f)$, where

$$V(f) = \int_M v(f) dv = \int_M \left[\sum_\alpha \sum_i \gamma_\alpha^i(x, f(x)) \frac{\partial f^\alpha}{\partial x^i}(x) + \Gamma(x, f(x)) \right] dv.$$

Here $\gamma \in C^r(M \times \mathcal{O}, T^*M \otimes \mathbf{R}^k)$ and $\Gamma \in C^r(M \times \mathcal{O}, \mathbf{R})$, where $\mathcal{O}$ is an open neighborhood of N in $\mathbf{R}^k$. A map $f \in L_1^2(M,N)$ is *$\tilde{E}$-minimizing* if $\tilde{E}(f) \le \tilde{E}(\phi)$ for any map $\phi \in L_1^2(M,N)$ with $(f - \phi) \in L_{1,0}^2(M,\mathbf{R}^k)$ (means those L_1^2 maps which are zero on ∂M). We assume that the metric on M is C^2 and $\gamma, \Gamma \in C^r$ for $r \ge 2$.

Let $\mathcal{O}$ be an open neighborhood of N in $\mathbf{R}^k$ such that the map $\Pi : \mathcal{O} \to N$ given by $\Pi(y) =$ nearest point in N to y, is a smooth fibration, and let N_0 be a compact subset of N. Since N_0 is a compact set, $\mathcal{O}$ contains a uniform neighborhood of N_0. By Schoen and Uhlenbeck [320], if f is $\tilde{E}$-minimizing on M and $f(x) \in N_0$ a.e., then f satisfies the Euler-Lagrange equations for E. These equations have the form

$$\Delta_M f - A(df, df) + \sum_{\alpha,i} B_{\alpha,i}(x, f(x)) \frac{\partial f^\alpha}{\partial x^i} + C(x, f(x)) = 0,$$

where A, B and C are smooth in their arguments and A is quadratic in df. We are not interested in the exact form of the lower order terms, but only in

$$\Delta_M f - A(df, df) = 0,$$

where A is the second fundamental form of N in $\mathbf{R}^k$.

The following theorems are proved by covering M with geodesic coordinate balls and showing regularity for $\tilde{E}$-minimizing maps on balls. Let $B_1 = B_1^m(0)$ be the unit ball in $\mathbf{R}^m$. For $\Lambda > 0$, let $\mathcal{F}_\Lambda$ be the class of functionals $\tilde{E}$ on B_1 with metric g_{ij} such that $g_{ij}(0) = \delta_{ij}$ and lower order terms satisfying, for $x \in B_1$, $f \in N_0$,

$$\sum_{i,j,k}\left|\frac{\partial g_{ij}(x)}{\partial x^k}\right| + |\gamma(x,f)| + |d_f\gamma(x,f)| + |\Gamma(x,f)|^{1/2}$$
$$+ |d_f\Gamma(x,f)|^{1/2} \le \Lambda. \tag{1.1}$$

If f is $\tilde{E}$-minimizing for $\tilde{E} \in \mathcal{F}_\Lambda$ and $f(x) \in N_0$ a.e., we say that $f \in \mathcal{H}_\Lambda$. The lower order terms are handled by showing that Λ is a dimensional constant which shrinks with the radius of a coordinate ball. Actually, if $\tilde{E}$ is a functional and $B_\sigma(p)$ is a geodesic ball in M of radius centered at $p \in M$, we define a functional $\tilde{E}^{p,\sigma}$ on B_1 by

$$\begin{aligned}\tilde{E}^{p,\sigma}(\phi) &= \int_{B_1}\Big(|d\phi|^2_{g_\sigma} + \sigma(d\phi\cdot\gamma(y,\phi)) + \sigma^2\Gamma(y,\phi)\Big)g_\sigma^{1/2}dy\\ &= \sigma^{2-m}\tilde{E}_{B_\sigma(p)}(f),\end{aligned} \tag{1.2}$$

where $\phi(y) = f(\sigma y)$ and $g_\sigma(y) = g(\sigma y)$. Since M and N_0 are compact, we can choose Λ so that $\tilde{E}^{p,\sigma} \in \mathcal{F}_\Lambda$ for all p and some $\sigma > 0$.

There are three main ingredients for the proofs of the following theorems. The first is the monotonicity (or scaling) inequality in Proposition 1.2.1 [320]. To study the behavior of df near a point $x \in \mathcal{S}_f$ (the singular set of f, see below), we take a small x-centered ball $B_{1/2}$ on which the metric g is C^1-close to a flat metric (the C^1 distance majorised by some $\Lambda > 0$ in (1.1)). On $B_{1/2}$ an energy minimizing map satisfying the following *monotonicity* (or *scaling*) *inequality.*

Proposition 1.2.1. *If $f \in \mathcal{H}_\Lambda$ for Λ sufficiently small, then*

$$\sigma^{2-m}E^x_\sigma(f) \le C[\rho^{2-m}E^x_\rho(f) + \Lambda\rho], \tag{1.3}$$

for $x \in B_{1/2}$, $0 < \sigma \le \rho \le 1/2$.

The number $m-2$ is the *scaling dimension.* It is computed from the relative weights of the coefficients of the metric in the energy density and the volume element in the integral $E(f)$.

The second main ingredient is Morrey's [272] growth lemma: *If $f \in L^2_1(B(x_0,R),N)$ and*

$$E_{B(x,\rho)}(f) \le C\rho^{m-2+2\alpha} \quad (0<\alpha<1)$$

for all $x \in B(x_0,R/2)$ and $\rho < R/2$, then $f \in C^\alpha(B(x_0,R/2),N)$, i.e., f is Hölder continuous on $B(x,R/2)$ with Hölder exponent α.

Let $f \in L^2_1(M,N)$ be an energy minimizing map whose image lies in a compact subset of N, with singular set of f defined as

$$\mathcal{S}_f = M - \{x \text{ having a neiborhood on which } f \text{ is continuous}\}.$$

We can now characterize the singular set as follows: Define the density of f at x by

$$\theta_f(x) = \liminf_{\rho\to 0}(E_{B(x,\rho)}(f)/\rho^{m-2}).$$

Then $\mathcal{S}_f = \{x \in M : \theta_f(x) > 0\}$. Roughly speaking, this shows that a singular point of f 'uses up' a positive amount of energy; that actually a restriction on $\mathcal{S}_f$. Hardt, Kinderlehrer, and Lin [172] proved: *If $f : (M, g) \to (N, h)$ is E-minimizing and N is simply connected, then at any point $x \in M$, $\theta_f(x)$ is bounded by a constant depending on (M, g) and (N, h).*

The third main ingredient is the Federer reduction, which is used in the proofs of Theorems 1.2.3 and 1.2.5. All the following theorems and results were obtained by Schoen and Uhlenbeck [320, 321].

Regularity Estimate. *There exists $\bar{\epsilon}(m, N_0) > 0$, depending on m and $N_0 \subset N$, such that if $f \in \mathcal{H}_\Lambda$, $\Lambda \le \epsilon$ and $E_1(f) \le \bar{\epsilon}$, then f is Hölder continuous on $B_{1/2}$ and $|f(x) - f(y)| \le c|x - y|^\alpha$ for $x, y \in B_{1/2}$ where $c, \alpha > 0$ depend on m and N_0.*

Theorem 1.2.2. *Let $f \in L_1^2(B_\sigma(a), N)$ be an $\tilde{E}$-minimizing map such that $f(x) \in N_0$ a.e. for some compact $N_0 \subset N$. If σ and $\sigma^{2-m}\tilde{E}(f)$ are sufficiently small, then f is Hölder continuous on $B_{\sigma/2}(a)$.*

Theorem 1.2.2 follows from the above regularity estimate by rescaling and a consequence of (1.2), namely given $\Lambda > 0$, there exists $\sigma_0 > 0$ such that for $0 < \sigma < \sigma_0$ and $p \in M$, if f is $\tilde{E}$-minimizing, then $\phi(y) = f(exp_p \sigma y)$ is $\tilde{E}^{p,\sigma}$-minimizing, where $\tilde{E}^{p,\sigma} \in \mathcal{F}_\Lambda$.

Since f is *Hölder* continuous on $B_{\sigma/2}(a)$ by Theorem 1.2.2, it follows that f is smooth in the interior of $B_{\sigma/2}(a)$ (this is a well-known result by Borchers and Garber [36]).

For $F \subset \mathbf{R}^m$, $s \ge 0$ the *Hausdorff measure* of F is defined by

$$\phi^s(F) = \inf\Big\{\sum r_i^s | F \subset \bigcup_i B_{r_i}(x_i)\Big\}.$$

We have by Federer [142]: (i) $\phi^s(F) = 0$ if and only if $\mathcal{H}^s(F) = 0$; (ii) the density inequality holds

$$\overline{\lim_{\lambda\to 0}} \lambda^{-s}\phi^s(F \cap B_\lambda^m(x)) \ge \text{const} > 0 \tag{1.4}$$

for ϕ^s a.e. $x \in F$.

Theorem 1.2.3. *If $f \in L_1^2(M, N)$ is $\tilde{E}$-minimizing with $f(x) \in N_0$ a.e. for a compact set $N_0 \subset N$, then dim$(\mathcal{S}_f \cap M) \le m - 3$ (m = dim M and dim F is the Hausdorff dimension of a set F). If $m = 3$, then $\mathcal{S}_f$ is a discrete set of points.*

Remark that a harmonic map is a weak solution to the Euler-Lagrange equations for E on $L_1^2(M, N)$. Let $f \in L_{1,\text{loc}}^2(\mathbf{R}^m, N)$ (the space of measurable maps whose restrictions to each compact set is in L_1^2), and assume that $\partial f/\partial r = 0$ a.e., where r denotes the radial coordinate. Then there is a map $\tilde{f} : S^{m-1} \to N$ such that $f(x) = \tilde{f}(x/|x|)$, and f is weakly harmonic if and only if $\tilde{f}$ is. Actually, $E(f|B_\sigma(0)) = (m-2)^{-1}\sigma^{m-2}E(\tilde{f})$. Furthermore, the map f has a singularity at 0 if and only if $\tilde{f}$ is non-constant. Such a homogeneous harmonic map f with an isolated singular point is called a *tangent map* (TM). It is said to be a *minimizing tangent map* (MTM) if f is $\tilde{E}$-minimizing on compact subsets of $\mathbf{R}^m$.

Proposition 1.2.4. *Let $f \in L_1^2(M, N)$ be an $\tilde{E}$-minimizing map and let $z \in \mathcal{S} \cap \mathit{int}M$. There exists a sequence $\sigma_i \in \mathbf{R}^+$, $\sigma_i \to 0$, such that the maps $f_i \in L_1^2(B_1(0), N)$, $f_i = f(\exp_z \sigma_i x)$, converge to $f \in L_1^2(B_1(0), N)$. The map f is a non-constant harmonic map satisfying $f(x) = \tilde{f}(x/|x|)$, where $\tilde{f} \in L_1^2(S^{m-1}, N)$ is harmonic.*

Theorem 1.2.5. *Assume that $l \geq 3$ such that every minimizing tangent map from $\mathbf{R}^j \to N$ is trivial for $3 \leq j \leq l$. If $f \in L_1^2(M, N)$ is an $\tilde{E}$-minimizing map with $f(x) \in N_0$ a.e., then $\dim(\mathcal{S} \cap \mathit{int}\, M) \leq m - l - 1$. If $m = l + 1$, then $\mathcal{S}$ is a discrete set of points, and if $m < l + 1$, then $\mathcal{S} = \emptyset$ and f is smooth.*

In order to prove Theorems 1.2.3 and 1.2.5, we require the following three lemmas (proofs are given in [320]). Let $\mathcal{H}_{\Lambda,B}$ be the set of maps $f \in \mathcal{H}_\Lambda$ with $E_1(f) \leq B$. Let $\bar{\mathcal{H}}_{\Lambda,B}$ be the closure of $\mathcal{H}_{\Lambda,B}$ taken in $L_1^2(B_1, N_0)$.

(L1) *Given $f \in \bar{\mathcal{H}}_{\Lambda,B}$ and $x_0 \in B_{1/2}$, there is a sequence $\lambda(i) \to 0$, $\lambda(i) \in (0, 1/2]$ such that the map $f_{x_0,\lambda(i)} \in L_1^2(B_1, N_0)$ defined by $f_{x_0,\lambda(i)}(x) = f(\lambda(i)(x - x_0))$ converge in L_1^2-norm on B_1^m to a harmonic map $f_0 \in \bar{\mathcal{H}}_{0,B}$ satisfying $\partial f_0/\partial r = 0$ a.e. on B_1. Furthermore, the convergence is uniform on compact subsets of $\bar{B}_1 \backslash \mathcal{S}_{f_0}$.*

(L2) *Let $f_0 \in L_{1,\text{loc}}^2(\mathbf{R}^l, N_0)$ with $l \geq 3$ be a harmonic map with an isolated singularity at 0 such that f_0 satisfies $\partial f_0/\partial r = 0$ a.e. Define $f \in L_{1,\text{loc}}^2(\mathbf{R}^m, N_0)$, $m \geq l$ by $f(x', x'') = f_0(x')$, $x' \in \mathbf{R}^l$, $x'' \in \mathbf{R}^{m-l}$. Assume that there is a sequence $f_i \in \mathcal{H}_{\Lambda_i,B}$ such that f_i converges to f in $L_1^2(B_1, N_0)$ and $\Lambda_i \to 0$. Then f and f_0 are E-minimizing on compact subsets of $\mathbf{R}^m$, $\mathbf{R}^l$, respectively. In particular, f_0 is a minimizing tangent map.*

(L3) *Suppose f_i is a sequence in $\mathcal{H}_\Lambda$ which converges weakly to f in $L_1^2(B_1^m, N)$ If $\mathcal{S}_i, \mathcal{S}$ are the singular sets of f_i, f, then we have*

$$\overline{\lim_{i\to 0}}\, \phi^s(\mathcal{S}_i \cap B_{1/2}^m) \leq \phi^s(\mathcal{S} \cap B_{1/2}^m), \quad \textit{for any } s \geq 0. \tag{1.5}$$

Proof (Proofs of Theorems 1.2.3 and 1.2.5). Assume that $f \in L_1^2(M, N)$ is $\tilde{E}$-minimizing with singular set $\mathcal{S} \subset \mathit{int}\, M$. Let $0 \leq s < m - 2$ with $\phi^s(\mathcal{S}) > 0$. Then by (1.4) we can select $p_0 \in \mathcal{S}$ such that

$$\lim_{\lambda_i \to 0} \lambda_i^{-s} \phi^s(\mathcal{S} \cap B^m_{\lambda_i/2}) > 0 \tag{1.6}$$

for a sequence $\lambda_i \to 0$, where B_λ is a ball in normal coordinates x centered at p_0. Consider the scaled maps $f_\lambda(x) = f(\lambda x)$. By (L1), we can select a subsequence of λ_i, still denote it by λ_i, such that f_{λ_i} converges weakly in $L^2_1(B^m_1, N)$, and strongly in $L^2_1(B^m_{1/2}, N)$ to a harmonic map f_0, where f_0 satisfies $\frac{\partial f_0}{\partial r} = 0$ a.e. If $\mathcal{S}_\lambda$ is the singular set of f_λ in B^m_1, we get $\mathcal{S}_\lambda \cap B^m_{1/2} = \{x/\lambda | x \in \mathcal{S} \cap B^m_{\lambda/2}\}$ and thus $\phi^s(\mathcal{S}_\lambda \cap B^m_{1/2}) = \lambda^{-s}\phi^s(\mathcal{S} \cap B^m_{\lambda/2})$. Hence, (1.6) implies $\lim_{\lambda_i \to 0} \phi^s(\mathcal{S}_{\lambda_i} \cap B^m_{\lambda_i/2}) > 0$. It follows from (L3) that $\phi^s(\mathcal{S}_0 \cap B^m_{1/2}) > 0$. Because $\frac{\partial f_0}{\partial r} = 0$ a.e., we have $\lambda \mathcal{S}_0 \subset \mathcal{S}_0$ for any $\lambda \geq 0$, and there are two possible cases: either $s \leq 0$, or we can select a point $x_1 \in \mathcal{S}_0 \cap \partial B^m_1$ by (1.4) such that $\overline{\lim}_{\lambda_i \to 0} \lambda^{-s} \phi^s(\mathcal{S}_0 \cap B^m_\lambda(x_1)) > 0$. We choose Euclidean coordinates centered at x_1 so that x^1 is radial at x_1. Repeating the above discussion at x_1 we get a radially independent harmonic map $f_1 \in L^2_{1,loc}(\mathbf{R}^m, N_0)$ with $\phi^s(\mathcal{S}_1 \cap B^m_1) > 0$. We have $\frac{\partial f_1}{\partial x^1} = 0$ a.e., since $\frac{\partial f_0}{\partial r} = 0$. If $s - 1 \leq 0$ we stop. Otherwise, there is a point $x_2 \in \mathcal{S}_1 \cap \partial B^{m-1}_1$, $\mathbf{R}^{m-1} = (0, x^2, \cdots, x^m)$ and we repeat the discussion at x_2. If we repeat this process n times, we get harmonic maps $f_j \in L^2_{1,\text{loc}}(\mathbf{R}^m, N_0)$ for $j = 1, \cdots, n$ such that $f_j|_{B^m_1} \in \bar{\mathcal{H}}_{\Lambda,B}$ for suitable B, by (L1), and $\frac{\partial f_j}{\partial r} = \frac{\partial f^j}{\partial x^\mu} = 0$ a.e. $\mu = 1, \cdots, j$. We also have $\phi^s(\mathcal{S}_j \cap B^m_1) > 0$ for $j = 1, \cdots, n$. We can repeat the argument until we have $s - n \leq 0$. In order to build f_n, we need to have $s - n + 1 > 0$. Since $s < m - 2$ and n is an integer, $n \leq m - 2$. If $n = m - 2$, $\{(x^1, \cdots, x^{m-2}, 0, 0)\} = \mathbf{R}^{m-2} \subset \mathcal{S}_n$, contradicting the fact that $\mathcal{H}^{m-2}(\mathcal{S}_n) = 0$. Then we have $n \leq m - 3$ and thus $\phi^t(\mathcal{S}_n \cap B^m_1) = 0$ for $t > m - 3$. Since $\phi^s(\mathcal{S}_n \cap B^m_1) > 0$, we have $s \leq m - 3$, and since s can be any number smaller then $dim\, \mathcal{S}$ we have verified $dim\, \mathcal{S} \leq m - 3$.

We only have trivial MTM from $\mathbf{R}^j \to N_0$, if we make more assumption for $j = 1, \cdots, l$. Then we can continue further. If $n = m - 3$, then we have $f_n \in L^2_{1,\text{loc}}(\mathbf{R}^m, N_0)$ such that $f_n|_{B^m_1} \in \bar{\mathcal{H}}_{\lambda,B}$ and $f_n(x', x'') = \tilde{f}_n(x'')$, for $x' \in \mathbf{R}^{m-3}$, $x'' \in \mathbf{R}^3$, where $\tilde{f}_n$ has an isolated singularity at $x'' = 0$. Consequently, by (L2) $\tilde{f}_n \in L^2_{1,\text{loc}}(\mathbf{R}^3, N_0)$ is an MTM and thus trivial by assumption. Therefore, we have $n \leq m - 4$. We can repeat the same arguments for $n = m - 4, \cdots, m - l$ and conclude that $n \leq m - l - 1$ which then implies $s \leq m - l - 1$ for any $s < dim\, \mathcal{S}$ and hence $dim\, \mathcal{S} \leq m - l - 1$.

Lastly, suppose we had $m = l + 1$ and $\mathcal{S} \neq \emptyset$. If $p_0 \in \mathcal{S}$ and $f_0 \in L^2_{1,\text{loc}}(\mathbf{R}^m, N_0)$ is a blown-up harmonic map at p_0, then the preceding discussion implies that f_0 has singular set $\mathcal{S}_0 = \{0\}$ and f_0 is an MTM. If there were a sequence $p_i \in \mathcal{S}$ with $p_i \to p_0$, then we could select $\lambda(i) = 4\, dist(p_i, p_0)$ and think of the scaled maps $f_{\lambda(i)} \in L^2_1(B^m_1, N_0)$. By the choice of $\lambda(i)$, we have $\mathcal{S}_{\lambda_i} \cap \partial B^m_{1/4} \neq \emptyset$ for each i. Because the limit f_0 has an isolated singularity at 0, this contradicts (L1). Consequently, $\mathcal{S}$ is discrete. The same argument shows that for $m = 3$ either $\mathcal{S} = \emptyset$ or $\mathcal{S}$ is discrete. Hence, the assertions of Theorems 1.2.3 and 1.2.5 hold true. □

Corollary 1.2.6. *If the sectional curvature of N is non-positive or if f(M) is contained in a strictly convex ball of N, then* $\mathcal{S}_f = \emptyset$*, i.e., any* $\tilde{E}$*-minimizing map* $f \in L_1^2(M, N)$ *is smooth.*

Proof. In order to apply Theorem 1.2.5, we want to show that any tangent map $\mathbf{R}^j \to N$ for $j \geq 3$ is trivial, i.e., any smooth harmonic map $f : S^{j-1} \to N$ is trivial. Since N has non-positive curvature, we can lift f to $\tilde{f} : S^{j-1} \to \tilde{N}$ where $\tilde{N}$ is the universal cover of N. Because the square of the distance function ρ to a point is strictly convex, $\rho^2 \circ f$ is a subharmonic function on S^{j-1} which is constant. Therefore, f is constant. Likewise, we can apply a similar argument if $f(M)$ is contained in a strictly convex ball of N. □

(1.2.B) For $m \geq 2$ an L_1^2-map may not be continuous, and there is no natural concept of its homotopy class. But, *an* L_1^2*-map* $f : (M, g) \to (N, h)$ *induces a conjugacy class of homomorphisms* $f_* : \pi_1(M) \to \pi_1(N)$*, and that action is preserved under weak limits in* $L_1^2(M, N)$. That was shown by Schoen and Yau [323] for $m = 2$. A detailed proof was given by Burstall [47] and White [402–404] for $m \geq 2$. The main idea is that even though f does not restrict to a continuous map on a given loop α in M, it does so restrict on almost every loop in a tubular neighborhood of α; their images are homotopic and thereby define $f_*[\alpha]$.

The fact is useful when $\pi_i(N) = 0$ for $i \geq 2$. Then the homotopy classes of maps $M \to N$ are parameterized by the conjugacy classes of homomorphisms $\pi_1(M) \to \pi_1(N)$.

1.2.3 Existence

The above regularity together with control of the π_1-action yields both new existence theorems and new proofs of known ones. In fact, the direct method provides an energy minimizing map in any class of L_1^2-maps with prescribed action on π_1. With further geometric assumptions such a map can be shown to be smooth. Supposing that M and N are compact, we discuss some of those as follows.

If every compact subset of U has a neighborhood U' on which there is a function $k : U' \to \mathbf{R}$ with positive definite Hessian Ddk, we say that $U \subset N$ is *convex supporting*. A theorem of Gordon [158] reads as follows: *Any harmonic map f from a compact manifold* (M, g) *to a convex supporting domain U is constant.* If k is a convex function on a neighborhood of the image $f(M)$ of the harmonic map f, then $k \circ f$ is subharmonic and thus constant. In general, it is sufficient that k has a positive semi-definite Hessian that is definite at some point of $f(M)$.

If the universal cover $(\tilde{N}, h)$ of (N, h) is convex supporting, then any harmonic map $\phi : S^{i-1} \to (N, h)$ is constant for $i \geq 3$. In fact, such a map lifts as a harmonic map $\tilde{\phi} : S^{i-1} \to (\tilde{N}, h)$. In particular, the assumptions of Theorem 1.2.5 are satisfied and every energy minimizing map $f : (M, g) \to (N, h)$ is smooth.

Here are a few examples of manifolds (N, h) whose universal covers are convex supporting.

Example 1. (N, h) has sectional curvature ≤ 0. More generally, $(\tilde{N}, h)$ has no focal point (i.e., no complete geodesic $\gamma : \mathbf{R} \to (\tilde{N}, h)$ has focal points along any perpendicular geodesic, where we consider γ as a submanifold of (N, h)). In this situation, the function $\psi(y) = dist^2(b, y)$ for a point $b \in \tilde{N}$ is smooth and strictly convex (cf. Eberlein [111] and Xin [423, 424]).

Example 2. (N, h) is a surface with no conjugate points (i.e., any two points of $(\tilde{N}, h)$ can be joined by exactly one geodesic). Then $(\tilde{N}, h)$ is convex supporting (cf. Burns [46]).

Applications of Theorem 1.2.5 and the direct method imply that in these cases *any map $f_0 : (M, g) \to (N, h)$ can be deformed into a harmonic map $f : (M, g) \to (N, h)$ whose energy is an absolute minimum in its homotopy class.* The case $Riem^N \leq 0$ was obtained by Eells and Sampson [129], while the other case $(\tilde{N}, h)$ with no focal points is due to Xin [423].

(1.2.C) The theorem of Morrey [270, 272] (in the beginning of Sect. 1.2.2) shows that when $dim\, M = 2$ a minimizing map has no singularity. The homotopy classes of maps of a surface M into a manifold N with $\pi_2(N) = 0$ are canonically identified with the conjugacy classes of homomorphism $\pi_1(M) \to \pi_1(N)$. Hence, we can apply Theorem 1.2.5 and the direct method: *Any map $f_0 : (M, g) \to (N, h)$ of a surface into a manifold N with $\pi_2(N) = 0$ can be deformed into a harmonic map which is an E-minimum in its conjugacy class* (cf. Schoen and Yau [323], Lemaire [245, 246, 248, 249], and Sacks and Uhlenbeck [310]).

If $dim\, M = 3$ and N is any surface not homeomorphic to S^2 or $P_2(\mathbf{R})$, *then with respect to any metrics, any map $f_0 : (M, g) \to (N, h)$ can be deformed into a harmonic map which is an E-minimum* (see Eells [113, 114] and White [404]). However, any harmonic map $f : S^2 \to (N, h)$ is conformal and constant.

1.2.4 Removable Singularities

A subset $E \subset (M, g)$ is *polar* (or 2-polar) if any L^2_1-function defined on a domain $V \subset M$ and harmonic on V-$(V \cap E)$ is harmonic on V. In general, E is *q-polar* $(1 < q < \infty)$ if 0 is the only distribution in $L^2_{-1}(M, \mathbf{R})$ with compact support in E, where $\frac{1}{p} + \frac{1}{q} = 1$. A set is *q-polar* if and only if every compact subset of E is *q-polar*. A compact set is q-polar if and only if $C_q(E) = 0$, where $C_q(E)$ is the *q-capacity* of E, i.e., the infimum of the L^q_1-norms of the smooth functions on M with values ≥ 1 on E.

Eells and Polking [126] proved that *if E is a polar set of (M, g) and $f \in L^2_1(M, N)$ a map weakly harmonic on $M - E$, then f is weakly harmonic on M.*

Let $B(y_0, \rho)$ be the geodesic ball of (N, h) with center y_0 and radius ρ and suppose that it is disjoint from the cut-locus of its centre and that $\rho < \pi/2\sqrt{Const_N}$, where $Riem^N \leq Const_N$. We call such a ball geodesically small. Meier [262] showed that *if* $f(M - E)$ *is a geodesically small ball of* (N, h), *then* f *is smooth and harmonic on* M.

(1.2.D) Sacks and Uhlenbeck [310] obtained that *any harmonic map from the 2-disk* $D^2 - \{0\}$ *to* (N, h) *of finite energy extends to a harmonic map on* D^2.

In contrast, Eells and Polking [126] applied the implicit mapping theorem to prove that *if* $f_0 : (M, g) \to (N, h)$ *is harmonic and* $E \subset M$ *is not q-polar with* $\frac{1}{p} + \frac{1}{q} = 1$ *and* $2 \leq m < p < \infty$, *then there is a map* $f \in L_1^2(M, N)$ *arbitrarily near* f_0 *which is harmonic on* $M - E$ *but not on* M.

Following Brezis, Coron, and Lieb [42], consider maps $\mathbf{R}^3 \to S^2$ with point singularities. Let $a_1, \cdots, a_p$ be p points in $\mathbf{R}^3$ and $d_1, \cdots, d_p$ be p non-zero integers such that $\sum d_i = 0$. Let

$$\mathcal{E} = \{f \in L_1^2(\mathbf{R}^3, S^2) \cap C^\infty(\mathbf{R}^3 - \cup\{a_i\}, S^2) : deg(f, a_i) = d_i\}$$

where $deg(f, a_i)$ is the Brouwer degree of f restricted to a small a_i-centered 2-sphere; $\mathcal{E} \neq \emptyset$, since $\sum d_i = 0$. Denote by $p_1, \cdots, p_q$ the list of the a_i with $d_i > 0$, with a_i repeated d_i times; and by $n_1, \cdots, n_q$ the list of the a_i with $d_i < 0$. Let $L = min_\sigma \sum_i d(p_i, n_{\sigma(i)})$, where σ is a permutation of the q-indices. Then Brezis, Coron, and Lieb [28, 40, 42] proved that *the infimum of the energy of* f *over* $\mathcal{E}$ *equals* $4\pi L$ *and is not realized by a map in* $\mathcal{E}$.

Bethuel and Zheng [33] showed the following statement concerning the definition of $\mathcal{E}$: *Every map in* $L_1^2(D^3, S^2)$ *can be* L_1^2*-approximated by maps that are smooth except at a finite number of points.*

1.3 Maps of Surfaces

We concentrate on the famous work of Sacks and Uhlenbeck [310, 311] about the existence of harmonic maps of surfaces. We then summarize some important results regarding harmonic diffeomorphisms, minimal surfaces, and surfaces of constant mean curvature.

1.3.1 Existence

Let (M, g) be a compact Riemann surface (not necessarily orientable), (M, ρ) be a compact Riemann surface (with a conformal class of metrics ρ), and (N, h) be a compact Riemannian manifold. Note that a harmonic map $f : (M, \rho) \to (N, h)$ is well defined as the harmonic map $f : (M, g) \to (N, h)$ for any g in the class ρ. As we have seen (1.2.C), Morrey's regularity theorem and the control of π_1-action imply that if N is a compact manifold with $\pi_2(N) = 0$, then any homotopy class of maps from $(M, \rho) \to (N, h)$ contains an energy minimizing harmonic map. We will discuss more as follows.

We approximate the energy functional E, whose critical points are harmonic maps, by a slightly different integral. Let

$$E_\alpha(f) = \int_M (1 + |df|^2)^\alpha dx \quad (\alpha > 1)$$

be the perturbed energy (for simplicity, we may choose a measure on M such that the area of M is 1). For $\alpha = 1$, $E_1(f) = 1 + E(f)$ has harmonic maps as critical points. For $\alpha > 1$, E_α behaves well. The Sobolev spaces of maps $L_1^{2\alpha}(M, N) = \{f \in L_1^{2\alpha}(M, \mathbf{R}^k) : f(x) \in N\} \subset C^0(M, N)$ is a C^2 separable Banach manifold for $\alpha > 1$.

Sacks and Uhlenbeck [310, 311] showed that *the critical maps of E_α in $L_1^{2\alpha}(M, N)$ are C^∞ if $\alpha > 1$*. They also obtained all the following theorems and results.

Theorem 1.3.1. *If N is compact and $\pi_2(N) = 0$, then there exists a minimizing harmonic map in every homotopy class of maps in $C^0(M, N)$.*

In order to prove the above theorem, we require the following three lemmas (cf. [310]).

(L1) *In every connected component of $L_1^{2\alpha}(M, N)$ $(\alpha > 1)$ the minimum value of E_α is taken on at some map $f_\alpha \in C^\infty(M, N)$, which also minimizes E_α in its connected component in $C^\infty(M, N)$. There exists a $K(= \max_{x \in M} |df(x)|)$ independent of α such that* $\min E_\alpha \le (1 + K^2)^\alpha$ *in that component.*

(L2) *Let $U \subset M$ be a open set and $f_\alpha : U \to N$ be a critical map of E_α and $E(f_\alpha) < K$, $\alpha \to 1$ and $f_\alpha \to f$ weakly in $L_1^2(U, \mathbf{R}^k)$. Then there exist a subsequence $\{f_\beta\}$ of $\{f_\alpha\}$ and a finite number of points $\{x_1, \cdots, x_l\}$, where l does not depend on U, such that $f_\beta \to f$ in $C^1(U - \{x_1, \cdots, x_l\}, N)$. Furthermore, $f : U \to N$ is a smooth harmonic map.*

(L3) *Let $f_\alpha : D(R) \to N$ be a sequence of critical maps of E_α for $\alpha \to 1$, and suppose that f_α converges weakly in $L_1^{2\alpha}(D(R), \mathbf{R}^k)$, where $D(R)$ is a disk with radius R and center the origin. Then there exists $\epsilon > 0$ such that if $E(f_\alpha) < \epsilon$, then $f_\alpha \to f$ in $C^1(D(R/2), N)$ and $f : D(R/2) \to N$ is a smooth harmonic map.*

Proof (Proof of Theorem 1.3.1). Let $f_\alpha : M \to N$ be a minimizing map for E_α in a fixed homotopy class with $E_\alpha(f_\alpha) < (1 + K^2)^\alpha$ as in (L1). Letting $M = U$ in (L2), we can select a subsequence $\beta \to 1$ such that f_β converges to f in $C^1(M - \{x_1, \cdots, x_l\}, N)$ with $f : M \to N$ is harmonic. We want to show that $f_\beta \to f$ in $C^1(M, N)$.

Let $D(\sigma)$ be a small disk centered at x_i in M with radius σ, where σ is sufficiently small such that $x_j \in D(\sigma)$ for $j \neq i$ (σ will be chosen later). Define a modified function $\tilde{f_\beta} : D(\sigma) \to N$ which coincides with f_β on the boundary of $D(\sigma)$ and with f in the center. Let μ be a smooth function which is equal to 1 on $r \geq 1$ and 0 on $r \leq 1/2$. Let *exp* be the exponential map on N.

$$\tilde{f_\beta}(x) = exp_{f(x)}\left(\mu(x/\sigma)exp^{-1}_{f^{-1}(x)} \circ f_\beta(x)\right). \tag{1.7}$$

Then $f_\beta \to f$ in $C^1(supp(\mu(x/\sigma)) \cap D(\sigma), N)$ and we get $\tilde{f_\beta} \to f$ in $C^1(D(\sigma), N)$. Notice that for $f \in L_1^{2\alpha}(M, N)$, $\tilde{E}_\alpha(f) = \int_{S^2}(1 + |df|^2)^\alpha dv - 1$. It follows from that

$$\lim_{\beta\to 1} \tilde{E}_\beta(\tilde{f_\beta}) = E(f|_{D(\sigma)}). \tag{1.8}$$

By hypothesis, $\pi_2(N) = 0$, so f_β is homotopic to $\tilde{f_\beta}$. Because f_β is a minimizing function for E_β in its homotopy class, $E_\beta(f_\beta|_{D(\sigma)}) \leq E_\beta(\tilde{f_\beta}|_{D(\sigma)})$. The equation (1.8) implies that

$$\overline{\lim_{\beta\to 1}} E_\beta(f_\beta|_{D(\sigma)}) \leq E(f|_{D(\sigma)}) \leq \sigma^2\pi||f||^2_{1,\infty}.$$

If we choose σ such that $\sigma^2\pi||f||^2_{1,\infty} < \epsilon/2$ from the beginning, we can utilize (L3) to get $f_\beta \to f$ in $C^1(D(\sigma), N)$ since $E_\beta(f_\beta|_{D(\sigma)}) < \epsilon$ for $\beta \to 1$. We can conclude $f_\beta \to f$ in $C^1(M, N)$. Since f_β minimizes $\tilde{E}_\beta$, f must minimize E in the same homotopy class. □

In order to prove Theorem 1.3.2, we need the following lemmas.

(L4) *Let f_α be a sequence of critical maps of E_α for $\alpha \to 1$, and $f_\alpha \to f$ weakly in $L_1^2(M, \mathbf{R}^k)$. Then either $f_\alpha \to f$ in $C^1(M, N)$ or there exists a non-trivial harmonic map $\tilde{f} : S^2 \to N$ with $\tilde{f}(S^2) \subset \bigcap_{\alpha\to 1} \overline{\bigcup_{\beta<\alpha} f_\beta(M)}$. Furthermore, $E(f) + E(\tilde{f}) \leq \overline{\lim}_{\alpha\to 1} E(f_\alpha)$.*

(L5) *Let $\epsilon_0 = \min E(f)$ for $f : S^2 \to N$, f harmonic and not a map to a point and $\epsilon_0 = \infty$ if this set of harmonic maps is empty. Then $E|_{E^{-1}[0,\epsilon_0)}$ satisfies a Morse theory for $M \neq S^2$, and $E|_{E^{-1}[0,2\epsilon_0)}$ satisfies a Morse theory for $M = S^2$.*

(L6) *If the domain $M = S^2$ and the universal covering space of N is not contractible, then there exist a K and a critical map of E_α with values in the range $(1, (1 + K^2)^\alpha)$ for $\alpha > 1$.*

Theorem 1.3.2. *If the universal covering space of N is not contractible, then there exists a non-trivial smooth conformal branched minimal immersion $f : S^2 \to N$.*

Proof. It follows from (L6) that we have a critical map f_α of $\tilde{E}_\alpha$ with $\epsilon < \tilde{E}_\alpha < K$. Then, by (L2), there is a subsequence, which we still call f_α, such that $f_\alpha \to f$ in

$C^1(S^2 - \{x_1, \cdots, x_l\}, N)$ as $\alpha \to 1$ and f is harmonic. If f is not a map to a point, we are done. If f is a point, due to $\epsilon < \tilde{E}_\alpha(f_\alpha)$, f_α must not converge to f at some point. By (L4) there exists a harmonic map $\tilde{f}$ with $\tilde{f}(S^2) \subset \bigcap_\alpha \overline{\bigcup_{\beta<\alpha} f_\alpha(S^2)}$. Remark that the image of a harmonic map from S^2 to N is a conformal branched minimal immersion. Combining this statement with (L5), we complete the proof of the theorem. Notice that the assumption on the universal covering space of N cannot be omitted, for if N has non-positive curvature then any harmonic map $f : S^2 \to N$ is constant. □

If $\pi_2(N) \neq 0$, then we have the following Theorem 1.3.3 and thus there exists a generating set for $\pi_2(N)$ consisting of conformal branched minimal immersion of spheres which minimize energy and area in their homotopy classes as in Corollary 1.3.4.

Theorem 1.3.3. *There exists a set of free homotopy classes $\Lambda_i \subset \pi_0 C^0(S^2, N)$ such that elements $\{\lambda \in \Lambda_i\}$ form a generating set for $\pi_2(N)$ acted on by $\pi_1(N)$ and each Λ_i contains a minimizing harmonic map $f_i : S^2 \to N$.*

Proof. Let Λ_i be the homotopy classes containing minimizing harmonic maps. Let $B \subset \pi_2(N)$ be the subgroup generated by elements $\{\lambda_i \in \Lambda_i\}$. Assume that the inclusion is proper. Choose a class Γ with elements $\gamma \in \Gamma, \gamma \notin B$, such that if $\#\Gamma' \leq \#\Gamma - \epsilon/2$, then the element $\{\gamma' \in \Gamma'\} \subset B$.

By hypothesis, Γ does not contain a minimizing harmonic map, so there exist Γ_1 and Γ_2 with $\pi_1(N)\gamma \subset \pi_1(N)\gamma_1 + \pi_1(N)\gamma_2$ and $\#\Gamma_1 + \#\Gamma_2 < \#\Gamma + \epsilon/2$. But Γ_1 and Γ_2 are not trivial, so $\#\Gamma_i \geq \epsilon$ for $i = 1, 2$. This implies $\#\Gamma_i < \#\Gamma - \epsilon/2$. By hypothesis, the sets $\pi_1(N)\gamma_i$ are both in B, so

$$\pi_1(N)\gamma \subset \pi_1(N)\gamma_1 + \pi_1(N)\gamma_2 \subset B.$$ □

Corollary 1.3.4. *There exists a set of free homotopy classes $\Lambda_i \subset \pi_0 C^0(S^2, N)$ such that elements $\{\lambda \in \Lambda_i\}$ form a generating set for $\pi_2(N)$ acted on by $\pi_1(N)$ and each Λ_i contains a conformal branched immersions of a sphere having the least area among maps S^2 into N which lie in Λ_i.*

Remark that Theorem 1.3.2 does not mean that each homotopy class of maps from S^2 to N contains a harmonic representative, although no counterexample has been produced. However, Futaki [151] proved that *certain homotopy classes from S^2 to a Hirzebruch surface (of real dimension 4) do not contain any energy minimizing harmonic map.* The idea is that $\pi_2(N) = \mathbf{Z} \oplus \mathbf{Z}$ and that two generators α, β of $\pi_2(N)$ have intersection properties such that a minimizing harmonic sphere in $\alpha + \beta$ would break into two separate spheres, one in α and one in β, one having bubbled off the other.

Problem: Let M and N be compact surfaces and $\mathcal{H} \in [M, N]$ a homotopy class. Do there exist metrics g and h on M and N such that $\mathcal{H}$ contains a harmonic map $f : (M, g) \to (N, h)$?

(1.3.A) The results of Lemaire [246, 247], Eells and Wood [133], Eells and Lemaire [121], and Adams [2,3] show that the answer to the above problem is yes – except the following cases (no):

(i) $M = T =$ the 2-torus, $N = S =$ the 2-sphere and $|deg(\mathcal{H})| = 1$, $deg(\mathcal{H})$ is the common degree of the maps $f \in \mathcal{H}$.
(ii) $M = T$, $N = P =$ the real projective plane and $\mathcal{H}$ is the class from (i) by composition with the covering map $\pi : S \to P$;
(iii) $M = P$, $N = S$ and $\mathcal{H}$ is the non-trivial homotopy class;
(iv) $M = P$, $N = P$ and $\mathcal{H}$ is the class from (iii) by composition with π.

Let $j_{\mathcal{H}}$ be the index of the subgroup $f_*(\pi_1(M))$ in $\pi_1(N)$. Edmonds [112] gave the following solution to the above problem for holomorphic maps. *Let M and N be compact orientable surfaces and $\mathcal{H}$ a homotopy class with $deg\,\mathcal{H} \neq 0$. For any complex structure on N, there is a complex structure on M relative to which $\mathcal{H}$ contains a holomorphic map if and only if $|deg\,\mathcal{H}| > j_{\mathcal{H}}$ or $f_* : \pi_1(M) \to \pi_1(N)$ is injective.*

1.3.2 Harmonic Diffeomorphisms

(1.3.B) Jost and Schoen [207] obtained the following result: *Let (M, g) and (N, h) be homeomorphic closed surfaces. Any diffeomorphism $\phi : M \to N$ is homotopic to a harmonic diffeomorphism f of least energy amongst all diffeomorphisms homotopic to ϕ.*

Coron and Hélein [93] showed that f is energy minimizing amongst all maps in the homotopy class of ϕ. The map f is unique if $genus(M) \geq 2$.

If $Riem^N < 0$, it was verified by Sampson [313] and Schoen and Yau [323] that the unique harmonic map in the class is a diffeomorphism.

Shibata and Sealey [325,337] in an attempt to prove (1.3.B), raised the following question. If $f \in C^0 \cap L_1^2(M, N)$ is a quasi-conformal homeomorphism for which $(f^*h)^{2,0}$ is holomorphic as a distribution on M, is f harmonic? If f were smooth that would be true (cf. [119]). This seems doubtful, as Jost proved in [202] that *any continuous map $\phi : (M, g) \to (N, h)$ is homotopic to a map $f \in C^0 \cap L_1^2(M, N)$ for which $(f^*h)^{2,0}$ is holomorphic.* Jost [202,204] also constructed a Lipschitz map $f : T \to S$ of degree 1, which by (1.3.A) is not harmonic.

Lelong-Ferrand [244] showed: *Let M, N be holomorphic surfaces of common genus ≥ 1. Then for any $c > 0$, $\{f \in L_1^2(M, N)$: f is a homeomorphism and $E(f) \leq c\}$ is equicontinuous.*

In contrast to (1.3.B), Calabi's construction [59] implies [121] that *there exist smooth metrics g on the 3-dimensional torus $\mathbf{T}^3$ such that any harmonic map f of $(\mathbf{T}^3, g)$ to the flat torus $(\mathbf{T}^3, g_0)$ has singularities* (i.e., points where the Jacobian J_f vanishes).

1.3.3 Minimal Surfaces

We note that any map $f : (M, \rho) \to (N, h)$ from a Riemann surface to a Riemannian manifold which is conformal and harmonic is minimal in the sense that it is an extremal of the area functional and is a branched immersion. If $M = S^2$, any harmonic map is conformal. If *genus(M)* ≥ 1, Sacks and Uhlenbeck [311, 385] proved that *a map f which is an extremal of E for all variations of the conformal structure on M and all deformations of f is conformal and harmonic.*

(1.3.C) This was used by Sacks and Uhlenbeck [311] and Schoen and Yau [323, 324] to prove that *if M is a surface, (N, h) is a manifold, and if* $\theta : \pi_1(M) \to \pi_1(N)$ *is an injective homomorphism, then there exist a conformal structure* ρ *on M and an area minimizing conformal harmonic map* $f : (M, \rho) \to (N, h)$ *such that* θ *is contained in the conjugacy class of the homomorphism* f_*. *When* $\pi_2(N) = 0$, *the homotopy class of f is determined by* θ.

The proof requires two minimizing processes. First, in the class of maps inducing θ, one looks for harmonic maps for given conformal structures ρ on M; then one minimizes the energy amongst the conformal structures. For simplicity, suppose that M is oriented and that the space of conformal structures is identified with the space of complex structures, which is not compact (cf. [110]). But if the structures ρ of a minimizing sequence leave all compact sets, the length of a closed geodesic of M tends to zero. Analysis of the energy shows that its image in N tends to a point, thus contradicting the fact that f_* is injective.

In general, a minimal map f obtained by (1.3.C) is a branched immersion. However, if $\dim N = 3$, a method of Osserman and Gulliver [167, 168, 283] and Alt [7] shows that *its Jacobian does not vanish, so that f is an immersion.*

(1.3.D) Minimal surfaces in the sphere S^3 were constructed by Lawson [241]: *For any compact surface M, except the projective plane, there exists a minimal immersion of M in* S^3, *for suitable* ρ *on M.*
If M is orientable, there exists a minimal embedding; and if the genus of M is not prime, the embedding is not unique.

The construction is made by successively reflecting geodesic polygonal minimal surfaces in S^3. On the other hand, there is no non-constant harmonic map from P (the real projective plane) to S^3 (cf. [120, 241]).

Lawson also showed that *a compact embedded minimal surface in* S^3 *separates it into two diffeomorphic components.* In all his examples, these components have equal volumes. But [215] produced *new minimal surfaces in* S^3 *separating it into two components of different volumes.*

Smyth [351] showed that every flat torus $\mathbf{T}^3$ contains minimally embedded surfaces of any high genus. (cf. Meeks [261], Micallef [263, 264] and Aviles [17].)

Bryant [44] proved that *any compact Riemann surface M can be conformally harmonically immersed in* S^4.

1.3.4 Surfaces of Constant Mean Curvature

For each p, (1.3.C) and (1.3.D) produce embedded minimal surfaces of genus p in S^3. Combining them with the canonical embedding $S^3 \to \mathbf{R}^4$ provides surfaces of constant mean curvature of every genus. Their Gauss maps $\gamma : M \to G_2^0(\mathbf{R}^4) = S_+ \times S_-$ decompose as $\gamma = (\gamma_+, \gamma_-)$ using the Riemannian product structure on the 2-spheres S_+ and S_-. The degree of γ_+ and γ_- is $1 - p$ (cf. [70]). Eells and Lemaire [120] showed that *for every oriented surface of genus p, the homotopy class of maps of degree* $1 - p$ *of* M *to* S *can be rendered harmonic by a Gauss map* $\gamma_\pm$.

Similarly, *the non-trivial homotopy class of maps from the Klein bottle K to S is represented harmonically via the Gauss map of Delaunay's nodoid in* $\mathbf{R}^3$ (cf. [120], using the identification $G_2^0(\mathbf{R}^3) = S$).

Given an oriented surface M of constant mean curvature in $\mathbf{R}^3$, its Gauss map is harmonic from M to S^2. Conversely, for any harmonic map γ from a simply connected surface to S^2 and for any positive number H, Kenmotsu [220] showed an explicit representation of a branched immersion $f : M \to \mathbf{R}^3$ with constant mean curvature H such that $\gamma_f = \gamma$.

The representation of f is derived by the following Weierstrass formula. Let γ_1 be the map from $M \setminus \gamma^{-1}(0,0,1)$ to $\mathbf{R}^2$ obtained by composing γ with the stereographic projection and put

$$Q = \left(\frac{-(1-\bar{\gamma}_1)^2}{H(1+|\gamma_1|^2)^2}, \frac{i(1+\bar{\gamma}_1)^2}{H(1+|\gamma_1|^2)^2}, \frac{-2\bar{\gamma}_1}{H(1+|\gamma_1|^2)^2} \right).$$

Then f is given by

$$f(z) = 2Re \int_0^z \bar{Q} \frac{\partial \bar{\gamma}_1}{\partial z} + c.$$

If M is not simply connected, the above formula does not provide a surface in $\mathbf{R}^3$ due to the difficulty with the periods of the integrals. In general, we have the following classical results:

(i) *Any isometric immersion* $f : S^2 \to \mathbf{R}^3$ *with constant mean curvature is an isometric embedding onto a Euclidean sphere* [189].
(ii) *Any compact isometrically embedded hypersurface* $f : (M^{n-1}, g) \to \mathbf{R}^n$ *of constant mean curvature is a Euclidean* $(n-1)$*-sphere* [4].

Wente [405] obtained that *certain tori* $\mathbf{T}^2$ *can be isometrically immersed in* $\mathbf{R}^3$ *with constant mean curvature.*

Kapouleas [213] proved that *for any* $p \geq 3$, *there are infinitely many orientable surfaces of genus* p *immersed in* $\mathbf{R}^3$ *with constant mean curvature.*

1.4 Maps of Kähler Manifolds

In this section, we discuss complex structures, negative curvature and rigidity, and complex variations. We focus on the complex-analyticity of harmonic maps and the strong rigidity of compact Kähler manifolds obtained by Siu [342, 343, 345, 346].

1.4.1 Complex Structures and Manifolds

Let M be an almost complex manifold, i.e., a real manifold with a field J of endormorphisms of $T(M)$ such that $J^2 = -id$. The real dimension of M is even, say $2m$ (m is the complex dimension). The operator J can be extended linearly to an operator (still denoted by J) on the complexified tangent space $T^{\mathbf{C}}(M)$ (with fibre $T_x(M) \otimes \mathbf{C}$ at x); we denote by $T'(M)$ and $T''(M)$ the bundles of eigenspaces of J on $T^{\mathbf{C}}(M)$ associated to the eigenvalues i and $-i$. Then $T'(M)$ (resp. $T''(M)$) is called the *holomorphic* (resp. *antiholomorphic*) tangent bundle; we have $T^{\mathbf{C}}(M) = T'(M) \oplus T''(M)$ and $T''(M) = \overline{T'(M)}$, the complex conjugate. This decomposition into complex types induces a dual decomposition $T^{*\mathbf{C}}(M) = T^{*'}(M) \oplus T^{*''}(M)$.

Suppose that M is equipped with a hermitian metric g i.e., a Riemannian metric g such that $g(JX, JY) = g(X, Y)$. The metric g extends to a complex bilinear form on $T^{\mathbf{C}}(M)$, and induces on $T'(M)$ the Hermitian form which associates to $X, Y \in T'_x(M)$ the number $g(X, \bar{Y})$. The *Kähler form* ω on (M, g, J) is the 2-form $\omega(X, Y) = g(X, JY)$. The manifold (M, g, J) is called *almost Kähler* if $d\omega = 0$. If the operator J is induced by a complex structure on M (i.e., it is multiplication by i in the charts of a holomorphic atlas), we talk about complex, Hermitian or *Kähler* manifolds.

We notice that if (M, g) is an orientable surface, it admits a compatible complex structure J such that (M, g, J) is Kähler. Then (M, J) is called a Riemann surface. A map $f : (M, g, J) \to (N, h)$ from a Riemann surface into a Riemannian manifold is harmonic iff it is so for any Hermitian metric on (M, J). The conformal class of these metrics is denoted by ρ, and we discuss harmonic maps from the Riemann surface (M, ρ) to (N, h).

Let $f : (M, g, J) \to (N, h, J)$ be a smooth map between almost hermitian manifolds. Its complexified differential $d^{\mathbf{C}} f : T^{\mathbf{C}}(M) \to T^{\mathbf{C}}(N)$ determines partial differentials by composition with the inclusions of $T'(M)$ and $T''(M)$ in $T^{\mathbf{C}}(M)$ and projections of $T^{\mathbf{C}}(N)$ on $T'(N)$ and $T''(N)$ as follows:

$$\partial f : T'(M) \to T'(N), \quad \bar{\partial} f : T''(M) \to T'(N);$$
$$\partial \bar{f} : T'(M) \to T''(N), \quad \bar{\partial} \bar{f} : T''(M) \to T''(N).$$

We get $\bar{\partial}\bar{f} = \overline{\partial f}, \partial\bar{f} = \overline{\bar{\partial} f}$, $d^{\mathbf{C}} f|_{T'(M)} = \partial f + \partial\bar{f}$ and $d^{\mathbf{C}} f|_{T''(M)} = \bar{\partial} f + \bar{\partial}\bar{f}$. A map f is holomorphic (resp. anti-holomorphic) iff $J \circ df = df \circ J$, i.e., $\bar{\partial} f = 0$ (resp. $J \circ df = -df \circ J$, i.e., $\partial f = 0$). A map f is $\pm$ holomorphic if it is holomorphic or antiholomorphic.

Let $e'(f) = |\partial f|^2$ and $e''(f) = |\bar{\partial} f|^2$ be the partial energy densities and $E'(f)$ and let $E''(f)$ be their integrals. Then $E(f) = E'(f) + E''(f)$ and f is holomorphic (resp. antiholomorphic) iff $E''(f) = 0$ (resp. $E'(f) = 0$). The relationship between $\pm$ holomorphic maps and harmonic maps can describe as follows.

Let (M, g, J) and (N, h, J) be almost hermitian manifolds. If M is *cosymplectic* (i.e., the codifferential of ω^M vanishes: $d^*\omega^M = 0$) and if N is (1, 2)-*symplectic* (i.e., the components of complex type (1, 2) of the 3-form $(d\omega^N)^{1,2} = 0$), then any $\pm$ holomorphic map $f : M \to N$ is harmonic.

Notice that these assumptions are satisfied if M and N are almost Kähler. In this case if M is compact, $E'(f) - E''(f)$ is constant on each homotopy class (by Lichnerowicz [252]) and for any variation f_t

$$\frac{\partial E'(f_t)}{\partial t} = \frac{\partial E''(f_t)}{\partial t} = \frac{1}{2}\frac{\partial E(f_t)}{\partial t}. \tag{1.9}$$

If both (M, g, J) and (N, h, J) are Kähler manifolds and if $\tau'(f)$ and $\tau''(f)$ denote the Euler-Lagrange operators of E' and E', then $\tau''(f) = \overline{\tau'(f)}$, $\tau(f) = \tau'(f) + \tau''(f)$ and τ, τ', τ'' vanish simultaneously.

(1.4.A) A vector bundle $B \to M$ is complex if the fibres are complex vector spaces. It is holomorphic if M is a complex manifold and the transition maps are holomorphic (as maps from the intersections of charts to $GL(\mathbf{C}^k)$). The operators $\bar{\partial}$ is well-defined on the sections of a holomorphic bundle, since it cancels the transition maps.

In the case where M is a Riemann surface and $B \to M$ is a complex bundle equipped with a connection, Koszul and Malgrange [232] proved that B admits a holomorphic bundle structure such that $\frac{\partial}{\partial\bar{z}} = \nabla_{\frac{\partial}{\partial\bar{z}}}$ on sections.

Wood [413] showed that *if $f : (M, g) \to (N, h)$ is a harmonic map from a Riemann surface to a Kähler manifold, then the complex bundle $f^{-1}T'(N) \to M$ admits a holomorphic structure such that ∂f is a holomorphic section of the holomorphic bundle $T'^*(M) \otimes f^{-1}T'(N)$. In fact, $\frac{\partial}{\partial\bar{z}}\partial f = \nabla_{\partial/\partial\bar{z}}\partial f = 0$.*

If $f : (M, g) \to (N, h)$ is a harmonic non $\pm$ holomorphic map from a Riemann surface to a Kähler manifold, then f is $\pm$ holomorphic on at most a discrete set. Since ∂f and $\bar{\partial} f$ are holomorphic sections of appropriate bundles, they vanish only at isolated points.

(1.4.B) Let B be a holomorphic vector bundle over the sphere S^2 (e.g., a complex bundle with Koszul-Malgrange [232] structure). By the Birkhoff-Grothendieck structure theorem, B decomposes into a direct sum of holomorphic line bundles

$$B = L_1 \oplus L_2 \oplus L_3 \cdots \oplus L_k.$$

The isomorphism classes of the L_i are uniquely determined; they can be arranged in order such that

$$c_1(L_1) \geq c_1(L_2) \geq \cdots \geq c_1(L_k),$$

where $c_1(L)$ denotes the first Chern class of L . Note that, by the Riemann-Roch theorem, for a line bundle L over S^2 the dimension of the space of holomorphic sections is $c_1(L) + 1$, if $c_1(L) \geq 0$, and 0, if $c_1(L) < 0$.

1.4.2 Complex-Analyticity and Rigidity

We discuss Siu's [342, 343, 345, 346] main results about complex-analyticity of harmonic maps and the strong rigidity of compact Kähler manifolds. We first introduce the concept of negative curvature.

Let (N, h) be a Kähler manifold and define a hermitian form

$$Q : (T'_y(N) \otimes T''_y(N)) \times (T'_y(N) \otimes T''_y(N)) \to \mathbf{C}$$

as follows [422]: For any $X, Y, Z, W \in T'_y(N)$, set

$$Q(X \otimes \bar{Y}, Z \otimes \bar{W}) = < R(X, \bar{Z})W, \bar{Y} >$$

where R is the curvature tensor of (N, h) and then extend the definition to any elements of $T'_y(N) \otimes T''_y(N)$ by requiring linearity in the first factor and conjugate linearity in the second. An element in $T'_y(N) \otimes T''_y(N)$ of the form $Z \otimes \bar{W}$ is called decomposable. Q is called *negative definite* (resp. *semi-negative*) of level k on N if $Q(A, A) < 0$ (resp. ≤ 0) for each non-zero element $A \in T'_y(N) \otimes T''_y(N)$ which can be expressed as a sum of at most k decomposable elements.

Following the terminology of Siu [343,345,346], the curvature tensor of a Kähler manifold is *strongly negative* (resp. *strongly semi-negative*) if Q is negative definite (resp. semi-definite) of level 2 at every point; and is *very strongly negative* (resp. *very strongly semi-negative*) if Q is negative definite (resp. semi-definite) at all levels at every point. We note that a Kähler manifold with strongly negative curvature tensor has negative sectional curvature and that it has negative holomorphic bisectional (cf. (1.14)) if and only if Q is negative definite of level 1.

Here are some examples. The open disc $D^n \subset \mathbf{C}^n$ with its Bergman metric has very strongly negative curvature by Siu [346]; the product $D^1 \times \cdots \times D^1$ has strongly semi-negative curvature by Jost and Yau [115]. In both cases, quotients by discrete groups of holomorphic isometries give examples of compact manifolds with the same properties. On the other hand, Mostow and Siu [273, 274] provided examples of compact manifolds with very strongly negative curvature which are not quotients of D^n by discrete groups.

If N is compact with $R^N \leq 0$, every homotopy class of maps contains a harmonic representative; thus the above result leads (Siu [344,345]) to the existence of holomorphic maps and the strong rigidity of compact Kähler manifolds. All the following theorems and results were obtained by Siu [342,343,345,346].

Theorem 1.4.1. *Let (M, g) and (N, h) be compact Kähler manifolds and let the curvature of N be strongly negative. If $f : (M,g) \to (N,h)$ is a harmonic map and the rank of the differential df of f over* **R** *is at least 4 at some point of M, then f is ± holomorphic.*

Theorem 1.4.1 follows from Theorem 1.4.3. If the curvature tensor $R_{\alpha\bar{\beta}\gamma\bar{\delta}}$ of a Kähler manifold N is strongly negative, then it is negative of order 2. The curvature condition of Theorem 1.4.1 can be weakened to the 'curvature of N is strongly semi-negative everywhere and is strongly negative at $f(a)$ for some $a \in M$ with $rank_{\mathbf{R}}\, df \geq 4$ at a'. The following strong rigidity theorem is a consequence of Theorem 1.4.1.

Theorem 1.4.2 (Strong Rigidity). *If N is a compact Kähler manifold of complex dimension at least two whose curvature is strongly negative, then a compact Kähler manifold of the same homotopy type as N must be either biholomorphic or conjugate biholomorphic to N.*

Theorem 1.4.3. *Suppose M and N are compact Kähler manifolds and the curvature of N is negative of order $k \geq 2$. If $f : M \to N$ is a harmonic map and the rank of the differential df of f over* **R** *is at least 2k at some point of M, then f is ± holomorphic.*

The curvature tensor $R_{\alpha\bar{\beta}\gamma\bar{\delta}}$ is said to be *strongly negative* (resp. *strongly semi–negative*) if

$$\sum_{\alpha,\beta\,\gamma,\delta} R_{\alpha\bar{\beta}\gamma\bar{\delta}}(A^\alpha \bar{B}^\beta - C^\alpha \bar{D}^\beta)\overline{(A^\delta \bar{B}^\gamma - C^\delta \bar{D}^\gamma)}$$

is positive (resp. non-negative) for any complex numbers A^α, B^α, C^α, D^α with $A^\alpha \bar{B}^\beta - C^\alpha \bar{D}^\beta \neq 0$ for at least one pair of indices (α, β).

The curvature tensor $R_{\alpha\bar{\beta}\gamma\bar{\delta}}$ is said to be *negative of order k* at a point p of N if it is strongly semi-negative at p and if it satisfies the following property. If $f : O \to N$ is a smooth map from an neighborhood O of 0 in $\mathbf{C}^k$ to N with $f(0) = p$ and $rank_{\mathbf{R}} df = 2k$ at 0 and if

$$\sum_{\alpha,\beta\,\gamma,\delta} R_{\alpha\bar{\beta}\gamma\bar{\delta}}\xi^{\alpha\bar{\beta}}_{\bar{i}j}\overline{\xi^{\delta\bar{\gamma}}_{\bar{i}j}} = 0,\ 1 \leq i,\ j \leq k,$$

where $\xi^{\alpha\bar{\beta}}_{\bar{i}j} = (\partial_{\bar{i}} f^\alpha)(0)\overline{(\partial_j f^\beta)(0)} - (\partial_{\bar{j}} f^\alpha)(0)\overline{(\partial_i f^\beta)(0)}$, the coordinates of $\mathbf{C}^k$ being w^i, $1 \leq i \leq k$, then either $\partial f = 0$ at 0 or $\bar{\partial} f = 0$ at 0.

To prove the above theorem, we require the following three lemmas (proofs are given in [343]).

(L1) *Suppose V is a vector space of dimension m over $\mathbf{C}$ and W is an $\mathbf{R}$-vector subspace of V and the dimension of W over $\mathbf{R}$ is $\leq 2m-2k$. Let E be the set of all bases of V over $\mathbf{C}$ and let F be the subset of E consisting of all bases of V over $\mathbf{C}$ such that $(\sum_{i=1}^{k} \mathbf{C}e_i) \cap W = 0$ for all $1 \leq i_1 < \cdots < i_k \leq m$. Then F is a dense open subset of E.*

(L2) *Let M and N be compact Kähler manifolds with $dim_{\mathbf{C}} M = m$ and $dim_{\mathbf{C}} N = n$ and $f : M \to N$ be a harmonic map with components f^{α} $(1 \leq \alpha \leq n)$ in a local coordinate system of N. Let w^i be a local coordinate system of M and let $\xi^{\alpha\bar{\beta}}_{\bar{i}\bar{j}} = (\partial_{\bar{i}} f^{\alpha})\overline{(\partial_j f^{\beta})} - (\partial_{\bar{j}} f^{\alpha})\overline{(\partial_{\bar{i}} f^{\beta})}$. If the curvature tensor $R_{\alpha\bar{\beta}\gamma\bar{\delta}}$ of N is strongly semi-negative, then*

$$\sum_{\alpha,\beta\,\gamma,\delta} R_{\alpha\bar{\beta}\gamma\bar{\delta}} \xi^{\alpha\bar{\beta}}_{\bar{i}\bar{j}} \overline{\xi^{\delta\bar{\gamma}}_{\bar{i}\bar{j}}} = 0, \ 1 \leq i, \ j \leq m.$$

(L3) *Let $f : M \to N$ be a harmonic map between two compact Kähler manifolds, and O be a non-empty open subset of M. If f is $\pm$ holomorphic on O, then f is $\pm$ holomorphic on M.*

Proof (Proof of Theorem 1.4.3). First we show that the assumption that $rank_{\mathbf{R}}\, df \geq 2k$ at some point of M, and thus at every point of some non-empty connected open subset O of M yields

$$\partial f = 0 \text{ or } \bar{\partial} f = 0 \text{ on } O. \tag{1.10}$$

It is sufficient to show that for every point a of O

$$\partial f = 0 \text{ at } a \text{ or } \bar{\partial} f = 0 \text{ at } a, \tag{1.11}$$

since the nowhere-vanishing of df on O implies that the two closed subsets $O \cap \{\partial f = 0\}$ and $O \cap \{\bar{\partial} f = 0\}$ are disjoint and by (1.11) their union is the connected set O, and so one of them equals to O.

Because the curvature tensor $R_{\alpha\bar{\beta}\gamma\bar{\delta}}$ is strongly semi-negative, (L2) implies that

$$\sum_{\alpha,\beta\,\gamma,\delta} R_{\alpha\bar{\beta}\gamma\bar{\delta}} \xi^{\alpha\bar{\beta}}_{\bar{i}\bar{j}} \overline{\xi^{\delta\bar{\gamma}}_{\bar{i}\bar{j}}} = 0\, at\ p, \ 1 \leq i, \ j \leq m, \ 1 \leq \alpha, \ \beta \leq n, \tag{1.12}$$

where $\xi^{\alpha\bar{\beta}}_{\bar{i}\bar{j}} = (\partial_{\bar{i}} f^{\alpha})(a)\overline{(\partial_j f^{\beta})(a)} - (\partial_{\bar{j}} f^{\alpha})(a)\overline{(\partial_i f^{\beta})(a)}$.

Let $p = f(a)$ and let $T_{M,a}$ (resp. $T_{N,p}$) be the tangent space of M at a (resp. N at p) when M (resp. N) is regarded as real manifold. Let K be the kernel of $df : T_{M,a} \to T_{N,p}$. The complex structure of M makes $T_{M,a}$ a vector space over $\mathbf{C}$. It follows that $dim_{\mathbf{R}} K \leq 2m - 2k$ since $rank_{\mathbf{R}}\, df \geq 2k$. By (L1) there exists a basis $e_1, \cdots, e_m$ of $T_{M,a}$ over $\mathbf{C}$ such that for $1 \le i_1 \le \cdots \le i_k \le m$ the intersection of K with the $\mathbf{C}$-vector subspace of $T_{M,a}$ spanned by $e_{i_1}, \cdots, e_{i_k}$ is the zero vector subspace. Select a holomorphic coordinate system (w_i) of an open neighborhood W of a in O such that $e_i = 2Re(\partial/\partial w_i)$. For $1 \le i_1 \le \cdots \le i_k \le m$, let $\Phi_{i_1,\cdots,i_k}$ be the restriction of f to

$$W \cap \{w_\mu = w_\mu(a),\ 1 \le \mu \le m,\ \mu \ne i_1, \cdots, i_k\}.$$

Then $rank_{\mathbf{R}} d\Phi_{i_1,\cdots,i_k} = 2k$ at a. Because $R_{\alpha\bar{\beta}\gamma\bar{\delta}}$ is negative of order k at every point of N, (1.12) implies that for $1 \le i_1 \le \cdots \le i_k \le m$ either $\partial_{\bar{j}} f^\alpha = 0$ at a for $1 \le \alpha \le n$ and $j = i_1, \cdots, i_k$ or $\partial_j f^\alpha = 0$ at a for $1 \le \alpha \le n$ and $j = i_1, \cdots, i_k$. Thus (1.11) follows from $k \ge 2$ since if $\partial_{\bar{j}} f^\alpha \ne 0$ for some $1 \le \alpha \le n$ and some $1 \le j \le m$, and $\partial_l f^\beta \ne 0$ at a for some $1 \le l \le m$ and some $1 \le \beta \le n$, then we choose $1 \le i_1 \le \cdots \le i_k \le m$ such that both j and l belong to the set $\{i_1, \cdots, i_k\}$. Hence, we can conclude the theorem from (1.11) and (L3). □

We next review four types of classical bounded symmetric domains denoted by $D^I_{m,n}$, D^{II}_n, D^{III}_n and D^{IV}_n and discuss their curvature tensors.

1. The domain $D^I_{m,n}$ is an open subset of $\mathbf{C}^{mn}$ and consists of all $m \times n$ matrices $Z = (z_{\alpha,\beta})$ with complex entries such that $I_n - Z^t\bar{Z}$ is positive definite, where I_n is the identity matrix of order n and $\bar{Z}^t$ is the transpose of the conjugate of Z. An invariant K$\ddot{a}$hler metric is given by the potential function

$$\begin{aligned}\Phi &= log\ det(I_n - \bar{Z}^t Z)^{-1}\\ &= \sum |z_{\alpha\beta}|^2 + \frac{1}{2}\sum \bar{z}_{\alpha\gamma} z_{\alpha\delta} \bar{z}_{\beta\delta} z_{\beta\gamma} + \text{higher order terms.}\end{aligned}$$

At $Z = 0$ the coordinates $(Z_{\alpha\beta})$ are normal coordinates such that the first order derivatives of the coefficients of the metric vanish at $Z = 0$. Thus the curvature tensor at $Z = 0$ is

$$R_{\alpha\gamma\overline{\beta\rho}\lambda\sigma\overline{\mu\tau}} = \frac{\partial^4\Phi}{\partial z_{\alpha\gamma}\partial\bar{z}_{\beta\rho}\partial z_{\lambda\sigma}\partial\bar{z}_{\mu\tau}} = \delta_{\alpha\beta}\delta_{\lambda\mu}\delta_{\gamma\tau}\delta_{\rho\sigma} + \delta_{\alpha\mu}\delta_{\beta\lambda}\delta_{\gamma\rho}\delta_{\sigma\tau}.$$

It follows that at $Z = 0$

$$\begin{aligned}&\sum R_{\alpha\gamma\overline{\beta\rho}\lambda\sigma\overline{\mu\tau}}\xi^{\alpha\gamma\overline{\beta\rho}}\,\overline{\xi^{\mu\tau\overline{\lambda\sigma}}}\\ &= \sum \delta_{\alpha\beta}\delta_{\lambda\mu}\delta_{\gamma\tau}\delta_{\rho\sigma}\xi^{\alpha\gamma\overline{\beta\rho}}\,\overline{\xi^{\mu\tau\overline{\lambda\sigma}}} + \sum \delta_{\alpha\mu}\delta_{\beta\lambda}\delta_{\gamma\rho}\delta_{\sigma\tau}\xi^{\alpha\gamma\overline{\beta\rho}}\,\overline{\xi^{\mu\tau\overline{\lambda\sigma}}}\\ &= \sum_{\alpha,\lambda,\gamma,\rho} \xi^{\alpha\gamma\overline{\alpha\rho}}\,\overline{\xi^{\lambda\gamma\overline{\lambda\rho}}} + \sum_{\alpha,\beta,\gamma,\sigma} \xi^{\alpha\gamma\overline{\beta\gamma}}\,\overline{\xi^{\alpha\sigma\overline{\beta\sigma}}}\\ &= \sum_{\gamma,\rho}|\sum_{\alpha}\xi^{\alpha\gamma\overline{\alpha\rho}}|^2 + \sum_{\alpha,\beta}|\sum_{\gamma}\xi^{\alpha\gamma\overline{\beta\gamma}}|^2.\end{aligned}$$

2. The domain D_n^{II} is an open set of $\mathbf{C}^{n(n-1)/2}$ and consists of all skew-symmetric $n \times n$ matrices $Z = (z_{\alpha,\beta})$ with complex entries such that $I_n - Z^t \bar{Z}$ is positive definite. D_n^{II} is a complex submanifold of $D_{n,n}^I$, which has an invariant Kähler metric induced from $D_{n,n}^I$. At $Z = 0$ the coordinates $(Z_{\alpha\beta})$ $(\alpha < \beta)$ are normal coordinates. Thus the second fundamental form of D_n^{II} in $D_{n,n}^I$ vanishes at $Z = 0$ and the curvature of D_n^{II} is the restriction of the curvature of $D_{n,n}^I$. Hence, at $Z = 0$

$$\sum_{\alpha<\gamma,\,\beta<\rho,\,\lambda<\sigma,\,\mu<\tau} R_{\alpha\gamma\overline{\beta\rho}\lambda\sigma\overline{\mu\tau}} \xi^{\alpha\gamma\overline{\beta\rho}} \overline{\xi^{\mu\tau\overline{\lambda\sigma}}}$$

$$= \sum_{\gamma,\rho=1}^{n} |\sum_{\alpha} \xi^{\alpha\gamma\overline{\alpha\rho}}|^2 + \sum_{\alpha,\beta=1}^{n} |\sum_{\gamma} \xi^{\alpha\gamma\overline{\beta\gamma}}|^2,$$

where $\xi^{\alpha\gamma\overline{\beta\rho}}$ is skew-symmetric in α, γ and skew-symmetric in β, ρ.

3. The domain D_n^{III} is an open set of $\mathbf{C}^{n(n+1)/2}$ and consists of all symmetric $n \times n$ matrices $Z = (z_{\alpha,\beta})$ with complex entries such that $I_n - Z^t \bar{Z}$ is positive definite. D_n^{III} is a complex submanifold of $D_{n,n}^I$, which has an invariant Kähler metric induced from $D_{n,n}^I$. At $Z = 0$, the coordinates $(Z_{\alpha\beta})$ $(\alpha \leq \beta)$ are normal coordinates. Thus the second fundamental form of D_n^{II} in $D_{n,n}^I$ vanishes at $Z = 0$ and the curvature of D_n^{II} is the restriction of the curvature of $D_{n,n}^I$. Hence, at $Z = 0$

$$\sum_{\alpha<\gamma,\,\beta\leq\rho,\,\lambda\leq\sigma,\,\mu\leq\tau} R_{\alpha\gamma\overline{\beta\rho}\lambda\sigma\overline{\mu\tau}} \xi^{\alpha\gamma\overline{\beta\rho}} \overline{\xi^{\mu\tau\overline{\lambda\sigma}}}$$

$$= \sum_{\gamma,\rho} |\sum_{\alpha} |\xi^{\alpha\gamma\overline{\alpha\rho}}|^2 + \sum_{\alpha,\beta} |\sum_{\gamma} \xi^{\alpha\gamma\overline{\beta\gamma}}|^2,$$

where $\xi^{\alpha\gamma\overline{\beta\rho}}$ is symmetric in α, γ and symmetric in β, ρ.

4. The domain D_n^{IV} is an open set of $\mathbf{C}^n$ and consists of all $z = (z_1, \cdots, z_n)$ in $\mathbf{C}^n$ such that $1 + |\sum_\alpha z_\alpha^2|^2 - 2\sum_\alpha |z_\alpha|^2 > 0$ and $\sum_\alpha |z_\alpha|^2 < 1$. An invariant Kähler metric is given by the potential function

$$\Phi = -log(1 + |\sum_{\alpha} z_\alpha^2|^2 - 2\sum_{\alpha} |z_\alpha|^2)$$

$$= 2\sum_{\alpha} |z_\alpha|^2 - |\sum_{\alpha} z_\alpha^2|^2 + 2(\sum_{\alpha} |z_\alpha|^2)^2 + \text{higher order terms.}$$

At $z = 0$, the coordinates $(z_{\alpha\beta})$ $(1 \leq \alpha \leq n)$ are normal coordinates. Thus the curvature is

$$R_{\alpha\gamma\bar{\rho}\beta\bar{\sigma}} = \frac{\partial^4 \Phi}{\partial z_\alpha \partial \bar{z}_\rho \partial z_\beta \partial \bar{z}_\sigma} = 4(\delta_{\alpha\rho}\delta_{\beta\sigma} + \delta_{\alpha\sigma}\delta_{\beta\rho} - \delta_{\alpha\beta}\delta\rho\sigma).$$

It follows that at $z = 0$

$$\begin{aligned}
&\sum R_{\alpha\bar{\rho}\beta\bar{\sigma}}\xi^{\alpha\bar{\rho}}\,\overline{\xi^{\alpha\bar{\beta}}}\\
&= 4\sum \delta_{\alpha\rho}\delta_{\beta\sigma}\xi^{\alpha\bar{\rho}}\,\overline{\xi^{\alpha\bar{\beta}}} + 4\sum \delta_{\alpha\sigma}\delta_{\beta\rho}\xi^{\alpha\bar{\rho}}\overline{\xi^{\sigma\bar{\beta}}} - 4\sum \delta_{\alpha\beta}\delta_{\rho\sigma}\xi^{\alpha\bar{\rho}}\,\overline{\xi^{\sigma\bar{\beta}}}\\
&= 4\sum \xi^{\alpha\bar{\alpha}}\overline{\xi^{\beta\bar{\beta}}} + 4\sum \xi^{\alpha\bar{\beta}}\overline{\xi^{\alpha\bar{\beta}}} - 4\sum \xi^{\alpha\bar{\rho}}\overline{\xi^{\bar{\rho}\alpha}}\\
&= 4\sum_{\alpha} |\xi^{\alpha\bar{\alpha}}|^2 + 2\sum_{\alpha\neq\beta} |\xi^{\alpha\bar{\beta}} - \xi^{\beta\bar{\alpha}}|^2.
\end{aligned}$$

The curvature tensors of these four types of bounded symmetric domains are strongly semi-negative, but not strongly negative. By Siu [343] we know that theses curvature tensors are adequately negative, and thus Theorem 1.4.3 implies Theorem 1.4.4. Theorem 1.4.5 in turn follows from Theorem 1.4.4.

Theorem 1.4.4. *Let M be a compact Kähler manifold and N be a compact quotient of a bounded symmetric domain of type* $D^{I}_{m,n}$ $(mn \geq 2)$, D^{II}_{n} $(n \geq 3)$, D^{III}_{n} $(n \geq 2)$, *or* D^{IV}_{n} $(n \geq 3)$. *Suppose that* $f : M \to N$ *is a harmonic map which is a submersion at some point of M. Then f is* $\pm$ *holomorphic.*

Theorem 1.4.5. *Let N be a compact quotient of a bounded symmetric domain of type* $D^{I}_{m,n}$ $(mn \geq 2)$, D^{II}_{n} $(n \geq 3)$, D^{III}_{n} $(n \geq 2)$, *or* D^{IV}_{n} $(n \geq 3)$. *Then any compact Kähler manifold of the same homotopy type as N must be biholomorphic or conjugate biholomorphic to N.*

We have the following application from Siu [345]. *Let* R^N *be strongly negative. If a homology class in* $H_{2k}(N, \mathbf{Z})$ *can be represented by the continuous image of a compact Kähler manifold of complex dimension* $\geq$*2, then it can be represented by a complex analytic subvariety of N.*

Let M be a compact submanifold of dimension $\geq$2 of a compact Kähler manifold (N, h) with R^N strongly negative (cf. Kalka [211], and Jost and Yau [210]), or of a compact irreducible quotient of a polydisc (cf. Mok [266, 267]). Then the deformations of M as a submanifold of N coincide with those of M as a complex manifold.

In the 1980s, Sampson [313–316] made great contributions on harmonic maps of Kähler manifolds. He used his elegant tensor techniques to show: *A complex Kähler manifold of dimension* >*1 cannot be minimally immersed in a space of constant negative curvature.* He also obtained the following results:

1. *If f is a harmonic map from a compact Kähler manifold* $(M, g_{i\bar{j}})$ *into a Kähler manifold* $(N, h_{\alpha\bar{\beta}})$ *of strongly semi-negative curvature, then the (2,0)-part of the pull-back* $f^*(ds_N{}^2)$ *with components* $\phi_{ij} = h_{\alpha\bar{\beta}} f_i^{\alpha} \bar{f}_j^{\bar{\beta}}$ *is holomorphic.*
2. (1) *If f is a harmonic map of a compact Kähler manifold into a Riemannian manifold of hermitian negative curvature, then* ϕ *is holomorphic, and we have*

$$f^{\alpha}_{i|\bar{j}} = 0, \quad R'_{\alpha\beta\gamma\mu} f^{\alpha}_{\bar{j}} f^{\beta}_{i} f^{\gamma}_{\bar{k}} f^{\mu}_{\bar{l}} g^{j\bar{k}} g^{i\bar{l}} = 0. \tag{1.13}$$

(2) *Under the conditions of* (1), *if the second equation of (1.13) holds, then* $d'f$ *maps* $T'_p(M)$ *onto an abelian algebra* $\underline{a} \subset \underline{P}^{\mathbf{C}}$, *where* $\underline{P}$ *is identified with* $T_{f(P)}(N)$. *(1.13) is satisfied iff f is harmonic and M is compact.*

3. (1) *Let N be a Riemannian locally symmetric space whose irreducible local factors are all of non-compact or Euclidean type. Then N has hermitian negative curvature.*
(2) (a) *The natural embedding of* $SP_n\mathbf{R}/U_n$ *in* $SL_{2n}\mathbf{R}/SO_{2n}$ *is totally geodesic.*
(b) *A harmonic map of a compact Kähler manifold into a locally symmetric space of bounded domain* $D^I{}_{m,n}$ *has rank* ≤ 2 *if* $m = 1$, *and* $\leq 2n$ *if* $m = 2$, $n \geq 2$.

Carlson and Toledo [61] also showed that *if (M, g) is a compact Kähler manifold and (N, h) is a locally symmetric space of non-compact type which is not locally Hermitian symmetric, then any harmonic map* $f : (M, g) \to (N, h)$ *has rank* $f < n$ *at every point.*

1.4.3 Complex Variations

Let $f : (M, g) \to (N, h)$ be a harmonic map from a compact Riemann surface to a Kähler manifold. Consider a variation of f with complex parameter $t \in \mathbf{C}$. By (1.9) we have

$$\frac{\partial E'(f)}{\partial \bar{t}} = \frac{E''(f)}{\partial \bar{t}}, \quad \frac{\partial^2 E(f)}{\partial t \partial \bar{t}} = 2\frac{\partial^2 E''(f)}{\partial t \partial \bar{t}}.$$

Siu and Yau [348] computed:

$$\begin{aligned} \left.\frac{\partial^2 E''}{\partial t \partial \bar{t}}\right|_{t=0} = \int_M \Big[&< |\nabla_t f_{\bar{z}}|^2 + |\nabla_{\bar{t}} f_{\bar{z}}|^2 + R^N(f_t, \bar{f_{\bar{t}}}) f_{\bar{z}}, \bar{f_{\bar{z}}} > \\ &+ < R^N(f_t, \bar{f_t}) f_{\bar{z}}, \bar{f_{\bar{z}}}) - 2Re(R^N(f_{\bar{t}}, \bar{f_{\bar{z}}}) f_t, \bar{f_{\bar{z}}} > \Big] \\ &\times \frac{i}{2} dz \wedge d\bar{z} \end{aligned} \tag{1.14}$$

where $< \cdot, \cdot >$ denotes the bilinear extension of h to $T^{\mathbf{C}}(N)$.

Remark that

$$< Re(R^N(f_t, \bar{f_{\bar{z}}}) f_{\bar{z}}, \bar{f_{\bar{z}}} > = -4|Re f_t|^2 |Re f_{\bar{z}}|^2 HBRiem^N(f_t, f_{\bar{z}}),$$

where $HBRiem^N$ denotes the holomorphic bisectional curvature of N

$$HBRiem^N(X, Y) = < R(X, JX)Y, JY > .$$

Suppose that for its Koszul-Malgrange complex structure (1.4.A), the bundle $f^{-1}T(N) \to M$ admits a non-zero holomorphic section v. Then a variation $f : \mathbf{C} \times M \to N$ of f can be constructed such that

$$\frac{\partial f}{\partial \bar{t}}\bigg|_{t=0} = 0, \quad \frac{\partial f}{\partial t}\bigg|_{t=0} = v.$$

Therefore, (1.14) reduces to

$$\frac{\partial^2 E''}{\partial t \partial \bar{t}}\bigg|_{t=0} = -4 \int_M |Re f_t|^2 |Re f_{\bar{z}}|^2 HBRiem^N(f_t, f_{\bar{z}}) \frac{i}{2} dz \wedge d\bar{z}.$$

This leads to the following result due to Siu and Yau [208, 347]:

(1.4.C) *Let $f : (M, g) \to (N, h)$ be a harmonic non $\pm$holomorphic map from a compact Riemann surface to a Kähler manifold. If $HBRiem^N > 0$ and if there is a non-trivial holomorphic variation $v \in \mathcal{C}(f^{-1}T(N))$, then E-index $f > 0$.*

In fact, the construction based on the variation v gives

$$\frac{\partial^2 E''(f)}{\partial t \partial \bar{t}}\bigg|_{t=0} < 0,$$

and gives a real variation of f along which the second derivative of E is negative.

An application of the Riemann-Roch theorem to the holomorphic vector bundle $f^{-1}T(N) \to M$ (with its Koszul-Malgrange structure) ensures the existence of holomorphic variations in various cases, e.g., Eells and Wood [136] showed: *Let (M, p) be a compact Riemann surface of genus $M = p$ and (N, h) a Kähler manifold with $HBRiem^N > 0$. If $f : M \to (N, h)$ is harmonic and non $\pm$holomorphic, then E-index $(f) \geq (f^*c_1(N)[M] + n(1 - p)$, where $c_1(N)$ is the first Chern class of T(N).*

If f is a map from a compact oriented surface M to a manifold N such that $H^2(N, \mathbf{Z}) = \mathbf{Z}$, the degree of *deg(f)* of f is the image of the generator of $H^2(N, \mathbf{Z})$ in $H^2(M, \mathbf{Z}) = \mathbf{Z}$ under the homomorphism induced by f.

For example, if $(N, h) = P_n(\mathbf{C})$ and $deg f \geq 0$, then

$$E - index(f) \geq deg(f)(n + 1) + n(1 - p).$$

In particular, let $n = 1$ and $p \geq 1$. Then for any integer d with $|d| \leq p - 1$, there exist (cf. [246]) a compact Riemann surface M of genus $M = p$ and a harmonic non $\pm$holomorphic map $f : M \to S^2 = P_1(\mathbf{C})$ of degree d. For $d \geq p/2$, f must have positive index.

Siu and Yau [347] used (1.4.C) to give a proof of a conjecture of Andreotti-Frankel: *Any compact Kähler manifold N with* $HBRiem^N > 0$ *is biholomorphically equivalent to the complex projective space* $P_n(\mathbf{C})$.

We explain very briefly here. Preceding considerations showed that (i) N can be assumed simply connected and $H^2(N, \mathbf{Z}) = \mathbf{Z}$; and (ii) the problem can be reduced to showing that a generator of the free part of $H_2(N, \mathbf{Z})$ can be represented by a rational curve, i.e., a holomorphic map from S^2 to N. Siu and Yau [187] apply a theorem of Sacks and Uhlenbeck [310] such that a set of generators of $\pi_2(N) = H_2(N, \mathbf{Z})$ can be represented by energy minimizing harmonic maps f_j of S^2. Applying the Birkhoff-Grothendieck decomposition of the bundle $f_j^{-1}TN$ and a computation of its Chern class, they establish the existence of a holomorphic variation of f_j, so that by (1.4.C) each f_j is $\pm$ holomorphic. Finally, the theory of deformations of curves is used to show that a given generator is represented by a single holomorphic map of S^2.

1.5 Maps into Groups and Grassmannians

In this section, we present a brief survey of recent developments of maps into Lie groups, maps into complex Grassmannians, maps into projective spaces, and Coulomb gauge fields. Since the methods are lengthy and technical, it is impossible to provide all the details.

1.5.1 Maps into Lie Groups

Let G be a compact Lie group endowed with a bi-invariant metric. Its Levi-Civita connection is given by $\nabla_X Y = \frac{1}{2}[X, Y]$, and its curvature by $R(X, Y)Z = \frac{1}{4}[[X, Y], Z]$. (By contrast, the connection $\nabla_X Y = [X, Y]$ has torsion $T(X, Y) = [X, Y]$ and curvature 0).

(1.5.A) Let $\mathcal{G}$ be the Lie algebra of G. The Mauer-Cartan form μ of G is the $\mathcal{G}$-valued 1-form on G given by $\mu(v) = v$ for all $v \in \mathcal{G}$ (viewed as a left-invariant vector field on G). We write $[\mu \wedge \mu](X, Y) = 2[\mu(X), \mu(Y)]$. Then $d\mu + \frac{1}{2}[\mu \wedge \mu] = 0$. The Jacobi identity ensures that $d(\mu \wedge [\mu \wedge \mu]) = 0$. We can regard μ as a connection 1-form on $T(G)$ with curvature 0.

If $f : M \to G$ is a map from a Riemann surface to G, we denote $\alpha = f^*\mu$ the pull-back of the Maurer-Cartan form. It is a $\mathcal{G}$-valued 1-form on M satisfying $d\alpha + \frac{1}{2}[\alpha, \alpha] = 0$. Then $\alpha = f^{-1}df$. The complexfication of α, still denoted here by α, can be split into complex types: $\alpha = \alpha' + \alpha''$, where $\alpha'' = \bar{\alpha}'$. Then $d\alpha = \bar{\partial}\alpha' + \partial\alpha''$ and

$$\bar{\partial}\alpha' + \partial\alpha'' + [\alpha'' \wedge \alpha'] = 0. \tag{1.15}$$

Since $*\alpha = -i\alpha' + i\alpha''$ (where $*$ is the Hodge operator), we have:

(1.5.B) A map $f : M \to G$ is harmonic if and only if $d^*(f^*\mu) = 0$, which is equivalent to

$$\bar{\partial}\alpha' = \partial\alpha''. \tag{1.16}$$

Using (1.15), we arrive at

$$\bar{\partial}\alpha' + \frac{1}{2}[\alpha'' \wedge \alpha'] = 0. \tag{1.17}$$

Taking conjugates, we see that (1.15) and (1.16) are equivalent to (1.17). An application of (1.5.B) by Uhlenbeck [387] is as follows: A non-constant harmonic map $f : S^2 \to G$ is *unstable* (G can be treated as a symmetric space). In fact, $*df^*$ is a closed $\mathcal{G}$-valued 1-form on S^2, and so there is a function $\psi : S^2 \to \mathcal{G}$ with $d\psi = *f^*\mu$. Then ψ can be viewed as a variation of f and its associated deformation (f_t) is energy-decreasing:

$$\frac{d^2}{dt^2}E(f_t)\Big|_{t=0} = -E(f).$$

Denoting $\alpha' = 2A_z dz$ and $\alpha'' = 2A_{\bar{z}} d\bar{z}$ in a complex chart of M, we recast (1.16) and (1.17) in the equivalent forms

$$\bar{\partial}A_z + \partial A_{\bar{z}} = 0,\ \bar{\partial}A_{\bar{z}} + [A_{\bar{z}}, A_z] = 0 = \partial A_{\bar{z}} - [A_{\bar{z}}, A_z].$$

We also have

$$E(f) = \frac{1}{2}\int_M |df|^2 dx = -2i\int_M trace\, A_z A_{\bar{z}} dz \wedge d\bar{z}.$$

Let $L = \partial + (1 - \lambda^{-1})A_z$ and $K = -(1 - \lambda)A_{\bar{z}}$ for any $\lambda \in \mathbf{C}^*$. Valli [393] proved that *if $f : M \to G$ is harmonic, then L and K satisfy the Lax equation* $\frac{\partial L}{\partial \bar{z}} = [K, L]$.

In particular, the spectrum of the problem $L\psi = \lambda\psi$ subject to the condition $\bar{\partial}\psi = K\psi$ varies holomorphically.

Hitchin [185, 186] used this information to emphasize its gauge theory and to study harmonic maps from the torus $T^2 \to SU(2) = S^3$. He reduced the study to an algebraic geometric setting involving a certain hyperelliptic curve C, called the *spectral curve* of the map. In this way new harmonic maps $f : T^2 \to SU(2)$ are obtained as well as certain deformations of these (cf. Toth [377–379]). For example,

(a) *genus* $(C) = 0$. If f is conformal harmonic, then it is a finite covering of the Clifford torus. If f is harmonic but not conformal, it is a finite covering of a rectangular torus.
(b) *genus* $(C) = 1$ contains the Gauss map of Delaunay's surfaces (cf. Eells and Lemaire [120]) and Hsiang-Lawson's [190] minimal tori in S^3 invariant under a circle action.
(c) *genus* $(C) = 3$ includes the Gauss map of Wente's [406] immersions.

Let M be a compact Riemann surface, and

$$\rho : \pi_1(M) \to PSL(\mathbf{C}^2) = SL(\mathbf{C}^2)/\text{centre}$$

an irreducible representation (in the sense that $\rho(\pi_1(M))$ fixes no point of $P_1(\mathbf{C})$). Using the reduced action of $\pi_1(M)$ on the hyperbolic 3-space RH^3 (via $PSL(\mathbf{C}^2)/SO(3)$) we form the associated flat bundle $\pi : W = \tilde{M} \times_\rho RH^3 \to M$, where $\tilde{M}$ is the universal cover of M; the fibres are isometric to RH^3.

The existence theorem in Sect. 1.2.3 was extended to sections of Riemannian fibration by C. M. Wood [414, 416] with growth restrictions if the fibres are not compact. Now, according to Donaldson [100] it is sufficient to obtain harmonic sections of π and thus solutions to the system of equations (1.18). For any section s of π he considers the principal $SO(3)$-bundle $P_s \to M$ associated to the pull-back $s^{-1}(T^B(W))$ of the vertical tangent bundle of W, with induced $SO(3)$-connection ∇_s; that is flat in the horizontal direction and restricts to the Poincaré metric on the fibres of W. Let $\sigma_s = \frac{1}{2} i(d^B s)$, where $d^B s$ is viewed as a $\mathcal{G}^{\mathbf{C}}(P_s)$-valued 1-form on M. Then

$$R_s + \frac{1}{2}[\sigma_s, \sigma_s] = 0; \; d_s\sigma = 0, \; d_s^*\sigma = 0 \tag{1.18}$$

if and only if s is a harmonic section of π. Here R_s denotes the curvature form of the connection ∇_s; d_s is the exterior differential on the appropriate vector bundle valued forms, and d_s^* its adjoint operator.

(1.5.C) A factorization theorem. Let $G = U(n)$ be the unitary group, whose Lie algebra contains the skew-hermitian $n \times n$ matrices, endowed with the bi-invariant metric which on $\mathcal{G}$ is given by $(u, v) = \mathit{trace}\,(uv^*)$, $v^* = \bar{v}^t$. The involution $K \to K^{-1}$ on $U(n)$ has fixed set $\{K \in U(n) : K^2 = 1\}$, which is identified with the Grassmannian $Grass(\mathbf{C}^n)$ of complex vector subspaces of $\mathbf{C}^n$ via the correspondence $P \mapsto P - P^\perp = K \in U(n)$, for each hermitian projection P onto the subspace $Im\,P \in Grass(\mathbf{C}^n)$, where $Grass(\mathbf{C}^n) = \bigcup_{r=0}^n G_r(\mathbf{C}^n)$, $G_r(\mathbf{C}^n)$ is the complex Grassmannian of r planes in $\mathbf{C}^n$. Each $G_r(\mathbf{C}^n)$ is a Kähler manifold (an irreducible hermitian symmetric space), totally geodesically embedded in $U(n)$. By this embedding, $G_r(C^n)$ is identified with $\{K \in U(n) : K^2 = I$ and its $(+1)$-eigenspace is r-dimensional$\}$.

For a map $f : M \to G_r(\mathbf{C}^n) \subset Grass(\mathbf{C}^n)$ of a Riemann surface into a complex Grassmannian, we associate the vector subbundle $\mathbf{f} \subset M \times \mathbf{C}^n$ of rank r, where for each $x \in M$, the fibre is the point $f(x) \in G_r(\mathbf{C}^n)$. The correspondence $f \mapsto \mathbf{f}$ is bijective. The bundle $\mathbf{f}^\perp$ is associated to the map $-f$ (where f and $-f$ are viewed as maps from M to $U(n)$, using the above embedding). *The map $f = P - P^\perp$ is holomorphic if and only if $P^\perp \cdot \bar{\partial} P = 0$, or equivalently if and only if the bundle P is a holomorphic subbundle of $M \times \mathbf{C}^n$.*

(1.5.D) The following result was obtained by Uhlenbeck [387] and refined by Valli [393, 394]. *Let M be a compact Riemann surface and $f : M \to U(n)$ a harmonic map. Suppose that $P : M \to Grass(\mathbf{C}^n)$ satisfies $P^\perp A_{\bar{z}} P = 0$ and $P \perp (\bar{\partial} P + A_{\bar{z}} P) = 0$. Then the map $\tilde{f} = f \cdot (P - P^\perp) : M \to U(n)$ is harmonic. Moreover, $E(\tilde{f}) - E(f) = Area(M) c_1(P)[M]$, where $c_1(P)$ is the first Chern class of P. In particular, if $M = S^2$, f is non-constant, and P can be chosen so that $E(\tilde{f}) - E(f) < 0$.* This requires to apply the Birkhoff-Grothendieck theorem in (1.4B) on the structure of holomorphic vector bundles over S^2. In fact, based on that theorem we can have a canonical choice of P to minimize $E(\tilde{f}) - E(f)$.

(1.5.E) A consequence is Valli's [393, 394] version of a theorem of Uhlenbeck [387]: *Associated to each harmonic map $f : S^2 \to U(n)$ is a sequence $f_0, f_1, \cdots, f_k$ of harmonic maps $S^2 \to U(n)$ with $f_0 = constant$,*

$$f_k = f, \; f_i = f_{i-1}(P_i - P_i^\perp) \quad (1 \le i \le k \le E(f)/4\pi)$$

and $E(f_i) - E(f_{i-1}) \ge 4\pi$. Hence, we have the canonical factorization $f = f_0(P_1 - P_1^\perp) \cdots (P_k - P_k^\perp)$. Each factor is holomorphic with respect to a specific connection.

Notice that $E(f) = -4\pi \sum_{i=1}^k c_1(P_i)$. Hence, the energy of any harmonic map $f : S^2 \to U(n)$ is an integral multiple of 4π. Actually, a similar integrality theorem holds for any compact group.

1.5.2 Maps into Complex Grassmannians

Let $f : M \to G_r(\mathbf{C}^n)$ be a map of a Riemann surface into a complex Grassmannian, and $\mathbf{f}$ the corresponding rank r subbundle of $M \times \mathbf{C}^n$ in (1.5.C). Thus there is an identification

$$f^{-1} T^{1,0}(G_r(\mathbf{C}^n)) \cong \mathbf{f}^* \otimes \mathbf{f}^{-1}$$

under which the (1, 0)-part of df corresponds to the second fundamental form of $\mathbf{f} \subset M \times \mathbf{C}^n$. The components of this second fundamental form may locally be written as

$$A'_f = \pi^{\perp} \circ \partial \circ \pi, \; A''_f = \pi^{\perp} \circ \partial \circ \pi,$$

where $\pi : M \times \mathbf{C}^n \to \mathbf{f}$ is the bundle projection and $\partial, \bar{\partial}$ are the flat differentiations in $M \times \mathbf{C}^n$. These induce connections in $\mathbf{f}$ and $\mathbf{f}^{\perp}$, and then Koszul-Malgrange holomorphic structures (1.4.A).

A map f is harmonic if and only the second fundamental form A'_f is holomorphic, i.e., $A'_f \circ \bar{\partial} f = \bar{\partial}_{f\perp} \circ A'_f$, where ∂_f (resp. $\bar{\partial}_{f\perp}$) denotes the ∂-operator on $\mathbf{f}$ (resp. $\mathbf{f}^{\perp}$).

(1.5.F) A special case of (1.5.E) is due to Burstall and Salamon [51] as follows. *Let $f : M \to G_r(\mathbf{C}^n)$ be a harmonic map and let $\alpha \subset \mathbf{f}, \beta \subset \mathbf{f}^{\perp}$ be holomorphic subbundles such that $A'_f(\alpha) \subset \beta$ and $A'_{f\perp}(\beta) \subset \alpha$. Then the map $\tilde{\mathbf{f}} = (\mathbf{f} \cap \alpha^{\perp}) \oplus \beta$ is harmonic. Moreover,*

$$E(\tilde{f}) - E(f) = -4\pi c_1(\alpha \oplus \beta).$$

In the technique of Burstall and Wood [52], $\tilde{f}$ is obtained from f by forward replacement of α by β. There is a dual notion of backward replacement of anti-holomorphic subbundles preserved by A'', f being obtained from $\tilde{f}$ by backward replacement of β by α. Such methods for producing new harmonic maps from old ones were used by Burstall and Wood [52], Burstall and Salamon [51], and Chern and Wolfson [71,72].

If $f : M \to G_r(\mathbf{C}^n)$ is harmonic, there is a holomorphic subbundle of f^{-1} which coincides with the image of A'_f almost everywhere. We call it the ∂-Gauss bundle and denote it by $G'(f)$. Likewise, A''_f defines the (anti-holomorphic) $\bar{\partial}$-Gauss bundle $G''(f)$.

Setting $\alpha = \mathbf{f}, \beta = G'(f)$ in (1.5.F), $G'(f)$ (and similarly $G''(f)$) correspond to harmonic maps into a Grassmannian $G_t(C^n)$, $t \leq r$. Therefore, we may iterate the procedure, defining bundles and thus harmonic maps by

$$G^{(s)}(f) = G'(G^{(s-1)}(f)), \; G^1(f) = G'(f), \quad s > 1,$$
$$G^{-s}(f) = G''(G^{-(s-1)}(f)), \; G^{(-1)}(f) = G''(f), \quad s > 1.$$

The sequence of harmonic maps $G^{(i)}(f)$ together with their second fundamental forms $A''_i = A'_{G^{(i)}f}$ is called a *harmonic sequence* in Wolfson [418].

Assume that each A'_i is an isomorphism on almost all fibres:

$$\mathbf{f} \xrightarrow{A'_0} G'(f) \xrightarrow{A'_1} G^{(2)}(f) \longrightarrow \cdots \longrightarrow G^{(s)}(f).$$

Thus we have

$$c_1(G^{(i)}) = c_1(G^{(i-1)}(f)) + r(2-2p) + ram(A'_{i-1}),$$

where $p = genus(M)$ and $ram(A'_{i-1})$ is the number of zeros of $dz \otimes A'_{i-1}$ counted with their multiplicities. From this, Wolfson [418] derived the estimate

$$(s+1)c_1(f) + r(2-2p)\frac{s(s+1)}{2} + \sum_{i=0}^{s} ram(A'_i) \leq E(f)$$

and obtained the following conclusion.

(1.5.G) *For any harmonic map $f : S^2 \to G_r(\mathbf{C}^n)$ there is some $G^{(i)}(f)$ with $i \geq 0$, for which*

$$rank\, G^{(i+1)}(f) < rank\, G^{(i)}(f);$$

hence by iteration, the harmonic sequence must terminate, i.e., there is a q for which

$$G^{(q)}(f) \neq 0, \;\; G^{(q+1)}(f) = 0,$$

(such maps have finite ∂-order). (1.5.G) may be interpreted as giving an explicit form of Uhlenbeck factorization in (1.5.E).

Applying the above theory, Wood [418] obtained the following theorem.

(1.5.H) Classification Theorem. *Let M be a closed Riemannian surface. There is a bijection between harmonic maps $f : M \to G_r(\mathbf{C}^n)$ of finite $\bar{\partial}$-order and subsets of subbundle $\mathbf{f} = \beta_0, \cdots, \beta_k$ of $M \times \mathbf{C}^n$ such that $\sum_{i=0}^{k} rank\, \beta_i = r$, where β_1 is a holomorphic subbundle of $Ker A'_{G'(\beta_0)^{\perp}}$, β_2 is a holomorphic subbundle of $Ker A'_{G'(G'(\beta_0)+\beta_1)^{\perp}}$, and so on.*

Twistorial approach. An alternate treatment of the factorization theorem for harmonic maps $S^2 \to G_r(\mathbf{C}^n)$ was given by Burstall and Salamon [51], building on methods of Burstall and Wood [49, 52, 417, 419]. This approach has two main components.

1. If $f : S^2 \to N$ is a harmonic map, then $f^{-1}T^{\mathbf{C}}(N)$ has the Birkhoff-Grothendieck decomposition into holomorphic subbundles:

$$f^{-1}T^{\mathbf{C}}(N) = n_1 L^{p_1} \oplus \cdots \oplus n_k L^{p_k},$$

where L is the Hopf bundle ($c_1(L) = 1$) and $p_1 > \cdots > p_k$. The integer $p_1 - p_k$ is the *length* of f. A non-constant harmonic map has length at least 4.
2. Burstall and Rawnsley [50] showed that *if $f : S^2 \to G/K$ is a harmonic map to an inner symmetric space, then there exists a parabolic subgroup P such that $K = K_p$ and a holomorphic map $\phi : S^2 \to (G/H, J_2^P)$ such that $\pi_P \circ \phi = f$, where J_2^P is the almost complex structure on G/H obtained by reversing the orientation of J_1^P on $Ker\, d\pi_P$, $\pi_P : G/H \to G/K_P$, J_1 and J_2*

are the CR structure on G/H. From this result, it follows that harmonic maps $S^2 \to G_r(\mathbf{C}^n)$ are covered by J_2-holomorphic maps into suitable twistor spaces, which are flag manifolds in the case at hand. By utilizing the geometry of these flag manifolds and further application of the Birkhoff-Grothendiek theorem, a sequence of harmonic maps is produced via (1.5.G) with strictly decreasing length and thus the factorization theorem is obtained. A key step in this approach is to encode the geometry of the flag manifolds into directed graphs.

1.5.3 Maps into Projective Spaces

(1.5.I) The following description is a reformulation of the theorem of Eells and Wood [136] (cf. [46, 98, 157]. Let $r = 1$ in (1.5.H). *There is a bijective correspondence between full isotropic harmonic maps* $f : M \to P_{n-1}\mathbf{C}$ *of* $\bar{\partial}$*-order k and holomorphic maps* $\phi : M \to P_{n-1}\mathbf{C}$ *of* ∂*-order* $\geq k$*, given by*

$$\phi = G^{(-r)}(f), \quad f = G^{(r)}(\phi).$$

The map f has finite $\bar{\partial}$-order if and only if f is complex isotropic. If $\phi : M \to P_{n-1}(\mathbf{C})$ is a full holomorphic map and $\phi_j : M \to G_{j+1}(\mathbf{C}^n)$ its jth associated curve, then for fixed k with $0 \leq k \leq n-1$, the map $f : M \to P_{n-1}\mathbf{C}$ defined by

$$f(z) = \phi_{k-1}^{\perp}(z) \cap \phi_k(z)$$

is a full complex harmonic map.

A holomorphic $\phi : M \to P_{n-1}(\mathbf{C})$ is *totally isotropic* if the associated curves ϕ_i and $\bar{\phi}_j$ are orthogonal for all $i, j \geq 0$ with $i + j < n - 1$. If n is even, then there are no full totally isotropic holomorphic maps $\phi : M \to P_{n-1}(\mathbf{C})$, as shown by an easy computation.

Considering $P_{n-1}(\mathbf{R})$ as the space of real points of $P_{n-1}(\mathbf{C})$, we recover the classification theorem of Calabi [58], which was the motivation of (1.5.I) and the whole twistor approach to the construction of harmonic maps of surfaces.

A map $f : M \to P_{n-1}(\mathbf{R})$ is *isotropic* if the composition $i \circ f : M \to P_{n-1}(\mathbf{C})$ is complex isotropic. There is a bijective correspondence between full isotropic harmonic maps $f : M \to P_{n-1}(\mathbf{R})$ and full totally isotropic holomorphic maps $\phi : M \to P_{n-1}(\mathbf{C})$. Particularly, such maps exist only for n odd, and the correspondence is described as in (1.5.I). Borchers and Garber [36] showed an iterative scheme for finding all such totally isotropic holomorphic maps.

Example 1. Any harmonic map $f : S^2 \to P_{n-1}(\mathbf{C})$ is complex isotropic (Zakrzewski [98] and Glaser and Stora [36]). For $n \geq 3$ there are harmonic non-holomorphic maps $S^2 \to P_{n-1}(\mathbf{C})$ of all degrees. For $n = 3$ these maps are full (Eells and Wood [136]).

Example 2. Any harmonic map $f : T^2 \to P_{n-1}(\mathbf{C})$ of $deg\, f \neq 0$ is complex isotropic. For $n \geq 3$ there are harmonic non $\pm$holomorphic maps $f : T^2 \to P_{n-1}(\mathbf{C})$ of all degrees [136] (in contrast to $n = 2$, cf. [124]).

Example 3. Let M be a closed Riemann surface of $genus(M) = p$. There are harmonic non-holomorphic maps $M \to P_{n-1}(\mathbf{C})$ of all degrees $\geq p + 1$, if $n \geq 3$ (cf. [136]).

1.5.4 Coulomb Gauge Fields

Let (N, h) be a Riemannian homogeneous space with G as transitive group of isometries. For each $v \in \mathcal{G}$, let $\tilde{v}$ be its associated vector field on N, and define the *moment map* $m{:}T(N) \to \mathcal{G}^*$ by $m(Y) \cdot v = - < Y, \tilde{v} >$.

(1.5.J) As a consequence of Noether's theorem, Pluzhnikov [300] and Rawnsley [306] showed that *if (M,g) is a Riemannian manifold and (N,h) a homogeneous space, then a map $f : (M, g) \to (N, h)$ is harmonic if and only of f^*m is a co-closed $\mathcal{G}^*$-valued 1-form on M, i.e., $d^*(f^*m) = 0$.*

Consider the group G as a homogeneous space, with G acting on itself by left translations. Then the moment map m can be identified with minus the Maurer-Cartan form μ (1.5.A). If $f : (M, g) \to G$ is a map from any manifold to G, then the pull-back $\alpha = f^*\mu$ is a $\mathcal{G}$-valued 1-form on M, a connection with curvature

$$d\alpha + \frac{1}{2}[\alpha \wedge \alpha] = 0. \tag{1.19}$$

Conversely, let α be a $\mathcal{G}$-valued 1-form on M such that (1.19) holds. Then α can be viewed as a flat connection 1-form on the product bundle $M \times G \to M$.

Assume that $Hom(\pi_1(M), G) = H^1(M, G) = 0$, and fix a base point $a \in M$. The correspondence $f \mapsto f^*\mu$ is a bijection between based maps $f : (M, a) \to (G, e)$ and $\mathcal{G}$-valued 1-forms α satisfying (1.19) (see Singer [341]).

In terms of gauge theory, we are given a trivial G-bundle over M and a connection $d + \tilde{\alpha}$, and ask when it is gauge equivalent to a connection $d + \alpha$ with $d^*\alpha = 0$. A gauge change is a map $f : (M, g) \to G$; with respect to it $\tilde{\alpha}$ changes to $\alpha = f^{-1}df + f^{-1}\tilde{\alpha} f$. Notice that $d^*\alpha = 0$ is equivalent to the Euler-Lagrange equation of the functional

$$E(f) = \int_M |f^{-1}df + f^{-1}\tilde{\alpha} f|^2 dx.$$

$G = U(1)$ gives a linear problem corresponding to the equations of Maxwell fields. Those α arising from f are then Coulomb gauge fields.

1.6 Harmonic Maps, Loop Groups and Integrable Systems

1.6.1 Introduction

Minimal surfaces, constant mean curvature surfaces, harmonic maps and many classical problems in differential geometry are actually connected with integrable systems. The theory of integrable system was developed after the discovery of the Korteweg-de Vries equations in the 1960s. Later, C. Gardner, J. Greene, M. Kruskal and R. Miura showed that this equation could be solved by the inverse scattering method. P. D. Lax [242] reinterpreted the method by his well-known equation and some progresses had been made in the 1970s. This theory then made contact with methods from algebraic geometry, loop group techniques and Grassmannians. About the same time, the twistor theory of R. Penrose was built independently, and then applied successfully by himself and R. S. Ward for constructing self-dual Yang-Mills connections and four-dimensional self-dual manifolds using complex geometry structures. In the 1980s, it became clear that integrable systems play an important part in differential geometry. This motivated Uhlenbeck [387] to construct harmonic maps on two-spheres with values in $U(n)$ using families of curvature-free connections depending on a complex 'spectral' parameter. Hitchin [186] also studied finite type tori into $SU(2)$ using similar method about the same time. An important contribution to these developments was the construction of an immersed constant mean curvature torus in $\mathbf{R}^3$ by Wente [406] in 1984.

In this section, we discuss maps into spheres, loop groups, and harmonic maps as integrable systems, based on the work of Frody and Wood [145], Guest [166] and Helein [178], where we can find all the developments and details on harmonic maps, loop groups and integrable systems.

1.6.2 Maps into Spheres

We first identify the n-dimensional sphere S^n with the group $SO(n+1)/SO(n)$ as follows. Select an orthonormal basis $(\epsilon_1, \cdots, \epsilon_{n+1})$ of R^{n+1}. Thus

$$SO(n) \cong \mathcal{R} = \{g \in SO(n+1) : g(\epsilon_{n+1}) = \epsilon_{n+1}\} = \left\{\begin{bmatrix} R & 0 \\ 0 & 1 \end{bmatrix} : R \in SO(n)\right\} \subset SO(n+1).$$

The subgroup $\mathcal{R}$ of $SO(n+1)$ is one of the two components of

$$\{g \in SO(n+1) : Ad_P(g) = PgP^{-1} = g\},$$

where $P = \begin{bmatrix} 1_n & 0 \\ 0 & -1 \end{bmatrix}$. An equivalence relation in $SO(n+1)$ is defined by $g\mathcal{R}g'$ iff $g^{-1}g' \in \mathcal{R}$ and the set of equivalence classes $\{[g] = g\mathcal{R} : g \in SO(n+1)\}$ is denoted by $SO(n+1)/SO(n)$. We note that the map $SO(n+1)/\mathcal{R} \to S^n$ given by $[g] \mapsto g(\epsilon_{n+1})$ is a diffeomorphism.

Let $f : \Omega \to S^n$ be a map on an open, simply connected subset Ω of $\mathbf{C}$. Then we can lift f to $SO(n+1)$, i.e., there is a map $F : \Omega \to SO(n+1)$ such that $F(z)(\epsilon_{n+1}) = f(z)$. Denote $F(z) = (e_1(z), \cdots, e_n(z), f(z))$, where $\{e_1, \cdots, e_n\}$ is an orthonormal basis of $T_f(S^n)$. We can express df as $df = \sum_{i=1}^n \psi^i e_i$, where $\psi_i = (df, e_i)$ and

$$de_i = \sum_{j=1}^n w_i^j e_j - \psi^i f, \quad i = 1, \cdots, n,$$

where $w_j^i = (de_j, e_i)$. The matrix

$$\alpha = \begin{bmatrix} 0 & w_2^1 & \cdots & w_n^1 & \psi^1 \\ w_1^2 & 0 & \cdots & w_n^2 & \psi^2 \\ \cdot & \cdot & \cdots & \cdot & \cdot \\ w_1^n & w_2^n & \cdots & 0 & \psi^n \\ -\psi^1 & -\psi^2 & \cdots & -\psi^n & 0 \end{bmatrix} \in T^*\Omega \otimes so(n+1)$$

satisfies $dF = F\alpha$. We have

$$0 = d^2F = d(dF) = d(F\alpha) = dF\alpha + Fd\alpha = F\alpha\wedge\alpha + Fd\alpha = F(d\alpha + \alpha\wedge\alpha),$$

which implies

$$d\alpha + \alpha \wedge \alpha = 0, \tag{1.20}$$

since F is invertible. If (1.20) holds, then there exists an $F : \Omega \to SO(n+1)$ such that

$$F(z_0) = F_0, \; dF = F\alpha,$$

where $F_0 \in SO(n+1)$ may be arbitrarily chosen. The process of passing from F to α can be viewed as linearization. The non-linearity of $F \in SO(n+1)$ is replaced by the linear equation $\alpha + \alpha^t = 0$ (assume $F_0 \in SO(n+1)$).

We denote

$$[\alpha \wedge \beta] = \alpha \wedge \beta + \beta \wedge \alpha = [\beta \wedge \alpha],$$

for 1-forms

$$\alpha = \begin{bmatrix} \alpha_1^1 & \cdots & \alpha_n^1 \\ \cdot & \cdots & \cdot \\ \alpha_1^n & \cdots & \alpha_n^n \end{bmatrix}, \quad \beta = \begin{bmatrix} \beta_1^1 & \cdots & \beta_n^1 \\ \cdot & \cdots & \cdot \\ \beta_1^n & \cdots & \beta_n^n \end{bmatrix} \in T^*\Omega \otimes gl(n+1, \mathbf{C}),$$

where

$$\alpha \wedge \beta = \sum_{i=1}^{n} \begin{bmatrix} \alpha_i^1 \wedge \beta_1^i & \cdots & \alpha_i^1 \wedge \beta_n^i \\ \cdot & \cdots & \cdot \\ \alpha_1^n \wedge \beta_1^i & \cdots & \alpha_i^n \wedge \beta_n^i \end{bmatrix}.$$

This can be generalized to arbitrary Lie algebra by putting

$$[\alpha \wedge \beta](X, Y) = [\alpha(X), \beta(Y)] - [\alpha(X), \beta(Y)].$$

Therefore, we can rewrite (1.20) as

$$d\alpha + \frac{1}{2}[\alpha \wedge \alpha] = 0.$$

If $f : \Omega \to S^n$ is harmonic, then we have

$$\Delta f + f|df|^2 = 0 \ \ (i.e., \ d(*df) || f). \tag{1.21}$$

Note that $df = \sum_{i=1}^n \psi^i e_i$ and $de_i = \sum_{j=1}^n w_i^j e_j - \psi^i f, \ i = 1, \cdots, n$. Therefore,

$$d(*df) = \sum_{j=1}^{n} \left[d(*\psi^j) + \sum_{i=1}^{n} w_i^j \wedge (*\psi^i) \right] e_j - \sum_{i=1}^{n} \psi^i \wedge (*\psi^i) f.$$

Hence, (1.21) becomes

$$d(*\psi^j) + \sum_{i=1}^{n} w_i^j \wedge (*\psi^i) = 0, \quad j = 1, \cdots, n,$$

which may be written as

$$d(*\alpha_1) + [\alpha_0 \wedge *\alpha_1] = 0.$$

Let's focus on α and ignore f for the moment. We need to study the system

$$d\alpha_0 + \frac{1}{2}[\alpha_0 \wedge \alpha_0] + \frac{1}{2}[\alpha_1 \wedge \alpha_1] = 0, \tag{1.22}$$

$$d\alpha_1 + [\alpha_0 \wedge \alpha_1] = 0, \tag{1.23}$$

$$d(*\alpha_1) + [\alpha_0 \wedge *\alpha_1] = 0. \tag{1.24}$$

We consider (1.23) and (1.24) as a non-linear Cauchy-Riemann system, in contrast to the linear one $d\beta = 0, \ d(*\beta) = 0$, arising from a harmonic function $f : \Omega \to \mathbf{R}$

by putting $\beta = df$. The idea to deal with (1.22)–(1.24) is to generalize the theory of the linear case, which leads to $d\beta_\sigma = 0$ for

$$\beta_\sigma = \frac{\sigma + \sigma^{-1}}{2}\beta + \frac{\sigma - \sigma^{-1}}{2i}(*\beta) = \sigma^{-1}\beta' + \sigma\beta'', \quad \sigma \in \mathbf{C}^*,$$

where $\beta' = \beta(\frac{\partial}{\partial z})dz$ and $\beta'' = \beta(\frac{\partial}{\partial \bar{z}})d\bar{z}$. We use two facts:

1. The Eqs. (1.23) and (1.24) are identical, one for α_1 and one for $*\alpha_1$, and linear in α.
2. The identity

$$[\cos(\theta)\alpha_1 + \sin(\theta)(*\alpha_1)) \wedge (\cos(\theta)\alpha_1 + \sin(\theta)(*\alpha_1))] = [\alpha_1 \wedge \alpha_1]$$

for all $\theta \in \mathbf{C}$, which is easy to check.

Notice that (1.22)–(1.24) are equivalent to

$$d\alpha_\sigma + \frac{1}{2}[\alpha_\sigma \wedge \alpha_\sigma] = 0, \quad \sigma \in \mathbf{C}^*,$$

where $\alpha_\sigma = \sigma^{-1}\alpha'_1 + \alpha_0 + \sigma\alpha''_1 = \frac{\sigma+\sigma^{-1}}{2}\alpha_1 + \alpha_0 + \frac{\sigma-\sigma^{-1}}{2i}(*\alpha_1)$. Hence, we obtain the following result.

Theorem 1.6.1. *Let $f : \Omega \to S^n$ be a map on a simply connected subset Ω of $\mathbf{C}$ with the lift $F : \Omega \to SO(n+1)$. Put $\alpha = F^{-1} \cdot dF$ and decompose $\alpha = \alpha_0 + \alpha_1$, based on the decomposition $so(n+1) = so(n+1)_0 \oplus so(n+1)_1$. Decompose further $\alpha_1 = \alpha'_1 + \alpha''_1$ with $\alpha'_1 = \alpha_1(\frac{\partial}{\partial z})dz$ and $\alpha''_1 = \alpha_1(\frac{\partial}{\partial \bar{z}})d\bar{z}$. Thus f is harmonic if and only if*

$$d\alpha_\sigma + \frac{1}{2}[\alpha_\sigma \wedge \alpha_\sigma] = 0 \tag{1.25}$$

for all $\sigma \in \mathbf{C}^*$, where $\alpha_\sigma = \sigma^{-1}\alpha'_1 + \alpha_0 + \sigma\alpha''_1$.

Conversely, each family of 1-forms α_σ on Ω of the form $\alpha_\sigma = \sigma^{-1}\alpha'_1 + \alpha_0 + \sigma\alpha''_1$ with coefficients fulfilling the above algebraic conditions, and which is a solution of (1.25), produces a S^1-family of harmonic maps in the following way: (i) Integrate $F_\sigma(z_0) = F_0$, $dF_\sigma = F_\sigma\alpha_\sigma$ on Ω for all $\sigma \in S^1$. (ii) Put $f_\sigma = [F_\sigma] = F_\sigma(\epsilon_{n+1})$. Then for any $\sigma \in S^1$, f_σ is harmonic into S^n.

1.6.3 Loop Groups

We may substitute $SO(n+1)$ by any compact Lie group G and $\mathcal{R} \cong SO(n)$ by a subgroup of G. Assume that G_τ is defined for an automorphism $\tau : G \to G$ such that $\tau^2 = 1_G$ by $G_\tau = \{g \in G : \tau(g) = g\}$ (τ acts like Ad_P) and let $(G_\tau)_0$ be the connected component of G_τ containing the identity. If $(G_\tau)_0 \subset \mathcal{R} \subset G_\tau$, then the previous results hold for a harmonic map $f : \Omega \to G/\mathcal{R}$. We can argue in a similar manner as follows.

We lift $f : \Omega \to G/\mathcal{R}$ to a map $F : \Omega \to G$ such that f is the composition of F with the projection $G \to G/\mathcal{R}$. Then let

$$\alpha = F^{-1}dF \in T^*\Omega \otimes \mathcal{G},$$

where $\mathcal{G}$ is the Lie algebra of G. We have the Cartan decomposition

$$\mathcal{G} = \mathcal{G}_0 \oplus \mathcal{G}_1 = \mathcal{H} \oplus \mathcal{G}_1,$$

where $\mathcal{G}_i = \{\xi \in \mathcal{G} : d\tau_1(\xi) = (-1)^i \xi\},\ i = 0, 1$. Observe that $\mathcal{H} = \mathcal{G}_0$ is the Lie algebra of $\mathcal{R}$. Split $\alpha = \alpha_0 + \alpha_1$ based on this decomposition. We set $\alpha_1' = \alpha_1(\frac{\partial}{\partial z})dz$, $\alpha_1'' = \alpha_1(\frac{\partial}{\partial \bar{z}})d\bar{z}$, and $\alpha_\sigma = \sigma^{-1}\alpha_1' + \alpha_0 + \sigma\alpha_1''$. Thus f is harmonic if and only if

$$d\alpha_\sigma + \frac{1}{2}[\alpha_\sigma \wedge \alpha_\sigma] = 0 \tag{1.26}$$

for all $\sigma \in \mathbf{C}^*$. Furthermore, if (1.26) holds, then we obtain an S^1-family of harmonic maps by integrating $F_\sigma(z_0) = F_0$, $dF_\sigma = F_\sigma\alpha_\sigma$, and letting $f_\sigma = [F_\sigma]$.

We next take a different approach by putting σ in the target. Let's define the loop groups as follows:

$$\Lambda G = \{\sigma \mapsto g_\sigma : \sigma \in S^1,\ g_\sigma \in G\},$$

$$\Lambda G^{\mathbf{C}} = \{\sigma \mapsto g_\sigma : \sigma \in S^1,\ g_\sigma \in G^{\mathbf{C}}\},$$

where $G^{\mathbf{C}}$ is the complexification of G (e.g. $G = SO(n+1)$ or $G = U(n)$),

$$SO(n+1)^{\mathbf{C}} = \{g \in GL(n+1, \mathbf{C}) : g^t g = 1_{n+1},\ det\, g = 1\},$$
$$U(n)^{\mathbf{C}} = GL(n, \mathbf{C}).$$

We also require the twisted loop groups

$$\Lambda G_\tau^{\mathbf{C}} = \{\sigma \mapsto g\sigma : \sigma \in S^1,\ g_\sigma \in G^{\mathbf{C}},\ \tau(g_\sigma) = g_{-\sigma}\},$$
$$\Lambda G_\tau = \Lambda G_\tau^{\mathbf{C}} \cap \Lambda G.$$

We endow these sets with a topology using the Fourier decomposition $g_\sigma = \sum_{k \in Z} \hat{g}_k \sigma^k$, and letting

$$||g_\sigma||_{H^s}^2 = \sum_{k \in Z} |\hat{g}_k|^2 (1 + k^2)^{s/2}$$

for $s > 1/2$. Then the pointwise product

$$[\sigma \mapsto g_\sigma] \cdot [\sigma \mapsto h_\sigma] = [\sigma \mapsto g_\sigma h_\sigma]$$

is continuous and gives these sets a 'Lie group' structure. The corresponding Lie algebras

$$\Lambda \mathcal{G}^{\mathbf{C}} = \{\sigma \mapsto \xi_\sigma : \sigma \in S^1,\ \xi_\sigma \in \mathcal{G}^{\mathbf{C}}\},$$

$\Lambda \mathcal{G}$, $\Lambda \mathcal{G}_\tau^{\mathbf{C}}$ have the Lie brackets

$$[[\sigma \mapsto \xi_\sigma], [\sigma, \eta_\sigma]] = [\sigma \mapsto [\xi_\sigma, \eta_\sigma]].$$

Notice that the twisting condition corresponds to $\tau(\xi_\sigma) = \xi_{-\sigma}$ for $[\sigma \mapsto \xi_\sigma] \in \Lambda \mathcal{G}_\tau^{\mathbf{C}}$. Using the Fourier decomposition $\xi_\sigma = \sum_{k \in Z} \hat{\xi}_k \sigma^k$, we see that it is equivalent to $\hat{\xi}_{2k} \in \mathcal{G}_0^{\mathbf{C}}$, $\hat{\xi}_{2k+1} \in \mathcal{G}_1^{\mathbf{C}}$ for all $k \in \mathbf{Z}$.

1.6.4 Harmonic Maps as Integrable Systems

Let $f : \Omega \to G/\mathcal{R}$ be a harmonic map. We know how to build from it a family of 1-forms

$$\alpha_\sigma = \sigma^{-1}\alpha_1' + \alpha_0 + \sigma \alpha_1'',$$

which we may regard as a 1-form with coefficients in $\Lambda \mathcal{G}^{\mathbf{C}}$. Moreover, we have

1. $\alpha_\sigma \in T^*\Omega \otimes \Lambda \mathcal{G}$ for $\sigma \in S^1$ (reality condition),
2. $\alpha_1', \alpha_1'' \in T^*\Omega \otimes \mathcal{G}_1^{\mathbf{C}}, \alpha_0 \in T^*\Omega \otimes \mathcal{G}_0$. That means $\alpha_\sigma \in T^*\Omega \otimes \Lambda \mathcal{G}_\tau^{\mathbf{C}}$.

We have a characterization of harmonic maps in terms of loop groups. We introduce more loop groups as follows.

$$\begin{aligned}
\Lambda^+ G_\tau^{\mathbf{C}} &= \{[\sigma \mapsto g_\sigma] \in \Lambda G_\tau^{\mathbf{C}} : g_\sigma = \sum_{k \geq 0} \hat{g}_k \sigma^k\} \\
&= \{[\sigma \mapsto g_\sigma] \in \Lambda G_\tau^{\mathbf{C}} : g_\sigma \text{ admits a holomorphic extension inside the disk} \\
&\qquad |\sigma| \leq 1\}, \\
\Lambda^- G_\tau^{\mathbf{C}} &= \{[\sigma \mapsto g_\sigma] \in \Lambda G_\tau^{\mathbf{C}} : g_\sigma = \sum_{k \leq 0} \hat{g}_k \sigma^k\} \\
&= \{[\sigma \mapsto g_\sigma] \in \Lambda G_\tau^{\mathbf{C}} : [\sigma \mapsto g_{\sigma^{-1}}] \in \Lambda^+ G_\tau^{\mathbf{C}}\}, \\
\Lambda_*^- G_\tau^C &= \{[\sigma \mapsto g_\sigma] \in \Lambda^- \mathcal{G}_\tau^C : g_\infty = 1\},
\end{aligned}$$

and their corresponding Lie algebra $\Lambda^+ \mathcal{G}_\tau^{\mathbf{C}}$, $\Lambda^- \mathcal{G}_\tau^{\mathbf{C}}$, etc. (Notice that the loops in $\Lambda^- G_\tau^{\mathbf{C}}$ can be extended holomorphically to $(\mathbf{C} \cup \{\infty\}) \cap \{|\sigma| \geq 1\}$, so that $g_\infty = 1$ makes sense). In particular, for any $[\sigma \mapsto \xi_\sigma] \in \Lambda \mathcal{G}_\tau^{\mathbf{C}}$, the splitting

$$\xi_\sigma = \sum_{k<0} \hat{\xi}_k \sigma^k + \sum_{k\geq 0} \hat{\xi}_k \sigma^k$$

implies that

$$\Lambda \mathcal{G}_\tau^C = \Lambda_*^- \mathcal{G}_\tau^{\mathbf{C}} \oplus \Lambda^+ \mathcal{G}_\tau^{\mathbf{C}}.$$

We denote by $[\xi_\sigma]_{\Lambda_*^- \mathcal{G}_\tau^{\mathbf{C}}}$ the component of $\xi_\sigma \in \Lambda \mathcal{G}_\tau^{\mathbf{C}}$ in $\Lambda_*^- \mathcal{G}_\tau^{\mathbf{C}}$ in this decomposition.

Theorem 1.6.2. *Let $F_\sigma : \Omega \to \Lambda G_\tau$ be a map satisfying*

$$[F_\sigma^{-1} dF_\sigma]_{\Lambda_*^- \mathcal{G}_\tau^{\mathbf{C}}} = \sigma^{-1} \beta \tag{1.27}$$

where $\beta \in T^*\Omega \otimes \mathcal{G}_1^{\mathbf{C}}$ with $\beta(\frac{\partial}{\partial \bar{z}}) = 0$. Then for any σ the map $z \mapsto [F_\sigma](z)$ is harmonic.

Any F_σ as in Theorem 1.6.2 is called an *extended lift* of a harmonic map.

Proof. Denote the family of 1-forms by $\alpha_\sigma = F_\sigma^{-1} dF_\sigma$. The twisting condition $F_\sigma \in \Lambda \mathcal{G}_\tau$ shows that α_σ is also twisted, i.e., $\alpha_\sigma = \sum_{k \in Z} \hat{\alpha}_k \sigma^k$ with $\hat{\alpha}_{2k} \in \mathcal{G}_0^{\mathbf{C}}$ and $\hat{\alpha}_{2k+1} \in \mathcal{G}_1^{\mathbf{C}}$. Equation (1.27) implies that $\hat{\alpha}_k = 0$ if $k < -1$. By the reality condition $F_\sigma \in \Lambda G$, we have $\alpha_\sigma \in \Lambda \mathcal{G}$ and thus

$$\sum_{k \geq -1} \overline{\hat{\alpha}_k} \sigma^{-k} = \sum_{k \geq -1} \hat{\alpha}_k \sigma^k$$

for $\sigma \in S^1$. Therefore, $\overline{\hat{\alpha}_{-k}} = \hat{\alpha}_k$, whence

$$\alpha_\sigma = \sigma^{-1} \overline{\hat{\alpha}_1} + \hat{\alpha}_0 + \sigma \hat{\alpha}_1 = \sigma^{-1} \beta + \hat{\alpha}_0 + \sigma \bar{\beta},$$

with $\hat{\alpha}_0 \in \mathcal{G}_0$. Since $\beta(\frac{\partial}{\partial \bar{z}}) = 0$ and thus $\bar{\beta}(\frac{\partial}{\partial z}) = 0$, this is precisely the decomposition in Theorem 1.6.1. But, in this case (1.26) holds trivially by the definition of α_σ. Hence, we can conclude the result from Theorem 1.6.1. □

Example 1. Let $f : \Omega \to SO(3)/SO(2) \cong S^2$ be conformal. Suppose that f is holomorphic, i.e.,

$$f \times \frac{\partial f}{\partial x} = \frac{\partial f}{\partial y}, \quad f \times \frac{\partial f}{\partial y} = -\frac{\partial f}{\partial x},$$

which means that

$$*(f \times df) = df. \tag{1.28}$$

Let $F = (e_1, e_2, f)$ be a lift of f, and let

$$\alpha = F^{-1}dF = \begin{bmatrix} 0 & w_2^1 & \psi^1 \\ w_1^2 & 0 & \psi^2 \\ -\psi^1 & -\psi^2 & 0 \end{bmatrix} = \begin{bmatrix} 0 & \alpha_2^1 & \alpha_3^1 \\ \alpha_1^2 & 0 & \alpha_3^2 \\ \alpha_1^3 & \alpha_2^3 & 0 \end{bmatrix}.$$

Then

$$df = e_1\psi^1 + e_2\psi^2,$$

and thus

$$f \times df = e_2\psi^1 - e_1\psi^2.$$

Consequently, (1.28) is equivalent to

$$*\,\psi^1 = \psi^2, \; *\psi^2 = -\psi^1,$$

and then to

$$(\psi^2)' = -i(\psi^1)', \; (\psi^2)'' = i(\psi^1)''.$$

Therefore, we have the representations

$$\alpha_1' = \begin{bmatrix} 0 & 0 & (\psi^1)' \\ 0 & 0 & (\psi^2)' \\ -(\psi^1)' & -(\psi^2)' & 0 \end{bmatrix} = (\psi^1)' \begin{bmatrix} 0 & 0 & 1 \\ 0 & 0 & -i \\ -1 & i & 0 \end{bmatrix},$$

and

$$\alpha_1'' = (\psi^1)'' \begin{bmatrix} 0 & 0 & 1 \\ 0 & 0 & i \\ -1 & -i & 0 \end{bmatrix}.$$

Put

$$A_0 = \begin{bmatrix} 0 & -1 & 0 \\ 1 & 0 & 0 \\ 0 & 0 & 0 \end{bmatrix}, \; A_+ = \begin{bmatrix} 0 & 0 & 1 \\ 0 & 0 & -i \\ -1 & i & 0 \end{bmatrix}, \; A_- = \begin{bmatrix} 0 & 0 & 1 \\ 0 & 0 & i \\ -1 & -i & 0 \end{bmatrix}.$$

Then

$$\alpha = (\psi^1)'A_+ + w_1^2 A_0 + (\psi^1)'' A_-.$$

Set

$$f = \frac{1}{1+|g|^2} \begin{bmatrix} g + \bar{g} \\ -i(g - \bar{g}) \\ 1 - |g|^2 \end{bmatrix},$$

where $g : \Omega \to \mathbf{C} \cup \{\infty\}$ is meromorphic, and select F such that

$$e_1 + ie_2 = \frac{1}{1+|g|^2}\begin{bmatrix} i(g^2-1) \\ g^2+1 \\ 2ig \end{bmatrix}.$$

We calculate

$$d'f = \frac{\partial f}{\partial z}dz = \frac{\frac{\partial g}{\partial z}dz}{1+|g|^2}(e_1 - ie_2) = (\psi^1)'(e_1 - ie_2),$$

$$d''f = \frac{\partial f}{\partial \bar{z}}d\bar{z} = \frac{\frac{\partial \bar{g}}{\partial \bar{z}}d\bar{z}}{1+|g|^2}(e_1 + ie_2) = (\psi^1)''(e_1 + ie_2),$$

and

$$w_1^2 = < de_1, e_2 > = \frac{i}{1+|g|^2}(\bar{g}dg - gd\bar{g}) = - * d(log(1+|g|^2)).$$

Consequently,

$$F = \frac{1}{1+|g|^2}\begin{bmatrix} 1-\frac{g^2+\bar{g}^2}{2} & i\frac{g^2-\bar{g}^2}{2} & g+\bar{g} \\ i\frac{g^2-\bar{g}^2}{2} & 1+\frac{g^2+\bar{g}^2}{2} & -i(g-\bar{g}) \\ -(g+\bar{g}) & i(g-\bar{g}) & 1-|g|^2 \end{bmatrix}$$

$$= \frac{1}{1+|g|^2}\left(\frac{g^2}{2}A_+^2 + gA_+ + 1 + \bar{g}A_- + \frac{\bar{g}}{2}A_-^2\right)$$

and

$$\alpha = \frac{d'g}{1+|g|^2}A_+ + \frac{i}{1+|g|^2}(\bar{g}dg - gd\bar{g})A_0 + \frac{d''g}{1+|g|^2}A_-.$$

Therefore, we may construct F_σ from F by replacing g by $\sigma^{-1}g$ to derive

$$F_\sigma = \frac{1}{1+|g|^2}\left(\frac{\sigma^{-2}g^2}{2}A_+^2 + \sigma^{-1}gA_+ + 1 + \sigma\bar{g}A_- + \frac{\sigma^2\bar{g}}{2}A_-^2\right).$$

Notice that f is deformed according to

$$f_\sigma = \begin{bmatrix} \frac{\sigma^{-1}+\sigma}{2} & -\frac{\sigma^{-1}-\sigma}{2i} & 0 \\ \frac{\sigma^{-1}-\sigma}{2i} & \frac{\sigma^{-1}+\sigma}{2} & 0 \\ 0 & 0 & 1 \end{bmatrix} f,$$

and so the action on S^1 is just a rotation.

Example 2. We try to find a rotationally symmetric harmonic map $f : \mathbf{C} \to S^2$ with

$$f(z) = f(x,y) = \begin{bmatrix} \sin\beta(x)\cos(y/a) \\ \sin\beta(x)\sin(y/a) \\ \cos\beta(x) \end{bmatrix} = f(x, y + 2\pi a)$$

for some function $\beta : \mathbf{R} \to \mathbf{R}$. So we have

$$\frac{\partial f}{\partial x} = \beta'(x)e_1, \quad \frac{\partial f}{\partial y} = \frac{\sin\beta(x)}{a}e_2,$$

where

$$e_1 = \begin{bmatrix} \cos\beta(x)\cos(y/a) \\ \cos\beta(x)\sin(y/a) \\ -\sin\beta(x) \end{bmatrix}, \; e_2 = \begin{bmatrix} -\sin(y/a) \\ \cos(y/a) \\ 0 \end{bmatrix}.$$

We can calculate

$$\frac{\partial}{\partial x}\left(f \times \frac{\partial f}{\partial x}\right) + \frac{\partial}{\partial y}\left(f \times \frac{\partial f}{\partial y}\right) = \left(\beta''(x) - \frac{\beta(x)\cos\beta(x)}{a^2}\right)e_2.$$

Thus f is harmonic if and only if β is a solution of the pendulum equation

$$\beta'' - \frac{\sin\beta\cos\beta}{a^2} = 0,$$

with conserved energy $(\beta')^2 - \frac{\sin^2\beta}{a^2} = E_0$. Keep in mind that the Hopf differential is $E_0(dz)^2$. If $E_0 > 0$, we may rescale by $z \mapsto rz$ and $a \mapsto \frac{a}{r}$ and assume that $E_0 = 1$ (likewise, if $E_0 < 0$ we may assume that $E_0 = -1$). Let $F = (e_1, e_2, f)$. Thus we get $F = e^{\frac{y}{a}A_3}e^{\beta A_2}$ and $dF = F\alpha$, with

$$\alpha = \begin{bmatrix} 0 & -\frac{\cos\beta}{a}dy & \beta' dx \\ \frac{\cos\beta}{a}dy & 0 & \frac{\sin\beta}{a}dy \\ -\beta' dx & -\frac{\sin\beta}{a}dy & 0 \end{bmatrix} = \frac{\cos\beta}{a}A_3 dy - \frac{\sin\beta}{a}dy + \beta' A_2 dx,$$

where

$$A_1 = \begin{bmatrix} 0 & 0 & 0 \\ 0 & 0 & -1 \\ 0 & 1 & 0 \end{bmatrix}, \quad A_2 = \begin{bmatrix} 0 & 0 & 1 \\ 0 & 0 & 0 \\ -1 & 0 & 0 \end{bmatrix}, \quad A_3 = \begin{bmatrix} 0 & -1 & 0 \\ 1 & 0 & 0 \\ 0 & 0 & 0 \end{bmatrix}.$$

(Note that $[A_i, A_j] = A_k$, $k = i + j \ (mod\,3)$). A solution of the pendulum equation can be derived by integrating ordinary differential equations in terms of vector fields. In this way we can eliminate β by setting

$$X = \frac{\sin\beta}{2a},\ Y = \frac{\beta'}{2},\ Z = -\frac{\cos\beta}{a},$$

so that

$$\frac{d}{dx}\begin{bmatrix} X \\ Y \\ Z \end{bmatrix} = \begin{bmatrix} -YZ \\ -XZ \\ 4XY \end{bmatrix}. \qquad \square$$

The extended lift F_σ of f is defined by $dF_\sigma = F_\sigma \alpha_\sigma$, and

$$\begin{aligned}\alpha_\sigma &= \sigma^{-1}\left[\frac{i\sin\beta}{2a}A_1 + \frac{\beta'}{2}A_2\right]dz + \frac{\cos\beta}{a}A_3 dy + \sigma\left[-\frac{i\sin\beta}{2a}A_1 + \frac{\beta'}{2}A_2\right]d\bar{z} \\ &= \sigma^{-1}(iXA_1 + YA_2)dz - ZA_3dy + \sigma(-iXA_1 + YA_2)d\bar{z}.\end{aligned}$$

This is equivalent to replacing (X, Y, Z) by $(\sigma^{-1}X, \sigma^{-1}Y, Z)$. Remark that

$$\frac{d}{dx}\begin{bmatrix} \sigma^{-1}X \\ \sigma^{-1}Y \\ Z \end{bmatrix} = \begin{bmatrix} -(\sigma^{-1}Y)Z \\ -(\sigma^{-1}X)Z \\ 2\overline{(\sigma^{-1}X)}(\sigma^{-1}Y) + (\sigma^{-1}X)\overline{(\sigma^{-1}Y)} \end{bmatrix}$$

for all $\sigma \in S^1$. Therefore, this one-parameter family of vector fields is embedded in the set of solutions $X, Y : \mathbf{R} \to \mathbf{C}$, $Z : \mathbf{R} \to \mathbf{R}$ of the system

$$\frac{d}{dx}\begin{bmatrix} X \\ Y \\ Z \end{bmatrix} = \begin{bmatrix} -YZ \\ -XZ \\ 2(\bar{X}Y + X\bar{Y}) \end{bmatrix}. \tag{1.29}$$

Any solution of (1.29) leads to a 1-form

$$\alpha_\sigma = \sigma^{-1}(iXA_1 + YA_2)dz - zA_3dy + \sigma(-i\bar{X}A_1 + \bar{Y}A_2)d\bar{z}$$

that solves the equation

$$d\alpha_\sigma + \frac{1}{2}[\alpha_\sigma \wedge \alpha_\sigma] = 0.$$

Furthermore, we have the following result.

Proposition 1.6.3. *Any solution $X, Y : \mathbf{R} \to \mathbf{C}, Z : \mathbf{R} \to \mathbf{R}$ of (1.29) is also a solution of*

$$d\eta_\sigma = [\eta_\sigma, \alpha_\sigma],$$

where

$$\eta_\sigma = \sigma^{-1}(XA_1 - iYA_2) + ZA_3 + \sigma(\bar{X}A_1 + i\bar{Y}A_2),$$

and conversely.

Proof. We obtain

$$d\eta_\sigma = \sigma^{-1}(dXA_1 - idYA_2) + dZA_3 + \sigma(d\bar{X}A_1 + id\bar{Y}A_2),$$

and

$$[\eta_\sigma, \alpha_\sigma] = \sigma^{-1}(-YZA_1 + iXZA_2)dx + 2(\bar{X}Y + \bar{Y}X)A_3dx + \sigma(-\bar{Y}ZA_1 - i\bar{X}ZA_2)dx.$$

Thus $d\eta_\sigma = [\eta_\sigma, \alpha_\sigma]$ holds if and only if

$$dX = -YZdx,\ dY = -XZdx,\ dZ = 2(\bar{X}Y + \bar{Y}X)dx.$$

Then we can conclude the result. □

We also notice that

$$\begin{aligned}\alpha_\sigma &= i\left[\sigma^{-1}(iXA_1 - iYA_2)dz + \frac{Z}{2}A_3\right]dz - i\left[\frac{Z}{2} + \sigma(\bar{X}A_1 + i\bar{Y}A_2)\right]d\bar{z}\\ &= i\eta_\sigma dz - \beta_\sigma,\end{aligned}$$

where

$$\beta_\sigma = iZA_3dx + 2i\sigma(\bar{X}A_1 + i\bar{Y}A_2)dx.$$

This decomposition $i\eta_\sigma dz = \alpha_\sigma + \beta_\sigma$ corresponds to

$$\Lambda so(3)_\tau^{\mathbf{C}} = \Lambda so(3)_\tau \oplus \Lambda_b^+ so(3)_\tau^{\mathbf{C}},$$

where $\Lambda_b^+ so(3)_\tau^{\mathbf{C}} = \{[\sigma \mapsto \xi_\sigma] \in \Lambda^+ so(3)_\tau^{\mathbf{C}} : \xi_0 = \hat{\xi}_0 \in b\}$ and $b = \{itA_3 : t \in \mathbf{R}\}$ is the Lie algebra of the subgroup

$$\mathcal{B} = \left\{\begin{bmatrix} \cosh t & -i\sinh t & 0\\ i\sinh t & \cosh t & 0\\ 0 & 0 & 1\end{bmatrix} : t \in \mathbf{R}\right\}$$

of $\mathcal{R}^{\mathbf{C}} \cong SO(2)^{\mathbf{C}}$. Thus if $r : \Lambda so(3)_\tau \oplus \Lambda_b^+ so(3)_\tau^{\mathbf{C}} \to \Lambda so(3)_\tau^{\mathbf{C}}$ is the corresponding projection, then we have $d\eta_\sigma = [\eta_\sigma, r(i\eta_\sigma dz)]$. This example shows that some harmonic maps can be obtained by integrating vector fields.

There is a similar way to construct a variety of harmonic maps as above by integrating a pair of vector fields on some finite-dimensional vector space

$$V = \Lambda^d \mathcal{G}_\tau = \left\{ [\sigma \mapsto \xi_\sigma] \in \Lambda \mathcal{G}_\tau : \xi_\sigma = \sum_{k=-d}^{d} \hat{\xi}_k \sigma^k \right\},$$

which is no longer a Lie algebra, where d is an odd number. Such maps are called harmonic maps of *finite type*.

Recall that constant mean curvature surfaces in $\mathbf{R}^3$ correspond to harmonic maps into S^2 by the Gauss map, that are not $\pm$ holomorphic. For constant mean curvature tori, all such harmonic maps are of finite type, which is a result of Pinkall and Sterling [299]. This result was extended to harmonic maps from a torus into Lie groups or symmetric spaces by Burstall, Ferus, Pedit, and Pinkall [53, 104].

Theorem 1.6.4 (Pinkall and Sterling). *All constant mean curvature immersions $T^2 \to \mathbf{R}^3$ are of finite type* (see the proof in [299]).

Notes. The idea of introducing a complex parameter is due to K. Pohlmeyer, and was used extensively by K. Uhlenbeck [387] and N. Hitchin [186]. This was followed by the works of A. Bobenko, F. Burstall, D. Ferus, F. Pedit, U. Pinkall, J. Dorfmeister, F. Helein, etc. The concept of loop groups was introduced by M. Sato, and was developed by G. Segal and G. Wilson in 1975. An algebraic theory of loop groups was constructed in the book written by Pressley and Segal [303] in 1988.

1.7 Harmonic Morphisms

Harmonic morphisms are morphisms which preserve the Laplace equation. This concept can be applied to Brelot harmonic spaces [91]: these are topological spaces where the notion of harmonic function is defined axiomatically, and they include Riemannian polyhedra (Eells and Fuglede [118]), Riemannian manifolds (Baird and Wood [23]), metric graphs (Urakawa [390, 391]) and Weyl spaces (Loubeau and Pantilie [258, 295]). Harmonic morphisms between Riemannian manifolds are maps which pull back (local) harmonic functions to (local) harmonic functions. A well-known result proved by Fuglede [148] and Ishihara [194] independently asserts that harmonic morphisms between Riemannian manifolds are harmonic maps which are horizontally (weakly) conformal. In recent decades, there were many new developments in the theory of harmonic morphisms. The reader is referred to the book "Harmonic Morphisms between Riemannian Manifolds" by Baird and Wood [23].

1.7.1 Morphisms of Euclidean Spaces

Let $f : V \to \mathbf{C}$ be a C^2 function on an open subset of Euclidean m-space $\mathbf{R}^m$ which is harmonic. Then f satisfies the Laplace equation:

$$\Delta f = \sum_{i=1}^{m} \frac{\partial^2 f}{\partial x_i^2} = 0, \ x = (x_1, \cdots, x_m) \in V. \tag{1.30}$$

Under what further conditions on f is the composition $\phi \circ f$ harmonic for any holomorphic map $\phi : U \to \mathbf{C}$ defined on an open subset of $\mathbf{C}$? By the chain rule,

$$\frac{\partial}{\partial x_i}(\phi \circ f) = \frac{d\phi}{dz}\frac{\partial f}{\partial x_i}, \ i = 1, \cdots, m;$$

differentiating both sides and summing, we get

$$\Delta(\phi \circ f) = \frac{d\phi}{dz}\Delta f + \frac{d^2\phi}{dz^2}\sum_{i=1}^{m}(\frac{\partial f}{\partial x_i})^2. \tag{1.31}$$

Because we can find holomorphic maps $z \mapsto \phi(z)$ with any prescribed values of their first and second derivatives $\frac{d\phi}{dz}$, $\frac{d^2\phi}{dz^2}$ at a point, we proceed as Jacobi did for $m = 3$ (cf. [195]) as follows.

(Stat1) *Let $f : V \to \mathbf{C}$ be a C^2 function on an open subset of $\mathbf{R}^3$. Then $\phi \circ f$ is harmonic for all holomorphic maps $\phi : U \to \mathbf{C}$ defined on open subsets of $\mathbf{C}$ iff f satisfies the further condition*

$$\sum_{i=1}^{m}\left(\frac{\partial f}{\partial x_i}\right)^2 = 0. \tag{1.32}$$

A continuous map $f : V \to \mathbf{C}$ defined on an open subset of $\mathbf{R}^m$ is called a *harmonic morphism* if for each $\phi : U \to \mathbf{R}$ is a harmonic function on an open subset U of $\mathbf{C}$ (with $f^{-1}(U)$ non-empty), then $\phi \circ f$ is harmonic. Let $f : V \to \mathbf{C}$ be a continuous map from an open subset V of $\mathbf{R}^m$. Then f is a harmonic morphism iff it is smooth and satisfies (1.30) and (1.32).

To find complex-valued harmonic morphisms, we need to solve (1.30) and (1.32). When $m = 2$, this pair of equations is easy to solve, and (1.32) is equivalent to $\frac{\partial f}{\partial x_2} = \pm i\frac{\partial f}{\partial x_1}$, which are the Cauchy-Riemann or conjugate Cauchy-Riemann equations. They imply that a map $f : V \to \mathbf{C}$ defined on an open subset of $\mathbf{R}^2 = \mathbf{C}$ is a harmonic morphism iff it is holomorphic or anti-holomorphic, or equivalently, it is weakly conformal.

We discuss harmonic morphisms in higher dimensions in three cases as follows.

(A) Complex-valued harmonic morphisms on $\mathbf{R}^3$. In order to solve the pair (1.30) and (1.32), Jacobi utilized the following idea implicitly:
(Stat2) *Let $\psi : \mathbf{R}^m \times \mathbf{C} \supset B \to \mathbf{C}$ $((x, z) \mapsto w)$ be*

(a) *A harmonic morphism in its first variable x, i.e.,*

$$\sum_{i=1}^{m} \frac{\partial^2 \psi}{\partial x_i^2} = 0 \text{ and } \sum_{i=1}^{m} \left(\frac{\partial \psi}{\partial x_i} \right)^2 = 0;$$

(b) *Holomorphic in its second variable z.*

Let $w_0 \in \mathbf{C}$ and assume that $d\psi \neq 0$ on $\psi^{-1}(w_0)$. Then any smooth local solution $f : V \to \mathbf{C}$ $(z = f(x))$, defined on an open subset V of $\mathbf{R}^m$, to the equation $\psi(x, z) = w_0$ is a harmonic morphism. (One can use the chain rule for the proof).

To apply the above (Stat2), put $\psi(x, z) = c_1(z)x_1 + c_2(z)x_2 + c_3(z)x_3 - d(z)$. Because it is linear, ψ is harmonic in x, i.e., $\sum_{i=1}^{3} \frac{\partial^2 \psi}{\partial x_i^2} = 0$. It is horizontally weakly conformal in x iff $\sum_{i=1}^{3} c_i^2 = 0$. It is holomorphic in z iff the c_i and d are holomorphic in z. Then one can write $\mathbf{c}(z) = (c_1(z), c_2(z), c_3(z))$ as

$$\mathbf{c}(z) = \frac{1}{2v(z)}(-2u(z), 1 - u(z)^2, i(1 + u(z)^2)),$$

for holomorphic functions u and v. We have shown part (a) of the following theorem.

Theorem 1.7.1 (Weierstrass formula). *Let $u, v : A \to \mathbf{C}$ be holomorphic functions defined on a domain of $\mathbf{C}$ (or a Riemann surface). Let $f : V \to \mathbf{C}$, $z = f(x_1, x_2, x_3)$ be a smooth solution to equation*

$$-2u(z)x_1 + (1 - u(z)^2) + i(1 + u(z)^2)x_3 = 2v(z) \tag{1.33}$$

defined on an open subset V of $\mathbf{R}^3$. Then

(a) *f is a harmonic morphism;*
(b) [20] *all harmonic morphisms from open subsets of $\mathbf{R}^3$ to $\mathbf{C}$ (or to Riemann surfaces) are given this way locally (up to composition with isometries of the domain and weakly conformal maps on the codomain);*
(c) [20] *the only harmonic morphism defined globally on $\mathbf{R}^3$ is orthogonal projection $z = x_2 + ix_3$ (up to composition with isometries of the domain and weakly conformal maps on the codomain).*

Proof. (b) Once one shows that any harmonic morphism can be factorized on suitable domains as the composition of a submersive harmonic morphism and a weakly conformal map, thus one can assume that the harmonic morphism is submersive. Then by a direct discussion or by Theorem 1.7.9, one observes that any harmonic morphism from an open subset of $\mathbf{R}^3$ to $\mathbf{C}$ (or to a Riemann surface) has fibres which are pieces of straight lines. Due to the horizontal conformality, these lines vary in a holomorphic manner, and so are determined by two holomorphic functions u and v. To show(c), one checks that some of the lines given by u and v will intersect somewhere unless u is constant (i.e., all lines are parallel). □

Example 1. (a) One obtains the orthogonal projection of (c) by putting $u = 0,\ v = \frac{1}{2}z$. Notice that $u = 0$ indicates that the fibres are pieces of straight lines parallel to the x_1-axis.

(b) On the extended complex plane with the Riemann sphere conformally identifying via stereographic projection

$$S^2 \ni (x_1, x_2, x_3) \mapsto \frac{1}{1+x_1} x_2 + i x_3 \in \mathbf{C} \cup \{\infty\}, \tag{1.34}$$

the choice $u(z) = z,\ v = 0$ gives the radial projection $\mathbf{R}^3 - \{0\} \to S^2,\ x \mapsto \frac{x}{|x|}$. Remark that $v = 0$ implies that the fibres are pieces of straight lines through the origin.

(B) Complex-valued harmonic morphisms on $\mathbf{R}^4$. First notice that any complex-valued holomorphic (resp. anti-holomorphic) function defined on a domain of $\mathbf{R}^4 = \mathbf{C}^2$ is a harmonic morphism. In fact, it is easy to see that the Cauchy-Riemann equations are equivalent to (1.30) and (1.32). More general solutions are given by a generalization of the Weierstrass-type construction for $\mathbf{R}^3$ as follows.

Let ϕ be a holomorphic function of three complex variables with $d\phi$ nowhere zero, and ν be a holomorphic function of one complex variable. Writing $x = (\zeta_1, \zeta_2) \in \mathbf{C}^2 = \mathbf{R}^4$, put

$$\psi(x, z) = \phi(z, \zeta_1 - \nu(z)\bar{\zeta}_2, \zeta_2 + \nu(z)\bar{\zeta}_1).$$

Then ψ is a harmonic morphism in x and holomorphic in z. Then (Stat2) implies part (a) of the following theorem.

Theorem 1.7.2. *Let $f : \mathbf{R}^4 \supset V \to \mathbf{C}\, (z = f(x))$ be a smooth submersive solution to the equation*

$$\psi(x, z) = 0. \tag{1.35}$$

Then

(a) *f is a (submersive) harmonic morphism;*
(b) [420] *each submersive harmonic morphism from an open subset of $\mathbf{R}^4$ to $\mathbf{C}$ (or a Riemann surface) is given this way locally (up to composition with isometries of $\mathbf{R}^4$ and conformal maps of the codomain);*
(c) [420] *any submersive harmonic morphism from $\mathbf{R}^4$ is holomorphic (up to precomposition with an isometry of $\mathbf{R}^4$).*

Actually, [395] allow us to verify (a) and (b) without the submersive condition. Notice that the construction of Theorem 1.7.1 can be viewed as a reduction to three dimensions of the construction in this theorem.

Example 2. Let $\phi(z, w_1, w_2) = w_1$ and $\nu(z) = z$. Then (1.35) becomes

$$\zeta_1 - z\bar{\zeta}_2 = 0,$$

and has solution

$$z = f(\zeta_1, \zeta_2) = \frac{\zeta_1}{\bar{\zeta}_2}.$$

This gives the harmonic morphism f. By the notion of harmonic morphism between Riemannian manifolds, we can interpret this as the composition of three harmonic morphisms

$$\mathbf{R}^4 - \{0\} \xrightarrow{\text{radial}} S^3 \xrightarrow{\overline{\text{Hopf}}} S^2 \xrightarrow{\text{stereo}} \mathbf{C} \cup \{\infty\},$$

where the first arrow denotes the radial projection, the second arrow denotes the $\overline{\text{Hopf}}$ map up to isometries, and the third arrow denotes the stereographic projection. Precomposing this map with the isometry which replaces $\bar{\zeta}_2$ by ζ_2 produces the harmonic morphism:

$$\mathbf{C}^2 - \{0\} \xrightarrow{\text{standard proj}} \mathbf{C}P^1 \xrightarrow{\text{standard identification}} \mathbf{C} \cup \{\infty\},$$

by $(\zeta_1, \zeta_2) \mapsto [\zeta_1, \zeta_2] \mapsto \frac{\zeta_1}{\zeta_2}$. If we select a more complicated formula for ϕ, e.g. quadratic in (w_1, w_2), obtain a few more solutions to (1.35), which may be described as a multi-valued harmonic morphism (cf. [165]).

The above constructions can be extended to find complex-valued harmonic morphisms from higher dimensional Euclidean spaces (cf. [21, 22, 421]), but we don't know whether all harmonic morphisms can be obtained by such methods.

(C) Harmonic morphisms of Euclidean spaces of any dimensions. We can apply a chain rule argument similar to that used in (Stat1) and obtain the following theorem.

Proposition 1.7.3. *A smooth map $f : V \to \mathbf{R}^n$ defined on an open subset V of $\mathbf{R}^m$ is a harmonic morphism iff it is harmonic, i.e.,*

$$\Delta f = 0 \tag{1.36}$$

and horizontally weakly conformal, i.e.,

$$\sum_{i=1}^{m} \frac{\partial f^\gamma}{\partial x^i}\frac{\partial f^\mu}{\partial x^i} = 0 \text{ and } \sum_{i=1}^{m} (\frac{\partial f^\gamma}{\partial x_i})^2 = \sum_{i=1}^{m} (\frac{\partial f^\mu}{\partial x_i})^2, \quad 1 \le \gamma, \ \mu \le m, \ \gamma \ne \mu. \tag{1.37}$$

Geometrically, (1.37) means that the gradients of the components f^γ of f are orthogonal and of the same length. Finding all harmonic morphisms between open subsets of Euclidean spaces remains an open problem. The following results are known: (a) [1] for maps given by polynomials, horizontal weak conformality (1.37) implies harmonicity (1.36); (b) [284, 291] all quadratic harmonic morphisms are given by orthogonal multiplications, or equivalently, Clifford systems.

Some value-distribution results can be found using the Brownian path-preserving characterization of harmonic morphisms. For instance, Duheille [107] proved that a non-constant harmonic morphism form $\mathbf{R}^m$ $(m > 3)$ to $\mathbf{R}^3$ cannot avoid three concurrent half-lines under certain conditions. For more details about harmonic morphisms of higher dimensional Euclidean spaces, refer to Baird and Wood [21].

1.7.2 Morphisms of Riemannian Manifolds

A smooth map $f : (M, g) \to (N, h)$ is called *weakly conformal* if for each point $p \in M$, either (a) p is a *branch point*, i.e., $df_p = 0$, or (b) p is a *regular point* if df_p maps the tangent space T_pM conformally into $T_{f(p)}N$, i.e., df_p is injective and there exists a number $\lambda(p) \neq 0$ such that

$$h(df_p(X), df_p(Y)) = \lambda(p)^2 g(X, Y), \ \ X, Y \in T_pM.$$

A *conformal map* is a weakly conformal map with no branch points. If we put $\lambda = 0$ at branch points, then $\lambda : M \to [0, \infty)$ is a continuous function called the *conformality factor* of f. Remark that λ^2 is a smooth since it equals $|df|^2/m$. A weakly conformal map preserves angles at regular points. For each $p \in M$, let df_p^* be the *adjoint* of df_p, i.e.,

$$g(X, df_p^*(Y)) = h(df_p(X), Y), \ \ X \in T_pM, \ Y \in T_{f(p)}N.$$

Proposition 1.7.4. *Let $f : (M, g) \to (N, h)$ be a smooth map. Then the following statements are equivalent:*

(a) *f is weakly conformal with conformality factor λ;*
(b) *$h(df_p(X), df_p(Y)) = \lambda(p)^2 g(X, Y), \ p \in M, X, Y \in T_pM$;*
(c) *$df_p^* \circ df_p = \lambda^2 Id_{T_p(M)}$;*
(d) *In local coordinates, $h_{\alpha\beta} f_i^\alpha f_j^\beta = \lambda^2 g_{ij}$.*

A smooth map $f : (M, g) \to (N, h)$ is called *horizontally weakly conformal* if for each point $p \in M$ either (a) p is a critical point, i.e., $df_p = 0$, or (b) p is a regular point if df_p maps the horizontal space $\mathcal{H}_p = (\ker df_p)^\perp$ conformally onto $T_{f(p)}N$, i.e., df_p is surjective and there exists a number $\lambda(p) \neq 0$ such that

$$h(df_p(X), df_p(Y)) = \lambda(p)^2 g(X, Y), \ \text{for} \ X, Y \in \mathcal{H}_p.$$

A *horizontally conformal map* is a horizontally weakly conformal map with no critical points. Set $\lambda = 0$ at critical points. Then $\lambda : M \to [0, \infty)$ is smooth since it equals $|df|^2/n$. For each $p \in M$, let $df_p^* : TN \to TM$ be the *adjoint* of df_p.

Proposition 1.7.5. *Let $f : (M, g) \to (N, h)$ be a smooth map. Then the following statements are equivalent:*

(a) *f is horizontally weakly conformal with dilation λ;*
(b) $h(df_p^*(X), df_p^*(Y)) = \lambda(p)^2 g(X, Y)$, *for* $p \in M$, $X, Y \in T_pM$;
(c) $df_p \circ df_p^* = \lambda^2 Id_{T_{f(p)}N}$;
(d) *In local coordinates,* $g^{ij} f_i^\alpha f_j^\beta = \lambda^2 h^{\alpha\beta}$.

A continuous map $f : M \to N$ is called a *harmonic morphism* if for each harmonic function $\phi : U \to \mathbf{R}$ defined on an open subset U of N with $f^{-1}(U)$ non-empty, the composition $\phi \circ f$ is harmonic on $f^{-1}(U)$. Fuglede [148] and Ishihara [194] independently proved the following theorem.

Theorem 1.7.6. *A smooth map $f : M \to N$ between Riemannian manifolds is a harmonic morphism iff it is both harmonic and horizontally weakly conformal.*

This is proved by computing the tension field of the composition of f with a harmonic function $\phi : N \supset U \to \mathbf{R}$, and showing that there is a harmonic function with prescribed (traceless) 2-jet (cf. [148, 194]).

Example 3. (a) The *Hopf maps* $S^3 \to S^2$, $S^7 \to S^4$, $S^{15} \to S^8$, $S^{2n+1} \to \mathbf{C}P^n$, $S^{4n+3} \to \mathbf{H}P^n$ are all Riemannian submersions up to scaling, i.e., horizontally conformal submersions with λ constant. One can verify that they are harmonic maps, thus they are harmonic morphisms.
(b) *Radial projection.* One can see that the map

$$\mathbf{R}^m - \{0\} \to S^{m-1}, \ x \mapsto \frac{x}{|x|}$$

is (i) harmonic; (ii) a horizontally conformal submersion with dilation $\lambda = \frac{1}{|x|}$. Thus it is a harmonic morphism.
(c) In [67] it is shown that all stable harmonic maps from a compact Riemannian manifolds to S^2 are harmonic morphisms. This is proved by computing the second variation to conclude that the non-negative curvature of S^2 forces the map to be horizontally weakly conformal.

1.7.3 Classification of Harmonic Morphisms

Twistor method. The harmonic morphisms constructed in Theorem 1.7.2 are holomorphic with respect to the almost Hermitian structure J on V which at any point $x = (\zeta_1, \zeta_2)$ has a (0,1)-tangent space spanned by

$$\frac{\partial}{\partial \bar{\zeta}_1} - \nu(f(x))\frac{\partial}{\partial \zeta_2} \text{ and } \frac{\partial}{\partial \bar{\zeta}_2} + \nu(f(x))\frac{\partial}{\partial \zeta_1}.$$

Clearly, J is parallel (constant) along the fibres, and this is equivalent to the integrability of J by Baird and Wood [23]. This is extended by the following result: A surface is *superminimal* with respect to an almost Hermitian structure J if its tangent space is closed under J and J is parallel along the surface.

Theorem 1.7.7. *Let $f : M^4 \to N^2$ be a non-trivial harmonic morphism from an orientable Einstein 4-manifold to a Riemann surface. Then f is holomorphic with respect to some integrable Hermitian structure J on M^4 and has superminimal fibres with respect to J.*

The above theorem was proved by Wood [420] for submersive harmonic morphisms, and was proved by Ville [395] for harmonic morphisms without the submersivity requirement.

A map is called *twistorial* if it is covered by a holomorphic map of suitable fibre bundles (cf. [296] for definition); for the setting of harmonic morphisms of Weyl spaces see [257, 295]. Theorem 1.7.7 says that any harmonic morphisms from an orientable Einstein 4-manifold to a Riemann surface is twistorial. Refer to [21, 22] for constructions of harmonic morphisms from some higher-dimensional manifolds by twistorial methods, and [164] for a different method of construction of complex-valued harmonic morphisms from symmetric spaces.

The fundamental equation. Baird and Eells [18] obtained the fundamental equation as follows.

Lemma 1.7.8 ([18]). *Let $f : M^m \to N^n$ be a horizontally weakly conformal map with dilation λ. Then*

$$\tau(f) = df\{-(n-2)\,\mathrm{grad}\ln\lambda - (m-n)k^{\mathcal{V}}\}, \tag{1.38}$$

at regular points, where $k^{\mathcal{V}}$ is the mean curvature of the fibres.

Therefore, f is harmonic, and so it is a harmonic morphism iff

$$(n-2)\mathcal{H}(\mathrm{grad}\ln\lambda) + (m-n)k^{\mathcal{V}} = 0,$$

at regular points.

Theorem 1.7.9 ([18]). *If $n = 2$, or grad λ is vertical at regular points, a horizontally weakly conformal map is harmonic, and so it is a harmonic morphism iff its fibres are minimal at regular points.*

A horizontally weakly conformal map is a *horizontally homothethic* map if grad λ is vertical. For instance, a Riemannian submersion is horizontally homothethic.

Example 4. (a) A Riemannian submersion is harmonic, and thus a harmonic morphism iff its fibres are minimal. The Hopf maps in Example 3 have minimal (totally geodesic) fibres, and thus are harmonic morphisms by Lemma 1.7.8.

(b) The projection $F \times_\eta N \to N$ of a warped product (cf. Sect. 5.3) onto its second factor is a horizontally homothetic map with totally geodesic fibres and integrable horizontal distribution. In particular, it is a harmonic morphism. Conversely, a horizontally homothetic map with totally geodesic fibres and integrable horizontal distribution is locally the projection of a warped product. Such maps are called *harmonic morphisms of warped product type* (or *umbilic harmonic morphisms* [45]). For instance, the radial projections of Example 3 are such maps.

For a submersion $f : M \to N$ of Riemannian manifolds, the connected components of its fibres form a smooth foliation by submanifolds of dimension $q = dimM - dimN = m - n$. A foliation is called *conformal* if it is locally given by a horizontally conformal submersion. Given a conformal foliation $\mathcal{F}$, one may ask under what condition it is given locally by harmonic morphisms, in which case one says that $\mathcal{F}$ *produces harmonic morphisms.* Let $k^{\mathcal{V}}$ be the mean curvature of the leaves of $\mathcal{F}$ and $k^{\mathcal{H}}$ the mean curvature of the *horizontal distribution* $\mathcal{H}$ (i.e., the distribution of subspaces orthogonal to the leaves). It is clear that for the foliation given by a horizontally conformal submersion of dilation λ, the vector field $k^{\mathcal{H}}$ is equal to the vertical component of $\operatorname{grad}\ln\lambda$. Then the fundamental equation immediately implies the following result.

Corollary 1.7.10. (a) *If $n = 2$, a conformal foliation gives harmonic morphisms iff its leaves are minimal (i.e. $k^{\mathcal{V}} = 0$).* (b) [45] *If $n \neq 2$, a conformal foliation gives harmonic morphisms iff*

$$Z = (n-2)k^{\mathcal{H}} - (m-n)k^{\mathcal{V}}$$

is locally a gradient field, equivalently, $Z^{\flat}$ is closed. (Here, $Z^{\flat}$ denotes the 1-form associated to Z by the 'musical' isomorphism $\flat : TM \to T^*M$ defined by the metric on M.)

The fundamental equation (1.38) for a horizontally weakly conformal map provides interesting information. An easy computation gives the interpretation of the mean curvature of the fibres as a Lie derivative of the volume form.

Proposition 1.7.11. *Let $\mathcal{F}$ be a foliation of dimension q on a Riemannian manifold and $v^{\mathcal{V}}$ be the vertical form on M which gives the volume form on each leaf (via some local orientation). Then for any horizontal vector field X,*

$$\mathcal{V}^*(\mathcal{L}_X(v^{\mathcal{V}})) = -q\, g(k^{\mathcal{V}}, X)v^{\mathcal{V}},$$

where $\mathcal{V}^*$ denotes the vertical part.

Applying the above proposition, the fundamental equation (1.38) can be expressed as

$$\mathcal{V}^*(\mathcal{L}_X(\lambda^{2-n}v^{\mathcal{V}})) = \lambda^{2-n}g(\tau(f), X)v^{\mathcal{V}}. \tag{1.39}$$

We can view $\mathcal{V}^* \circ \mathcal{L}_X$ as a Bott partial connection on the vertical bundle $\mathcal{V}$, so that a horizontally weakly conformal map f is a *harmonic morphism* iff at each regular point the form $\lambda^{2-n}v^{\mathcal{V}}$ *is parallel in all horizontal directions with respect to the Bott partial connection* (cf. [23] for more details).

1.7.4 Morphisms with One-Dimensional Fibres

When the foliation has one-dimensional fibres, the fundamental equation (1.39) for a horizontally weakly conformal map can be re-expressed as follows. Let $W = \lambda^{n-2}V$, where V is the positive unit tangent vector field to the fibres (with respect to some local orientation), and let θ be the dual vertical 1-form such that $\theta(W) = 1$ and $ker\,\theta = \mathcal{H}$. Thus $\theta = \lambda^{2-n}V^b$, and one has that, for each horizontal vector field X,

$$g(\tau(f), X) = \mathcal{L}_X\theta(W) = -d\theta(W, X) = -(\mathcal{L}_W\theta)(X) \qquad (1.40)$$

at all regular points.

Proposition 1.7.12. *A horizontally weakly conformal map f is a harmonic morphism iff at each regular point* $\mathcal{L}_W(\theta) = 0$*, or equivalently, its horizontal distribution* $\mathcal{H}$ *is invariant under the flow of W.*

Hence, a harmonic morphism is locally a principal bundle with group **R** or S^1 and connection form θ at regular points. Moreover, one can state that the metric g is of the form

$$g = \lambda^{-2}f^*h + \lambda^{2n-4}\theta^2. \qquad (1.41)$$

In fact, the first term describes the horizontal conformality with dilation λ, and the second term is the square of the volume form on the fibres.

Theorem 1.7.13 (Bryant [43, 45]). *Let* (M^{n+1}, N^n, S^1) *be a principal bundle with projection* $f : M^{n+1} \to N^n$ *and endowed with a principal connection* $\mathcal{H} \subset TM$*. Let h be a Riemannian metric on* N^n *and* λ *a smooth positive function on* M^{n+1}*. Define a Riemannian metric on* M^{n+1} *by (1.41), where* θ *is the connection form of* $\mathcal{H}$*. Then* $f : (M^{n+1} g) \to (N^n, h)$ *is a submersive harmonic morphism. Conversely, any submersive harmonic morphism with one-dimensional fibres is locally of this form (up to isometries).*

We now discuss under what condition the fundamental vector field W is a Killing vector field, i.e., $L_W g = 0$. Remark that (1.40) equals $-\lambda^{4-2n}(\mathcal{L}_W g)(W, Y)$, and that for any horizontally conformal submersion the other two components of $\mathcal{L}_W g$ are given by

$$(\mathcal{L}_W g)(W, W) = W(\lambda^{2n-4}) \text{ and } (\mathcal{L}_W g)(X, Y) = -W(ln\,\lambda^2)g(X, Y).$$

Proposition 1.7.14 ([23]). *Let* $f : (M^{n+1}, g) \to (N^n, h)$ $(n \geq 1)$ *be a horizontally conformal submersion with one-dimensional fibres,* λ *be its dilation, and* W *be its fundamental vertical vector. Then* W *is Killing iff* f *is a harmonic morphism with grad* λ *horizontal, i.e., with dilation constant along the fibre components.*

Such maps are called *harmonic morphisms of Killing type.*

Proposition 1.7.15 ([45]). *Let* $\mathcal{F}$ *be a one-dimensional Riemannian foliation on a Riemannian manifold* M^{n+1} $(n \neq 2)$*. Then* $\mathcal{F}$ *provides harmonic morphisms iff its leaves are tangent to (locally defined) Killing fields.*

Example 5. The Hopf polynomial map $\mathbf{R}^4 \to \mathbf{R}^3$ defined by

$$\mathbf{C}^2 \ni (z, w) \mapsto (|z|^2 - |w|^2, 2\bar{z}w) \in R \times \mathbf{C}$$

is a harmonic morphism of Killing type. Its fibres are spanned by the Killing field corresponding to the isometric action

$$t \cdot (z, w) = (e^{it}z,\ e^{it}w),\ t \in \mathbf{R}.$$

It restricts to the Hopf map $S^3 \to S^2$.

1.7.5 Relationship with Two Equations of Mathematical Physics

We narrate that harmonic morphisms are connected with two equations of Mathematical Physics: *the monopole equation* and the *Beltrami fields equation.* In the following result, assertion (b) was obtained by Jones and Tod [200], and assertion (c) by Gibbons and Hawking [155].

Theorem 1.7.16. *Let* N^3 *be an oriented constant curvature 3-manifold, put* $M^4 = \mathbf{R} \times N^3$*, and let* $f : M^4 \to N^3$ *be the projection onto the second factor. Define a Riemannian metric* g *on* M^4 *by*

$$g = u f^*h + u^{-1}(dt + A)^2,$$

where u *is a positive smooth function and* A *is a 1-form on* N^3*. Then*

(a) *f is a harmonic morphism of Killing type;*
(b) (M^4, g) *is self-dual (resp. anti-self-dual) if u and A are related by the monople equation*

$$du = \star dA \ \ (resp.\ du = -\star dA); \tag{1.42}$$

(c) $(M^4,\ g)$ *is Einstein iff (1.42) holds and* $(N^3,\ h)$ *is flat, in which case g is Ricci-flat and self-dual.*

Remark that the connection form θ is given by $\theta = dt + A$ and dA is the curvature of the connection form given by θ (cf. [296, 297]).

Example 6. For $a \geq 0$, define a harmonic function on $\mathbf{R}^3 - \{0\}$ by

$$u_a(x) = \frac{1}{4}\left(\frac{1}{|x|} + a\right).$$

This construction provides the *Hawking Tau-Nut metric* g_a *on* $\mathbf{R}^4$ $(a > 0)$ or the standard metric g_0 $(a = 0)$, and the harmonic morphism is the Hopf polynomial map $(\mathbf{R}^4, g_a) \to (\mathbf{R}^3, h)$.

Notice that the metric g_a can be extended to the whole of $\mathbf{R}^4$; indeed it is given by the explicit formula

$$g_a = (a|x|^2 + 1)g_0 - \frac{a(a|x|^2 + 2)}{a|x|^2 + 1}(-x_2dx_1 + x_1dx_2 - x_4dx_3 + x_3dx_4)^2,$$

(cf. [243] for g_1). Pantilie characterizes the above example as follows.

Theorem 1.7.17 ([294]). *Let* $f : M^4 \to N^3$ *be a surjective harmonic morphism between complete simply-connected Einstein manifolds which has exactly one critical point. Then we necessarily are in the case of Example 6 for some* $a \geq 0$ *(up to homotheties).*

We now describe a second construction that provides morphisms that are neither of Killing type nor of warped product type, which are called 'type 3'.

Theorem 1.7.18 ([60, 296]). *Let* $(N^3,\ h)$ *be an oriented constant curvature 3-manifold and let A be a 1-form on* N^3*. Define a Riemannian metric on* $(0,\ \infty) \times N^3$ *by*

$$g = \rho h + \rho^{-1}(d\rho + A)^2, \quad \rho \in (0,\ \infty). \tag{1.43}$$

Then

(a) *f is harmonic morphism.*
(b) [297] *g is self-dual (resp. anti-self-dual) if the following Beltrami fields equation holds on* N^3*:*

$$dA = - \star A \ \ (resp.\ dA = \star A). \tag{1.44}$$

(c) [296] *g is Einstein iff (1.44) holds and h has constant curvature 1/4, in which case g is Ricci-flat and self-dual.*

Example 7. Let $A = \alpha i^*(-x^2dx^1 + x^1dx^2 - x^4dx^3 + x^3dx^4$ ($\alpha \in \mathbf{R}$), and consider the standard inclusion map $i : S^3 \hookrightarrow \mathbf{R}^4$ where S^3 is the 3-sphere. Then A induces the Beltrami fields equation, g is the Eguchi-Hansen metric ($\alpha \neq 0$) on $\mathbf{R}^4$ or the standard metric ($\alpha = 0$), and the harmonic morphism is the radial projection $(\mathbf{R}^4 - \{0\}, g) \to (S^3, h)$, $x \mapsto \frac{x}{|x|}$.

Pantilie [294] proved that any harmonic morphism from an Einstein 4-manifold to a Riemannian 3-manifold must be of Killing type, warped-product type, or is such up to homotheties as Theorem 1.7.18(c). For maps from self-dual manifolds, a fourth type of harmonic morphism is introduced (cf. [297]).

1.8 Maps of Singular Spaces

Eells' final monograph, *Harmonic Maps between Riemannian Polyhedra* [118], co-authored with the Danish mathematician B. Fuglede, on harmonic theory with singular domains and targets. The reader is referred to this book for details. In this section, we focus on the spectral geometry of V-manifolds, harmonic maps of V-manifolds based on the results obtained by Chiang [73,74] in 1990, and harmonic maps on spaces with conical singularities by Chiang and Ratto [79] in 1992.

1.8.1 Spectral Geometry of V-Manifolds

Let $(M, \mathcal{F})$ be a (C^∞) V-manifold, and U be an open subset of M. By a *V-chart* on M over U we mean a triplet $\{\tilde{U}, G, \pi\}$ consisting of (i) a connected open subset $\tilde{U}$ of $\mathbf{R}^m$, (ii) a finite group G of diffeomorphisms of $\tilde{U}$, with the set of fixed points of codimension ≥ 2, and (iii) a continuous map of $\pi : \tilde{U} \to U$ such that $\pi \circ \sigma = \pi$ for all $\sigma \in G$ and such that π induces a homeomorphism of $\tilde{U}/G$ onto U. U is called the *support* of the V-chart, and π is called the *projection* onto U.

Let $\{\tilde{U}, G, \pi\}$ and $\{\tilde{U}', G', \pi'\}$ be two V-charts for U, U', respectively, and $U \subset U'$. By an injection $\lambda : \{\tilde{U}, G, \pi\} \to \{\tilde{U}', G', \pi'\}$ we mean an isomorphism λ from U onto an open subset of U' such that for any $\sigma \in G$ there exists a $\sigma' \in G'$ satisfying the relations $\lambda \cdot \sigma = \sigma' \cdot \lambda$ and $\pi = \pi' \cdot \lambda$. Clearly, σ' is then uniquely determined by σ, and $\sigma \mapsto \sigma'$ defines an isomorphism of G onto a subgroup of G'.

A (C^∞) *V-manifold* $(M, \mathcal{F})$ consists of a Hausdorff space M with an atlas $\mathcal{F}$ of V-charts satisfying the following conditions: (a) If $\{\tilde{U}, G, \pi\}$ and $\{\tilde{U}', G', \pi'\}$ are two V-charts over U, U', respectively, in M such that $U \subset U'$, then there exists an injection $\lambda : \{\tilde{U}, G, \pi\} \to \{\tilde{U}', G', \pi'\}$. (b) The supports of V-charts in $\mathcal{F}$ form a basis for the open sets in M.

Let $(M, \mathcal{F})$ be a V-manifold and $p \in M$. Take a chart $\{\tilde{U}, G, \pi\} \in \mathcal{F}$ such that $p \in \pi(\tilde{U})$ and choose $\tilde{p} \in \tilde{U}$ such that $\pi(\tilde{p}) = p$. The isotropy subgroup $G_{\tilde{p}}$ of G at $\tilde{p}$ is the set of all $\sigma \in G$ such that $\sigma \tilde{p} = \tilde{p}$, and is uniquely determined by p. Therefore, $G_{\tilde{p}}$ is called the *isotropy group* of p. The singular set $\mathcal{S}$ of M consists of all singular points of M, i.e., the points of M with non-trivial isotropy groups. Let $(\tilde{x}^1, \cdots, \tilde{x}^m)$ be a coordinate system around $\tilde{p}$ and consider the system $\tilde{y}^i = \frac{1}{|G_{\tilde{p}}|}\sum l_{ij}(\sigma^{-1})\tilde{x}^j \cdot \sigma$, with

$$l_{ij}(\sigma) = \left[\frac{\partial \tilde{x}^i \circ \sigma}{\partial \tilde{x}^j}\right]_{\tilde{p}}, \quad |G_{\tilde{p}}| = \text{order of } G_{\tilde{p}}.$$

Then the $\{\tilde{y}^i\}$ are a new coordinate system around $\tilde{p}$ and $G_{\tilde{p}}$ operates linearly in the $\tilde{y}$-system. After this suitable C^∞ change of coordinate around $\tilde{p}$, $G_{\tilde{p}}$ becomes a finite group of linear transformations. The fixed point set of any $\sigma \in G_{\tilde{p}}$ is defined by linear equations in $\tilde{y}$, and consequently the fixed point set of $\sigma \in G_{\tilde{p}}$ in $\tilde{U}$ is the intersection of $\tilde{U}$ with a linear space. Therefore, $\pi^{-1}\mathcal{S}$ is locally a finite union of linear spaces intersected with $\tilde{U}$. Hence, $\mathcal{S}$ is a V-submanifold of codimension ≥ 2 of M. Clearly, $M - \mathcal{S}$ is an ordinary manifold.

The main difficulties arise from the complicated behavior near the singular locus of V-manifolds. Therefore, a different method than the usual one is required to deal with Sobolev chains, Laplacian and the heat operator.

We fix a paracompact V-manifold M with defining atlas $\mathcal{F}$. A *smooth* map $f : M \to N$ from a V-manifold into an ordinary manifold N is defined as follows: For any $\{\tilde{U}, G, \pi\} \in \mathcal{F}$ there corresponds a G-invariant smooth map $f_{\tilde{U}}^G = \frac{1}{|G|}\sum_{\sigma\in G} f_{\tilde{U}} \circ \sigma : \tilde{U} \to N$ such that $f_{\tilde{U}}^G = f \circ \pi$ and $f_{\tilde{U}}^G = f_{\tilde{U}'}^{G'} \circ \lambda$ for any injection $\lambda : \{\tilde{U}, G, \pi\} \to \{\tilde{U}', G', \pi'\}$, where $f_{\tilde{U}} : \tilde{U} \to N$ is an ordinary smooth map. Considering $\mathbf{R}$ as a 1-dimensional manifold, we define a smooth function on a V-manifold as a smooth map $M \to \mathbf{R}$. It is not hard to extend the definition to maps of M into another V-manifold.

If $\tilde{\rho}$ is a smooth function on the chart $\{\tilde{U}, G, \pi\} \in \mathcal{F}$ over U, then the G-average $\tilde{\rho}^G$ defines a smooth function ρ on U. If $\tilde{\rho} > 0$ on $\tilde{U}$, then certainly $\rho > 0$ on U. In this way we can follow the usual procedure to construct a partition of unity on M and obtain the following proposition.

Proposition 1.8.1. *If $\{U_\alpha\}$ is a locally finite covering of a paracompact V-manifold M by open sets which are supports of V-charts, then there is a smooth partition of unity $\{\rho_\alpha\}$ such that each ρ_α has compact support contained in U_α.*

We next discuss tensors. If $\tilde{\psi}$ is a tensor field of some fixed type (r, s) on $\tilde{U}$ over U, then the composition with $\sigma \in G$ defines a new tensor $\tilde{\psi}^\sigma$ of the same type on $\tilde{U}$ and we may form the G-average $\tilde{\psi}^G = \frac{1}{|G|}\sum_{\sigma\in G} \tilde{\psi}^\sigma$, which is G-invariant. We interpret this as a tensor field ψ of type (r, s) on the support. Then a global smooth tensor field ψ of type (r, s) on M has a smooth G-invariant lift over the support of any chart in the atlas $\mathcal{F}$. In general, we make calculations in the V-chart $\tilde{U}$.

Let us put a Riemannian metric $g_{\tilde{U}} = g_{ij} d\tilde{x}^i d\tilde{x}^j$ on $\tilde{U}$. By taking the G-average if necessary, we can assume that $g_{\tilde{U}}$ is G-invariant. Thus the transformations $\sigma \in G$ are isometries for $g_{\tilde{U}}$. By using the standard partition of unity construction, we can patch all such local invariant metrics together into a global metric tensor field of type (0, 2) on the V-manifold M, which we call a Riemannian metric on M.

If $\tilde{\psi}$ is a tensor field on $\tilde{U}$ and D is the covariant derivative given by our G-invariant metric, then $D(\tilde{\psi}^\sigma) = (D\tilde{\psi})^\sigma$, where the superscript indicates the new field obtained by applying the isometry $\sigma \in G$. If $\tilde{\psi}$ is G-invariant, then so is $D\tilde{\psi}$ and they define tensor fields on the support U of the V-chart $\{U, G, \pi\} \in \mathcal{F}$ which are compatible with the injections. If ψ is a global tensor field on M, we may therefore define its covariant derivative $D\tilde{\psi}$, which is of course calculated locally by the usual formula.

Let u be a smooth function on a compact V-manifold M and fix a chart $\{\tilde{U}, G, \pi\} \in \mathcal{F}$ over $U \subset M$. The lift $\tilde{u} = u \circ \pi$ is G-invariant in $\tilde{U}$ and so is $d\tilde{u}$, and accordingly we can define a global smooth 1-form du on M. The laplacian Δu of u is the trace of the tensor field Ddu of type (0, 2). In $\tilde{U}$ this is given by the usual formula, namely

$$\Delta \tilde{u} = g^{ij}\tilde{u}_{ij} = \frac{1}{\sqrt{g}}\frac{\partial}{\partial \tilde{x}^i}\left(\sqrt{g}g^{ij}\frac{\partial \tilde{u}}{\partial \tilde{x}_j}\right)$$

$$= g^{ij}\frac{\partial^2 \tilde{u}}{\partial \tilde{x}^i \partial \tilde{x}^j} + \sum_j B^j \frac{\partial \tilde{u}}{\partial \tilde{x}^j}, \quad B = \frac{1}{2g}\sum \frac{\partial g}{\partial \tilde{x}^i}g^{ij} + \sum \frac{\partial g_{ij}}{\partial \tilde{x}^i}, \tag{1.45}$$

where g_{ij} is the G-invariant metric in $\tilde{U}$.

From now on we assume that the V-manifold M is compact. Let $\{\tilde{U}_\alpha\}$ be a finite covering of M such that $\{\tilde{U}_\alpha, G_\alpha, \pi_\alpha\} \in \mathcal{F}$ and fix it for the ensuing discussion. For any $p \in M$ there exists a neighborhood $\tilde{U}_\alpha$ of $\tilde{p}$ such that $\{\tilde{U}_\alpha, G_\alpha, \pi_\alpha\} \in \mathcal{F}$; let $\tilde{x}_\alpha$ be a local coordinate about $\tilde{p}$. Let $\mathbf{T}^m = \mathbf{R}^m / 2\pi\mathbf{Z}^m$ denote the m-torus. Each $\tilde{U}_\alpha$ can be provided with an isomorphism $\tau_\alpha : \tilde{U}_\alpha \to U'_\alpha$, where U'_α is an open subset of $\mathbf{T}^m$. Then $x'_\alpha = \tilde{x}_\alpha \circ \tau_\alpha^{-1}$ is a local coordinate on $\mathbf{T}^m$ (x' is determined up to $2\pi\mathbf{Z}^m$ by the canonical coordinate of $\mathbf{R}^m$). We are taking charts in $\mathbf{T}^m$ instead of $\mathbf{R}^m$ because we want to use the Fourier transform on $\mathbf{T}^m$ and it is much simpler for our purposes than $\mathbf{R}^m$. For convenience, we may choose τ_α such that $\{U'_\alpha\}$ are disjoint in $\mathbf{T}^m$. Let $\{\tilde{\rho}_\alpha(\tilde{x})\}$ be a partition of unity with compact support subordinate to the finite covering $\{\tilde{U}_\alpha\}$ of M such that $\{\tilde{U}_\alpha, G_\alpha, \pi_\alpha\} \in \mathcal{F}$, i.e., $\tilde{\rho}_\alpha(\tilde{x}) = \rho_\alpha(x) \circ \pi_\alpha$ in Proposition 1.8.1. For $u,\ v \in C^\infty(M),\ t = 1, 2, \cdots$, we define

$$(u, v)_t = \int_M (1+\Delta)^t u \cdot v\, dx = \sum_\alpha \frac{1}{|G_\alpha|}\int_{\tilde{U}_\alpha} \tilde{\rho}_\alpha(\tilde{x})(1+\Delta)^t \tilde{u} \cdot \tilde{v}\, d\tilde{x}_\alpha$$

$$= \sum_\alpha \sum_{[\mu]\le t} \binom{t}{[\mu]} \frac{1}{|G_\alpha|}\int_{\tilde{U}_\alpha} \tilde{\rho}_\alpha(\tilde{x}) < D^\mu_{\tilde{x}_\alpha}\tilde{u}_\alpha, D^\mu_{\tilde{x}_\alpha}\tilde{v}_\alpha > d\tilde{x}_\alpha, \tag{1.46}$$

where $\Delta_\alpha = (-)(\frac{\partial}{\partial \tilde{x}_\alpha})^2 = (-)\sum_{j=1}^m (\frac{\partial}{\partial \tilde{x}_j})^2$ (Note that there is a $-$ or $+$ sign convention for the laplacian Δ.),

$$D^{\mu}_{\tilde{x}_\alpha} = \left(\frac{\partial}{\partial \tilde{x}_\alpha}\right)^{\mu} = (\partial/\partial\tilde{x}_1)^{\mu_1}\cdots(\partial/\partial\tilde{x}_m)^{\mu_m},$$

$\tilde{u}_\alpha = u_\alpha \circ \pi_\alpha : \tilde{U}_\alpha \to \mathbf{R}$ is the G_α-invariant smooth functions, $u_\alpha = u|_{U_\alpha}$, $< \cdot, \cdot >$ is the ordinary inner product. For $t = 0$, we observe that

$$(u, v)_0 = \sum_\alpha \frac{1}{|G_\alpha|}\int_{\tilde{U}_\alpha} \tilde{\rho}_\alpha(\tilde{x}) < \tilde{u}_\alpha, \tilde{v}_\alpha > d\tilde{x}_\alpha. \tag{1.47}$$

We first set $H^t(M)$ to be the completion of $C^\infty(M)$ with respect to the norm $|u_t| = (u,u)_t^{1/2}$. Then $H^t(M)$ is a Hilbert space. We next set $H^{-t}(M)$ to be the dual of $H^t(M)$ with respect to the product (1.47), with $H^{-0}(M)$ identified to $H^0(M)$. Therefore, we clearly have the injection

$$\cdots H^t(M) \subset \cdots \subset H^1(M) \subset H^0(M) \subset H^{-1}(M) \subset \cdots \subset H^{-t}(M)\cdots \tag{1.48}$$

where $t \geq 0$, $H^0(M) = L^2(M)$. If $\lambda : (\tilde{U}_\alpha, \tilde{x}) \to (\tilde{U}'_\alpha, \tilde{x}')$ is an injection for any two V-charts $\{\tilde{U}_\alpha, G_\alpha, \pi_\alpha\}$, $\{\tilde{U}'_\alpha, G'_\alpha, \pi\} \in \mathcal{F}$ over the compact V-manifold M, then the norm $|u(x)|_t$ from the product

$$\begin{aligned}
(u,v)_t &= \sum_\alpha \sum_{[\mu]\leq t} \binom{t}{[\mu]} \frac{1}{|G_\alpha|}\int_{\tilde{U}_\alpha} \tilde{\rho}_\alpha(\tilde{x}) < D^{\mu}_{\tilde{x}_\alpha}, D^{\mu}_{\tilde{x}_\alpha}\tilde{v}_\alpha > d\tilde{x}_\alpha \\
&= \sum_\alpha \sum_{[\mu]\leq t} \binom{t}{[\mu]} \frac{1}{|G_\alpha|}\int_{\tilde{U}_\alpha} \tilde{\rho}_\alpha(\tilde{x}) < \frac{\partial^\mu \tilde{x}'\circ\lambda}{\partial \tilde{x}^\mu_j}\frac{\partial^\mu \tilde{u}_\alpha}{\partial(\tilde{x}'_i\circ\lambda)^\mu}, \frac{\partial^\mu \tilde{x}'_i\circ\lambda}{\partial \tilde{x}^\mu_j}\frac{\partial^\mu \tilde{v}_\alpha}{\partial(\tilde{x}'_i\circ\lambda)^\mu} > d\tilde{x}_\alpha \\
&= \sum_\alpha \sum_{[\mu]\leq t} \binom{t}{[\mu]} \frac{1}{|G_\alpha|}\int_{\tilde{U}_\alpha} \tilde{\rho}_\alpha(\tilde{x}) \left| det\left(\frac{\partial^\mu \tilde{x}'_i\circ\lambda}{\partial \tilde{x}^\mu_j}\right)\right|^2 \\
&\quad < D^{\mu}_{\tilde{x}_\alpha\circ\lambda}, D^{\mu}_{\tilde{x}_\alpha\circ\lambda}\tilde{v}_\alpha > det\left(\frac{\partial \tilde{x}_\alpha}{\partial \tilde{x}'_\alpha}\circ\lambda\right) d\tilde{x}'_\alpha\circ\lambda
\end{aligned}$$

is equivalent to the norm $|u(x'\circ\lambda)|_t$ (notation abuse) based on the product

$$(u,v)_t = \sum_\alpha \sum_{[\mu]\leq t} \binom{t}{[\mu]} \frac{1}{|G_\alpha|}\int_{\tilde{U}_\alpha} \tilde{\rho}_\alpha(\tilde{x}'\circ\lambda) < D^{\mu}_{\tilde{x}'_\alpha\circ\lambda}, D^{\mu}_{\tilde{x}'_\alpha\circ\lambda}\tilde{v}_\alpha > d(\tilde{x}'_\alpha\circ\lambda)$$

under the assumption that M is orientable, i.e., $det(\frac{\partial \tilde{x}_\alpha}{\partial \tilde{x}'_\alpha\circ\lambda}) > 0$. So the Hilbert space structure on $H^t(M)$ is determined up to equivalence of norms by the choice of V-charts, and therefore the Hilbertian structure does not depend on the choice of V-charts. The chain structure of (1.48) is called the *Sobolev chain* of the compact V-manifold M. Since the Sobolev chain on $\mathbf{T}^m$ is well-known, we are going to make use of that fact by embedding the Sobolev chain of M into Sobolev chain of $\mathbf{T}^m$ as follows: We define a map $\rho : C^\infty(M) \to C^\infty(\mathbf{T}^m)$ by

$$\rho(u) = \sum_\alpha (\rho_\alpha \cdot u_\alpha) \circ \pi_\alpha \circ \tau_\alpha^{-1}, \tag{1.49}$$

where $\{\rho_\alpha\}$ is a partition of unity subordinate to the finite covering $\{U_\alpha\}$ of M, $\tilde{U}_\alpha = u_\alpha \circ \pi_\alpha : \tilde{U}_\alpha \to \mathbf{R}$ is a G_α-invariant smooth function in $\tilde{U}_\alpha$ for each α. It is clear that ρ is linear and continuous with respect to the C^∞ topology. Since $C^\infty(M)$ and $C^\infty(\mathbf{T}^m)$ are dense in $H^t(M)$ and $H^t(\mathbf{T}^m)$, we may consider the extension $\rho : H^t(M) \to H^t(\mathbf{T}^m)$ for $t \in \mathbf{Z}$. This ρ allows us to identify the chain of (1.48) with norm

$$(u, v)_t = \sum_\alpha \sum_{[\mu]\le t} \binom{t}{[\mu]} \frac{1}{|G_\alpha|} \int_{\tilde{U}_\alpha} \tilde{\rho}_\alpha(\tilde{x}) < D^\mu_{\tilde{x}_\alpha}, D^\mu_{\tilde{x}_\alpha} \tilde{v}_\alpha > d\tilde{x}_\alpha$$

to a closed sub-chain $\{H^t_{M'}, M' = \bigcup U'_\alpha\}$ of $\{H^t(\mathbf{T}^m)\}$ with the norm

$$(u', v')_t = \sum_\alpha \sum_{[\mu]\le t} \binom{t}{[\mu]} \frac{1}{|G_\alpha|} \int_{U'_\alpha} \rho'_\alpha(x') < D^\mu_{x'_\alpha} u'_\alpha, D^\mu_{x'_\alpha} v'_\alpha > dx'_\alpha,$$

thanks to the completeness of Hilbert subspaces for all $t \in \mathbf{Z}$, where $\rho'_\alpha = \rho_\alpha \circ \pi_\alpha \circ \tau_\alpha^{-1}$, $u'_\alpha = u_\alpha \circ \pi_\alpha \circ \tau_\alpha^{-1}$. This identification is an isomorphism for the Hilbert structure, but in general, it does not preserve the metric since $(u, v)_t \neq (\rho(u), \rho(v))_t$, where the right-hand side involves the derivatives of the partition of unity, but the left-hand side does not. It is possible to extend the chain (1.48) to real values $t \in \mathbf{R}$, but that will not be used here.

Following [312] we shall study the properties of the Sobolev chain $\{H^t(M)\}$ of the compact V-manifold M. From the identification $\rho : H^t(M) \cong H^t_{M'}$ for $t \in \mathbf{Z}$, we may study the Sobolev closed subchain $\{H^t_{M'}\}$ of $\{H^t(\mathbf{T}^m)\}$ instead, which is derived from the Sobolev subchain $\{\hat{H}^t_{M'}\}$ of $\{\hat{H}^t(\mathbf{T}^m)\}$. For each summable function u' defined on $\mathbf{T}^m$ we associate a Fourier transform $\hat{u}$ over the dual space $\hat{\mathbf{R}^m}$ defined by

$$\hat{u}'_{\xi'} = \frac{1}{(2\pi)^{m/2}} \int_{\mathbf{T}^m} u'(x') e^{-ix'\xi'} dx',$$

where $x'\xi' = x'^1\xi'_1 + \cdots + x'^m\xi'_m$. Consider the function $P_\xi = 1 + |\xi'|^2$, $|\xi'|^2 = (\xi'^2_1 + \cdots + \xi'^2_m)^{1/2}$ over the dual space $\widehat{\mathbf{R}^m}$ of $\mathbf{R}^m$. Let $\hat{H}^t_{M'}$ denote the Hilbert space formed by the function $\hat{u}' : \mathbf{Z}^m \to \mathbf{R}$ such that

$$|\hat{u}'| = \sum_k P^t_k |\hat{u}'_k|^2 < \infty, \; \hat{u}'_k = \hat{u}'_k | M'.$$

The scalar product is given by $(\hat{u}', \hat{v}')_t = \sum P_k^t \hat{u}'_k \hat{v}'_k$. If $s \leq t$, then $|\hat{u}'|_t \geq |\hat{u}'|_s$, and we have a continuous injection $\hat{H}^t_{M'} \subset \hat{H}^s_{M'}$. We observe that $\hat{H}^0_{M'}$ and $H^0_{M'} (= L^2_{M'})$ are isomorphic via the correspondence

$$\hat{u}' \longleftrightarrow \sum_k \hat{u}'_k e^{ikx},$$

thanks to the Riesz-Fischer theorem. Recall that $H^t_{M'}$ is the image of $\hat{H}^t_{M'}$ in $H^0_{M'}$ by this correspondence ($t \geq 0$), and $H^t_{M'}$ are provided with the scalar product $(\cdot, \cdot)_t$ induced by the product $(\hat{\cdot}, \hat{\cdot})_t$ from $\hat{H}^t_{M'}$. Put H^{-t} = dual of H^t via $(\cdot, \cdot)_0$. For $t = 0, 1, 2, \cdots$, the norm of $H^t_{M'}$ is computed as follows: Take $u', v' \in C^\infty_{M'}$; then

$$\begin{aligned}(u', v')_t &= \int_{M'} (1 + \Delta')^t u' \cdot v' \, dt \\ &= \sum_\alpha \int_{U'_\alpha} \rho'_\alpha(x')(1 + \Delta'_\alpha)^t u' \cdot v' dx' \\ &= \sum_\alpha \sum_{[\mu] \leq t} \binom{t}{[\mu]} \int_{U'_\alpha} \rho'_\alpha(x') < D^\mu_{x'_\alpha} u'_\alpha, D^\mu_{x'_\alpha} v'_\alpha > dx'_\alpha.\end{aligned}$$

Since $[(1 + \widehat{\Delta'})^t u']_k = P_k^t \hat{u}'_k$, we have $(u', v')_t = (\hat{u}', \hat{v}')_t$. Actually, the Sobolev chain $\{H^t_{M'}\}$ and the Sobolev chain $\{\hat{H}^t_{M'}\}$ are isomorphic for $t \in \mathbf{Z}$. Following [312], the fundamental theorems of Rellich and Sobolev hold for the Sobolev chain $\{\hat{H}^t(T^m)\}$, hence also hold for our closed subchain $\{H^t_{M'}\}$. By the foregoing identification $\rho : H^t(M) \cong H^t_{M'}$, we can obtain the following two theorems for the Sobolev chain $\{H^t(M)\}$ of the compact V-manifold M.

Theorem 1.8.2 (Rellich). *Let M be a compact V-manifold. If $s < t$, then the inclusion $H^t(M) \subset H^s(M)$ is a compact injection.*

Theorem 1.8.3 (Sobolev). *For $t > m/2$ we have a continuous injection $H^{\nu+t}(M) \subset C^\nu(M)$, where $C^\nu(M)$ is the space of the functions of class C^ν on the compact V-manifold M provided with norm by* $\sup_{[\mu] \leq \nu} |D^\mu_x|$.

For any G-invariant smooth function $\tilde{u}$ in a V-chart with $\{\tilde{U}, G, \pi\} \in \mathcal{F}$ over the compact V-manifold M, the laplacian is given by (1.45):

$$\begin{aligned}\Delta \tilde{u} &= g^{ij} \frac{\partial^2 \tilde{u}}{\partial \tilde{x}^i \partial \tilde{x}^j} + \sum_j B^j \frac{\partial \tilde{u}}{\partial \tilde{x}^j} \\ &= a_2(\tilde{x}) D^2_{\tilde{x}_\alpha} \tilde{u} + a_1(\tilde{x}) D^1_{\tilde{x}_\alpha} \tilde{u},\end{aligned}$$

which is a linear differential operator of order 2. Let $\tilde{\xi} = \tilde{\xi}_j d\tilde{x}^j$ be a co-vector at $\tilde{x}$ in $\tilde{U}$. Then the 2-symbol of Δ is

$$a_2(\tilde{x}, \tilde{\xi}) = a_2(\tilde{x})(i\tilde{\xi})^2 = g^{hk}(i\tilde{\xi}_h)(i\tilde{\xi}_k) = -|\tilde{\xi}|^2 \neq 0$$

for any $\tilde{x}, \tilde{\xi} \neq 0$. Thus, the laplacian $\Delta : C^\infty(M) \to C^\infty(M)$ is a *strongly elliptic differential operator* of order 2. Since $C^\infty(M)$ is dense in both $H^t(M)$ and $H^{t-2}(M)$ for $t \in \mathbf{Z}$, we may consider the extension $\Delta : H^t(M) \to H^{t-2}(M)$.

We can apply Theorems 1.8.2 and 1.8.3 to prove the following results (see the proofs in [73]).

Theorem 1.8.4. *Let M be a compact V-manifold, and $\Delta u = v$ in $H^{-\infty}(M)$ such that $v \in H^t$. Then $u \in H^{t+2}$. In particular, if $v \in C^\infty(M)$, then $u \in C^\infty(M)$.*

Theorem 1.8.5. *For any $\lambda \in \mathbf{R}$, Ker$(\Delta - \lambda)$ has finite dimension, and is contained in $C^\infty(M)$. The set of $\lambda \in \mathbf{R}$ such that Ker$(\Delta - \lambda) \neq 0$ has no limit point, i.e., the eigenvalues $\{\lambda_j\}$ of Δ are discrete. Moreover, we have $0 = \lambda_0 < \lambda_1 \leq \lambda_2 \leq \cdots$ and $\lim_{j\to\infty} \lambda_j = \infty$.*

Theorem 1.8.6. *Let M be a compact V-manifold. If we consider the k-th power of the laplacian Δ^k, $k = 1, 2, \cdots$, then there exists a k-th iterate Green's function $G_x^{(k)} \in H^{-\nu+2k}(M)$ $(\nu > m/2)$ such that $\Delta^k G_x^{(k)} = \delta_x$. Furthermore, all $G_x^{(k)}$ are C^∞ on $M - \{x\}$.*

Theorem 1.8.7. *Let M be a compact V-manifold. There exists an orthonormal basis $\{\phi_j\}$ of $H_0(M)(= L^2(M))$ consisting of the eigenfunctions of Δ corresponding to the eigenvalues λ_j, i.e., $\Delta\phi_j = \lambda_j\phi_j$ where $0 = \lambda_0 < \lambda_1 \leq \lambda_2 \cdots$. Here the λ_j have finite multiplicity and tend to ∞ as $j \to \infty$.*

Theorem 1.8.8. *For any $t > 0$ the series of the heat kernel $H(x, y; t) = \sum_0^\infty e^{-\lambda_j t}\phi_j(x)\phi_j(y)$ on a compact V-manifold M converges uniformly along with its derivatives of all orders. Consequently, for $t > 0$, $H(x, y; t)$ is a smooth function of all three arguments. Moreover, it is symmetric and is a solution of the heat equation.*

1.8.2 Maps of V-Manifolds

Let (M^m, g) be a smooth compact V-manifold without boundary and (N^n, h) be a smooth Riemannian manifold. By Sect. 1 of [73], the compact V-manifold M admits a finite triangulation $T = \bigcup s_\alpha$ such that each simplex s_α is contained in the support U_α of a V-chart $\{\tilde{U}_\alpha, G_\alpha, \pi_\alpha\} \in \mathcal{F}$ on M and is the homeomorphic projection of a regular simplex $\tilde{s}_\alpha$ in $\tilde{U}_\alpha$. For technical reason we may assume that N is smoothly and isometrically embedded in some Euclidean space $\mathbf{R}^k$. For a map $f \in C^2(M, N)$ the energy functional of f is given by

$$E(f) = \int_M |df|^2 dv = \sum_\alpha \int_{s_\alpha} |df|^2 dx_\alpha = \sum_\alpha \frac{1}{|G_\alpha|} \int_{\tilde{s}_\alpha} |d\tilde{f}|^2 d\tilde{x}_\alpha, \quad (1.50)$$

where $d\tilde{x}_\alpha$ denotes the volume form with respect to the G_α-invariant metric (g_{ij}) in $\tilde{U}_\alpha$, and $\tilde{f} : \tilde{U}_\alpha \to N$ is the G_α-invariant lift of f. The Euler-Lagrange equations describing the critical points of the energy functional (1.50) are

$$\begin{aligned}\tau(\tilde{f})^\gamma &= g^{ij}\tilde{f}^\gamma_{i|j} = \Delta\tilde{f}^\gamma + g^{ij}\Gamma'^\gamma_{\alpha\beta}\tilde{f}^\alpha_i\tilde{f}^\beta_j \\ &= \tilde{g}^{ij}\left[\left(\frac{\partial^2\tilde{f}^\gamma}{\partial x^i\partial x^j} - \Gamma^k_{ij}\tilde{f}^\gamma_k\right) + \Gamma'^\gamma_{\alpha\beta}\tilde{f}^\alpha_i\tilde{f}^\beta_j\right] = 0, \quad 1 \le \gamma \le n, \quad (1.51)\end{aligned}$$

where Γ^k_{ij} and $\Gamma'^\gamma_{\alpha\beta}$ are the Christoffel symbols with respect to (g_{ij}) and $(h_{\alpha\beta})$ on $\tilde{U}$ and N, respectively. The map $f : M \to N$ is *harmonic* iff the tension field of $\tilde{f}$ satisfies (1.51). The divergence theorem on a compact V-manifold M plays a fundamental role in the construction of harmonic maps of V-manifolds; for the proof, see [73] (pp. 320–321).

Chiang obtained the following results (see the proofs in [73]), which are required to prove Theorem 1.8.12.

Proposition 1.8.9. *Let $H^G(\tilde{x}, \tilde{y}; t)$ be the G-invariant heat kernel and K^G $(\tilde{x}, \tilde{y}, ; t)$ be the G-invariant fundamental solution of the heat equation in $\tilde{U}$. For $\tilde{x} \in \tilde{\bar{V}}$, $y \in \tilde{V}$ $(\tilde{V} \subset \tilde{\bar{V}} \subset \tilde{U})$, the G-invariant function $\tilde{g}^G(\tilde{x}, \tilde{y}; t) = H^G(\tilde{x}, \tilde{y}; t) - K^G(\tilde{x}, \tilde{y}, ; t)$ is bounded and tends uniformly to zero as $t \to 0$. The same holds for the space derivatives of all orders.*

Proposition 1.8.10. *Let M be a compact V-manifold and $W(x,t) = \int_M H(x, y; t)$ $w(y)dy$ be the integral operator associated to the heat kernel on M. For any $w \in C^1(M)$ we have $|W_i(x,t)| \le \mathrm{const} \cdot \sup_{x\in M} |w_i(x)|$, and $W_i(x,t) \to w_i(x)$ as $t \to 0$.*

Proposition 1.8.11. *For $u(y,t) \in C^0(M)$ we have*

$$\int_0^t d\tau \int_M H(x, y; t-\tau)u(y,\tau)dy| \le \mathrm{const} \cdot \sup_{y\in M} |\bar{u}(y)|,$$

$$\int_0^t d\tau \int_M H_i(x, y, t-\tau)u(y,\tau)dy| \le \mathrm{const} \cdot \sup_{y\in M} |\bar{u}(y)|,$$

where $|\bar{u}(y)| = \sup_{0\le\tau\le T} |u(y,\tau)|$.

By combining all the results on the spectral geometry of V-manifolds, the above propositions, and Eells-Sampson's techniques [129], Chiang [73, 74] proved the following theorem.

Theorem 1.8.12. (1) *Let M be a compact V-manifold, N be a Riemannian manifold with non-positive Riemannian curvature and $f : M \to N$ be a continuous differentiable map. Let f_t be a solution of $\tau(f) = \frac{\partial f_t}{\partial t}$ which reduces to f at $t = 0$ (assume that N is compact). As t tends to ∞, f is homotopic to a harmonic map*

f' for which $E(f') \leq E(f)$. In particular, every $f \in C^0(M,N)$ is homotopic to a harmonic map. (2) *We can construct a harmonic map $f : M \to M'$ from a compact V-manifold M into another V-manifold M' via factorization through a compact Riemannian manifold N with non-positive curvature. Let $f_1 : M \to N$ be a continuous map and $f_2 : N \to M'$, such that $f = f_2 \circ f_1$. By* (1), *f_1 is homotopic to a harmonic map, still denoted by f_1. If $f_2 : N \to M'$ is totally geodesic, then $f = f_2 \circ f_1 : M \to M'$ is a harmonic map of V-manifolds.*

1.8.3 Maps on Spaces with Conical Singularities

Since the structures of spaces with conical singularities are different than those of V-manifolds, we dealt with harmonic maps on spaces with conical singularities (see Chiang and Ratto [79]) differently than harmonic maps of V-manifolds.

Let M be an m-dimensional Riemannian manifold with metric g; the metric cone $C(M)$ on M is the space $M \times (0, +\infty)$, equipped with the metric $r^2 g_M + dr^2$. Set

$$C_{u,v}(M) = \{(x,r) \in C(M) | u < r < v\}$$

and $C^*(M) = C(M) \cup p$, the completion of $C(M)$ by adding the vertex p. We shall consider the following class of spaces.

Definition 1.8.13. A compact metric space $X^{m+1} = X$, $m \geq 1$, is a space with *conical singularities* if there exist points $p_j \in X$, $j = 1, \cdots, k$ such that $X - \bigcup_{j=1,\cdots,k} p_j$ is an open (m + 1)-dimensional Riemannian manifold and each p_j has a neighborhood U_j such that $U_j - p_j$ is isometric to $C_{0,v_j}(M_j)$, for some v_j and compact (not necessarily connected) manifolds M_j. We denote $\Sigma = \bigcup_{j=1,\cdots,k} p_j$.

The standard de Rham-Hodge theory admits an extension to spaces with conical singularities, provided that the usual cohomology and homology are replaced by L^2-cohomology and intersection homology respectively (cf. Cheeger, Goresky, and MacPherson [65]). A related type of L^2 analysis concerns the spectral theory of Δ, which leads to a detailed study of wave and heat operators on X, as carried out in Cheeger [64]. In particular, a smooth symmetric kernel is associated with the heat operator $e^{-\Delta t}$, a fact that will be the starting point of our analysis.

From standard spectral theory (cf. Dunford and Schwartz [108]) we obtain the following formal expression for the kernel of the operator $f(\Delta)$:

$$k_f(r_1, x_1, r_2, x_2) = \sum_i k_f(r_1, r_2, \nu_i)\phi_i(x_1)\phi_i(x_2), \tag{1.52}$$

where

$$k_f(r_1, r_2, \nu_i) = (r_1 r_2)^\alpha \int_0^\infty f(\lambda^2) J_{\nu_i}(\lambda r_1) J_{\nu_i}(\lambda r_2) \lambda \, d\lambda. \tag{1.53}$$

Here $\{\phi_i\}$ is an orthonormal basis of eigenfunctions of $\triangle^M$, and J_{ν_i} are Bessel functions. In many interesting cases the above series converges uniformly; in particular, we have:

Proposition 1.8.14. *If $f(\triangle) = e^{-\Delta t}$, (1.52) converges uniformly on $C_{0,a}(M) \times C_{0,a}(M) \times [b,c]$ for all $a, b, c > 0$. The associated kernel, denoted by $H(r_1, x_1, r_2, x_2; t)$, is smooth, bounded and symmetric. Similar properties hold for the iterated Green's function G^k on $C_{0,a}(M) \times C_{0,a}(M)$ (corresponding to $f(\triangle) = \triangle^{-k}$, $k \geq 1$).*

The distance function on $C(M)$ is given by

$$\rho^2((r_1, x_1), (r_2, x_2)) = \begin{cases} r_1^2 + r_2^2 - 2r_1r_2 \cos(\theta(x_1, x_2)), & \text{if } \theta(x_1, x_2) \leq \pi, \\ (r_1 + r_2)^2, & \text{if } \theta(x_1, x_2) > \pi, \end{cases}$$

where θ denotes distance between points in M. Writing v, z for two arbitrary points in $C(M)$, the parametrix for the heat equation is

$$P(v, z, t) = \frac{1}{(2\sqrt{\pi})^{m+1} t^{(m+1)/2}} exp\{-\rho^2(v, z)/4t\}.$$

Let $K = K(v, z, t)$ be the fundamental solution of the heat equation associated with P. The following estimates follow by simple modification of the arguments of the nonsingular case (see Friedman [147] and Pogorzelski [301]).

Lemma 1.8.15. *For each $0 < \mu < 1$,*

$$K(v, z, t) \leq Bt^{-\mu}[\rho(v, z)]^{2\mu-m-1},$$
$$(\partial K/\partial v_i)(v, z, t) \leq Bt^{-\mu}[\rho(v, z)]^{2\mu-m-2},$$
$$(\partial^2 K/\partial v_i \partial v_j)(v, z, t) \leq Bt^{-\nu}[\rho(v, z)]^{2\mu-m-3},$$

for some constant $B > 0$.

A map $f : X \to N$ from a space with conical singularities X into a compact Riemannian manifold N is *harmonic* if it is continuous and its restriction to $X - \Sigma$ is harmonic in the usual sense. By combining all the above arguments, Chiang and Ratto [79] obtained the following theorems.

Theorem 1.8.16. *Let X be a space with conical singularity and N be a compact Riemannian manifold. Assume Riem $N \leq 0$. Then any continuous map $f_0 : X \to N$ is homotopic to a harmonic map.*

We may construct a harmonic map $f : X_1 \to X_2$ between two spaces with conical singularities via factorization through a Riemannian manifold N of non-positive curvature.

Let C be a (possibly singular) complex projective algebraic curve (i.e., C is the locus of common zeros of a finite set of homogeneous polynomials on $\mathbf{C}^{n+1}$). To avoid trivialities, we can assume that C has only isolated singular points, whose union is denoted by Σ. The inclusion $C \to \mathbf{C}P^n$ induces a Kähler metric ω on $C - \Sigma$.

Theorem 1.8.17. *Let $f_0 : C \to N$ be a continuous map into a compact manifold N with Riem $N \leq 0$. Then f_0 is homotopic to a continuous maps whose restriction $f : (C - \Sigma, \omega) \to N$ is harmonic.*

Let W, Y be compact Riemann surfaces. Suppose that there exists a non-constant $\pm$ holomorphic map (i.e., $\phi : W \to \mathbf{C}P^n$ is $\pm$ holomorphic and $\phi(W) \subset C$) of degree d_ϕ. Let $f : C \to Y$ be a non-constant map of degree d_f which is harmonic on $C - \Sigma$. Applying the main theorem of Eells and Wood [133] to $f \circ \phi$ and using the characterization $\pm$ holomorphic = weakly conformal, it is not hard to deduce the following:

If $e(W) + |d_f d_\phi e(Y)| > 0$, then f is $\pm$ holomorphic on $C - \Sigma$. (Here $e(W)$ and $e(Y)$ denote the corresponding Euler characteristics.)

1.9 Transversally Harmonic Maps

Harmonic maps between foliated Riemannian manifolds with one manifold foliated by points were first studied by Eells and Verjovsky [132] and by Kacimi Alaoui and Gomez [138]. Transversally harmonic maps between manifolds with Riemannian foliations were first defined by Konderak and Wolak [230, 231] in 2003. In this section, we first review foliations, and then discuss the results obtained in [230,231].

1.9.1 Foliations

Let $\mathcal{F}$ be a foliation on a Riemannian n-manifold (M, g). Then $\mathcal{F}$ is defined by a cocycle $\mathcal{U} = \{U_i, f_i, g_{ij}\}_{i\in I}$ modeled on a q-manifold N_0, where

1. $\{U_i\}_{i\in I}$ is an open covering of M,
2. $f_i : U_i \to N_0$ are submersions with connected fibres,
3. $g_{ij} : N_0 \to N_0$ are local diffeomorphisms of N_0 such that $f_i = g_{ij} f_j$ on $U_i \cap U_j$.

The connected components of the trace of any leaf of $\mathcal{F}$ on U_i consist of fibres of f_i. The open subsets $N_i = f_i(U_i) \subset N_0$ form a q-manifold $N = \amalg N_i$, which can be considered to be a transverse manifold to the foliation $\mathcal{F}$. The pseudogroup $\mathcal{H}_N$ of local diffeomorphisms of N generated by g_{ij} is called the holonomy pseudogroup of the foliated manifold $(M, \mathcal{F})$ defined by the cocycle $\mathcal{U}$. If the foliation $\mathcal{F}$ is Riemannian for the Riemannian metric g, then it induces a Riemannian metric $\bar{g}$

on N such that the submersions f_i are Riemannian submersions and the elements of the holonomy group are isometries.

Let $\phi : U \to \mathbf{R}^p \times \mathbf{R}^q$, $\phi = (\phi^1, \phi^2) = (x_1, \cdots, x_p, y_1, \cdots, y_q)$ be an adapted chart on a foliated manifold $(M, \mathcal{F})$. Then on U the vector fields $\frac{\partial}{\partial x_1}, \cdots, \frac{\partial}{\partial x_p}$ span the bundle $T\mathcal{F}$ tangent to the leaves of $\mathcal{F}$, the equivalence classes of $\frac{\partial}{\partial y_1}, \cdots, \frac{\partial}{\partial y_q}$ denoted by $\frac{\bar{\partial}}{\partial y_1}, \cdots, \frac{\bar{\partial}}{\partial y_q}$, span the normal bundle $N(M, \mathcal{F}) = TM/T\mathcal{F}$, which is isomorphic to the subbundle $T\mathcal{F}^\perp$. This bundle and the others considered here are naturally foliated by foliations whose leaves are covering spaces of leaves of $\mathcal{F}$ and whose defining cocycles can be derived in the obvious way from the cocycle $\mathcal{U}$, cf. [408]. In the non-Riemannian case we can take any subbundle Q complementary to $T\mathcal{F}$ and for simplicity we shall denote it by the same symbol.

The sheaf $\Gamma_b(T\mathcal{F}^\perp)$ of foliated sections of the vector bundle $T\mathcal{F}^\perp \to M$ may be described as follows: If U is an open subset of M, then $X \in \Gamma_b(U, T\mathcal{F}^\perp)$ if and only if for each local Riemannian submersion $\phi : U \to \bar{U}$ defining $\mathcal{F}$, the restriction of X to U is projectable via the map ϕ to a vector field $\bar{X}$ on $\bar{U}$.

Definition 1.9.1 (Molino [268]). A *basic partial connection* $(M, \mathcal{F}, g)$ is a sheaf operator D such that for each open subset U of M

$$D : \Gamma_b(U, T\mathcal{F}^\perp) \times \Gamma_b(U, T\mathcal{F}^\perp) \to \Gamma_b(U, T\mathcal{F}^\perp)$$

and for any $X, Y, Z \in \Gamma_b(U, T\mathcal{F}^\perp)$ and any $f, h \in C_b^\infty(U)$:

1. $D_{fX+hY}Z = fD_XY + hD_XZ$,
2. D_X is $\mathbf{R}$-linear,
3. $D_X fY = X(f)Z + fD_XY$ (the transversal Leibniz rule).

Let ∇ be the Levi-Civita connection of g; then for any open subset U of M and $X, Y \in \Gamma_b(U, T\mathcal{F}^\perp)$ we define D by

$$D_XY = (\nabla_XY)^\perp$$

where $(\nabla_XY)^\perp$ is a local foliated section of $T\mathcal{F}^\perp$. It is easy to check that D is a basic partial connection on $(M, g, \mathcal{F})$. Let $\phi : U \to \bar{U}$ be a Riemannian submersion defining the foliation $\mathcal{F}$ on an open set U. Let us assume that $X, Y \in \Gamma_b(U, T\mathcal{F}^\perp)$, and let $\bar{X}, \bar{Y}$ be the corresponding push-forward vector fields via the map ϕ. Then there is a well-known property of Riemannian foliations from [200] that

$$d\phi(D_XY) = \nabla^{\bar{g}}_{\bar{X}}\bar{Y}$$

where $\nabla^{\bar{g}}$ is the Levi-Civita connection of the metric $\bar{g}$. For more details about foliations and Riemannian foliations, see Molino [268] and Tondeur [376].

The operator D can be defined using the induced metric on the normal bundle via the well-known formula for the Levi-Civita connection restricted to normal vectors.

Foliated semi-Riemannian (resp. Minkowskian, Lorentzian) metrics in the normal bundle define basic partial connections in the standard way [92, 408].

Let $(M_1, \mathcal{F}_1, g_1)$ and $(M_2, \mathcal{F}_2, g_2)$ be two foliated Riemannian manifolds. Let ∇^i be the Levi-Civita connections of the respective metrics and D^i be the induced basic partial connections on the orthogonal complement bundles $T\mathcal{F}_i^{\perp} \to M_i$, $i = 1, 2$. Suppose that $f : M_1 \to M_2$ is a smooth foliated leaf preserving map, i.e., $df(T\mathcal{F}_1) \subset T\mathcal{F}_2$. Then there are given natural bundle maps

$$I_i : T\mathcal{F}_i^{\perp} \to TM_i, \ \ \Pi_i : TM_i \to T\mathcal{F}_i^{\perp} \text{ for } i = 1, 2,$$

where I_i is the inclusion of $T\mathcal{F}_i^{\perp}$ in TM_i and Π_i is the orthogonal projection of TM_i onto $T\mathcal{F}_i^{\perp}$. Let X be a local foliated section of $T\mathcal{F}_1^{\perp} \to M_1$; then $\Pi_2 df(X)$ is a foliated section of the bundle $f^{-1}T\mathcal{F}_2^{\perp}$. Therefore, $\Pi_2 df I_1$ is a foliated section of the bundle $(T\mathcal{F}_1^{\perp})^* \otimes f^{-1}T\mathcal{F}_2^{\perp}$.

We define the transversal second fundamental form as the covariant derivative $D(\Pi_2 df I_1)$, which is a global section of the bundle

$$(T\mathcal{F}_1^{\perp})^* \otimes (T\mathcal{F}_1^{\perp})^* \otimes f^{-1}T\mathcal{F}_2^{\perp} \to M_1,$$

where D is the connection on the bundle $(T\mathcal{F}_1^{\perp})^* \otimes f^{-1}T\mathcal{F}_2^{\perp} \to M_1$ induced by D^1 and D^2. The trace of the transversal second fundamental form is called the transversal tension field of f, and it is denoted by $\tau_b(f)$. If $X_{1x}, \cdots, X_{q_1 x}$ is an orthonormal basis of the space $T_x\mathcal{F}_1^{\perp}$, then

$$\tau_b(f)_x = trace_{T\mathcal{F}_1^{\perp}} D(\Pi_2 df I_1) = \sum_{\alpha=1}^{q_1} D(\Pi_2 df I_1)(X_\alpha, X_\alpha) \tag{1.54}$$

is a section of the bundle $f^{-1}T\mathcal{F}_2^{\perp} \to M_1$. For more details about transversal tension fields, see [230, 231].

1.9.2 Definition and Examples

Let $(M_1, \mathcal{F}_1, g_1)$ and $(M_2, \mathcal{F}_2, g_2)$ be two foliated Riemannian manifolds. Let $f : (M_1, \mathcal{F}_1) \to (M_2, \mathcal{F}_2)$ be a smooth foliated leaf preserving map. Let $U_i \subset M_i$ be open subsets and let $\phi_i : (U_i, g_i) \to (\bar{U}_i, \bar{g}_i)$ be Riemannian submersions on U_i which define locally the Riemannian foliations $\mathcal{F}_i$ for $i = 1, 2$. Suppose that $f(U_1) \subset U_2$. Let $X_1, \cdots, X_{q_1}$ and $Y_1, \cdots, Y_{q_2}$ be two local bases of foliated sections of $T\mathcal{F}_1{}^{\perp}$ and $T\mathcal{F}_2{}^{\perp}$ over U_1 and U_2, respectively. Then $X_1, \cdots, X_{q_1}$ project via the map ϕ_1 to frame sections $\bar{X}_1, \cdots, \bar{X}_{q_1}$, and $Y_1, \cdots, Y_{q_2}$ project via the map ϕ_2 to frame sections $\bar{Y}_1, \cdots, \bar{Y}_{q_2}$. Then there exists the unique map $\bar{f} : \bar{U}_1 \to \bar{U}_2$ such that

$$
\begin{array}{ccc}
U_1 & \xrightarrow{f} & U_2 \\
{\scriptstyle \phi_1}\big\downarrow & & {\scriptstyle \phi_2}\big\downarrow \\
\bar{U}_1 & \xrightarrow{\bar{f}} & \bar{U}_2
\end{array}
$$

Diagram 1.9.1.

commutes.

There is a close relationship between the transversal tension field of f and the tension fields of the induced maps $\bar{f}$ obtained by using the local submersions defining the foliations $\mathcal{F}_1$ and $\mathcal{F}_2$. By Diagram 1.9.1 and (1.54),

$$d\phi_2(\tau)_b(f)_x = \tau(\bar{f})_{\phi_1(x)} \tag{1.55}$$

holds for each of the foliations defining local submersions $\phi_i : U_i \to \bar{U}_i$, $i = 1, 2$, such that $f(U_1) \subset U_2$. Hence, we arrive at the following theorem.

Theorem 1.9.2 ([230]). *Let $f : (M_1, \mathcal{F}_1) \to (M_2, \mathcal{F}_2)$ be a smooth foliated map between two foliated Riemannian manifolds. Then f is transversally harmonic if and only if the induced map $\bar{f}$ is harmonic in each $\bar{U}_1$.*

Proof. The assertion follows from Diagram 1.9.1 and (1.55).

The definitions of transversally harmonic maps do not depend on the choices of local Riemannian submersions defining the Riemannian foliations. □

There is no if and only if relationship between harmonicity and transversal harmonicity. The following example shows that there are transversally harmonic maps which are not harmonic maps.

Example 1. Let (B_1, g_1), (B_2, g_2), (F_1, h_1) and (F_2, h_2) be Riemannian manifolds. Consider the foliations on $B_1 \times F_1$ and $B_2 \times F_2$ given by the projections on the first component $\pi_1 : B_1 \times F_1 \to B_1$ and $\pi_2 : B_2 \times F_2 \to B_2$, respectively. The projections π_1 and π_2 are Riemannian submersions, and the foliations are also Riemannian. Let $f : B_1 \times F_1 \to B_2 \times F_2$ be a smooth map which preserves the leaves of the foliations. Then f must be of the form $f(x, y) = (f_1(x), f_2(x, y))$, $x \in B_1$ and $y \in F_1$, where $f_1 : B_1 \to B_2$ and $f_2 : B_1 \times F_1 \to F_2$ are smooth. For the product Riemannian metrics on $B_1 \times F_1$ and $B_2 \times F_2$, the tension field of f can be expressed as

$$\tau(f) = (\tau(f_1), \tau(f_2|_{B_1}) + \tau(f_2|_{F_1})), \tag{1.56}$$

where $\tau(f_1)$ is the tension field at x of $f_1 : B_1 \to B_2$, $\tau(f_2|_{B_1})$ is the tension field at x of the map $x \mapsto f_2(x, y)$ while y is fixed, and $\tau(f_2|_{F_1})$ is the tension field at y of the map $y \mapsto f_2(x, y)$ while x is fixed. On the one hand, by (1.56) the harmonicity of $f = (f_1, f_2)$ is equivalent to f_1 is harmonic and $\tau(f_2|_{B_1}) + \tau(f_2|_{F_1}) = 0$, i.e., the vertical and horizontal contributions to the tension field annihilate each other. On the other hand, if f_1 is harmonic and $f_2|_{B_1}$, $f_2|_{F_1}$ are harmonic for $x \in B_1, y \in F_1$, then f is harmonic. Hence, it follows that there are maps f which are transversally harmonic, but not harmonic.

There are also harmonic maps which are not transversally harmonic maps either. In Example 2 below we construct a harmonic map that is not transversally harmonic using a warped product of two manifolds. By O'Neill [160], a warped product can be defined on Riemannian manifolds. Let (B, g) and (F, h) be Riemannian manifolds and $\alpha : B \to \mathbf{R}$ be a smooth function. On the product manifold $B \times F$ we define the metric tensor $k = g \oplus e^{2\alpha} h$. Let ∇^g, ∇^h be the Levi-Civita connections on (B, g) and (F, h), respectively. Let X, Y be vector fields on B and V, W be vector fields on F. The Levi-Civita connection ∇^k on $B \times F$ can be related to those of B and F as follows:

$$\begin{aligned} \nabla^k_X Y &= \nabla^g_X Y, \\ \nabla^k_X V &= \nabla^k_V X = X(\alpha) V, \\ \nabla^k_V W &= -h(V, W) grad_g \alpha + \nabla^h_V (W). \end{aligned} \tag{1.57}$$

Let $(B_i, g_i), (F_i, h_i), i = 1, 2$ be four Riemannian manifolds such that the following diagram

$$\begin{array}{ccc} B_1 \times F_1 & \xrightarrow{f} & B_2 \times F_2 \\ \pi_1 \downarrow & & \pi_2 \downarrow \\ B_1 & \xrightarrow{f_1} & B_2 \end{array}$$

Diagram 1.9.2.

commutes. The construction of the tension field of a map, Theorem 1.9.2 and Diagram 1.9.2 yield the following result.

Corollary 1.9.3. *The map f is transversally harmonic if and only if* $\tau(f_1) = 0$.

By a tedious computation [230],

$$\begin{aligned} \tau(f) = \tau(f_1) + \tau(f_2|_{B_1}) - ||df_2|_{B_1}||^2 (grad_{g_2} \alpha_2) \circ f_1 + e^{-2\alpha_1} \tau(f_2|_{F_1}) \\ - e^{-2\alpha_1} ||df_2|_{F_1}||^2 (grad_{g_2} \alpha_2) \circ f_1 + dim\, F_1 e^{-2\alpha_1} df_2 (grad_{g_1} \alpha_1). \end{aligned} \tag{1.58}$$

If the warping map $\alpha_1 = 0$, then (1.58) reduces to

$$\tau(f) = \tau(f_1) + \tau(f_2|_{B_1}) + \tau(f_2|_{F_1}) - ||df_2||^2 (grad_{g_2} \alpha_2) \circ f_1, \tag{1.59}$$

where the Hilbert-Schmidt norm $||df||_2$ is taken with respect to the $g_1 \oplus h_1$ metric on $B_1 \times F_1$ and h_2 on F_2.

Example 2. Let $B_1 = B_2 = F_1 = F_2 = \mathbf{R}$, $\alpha_1 = 0$, $\alpha_2(x) = x$, $f_1(x) = x$, $f_2(x) = x^2$, $f_2(x, y) = \sqrt{2} y$ and $f(x, y) = (f_1(x), f_2(x, y))$. The foliations on $B_1 \times F_1$ and $B_2 \times F_2$ are given by the projections on B_1 and B_2 respectively.

Then f is a smooth leaf preserving map from $B_1 \times F_1$ to $B_2 \times_{e^{2x}} F_2$. By (1.59) we have $\tau(f) = 0$, but $\tau(f_1) = 2 \neq 0$. Hence, f is harmonic, but not transversally harmonic.

1.9.3 Suspension Constructions

Let (F, g) be a Riemannian manifold and $Isom(F, g)$ its group of isometries. Let P be a smooth manifold. Denote its fundamental group $\pi_1(P)$ by G. Let h be a representation of the group G into $Isom(F, g)$. Let $\tilde{P}$ be the universal covering of P and $\tilde{P} \times F$ be the Cartesian product. G acts on this product via the deck transformation h on F:

$$(p, v) \cdot \gamma = (p \cdot \gamma, h(\gamma(v))).$$

The product $\tilde{P} \times F$ is equipped with the Riemannian metric $g_P \times g$, where g_P is a Riemannian metric lifted from P to $\tilde{P}$. The action of G on $\tilde{P} \times F$, denoted by the same letter h, is isometric for this Riemannian metric.

The action of the group G is totally discontinuous and the quotient manifold $(\tilde{P} \times F)/h$ is denoted by $M(P, F; h)$. It is a fibre bundle over P with the standard fibre F, which admits a foliation $\mathcal{F}_M$ transverse to the fibres. Its leaves are covering spaces of P. In the induced Riemannian metric, the foliation by the fibres is totally geodesic (cf. [180]), and the foliation $\mathcal{F}$ is Riemannian.

Let (P_1, g_1), (P_2, g_2), (F_1, h_1) and (F_2, h_2) be four Riemannian manifolds. Denote by G_i the fundamental group of the manifold P_i and let h_i be a representation of G_i into the group $Isom(F_i, g_i)$ of isometries of the Riemannian manifolds (F_i, g_i), $i = 1, 2$. Let $f : P_1 \to P_2$ be a smooth map and $\tilde{f} : \tilde{P}_1 \to \tilde{P}_2$ be its lift to universal coverings of P_1 and P_2. Then $\tilde{f}$ is (G_1, G_2)-equivariant. Let $\pi_1(f) : G_1 \to G_2$ be the map induced by f on the fundamental groups of the manifolds. Consider a map $\phi : F_1 \to F_2$ which is $(G_1, h_1; G_1, h_2)$-equivariant, i.e., $\phi(v \cdot h_1(\gamma)) = \phi(v) \cdot h_2(\pi_1(f)(\gamma))$ for $v \in F_1$, $\gamma \in G_1$. The map $\tilde{\psi} : \tilde{P}_1 \times F_1 \to \tilde{P}_1 \times F_2$ is defined by $\tilde{\psi} = (\tilde{f}, \phi)$ and is also (G_1, G_2)-equivariant. Hence, it induces a map $\psi : M_1(P_1, F_1; h_1) \to M_2(P_2, F_2; h_2)$.

Thus, we have the following commutative diagram:

$$\begin{array}{ccc} \tilde{P}_1 \times F_1 & \xrightarrow{\tilde{f} \times \phi} & \tilde{P}_2 \times F_2 \\ \downarrow & & \downarrow \\ M_1 & \xrightarrow{\psi} & M_2 \end{array}$$

Diagram 1.9.3.

Using this diagram involving suspension, we can construct transversally harmonic maps as follows.

Lemma 1.9.4 ([139]). *The map $\phi : F_1 \to F_2$ is harmonic if and only if ψ is transversally harmonic.*

Remark that the map $\tilde{f} \times \phi$ is transversally harmonic if and only if ψ is transversally harmonic. We know that the transversal harmonicity of $\tilde{f} \times \phi$ is equivalent to the harmonicity of ϕ.

Example 3. Let $P_1 = P_2 = S^1$ be the unit circle and $f : P_1 \to P_2$ be a smooth map. Since $\pi_1(S^1) = \mathbf{Z}$, there exists $n \in \mathbf{Z}$ such that $\pi_1(f)(m) = mn$, $m \in \mathbf{Z}$. In fact, f is smoothly homotopic to the map $f_c : S^1 \to S^1$ given by $f_c(z) = z^n$, $z \in S^1 \subset \mathbf{C}$. Let $\phi : \mathbf{R} \to \mathbf{R}$ be a map such that $\phi(y) = \alpha y + \beta$ for some $\alpha, \beta \in \mathbf{R}$. It is obvious that ϕ is harmonic. Then we define two homomorphisms $h_i : \mathbf{Z} \to Isom(\mathbf{R})$, $i = 1, 2$ by $h_1(m)x = x + \gamma nm$ and $h_2(m)x = x + \alpha\gamma m$ for some $\gamma \in \mathbf{R}$. Since $\tilde{S}^1 = \mathbf{R}$, we have that

$$M_1 = (\mathbf{R} \times \mathbf{R})/h_1, \quad M_2 = (\mathbf{R} \times \mathbf{R})/h_2.$$

Then we obtain that the map $\psi : M_1 \to M_2$, $\psi([x, y]) = [\tilde{f}(x), \alpha y + \beta]$ is transversally harmonic. If we consider the particular case of the map $f = f_c$, then the suspension map is given by $\psi([x, y]) = [2xn\pi, \alpha y + \beta]$.

Example 4. Let $P_1 = P_2 = S^1$ and $F_1 = F_2 = S^1$ with their standard Riemannian structures. Assume that there is given a smooth map $f : P_1 \to P_2$ such that $\pi_1(f)(m) = mn$, $n \in \mathbf{Z}^*$. If there is a map $\phi : F_1 \to F_2$ such that $\phi(z) = wz^k$, $w \in S^1 \subset \mathbf{C}$, then ϕ is harmonic for any $k \in \mathbf{Z}$. Let the representations $\mu_i : \mathbf{Z} \to Isom(F_i)$, $i = 1, 2$ be given by $\mu_1(m)z = q^{mn}z$ and $\mu_2(m)z = q^{km}z$, $q \in S^1 \subset \mathbf{C}$. Then the map ψ is equivariant with respect to μ_1 and μ_2 and we obtain a transversally harmonic map $\psi : M_1 \to M_2$ such that $\psi([x, z]) = [\tilde{f}(x), wz^k]$, where $M_1 = (\mathbf{R} \times S^1)/\mu_1$, $M_2 = (\mathbf{R} \times S^1)/\mu_2$. In particular, if $f(z) = f_c(z) = z^n$, we have that ψ is a local diffeomorphism which is harmonic. The ψ is also a harmonic morphism. Furthermore, $\psi([x, z]) = [2xn\pi, wz^k]$.

Example 5. Let $P_1 = P_2 = S^1$ and $F_1 = S^3$ and $F_2 = S^2$ with their standard Riemannian structures. Suppose that there is a given smooth map $f : S_1 \to S_2$ such that $\pi_1(f)(m) = mn$, $n \in \mathbf{Z}$. Let $\phi : S^3 \to S^2$ be the Hopf fibration. We consider the representations $\mu_i : Z \to Isom(f_i)$, $i = 1, 2$ given by $\mu_1(m)(z_1 \, z_2) = (q_1^{nm} z_1, \, q_2^{nm} z_2)$, where $(z_1, z_2) \in S^3 \subset \mathbf{C}^2$, $q_1, q_2 \in S^1 \subset \mathbf{C}$ and $\mu_2(m)(a, z) = (a, \, q_1^m z \bar{q}_2^m)$ for $(a, z) \in S^2 \subset \mathbf{R} \times \mathbf{C}$. It is easy to see that μ_1 and μ_2 act by isometries. Then the map ϕ is equivariant with respect to the actions μ_1 and μ_2. Thus we obtain a transversally harmonic map $\psi : M_1 \to M_2$, where $M_1 = (\mathbf{R} \times S^3)/\mu_1$ and $M_2 = (\mathbf{R} \times S^2)/\mu_2$. The map ψ is given by $\psi([x, (z_1, z_2)] = [\tilde{f}(x), (|z_1|^2 - |z_2|^2, 2z_1\bar{z}_2)]$. In particular, if $f(z) = f_c(z) = z^n$, f_c is a local morphism which is harmonic. Because the Hopf fibration ϕ is a harmonic morphism, ψ is also a harmonic morphism. In this case, ψ is given by $\psi([x, (z_1, z_2)]) = [2xn\pi, (|z_1|^2 - |z_2|^2, 2z_1\bar{z}_2)]$.

Chapter 2
Wave Maps

2.1 Introduction

Wave maps are harmonic maps on Minkowski spaces, and they were studied in the early 1990s. In these last two decades, there were many new developments concerning local well-posedness (LWP), global well-posedness (GWP) and global regularity of wave maps. Klainerman and Machedon [222–224] and Klainerman and Selberg [227] investigated the general Cauchy problem in any dimension greater or equal than two for regular data and obtained the almost optimal LWP. In the difficult case of dimension 2, Christodoulou and Tahvildar-Zadeh [90] studied the regularity of spherically symmetric wave maps by imposing a convexity condition for the target manifold. Shatah and Tahvildar-Zadeh [334] also studied the optimal regularity of equivariant wave maps into two-dimensional rotationally symmetric and geodesically convex Riemannian manifolds. The study of the general wave maps problem incorporated methods that exploited the null-form structure of the wave map system such as that of Grillakis [160, 161] as well as the geometric structure of the equation as done by Struwe [352, 353, 356]. Keel and Tao studied the one (spatial) dimensional case in [216].

Tataru [366, 367], following Tao [360, 361], has used new techniques, which allow one to treat the Cauchy problem with critical data. Their methods rely on harmonic analysis such as adapted frequency, in conjunction with gauge theoretic geometric methods. Tao [360, 361] established the global regularity for wave maps from $\mathbf{R}^{1+m}$ into the sphere S^n for $m \geq 5$, and $m = 2, 3, 4$; the latter low dimensional case is much harder than the former high dimensional case. Similar results were obtained by Klainerman and Rodnianski [226] for target manifolds that admit a bounded parallellizable structure.

Nahmod, Stefanov and Uhlenbeck [275] studied the Cauchy problem of wave maps from $\mathbf{R}^{1+m}$ into a (compact) Lie group (or Riemannian symmetric spaces) when $m \geq 4$ and established global existence and uniqueness provided the Cauchy initial data are small in the critical norm. About the same time, Shatah and Struwe

Y.-J. Chiang, *Developments of Harmonic Maps, Wave Maps and Yang-Mills Fields into Biharmonic Maps, Biwave Maps and Bi-Yang-Mills Fields*, Frontiers in Mathematics,
DOI 10.1007/978-3-0348-0534-6_2,

obtained similar results, in the case when the target is any complete Riemannian manifold with bounded curvature.

Recently, Kenig, Merle and Duyckaerts [109, 218, 219] have studied GWP, scattering and finite time blow-up. Kenig and Merle have developed a method called the "concentration compactness/rigidity theorem" method. The ideas used here are natural extensions of many authors to study critical non-linear elliptic problems (e.g. Yamabe problems and harmonic maps). For more detailed developments of wave maps, please read Shatah and Struwe [332] and Tataru [366].

Let $\mathbf{R}^{1+m}$ be the $m+1$ dimensional Minkowski space with the metric $(g_{ij}) = diag(-1, 1, \cdots, 1)$ and the coordinates $x^0 = t, x^1, x^2, \cdots, x^m$, $(N\ h_{\alpha\beta})$ be an n-dimensional Riemannian manifold, and $f : (\mathbf{R}^{1+m}, g_{ij}) \to (N, h_{\alpha\beta})$ be a map. A wave map is a harmonic map on the Minkowski space $\mathbf{R}^{1+m}$ with the energy functional (see (1))

$$\begin{aligned} E(f) &= \frac{1}{2}\int_{\mathbf{R}^{1+m}} \left(-|f_t|^2 + |\nabla_x f|^2\right) dt\,dx \\ &= \frac{1}{2}\int_{\mathbf{R}^{1+m}} h_{\alpha\beta}\left(-f_t^\alpha f_t^\beta + \sum_{i=1}^m f_i^\alpha f_i^\beta\right) dt\,dx_i. \end{aligned} \tag{2.1}$$

The Euler-Lagrange equation describing the critical point of (2.1) is, by (3),

$$\tau_{\Box}^{\alpha}(f) = \Box f^\alpha + \Gamma'^{\alpha}_{\beta\gamma}\Big(-f_t^\beta f_t^\gamma + \sum_{i=1}^m f_i^\beta f_i^\gamma\Big) = 0, \tag{2.2}$$

where $\Box = -\frac{\partial^2}{\partial t^2} + \Delta_x$ is the wave operator on $\mathbf{R}^{1+m}$, and $\Gamma'^{\alpha}_{\beta\gamma}$ are the Christoffel symbols of N. The map f is a wave map iff the wave field $\tau_{\Box}^{\alpha}(f)$ (i.e., tension field on the Minkowski space) vanishes identically. The wave map equation is invariant under the dimensionless scaling $f(t, x) \mapsto f(\lambda t, \lambda x)$, $\lambda \in \mathbf{R}$. However, the energy is scale invariant only in dimension $m = 2$.

We can rewrite (2.2) as the Cauchy problem

$$\begin{aligned} D^\alpha \partial_\alpha f(t, x) &= 0, \quad (t, x) \in \mathbf{R}^{1+m} \\ f(0, x) &= f_0(x), \ f_t(0, x) = f_1(x), \quad x \in \mathbf{R}^m, \end{aligned} \tag{2.3}$$

where D is the connection on the pull-back bundle f^*TN, the initial data $f(0, x) = f_0(x) \in N$, and $f_t(0, x) = f_1(x) \in T_{f_0(x)}N$ for $x \in \mathbf{R}^m$. For given initial data in Sobolev spaces $(f_0, f_1) \in H^s(\mathbf{R}^m) \times H^{s-1}(\mathbf{R}^m)$, one looks for a solution $f \in C([-T, T]; H^s(\mathbf{R}^m))$, $f_t \in C([-T, T]; H^{s-1}(\mathbf{R}^m))$ with a lifespan T, which depends on the initial data or on the size of the initial data. This is rather easy to solve if s is large enough, but it becomes increasing difficult as s decreases.

The answer is easy for $s > \frac{m}{2}$ since the H^s functions are continuous, and thus locally the image of f is contained in the domain of a local map for N. Then we can measure the regularity of f by using the corresponding local coordinates on N. For low s we have difficulties even with the definition of the Sobolev spaces. What does it mean of $H^s(\mathbf{R}^m)$ for functions that take values into a manifold? If $s \leq \frac{m}{2}$ the problem becomes non-local and the answer may depend on the global properties of the manifold N. Suppose that N is embedded isometrically into R^k; then we might use this to define the space of H^s functions with values in N for all $s \geq 0$. In order to do this, we need to understand whether these spaces depend on the isometric embedding or not. Topological information is missing when $s < m/2$. For example, when $N = S^m$, one needs to know how many times does an S^m-valued H^s function wrap around the sphere. By Brezis and Nirenberg [41], this rotation number is well defined for $s = \frac{m}{2}$, but not for $s < \frac{m}{2}$. Scaling provides additional information. We can balance the size of the initial data and the lifespan of the solution by rescaling the equation. The initial data space is scale invariant if $s = \frac{m}{2}$.

2.1.1 Local Theory

Since the wave map equation looks like a semi-linear wave equation in local coordinates, we can try to treat it as such. The usual approach is to consider the non-linear term as a small perturbation of the governing linear operator. The equation (2.3) takes the form

$$\Box f = N(f, \nabla f), \ f(0) = f_0, \ \partial_t f(0) = f_1, \tag{2.4}$$

where N is a quadratic form, and the initial data (f_0, f_1) are prescribed in some Sobolev space. Then the idea is to consider the non-linear term as a small perturbation with respect to the linear equation and to treat the non-linearity by a fixed point argument.

We can rewrite (2.4) as

$$f = K(f_0, f_1) + \Box^{-1} N(f) \tag{2.5}$$

where $K(f_0, f_1)$ is the solution operator for the free wave equation.

The question of the local well-posedness for initial data $(f_0, f_1) \in H^s(\mathbf{R}^m) \times H^{s-1}(\mathbf{R}^m)$ reduces to that of finding a fixed point for small time in Banach spaces X, Y, for which

$$K : H^s \times H^{s-1} \to X, \ \Box^{-1} : Y \to X, \ N : X \to Y.$$

For $s > \frac{m}{2} + 1$ this can be studied by energy methods, taking

$$X = \{f : f \in L^\infty(H^s), \ df \in L^\infty(H^{s-1})\}, \ Y = L^1(H^{s-1}).$$

Recall that the wave map system is invariant under the scaling $f_\lambda(t, x) = f(\lambda t, \lambda x)$. The $\dot{H}^s$ norm of the scaled initial data scales as

$$||f_\lambda(0, \cdot)||_{\dot{H}^s} = \lambda^{s-m/2} ||f(0, \cdot)||_{\dot{H}^s},$$

and consequently, the Sobolev norm with exponent $s_c = m/2$ is invariant under scaling. One has the following equivalences between local well-posedness and global well-posedness for different exponents.

1. $s > m/2$ 'small data, large time' is equivalent to 'large data, small time'.
2. $s = m/2$ 'small data, small time' is equivalent to 'small data, large time'.
3. $s < m/2$ 'small data, small time' is equivalent to 'large data, large time'.

In the case 1, we can attempt to prove local well-posedness for $s > m/2$. As s decreases toward $m/2$ we get better clues regarding the lifespan of solutions. For $s = m/2$ a local result yields a global result, but we need to distinguish between small and large data. In the case 2, under reasonable hypotheses on N we may expect global well-posedness for small data and $s = m/2$. In the case 3, we may expect ill-posedness for $s < m/2$.

Note that the energy methods allow us to obtain local well-posedness only for $s > s_c + 1 = \frac{m}{2} + 1$. We can take advantage of the special structure of the nonlinearity in our problem, which can be written as $N(f) = \Gamma(u)Q_0(f, \psi)$, where

$$Q_0(u, v) = \partial_\alpha u \, \partial_\alpha v = u_t v_t - \nabla u \cdot \nabla v.$$

This allows us to establish local well-posedness up to the critical scaling, which was utilized by Klainerman and Machedon [224] ($m \geq 3$), and Klainerman and Selberg [227] ($m = 2$). They used the spaces

$$(X, Y) = X^{s,b} \times X^{s-1,b-1+}$$

to establish local well-posedness in H^s for $s > s_c$. The wave-Sobolev space $X^{s,b}$ is defined by

$$X^{s,b} = \{f : ||(1 + |\xi|^2)^{s/2}(1 + |\tau^2 - |\xi|^2|)^{b/2} \hat{f}||_{L^2} < \infty\},$$

compared to

$$H^s = \{f : ||(1 + |\xi|^2)^{s/2} \hat{f}||_{L^2} < \infty\}.$$

More details will be provided in subsequent sections.

2.1.2 Global Theory

If $f : \mathbf{R}^{1+m} \to \mathbf{R}$ is a smooth solution to the linear scalar wave equation $\Box u = 0$, and $\gamma : \mathbf{R} \to N$ is a geodesic in the target manifold, then the map $f(t, x) = \gamma(u(t, x)) : \mathbf{R}^{1+m} \to N$ is a global smooth wave map satisfying

$$D^\alpha \partial_\alpha \gamma(u) = \Box u \, \dot{\gamma}(u) + \partial_\alpha u \, D^\alpha \gamma(u) = 0.$$

In the $1+2$ dimensional case global regularity for spherically symmetric wave maps $f(t,x)=f(t,|x|)$ was proved by Christodolou and Tahvildar-Zadeh [90] for a geodesically convex target N. This result was extended by Struwe [355] to include the sphere and then to any Riemannian manifold without boundary [356].

For a rotationally symmetric target N, this is similar to the preceding results that were obtained for co-rotational (equivariant) wave maps into geodesically convex targets by Shatah and Tahvildar-Zadeh [334], and then was extended to more general targets by Grillakis [160] and Struwe [356]. The numerical work of Bizon et al. [34] showed blow up in the case $N = S^2$.

Under smallness assumption on the data we can have global well-posedness in the Besov space $\dot{B}^{m/2,1} \times \dot{B}^{m/2-1,1}$. This was first established by Tataru [363] in high dimensions $m \geq 4$, and then in low dimensions $m = 2, 3$ [365]. He showed that the Cauchy problem

$$\Box f^{\alpha} + \Gamma'^{\alpha}_{jk}\partial^i f^j \partial_i f^k = 0,$$
$$f(0,\cdot) = f_0, \quad f_t(0,\cdot) = f_1,$$

with initial data fulfilling

$$||(f_0, f_1)||_{\dot{B}^{2,1}_{m/2}\times \dot{B}^{2,1}_{m/2-1}} < \epsilon,$$

has a global solution, which is unique and a limit of smooth solutions; moreover, the solution depends in a Lipschitz continuity on the initial data.

Under the hypothesis of smallness of the initial data in the Besov space, Tataru was able to work in local coordinates thanks to the embedding $\dot{B}^{m/1,1} \hookrightarrow L^{\infty}$, which prevents the solution from exiting the chart domain. This property allows us to ignore the geometry of the target manifold.

The space $\dot{B}^{s,p}$ can be defined by decomposing the functions in Fourier space and applying Littlewood-Paley theory. If $f = \sum_{k\in\mathbf{Z}} P_k f$ in the Littlewood-Paley decomposition, then for $1 \leq p \leq \infty$ we have

$$||f||_{\dot{B}^{s,p}} = \left(\sum_{k\in\mathbf{Z}} (2^{sk}||P_k f||_{L^2})^p\right)^{1/p} = ||\{2^{sk}||P_k f||_{L^2}\}_{k\in\mathbf{Z}}||_{l^p}.$$

Therefore,

$$||f||_{\dot{B}^{m/2,1}} = ||\{2^{mk/2}||P_k f||_{L^2}\}_{k\in\mathbf{Z}}||_{l^1} \text{ vs. } ||f||_{\dot{H}^{m/2}} = ||\{2^{mk/2}||P_k f||_{L^2}\}_{k\in\mathbf{Z}}||_{l^2}.$$

The techniques used in dimension $m \geq 4$ and $m = 2, 3$ are different, because of the lack of suitable Strichartz estimates in lower dimensions. The latter is compensated by the use of 'null-frame spaces' $L^p_{t_w}(L^q_{x_w})$, in which the $L^2(L^{\infty})$ Strichartz estimates hold, allowing us to close the estimates in the space $\dot{X}^{s,b,1}$.

Global regularity for general small data in the critical Sobolev space $\dot{H}^{m/2}$ was first shown by Tao [360] for $m \geq 5$ (see Sect. 2.4) using only Strichartz estimates, while dealing with the logarithmic divergence that arises in the l^2 space. We cannot utilize the formulation in local coordinates here, though Tao studied the case $S^n \subset \mathbf{R}^{n+1}$, and used the extrinsic formulation to write the equations as

$$\Box f = -f(\partial_\alpha f \cdot \partial^\alpha f). \tag{2.6}$$

After decomposing in frequency $f = \sum_{k \in \mathbf{Z}} P_k f$, he dealt with all the intersections in non-linearity with the exception of $f_{low} \partial_\alpha f_{low} \partial^\alpha f_{high}$, which is then 'gauge away' by a microlocal gauge.

To observe the importance of the gauge in the wave map equation, let $e_1, \cdots, e_n$ be an orthonormal frame in TN. For a smooth map $f : \mathbf{R}^{1+m} \to N$ we can pull-back this frame to the pull-back bundle $f^*(TN)$. Then the derivatives of the map f can be written in this frame as $\partial_\alpha f = e\psi_\alpha$ and the covariant derivative has the form $D_\alpha = \partial_\alpha + A_\alpha$, where $A = A_\alpha dx^\alpha$ is a matrix-valued 1-form. By the wave map equation, the definition of the connection form A, and the zero-torsion identity $D_\alpha \psi_\beta = D_\beta \psi_\alpha$, we can write the derivative formulation of the wave map system as

$$\begin{aligned}
&\partial_\alpha f = e\psi_\alpha,\ (f^*\nabla)_\alpha = eA_\alpha,\\
&D^\alpha \psi_\alpha = 0,\ D_\alpha \psi_\beta - D_\beta \psi_\alpha = 0,\\
&F_{\alpha\beta} = \partial_\alpha A_\beta - \partial_\beta A_\alpha + [A_\alpha, A_\beta].
\end{aligned}$$

This system is undetermined, since we have freedom of choice of the frame e.

The above derivative formulation was used by Klainerman and Rodnianski [226, 228] to generalize Tao's result to more general targets, as well as by Shatah and Struwe [333] and Nahmod, Stefanov and Uhlenbeck [275] to give alternate proofs for dimension $m \geq 4$. The result of Shatah and Struwe is very interesting, since it simplifies the proof significantly with only Strichartz estimates, but no microlocalization. In the previous two papers, the Coulomb gauge $\sum \partial_i A_i$ was used, which gives elliptic equations for the connection form A:

$$\triangle A_\beta + \partial_i [A_i, A_\beta] = \partial_i F_{i\beta} = \partial_i (R(\partial_i f, \partial_\beta f)), \quad 0 \leq \beta \leq n. \tag{2.7}$$

For the low dimensional cases $m = 2, 3, 4$, the global regularity in the critical Sobolev norm was also obtained by Tao in [360], where he examined S^n and analyzed the equation (2.6). Using microlocolization and Tataru's null-frame spaces to compensate for the missing Strichartz estimates, Tao tries to control all the interactions except one, which is again gauged away by a microlocal gauge. This paper will be discussed in Sect. 2.5, which is very technical. The solution space looks like ($\sim 2^k$ frequency terms)

$$\begin{aligned}
||f||_{S[k]} &= ||\nabla_{x,t} f||_{L_t^\infty \dot{H}_x^{m/2-1}} + ||\nabla_{x,t} f||_{\dot{X}_k^{m/2-1,1/2,\infty}}\\
&+ \sup_{\pm} \sup_{l>10} \left(\sum_{\kappa \in K_l} ||P_{k,\pm\kappa} Q^{\pm}_{<k-2l} f||^2_{S[k,\kappa]} \right)^{1/2}
\end{aligned}$$

where

$$||f||_{S[k,\kappa]} = 2^{mk/2}||f||_{NFA^*[\kappa]} + |\kappa|^{-1}2^{k/2}||f||_{PW[\kappa]} + 2^{mk/2}||f||_{L_t^\infty L_x^2}.$$

Krieger [233] generalized Tao's low dimensional result to the hyperbolic space $\mathbf{H}^2$ for $m = 3$, then for $m = 2$ [234]. He utilized the same functional spaces, while using the Coulomb gauge at the beginning similarly to the work of Shatah-Struwe [333] and Namod-Stefanov-Uhlenbeck [275] (rather than after microlocalization, as was done by Tao and Klainerman-Rodnianski). For usual targets which can be 'uniformly isometrically embedded' into some Euclidean space $\mathbf{R}^k$, the global regularity for small data in the critical Sobolev space was shown by Tataru [366] by applying similar techniques.

2.1.3 *Stability*

The orbital stability for the geodesic wave map $f = \gamma(u)$ in $\mathbf{R}^{1+3}$ was studied by Sideris [338], where a global smooth solution to the wave maps system is constructed by a perturbation of the geodesic wave map. This perturbed solution stays in a tubular neighborhood around the geodesic for all time. The spaces used are of $X^{s,b}$ type with $s > 10$, and the techniques are energy estimates based on embeddings of $X^{s,b}$ spaces and bilinear null-form estimates in these spaces established by Klainerman in his earlier papers. The global geometry of the target manifold is not involved because of the high regularity.

Krieger [235] obtained stability of spherically symmetric and geodesic wave maps $f : \mathbf{R}^{1+2} \to \mathbf{H}^2$. The stability of spherical maps is based on the asymptotic behavior of such maps established by Christodoulou and Tahvildar-Zadeh [90]. The stability is in the sense of the closeness of the (Coulomb) gauged derivative components of the perturbed map to the spherical symmetric one in the L^2 sense. It is not clear from this result whether the map itself (not the gauged components) stays close to the spherical symmetric map.

It is possible to deduce the stability of the geodesic wave map using the exponential map on the target manifold N by comparing the 'difference' between the perturbed map and the geodesic map. In the Fermi chart, in a neighborhood of the geodesic γ, the 'difference equations' appear similarly to the wave map system in local coordinates. To explain this we focus on $f : \mathbf{R}^{1+m} \to S^2$. Let $f(t,x) = \gamma(u(t,x)) : \mathbf{R}^{1+m} \to S^2$ be a geodesic map, where $\gamma : \mathbf{R} \to S^2$ is a geodesic and $u : \mathbf{R}^{1+m} \to \mathbf{R}$ is a free wave map, i.e., $\Box u = 0$.

For convenience, we define $\rho(t,x) = \gamma(u(t,x) + s(t,x))$. Thus the perturbed map is $\psi = \exp_\rho \vec{v}$, where $\vec{v} = v\vec{n}$. Differentiating the expression of ψ in the direction of x_α $(\alpha = 0, 1, \cdots, m)$ we have

$$\partial_\alpha \psi = \partial_\alpha v \vec{n}_\psi + \cos(\partial_\alpha u + \partial_\alpha s)\vec{t_\psi}.$$

(Think of $S^2 \hookrightarrow \mathbf{R}^3$ and the family of geodesics $\Psi(v,r) = (\cos v \cos r, \cos v \sin r, \sin v)$, then $|\partial_t \Psi(v,r)|^2 = \cos^2 v$). Therefore, the wave map system for ψ is

$$0 = D^\alpha \partial_\alpha \psi = \Box v \vec{n} + \cos v\, Q_0(v, u+s)\nabla_t \vec{n} + (\cos v\, \Box s + Q_0(\cos v,\, u+s))\vec{t},$$
$$+ \cos v\, Q_0(v, u+s)\nabla_{\vec{n}} \vec{t} + \cos v\, Q_0(u+s, u+s)\nabla_{\vec{t}} \vec{t}.$$

The $s - v$ system has the form [159]:

$$\Box v + \cos v[(\Gamma^1_{12}(\psi) + \Gamma^1_{21}(\psi))\, Q_0(v, u+s) + \Gamma^1_{22}\, Q_0(u+s, u+s)] = 0,$$
$$\cos v\, \Box s + \cos v\, [(\Gamma^2_{12}(\psi) + \Gamma^2_{21}(\psi))\, Q_0(v, u+s) + \Gamma^2_{22}(\psi)\, Q_0(u+s, u+s)]$$
$$+ Q_0(\cos v, u+s) = 0,$$
$$\Box u = 0,$$

where the second equation can be simplified (v is small, $\cos v \neq 0$) as follows:

$$\Box s + (\Gamma^2_{12}(\psi) + \Gamma^2_{21}(\psi))Q_0(v, u+s) + \Gamma^2_{22}(\psi)Q_0(u+s, u+s) + \tan v\, Q_0(\cos v, u+s) = 0.$$

These equations can be recast as $\Box V = \Gamma(V)Q_0(V, V)$, which is similar to Tataru's global well-posedness result in Besov spaces. We obtain global existence for the perturbed map if and only if $||(v,s)||_{\dot{B}^{m/1,1}}$ is small, which remains close to the geodesic wave map for all time. We have (strong) pointwise asymptotic stability thanks to the embedding $\dot{B}^{m/1,1} \hookrightarrow L^\infty$.

Finally, we discuss briefly two recent blow-up results for the large-data wave maps $f : \mathbf{R}^{1+2} \to S^2$, which are consistent with the numerical evidence provided by Bizon et al. [34]. The first result is due to Rodnianski and Sterbenz [307], who found a set of initial data that led to the development of singularities in finite time. This was done by regarding an n-equivariant wave map as a perturbation of a self-similar scaled harmonic map for $n \geq 4$. The second result was obtained by Krieger, Schlag and Tataru [236], who constructed a 1-equivariant map as a perturbation of a time-scaled harmonic map. In this result, the blow-up rate can be controlled arbitrarily slow, while the initial data leading to this blow up are not generic.

2.2 Geometric Aspects

2.2.1 General Results

Let $f = (f^1, \cdots, f^n) : \mathbf{R}^{1+m} \to N \subset \mathbf{R}^k$ be a wave map to a Riemannian manifold N isometrically embedded in $\mathbf{R}^k$ with second fundamental form A; f satisfies

$$\Box f = A(f)(Df, Df) \perp T_f N, \tag{2.8}$$

where $Df = (\partial_t f, \nabla f)$ is the vector of time-space derivatives of f, and $\Box = -\frac{\partial^2}{\partial t^2} + \Delta$ is the wave operator. We want to study the well-posedness of the Cauchy problem with initial data

$$(f, f_t)|_{t=0} = (f_0, f_1) : \mathbf{R}^m \to TN, \tag{2.9}$$

i.e., $f_0(x) \in N \subset \mathbf{R}^k$ and $f_1(x) \in T_{f_0(x)}N \subset TN$. For simplicity, we assume that N is compact.

The treatment of (2.8) can be illustrated under weak regularity assumptions on the solution and on the initial data. Because N is compact, the equation can be treated in the sense of distributions if $f \in L^2_{\text{loc}}(\mathbf{R}^{1+m}; N)$ and $Df \in L^2_{\text{loc}}(\mathbf{R}^{1+m})$. For given initial data

$$(f_0, f_1) \in (H^s \times H^{s-1})(\mathbf{R}^m; TN), \ s \geq 1, \tag{2.10}$$

one may ask the following questions:

- Local well-posedness: For what values of s does the initial value problem of (2.8)–(2.10) have a unique local solution $f \in H^s$?
- Global well-posedness: For what values of s does the solution extend to all time?
- Global regularity: Does the solution preserve the regularity of the initial data?

We can obtain answers to the above questions by applying dimensional analysis. If we consider the dimension of each coordinate x^α as 1, and the map f as dimensionless (i.e. the dimension of f is 0), then the H^s energy norm on $\mathbf{R}^m$ has dimension $m - 2s$. Thus the question is critical in $H^{m/2}$, subcritical in H^s for $s > m/2$, and supercritical in H^s for $s < m/2$. Classical energy estimates for the equation $\Box f = k(f, Df)$ (k is a function of f and Df), and thus for wave maps, imply local well-posedness of the Cauchy problem in H^s for $s > \frac{m}{2}+1$. However, if we use the geometric structure of the wave map system, the result can be improved nicely.

To investigate how the wave map leads to local well-posedness in H^s for $s < \frac{m}{2} + 1$, we can use the above geometric structure of the wave map system in (2.8). All the following theorems and results were obtained by Shatah and Struwe [329–333, 353, 354].

Theorem 2.2.1. *Consider wave maps $f : \mathbf{R}^{1+m} \to N$ from a Minkowski space into a Riemannian manifold and assume $m \leq 3$. Then for any data $(f_0, f_1) \in (H^2 \times H^1)(\mathbf{R}^m; TN)$, there exists a unique local solution f of class H^2. If $m = 1$, the solution extends uniquely for all time. If $(f_0, f_1) \in H^s \times H^{s-1}$, $s > 2$, then so does f.*

Proof. We begin with (2.8), which implies the energy-momentum conservation

$$0 = (\Box f, f_t) = \frac{1}{2}\partial_t |Df|^2 - div(\nabla f, f_t),$$

with the energy

$$E(f_t) = \frac{1}{2}||Df(t)||^2_{L^2(\mathbf{R}^m)} = \frac{1}{2}||Df(0)||^2_{L^2(\mathbf{R}^m)}. \tag{2.11}$$

We compute

$$\Box(\partial f) = \partial[A(f)(\partial_a f, \partial^a f)] = dA(f)(\partial f, \partial_a f, \partial^a f) + 2A(f)(\partial_a \partial f, \partial^a f).$$

Applying the fact $(f_t, A(f)(\cdot,\cdot)) = 0$ to the above equation, we have

$$(\partial f_t, A(f)(\partial_a \partial f, \partial^a f)) = -(f_t, dA(f)(\partial f, \partial_a \partial f, \partial^a f)).$$

Then we derive

$$\frac{d}{dt}E(\partial f(t)) = \int_{R^m} (\Box \partial f, \partial f_t)dx \leq C||dA(f)||_{L^\infty} \cdot \int_{R^m} |Df(t)|^3|D^2 f(t)|dx. \tag{2.12}$$

We obtain from Sobolev's embedding that

$$\int_{R^m} |Df(t)|^3|D^2 f(t)|dx \leq C||Df(t)||^{4-\theta}_{L^2}||D^2 f(t)||^{\theta}_{L^2},$$

where $\theta = 2, 3, 4$ if $m = 1, 2, 3$, respectively. Thus, we have a Gronwall inequality

$$\frac{d}{dt}||D^2 f(t)||^2_{L^2} \leq C||D^2 f(t)||^{\theta}_{L^2}, \tag{2.13}$$

which implies a local-in-time H^2 a priori bound. If $m = 1$, we get $\theta = 2$ and the H^2 bound is global. Since $H^2(\mathbf{R}^m) \subset C^0(\mathbf{R}^m)$ for $m \leq 3$, these a priori estimates imply local well-posedness of the Cauchy problem in H^2.

We next show that the energy inequality implies the uniqueness of H^2 solutions. Remark that $(f_t, A(f)(\cdot,\cdot)) = 0$, and for H^2 solutions $f, g : \mathbf{R}^m \to N$ we obtain

$$\begin{aligned}
&\frac{1}{2}\frac{d}{dt}||D(f-g)(t)||^2_{L^2} \leq \int_{\mathbf{R}^m} (f_t - g_t, A(f)(Df, Df) - A(g)(Dg, Dg))dx \\
=& \int_{\mathbf{R}^m} \Big[\Big(f_t, A(f)(Df, Df) - A(g)(Dg, Dg)\Big) - \Big(g_t, A(f)(D(f, Df) - A(g)(Dg, Dg))\Big)\Big]dx \\
\leq& C\int_{\mathbf{R}^m} |A(f) - A(g)|\,|Df - Dg|\Big(||Df||^2 + ||Dg||^2\Big)dx.
\end{aligned} \tag{2.14}$$

It follows from (2.14) that

$$\frac{d}{dt}||D(f-g)(t)||_{L^2} \leq C||(|f-g|)(|Df|^2 + |Dg|^2)||_{L^2}. \tag{2.15}$$

Therefore, by Sobolev's embedding theorem, we have

$$\frac{d}{dt}||D(f-g)(t)||_{L^2} \le C||f-g||_{L^6}(||Df||^2_{L^6} + ||Dg||^2||_{L^6}) \le C||D(f-g)||_{L^2},$$

and thus the Gronwall inequality yields the uniqueness. □

Let $f : \mathbf{R}^{1+m} \to N$ be a wave map. The above theorem can be generalized to dimension $m > 3$ to give $H^{\frac{m+1}{2}}$ local well-posedness and can be improved for $m = 1$ or 2. The reason is that for $m > 3$, (2.15) can be expressed as

$$\frac{d}{dt}||D(f-g)(t)||_{L^2} \le C||D(f-g)||_{L^2}\Big(||Df||^2_{L^{2m}} + ||Dg||^2_{L^{2m}}\Big),$$

and then the uniqueness follows, if $Df \in H^{\frac{m-1}{2}} \hookrightarrow L^{2m}$. To find a priori bounds for the solution $f : \mathbf{R}^{1+m} \to N$ in $H^{\frac{m+1}{2}}$, we use energy estimates. Since $\frac{m+1}{2}$ may not be an integer, we must apply interpolation to obtain our estimates. Let $\mathcal{D}_a$ be the covariant derivative in the pull-back bundle $f^{-1}TN$. We consider the intrinsic form of the wave map system $\mathcal{D}^a\partial_a f = 0,\ 0 \le a \le m$. We define $V_a = \partial_a f,\ 0 \le a \le m$ as a family of vector fields along f satisfying the Hodge system of equations

$$\mathcal{D}^a\partial_a f = 0,\ \ \mathcal{D}_a V_b - \mathcal{D}_b V_a = 0.$$

We now consider the abstract linear Hodge system of $f : \mathbf{R}^{1+m} \to N$,

$$\mathcal{D}^a\partial_a f = F,\ \ \mathcal{D}_a V_b - \mathcal{D}_b V_a = G_{ab},\ \ 0 \le a, b \le m, \tag{2.16}$$

for vector fields $V = \{V_a\}_{0\le a\le m}$, F, and $G = \{G_{ab}\}_{0\le a,b\le m} \in X^s$. For $m \ge 2$ we need the following lemma (proof in [332]) to prove Theorem 2.2.2.

(L1) *For $0 \le s \le 1$ solutions to (2.16) satisfy*

$$||V(t)||_{H^s} \le C\Big[\int_0^t (||F(\tau)||_{H^s} + ||G(\tau)||_{H^s})d\tau + ||V(0)||_{H^s}\Big], \tag{2.17}$$

where C depends on $||f||_{H^\mu}$, $\mu = \frac{m+1}{2}$ for $m \ge 3$, and $\mu > 3/2$ for $m = 2$.

Theorem 2.2.2. *Let $f : \mathbf{R}^{1+m} \to N$ be a wave map from a Minkowski space into a Riemannian manifold. If $(f_0, f_1) \in H^{(m+1)/2} \times H^{(m-1)/2}(\mathbf{R}^m; TN)$ for $m \ge 3$ and $(f_0, f_1) \in H^s \times H^{s-1}(\mathbf{R}^2; TN)$ for some $s > 3/2$ and for $m = 2$, then there exists a local solution f to (2.8), (2.9) of class $H^{(m+1)/2}$. If $(f_0, f_1) \in H^s \times H^{s-1}$ for some $s > \frac{m+1}{2}$, then f is also in H^s.*

Proof. We first show uniqueness. Let f, g be local $H^{\frac{m+1}{2}}$ solutions of (2.8) with

$$(f, f_t)|_{t=0} = (g, g_t)|_{t=0} \in H^{\frac{m+1}{2}} \times H^{\frac{m-1}{2}}(\mathbf{R}^m; TN)$$

and let $h = f - g$. It follows from (2.15) that for $m \geq 3$ we have

$$\frac{d}{dt}||D(h)(t)||_{L^2} \leq C|||h|(|Df|^2 + |Dg|^2)||_{L^2} \tag{2.18}$$

$$\leq C(||Df||^2_{L^{2m}} + ||Dg||^2_{L^{2m}})||h||_{L^\alpha},$$

where $\frac{1}{\alpha} = \frac{1}{2} - \frac{1}{m}$. Applying Sobolev's embedding, $H^{\frac{m+1}{2}}(\mathbf{R}^m) \hookrightarrow W^{1,2m}(\mathbf{R}^m)$ and $H^1(\mathbf{R}^m) \hookrightarrow L^\alpha(\mathbf{R}^m)$, we obtain that

$$\frac{d}{dt}||Dh||_{L^2} \leq C(||Df||^2_{H^{\frac{m+1}{2}}} + ||D\bar{g}||^2_{H^{\frac{m+1}{2}}})||Dh||_{L^2}.$$

This yields uniqueness if $m \geq 3$.

When $m = 2$, we apply the Brezis-Wainger inequality

$$||u||_{L^\infty} \leq C||u||_{H^1} log(2 + ||\nabla u||_{H^{1/2}}/||u||_{H^1})$$

and (2.15) to get

$$\frac{d}{dt}||Dh||_{L^2} \leq C(||Df||^2_{L^4} + ||Dg||^2_{L^4})||h||_{L^\infty}$$

$$\leq C(||Df||^2_{H^{1/2}} + ||Dg||^2_{H^{1/2}})||h||_{H^1}\, log\left[2 + \frac{||Df||_{H^{1/2}} + ||Dg||_{H^{1/2}}}{||h||_{H^1}}\right]$$

$$\leq C(f,g)||h||_{H^1}\left[1 + log\left(\frac{C(f,g)}{||h||^2_{H^1}}\right)\right].$$

Furthermore, we have

$$\frac{1}{2}\frac{d}{dt}||h||^2_{L^2} = \int_{\mathbf{R}^m} h_t h dx \leq ||h_t||_{L^2}||h||_{L^2},$$

and thus

$$\frac{d}{dt}||h(t)||_{L^2} \leq ||h_t||_{L^2} \leq ||Dh||_{L^2}.$$

Hence,

$$\frac{d}{dt}||h||_{H^1} \leq C||h||_{H^1}\left(1 + log(\frac{C}{||h||_{H^1}})\right)$$

where $C = C(f,g) \geq 1$, i.e., $\frac{d}{dt} log\left(1 + log(\frac{C}{||h||_{H^1}})\right) \geq -C$. So Gronwall inequality yields the uniqueness.

We next show some a priori bounds. For a wave map $f : \mathbf{R}^{1+m} \to N$, let $l = [\frac{m-1}{2}] = \sup\{i \in \mathbf{N} | i \leq \frac{m-1}{2}\}$ and let $D^l \partial_a f$ be any l-fold covariant derivative of $V = \partial_a f$; then

$$D^a D^l \partial_a f = D^l D^a \partial_a f + \sum_{l_1+l_2=l-1} D^{l_1} B_{l_1,l_2}(Df, Df) D^{l_2+1} f,$$

$$D_a D^l \partial_b f - D_b D^l \partial_a f = D^l (D_a \partial_b f - D_b \partial_a f) + \sum_{l_1+l_2=l-1} D^{l_1} \tilde{B}_{l_1 l_2}(Df, Df) D^{l_2+1} f,$$

where the coefficients of the bilinear forms $B_{l_1 l_2}$, $\tilde{B}_{l_1 l_2}$ depend smoothly on f, i.e.,

$$D^a D^l \partial_a f = D^{l-1} T(Df, Df, Df),$$
$$D_a D^l \partial_b f - D_b D^l \partial_a f = D^{l-1} \tilde{T}(Df, Df, Df),$$

with smooth tri-linear forms $T = T(f)$, $\tilde{T} = \tilde{T}(f)$. Remark that for $m = 2$, we have $l = 0$ and $T = \tilde{T} = 0$. Putting $s = \frac{m-1}{2} - l \in \{0, \frac{1}{2}\}$ and applying (L1) for $m \geq 3$ we get

$$||Df(t)||_{H^{\frac{m-1}{2}}} \leq C \int_0^t \left(||T(Df, Df, Df)||_{H^{\frac{m-3}{2}}} ||\tilde{T}(Df, Df, Df)||_{H^{\frac{m-3}{2}}} \right) d\tau$$
$$+ C||Df(0)||_{H^{\frac{m-1}{2}}} \leq C \int_0^t \left(1 + ||Df||_{H^{\frac{m-1}{2}}}^{\frac{m+3}{2}} \right) d\tau + C||Df(0)||_{H^{\frac{m-1}{2}}},$$

and we obtain the a priori bound. For $m = 2$, we have a bound on the H^μ norm for $\mu \geq s/2$ from (L1). □

For $m = 1$ we use the characteristic coordinates $\xi = t + r$, $\eta = t - r$ to rewrite the wave map system for $f : \mathbf{R} \times \mathbf{R} \to N \hookrightarrow \mathbf{R}^k$ as

$$- f_{\xi\eta} = A(f)(\partial_\xi f, \partial_\eta f) \perp T_f N. \tag{2.19}$$

These characteristic coordinates permit us to use the (L^1, L^∞) estimate to prove global well-posedness in the class of finite energy solutions to (2.8) or (2.19).

In the Minkowski space $\mathbf{R} \times \mathbf{R}^m$, consider the light cone

$$K^\pm(z_0) = \{z = (t, x) \in \mathbf{R} \times \mathbf{R}^m; |x - x_0| < \pm(t - t_0)\}$$

through any point $z_0 = (x_0, t_0)$. Because we are more interested in evolution and singularities in forward time, our estimates are usually done on the backward light cone $K^-(z_0)$ with vertex at $z_0 = (x_0, t_0)$ and with lateral boundary

$$M^-(z_0) = \{z \in \mathbf{R} \times \mathbf{R}^m; |x - x_0| = t_0 - t\}$$

and horizontal sections

$$D(t; z_0) = K^-(z_0) \cap (\{t\} \times \mathbf{R}^m), \ t < t_0.$$

Moreover, for $s < t < t_0$ we denote by

$$K_s^t(z_0) = K^-(z_0) \cap [s,t] \times \mathbf{R}^m,$$
$$M_s^t(z_0) = M^-(z_0) \cap [s,t] \times \mathbf{R}^m,$$

the truncated cone and its mantle. For convenience, we put $K(z_0) = K_0^{t_0}(z_0)$ and $M(z_0) = M_0^{t_0}(z_0)$. Recall that

$$\partial_t \left(\frac{|f_t|^2}{2} + \frac{|\nabla f|^2}{2} \right) - div(\nabla f f_t) = 0.$$

Integrating the above equation over a truncated cone $K_s^t(z_0)$, we have the local energy identity

$$E(f; D(t; z_0)) + Flux(f; M_t^s(z_0)) = E(f; D(s; z_0)),$$

where $E(f; D(t, z_0)) = \frac{1}{2} \int_{D(t,z_0)} |Df|^2 dx$ and

$$\begin{aligned} Flux(f; M_x^t(z_0)) &= \frac{1}{\sqrt{2}} \int_{M_s^t(z_0)} \left(\frac{|Df|^2}{2} - < \frac{x - x_0}{|x - x_0|} \cdot \nabla f, f_t > \right) d\sigma \\ &= \frac{1}{2\sqrt{2}} \int_{M_s^t(z_0)} \left| \nabla f - \frac{x - x_0}{|x - x_0|} f_t \right|^2 d\sigma, \end{aligned}$$

which involves the tangential derivative of f on the lateral boundary of $K(z_0)$. If we introduce new coordinates on $M_s^t(z_0)$ via $y \mapsto (t_0 - |y|, x_0 + y)$ and if we denote $g(y) = f(t_0 - |y|, x_0 + y)$, then

$$Flux(f; M_s^t(z_0)) = \frac{1}{2} \int_{B_s \setminus B_t(0)} |\nabla g|^2 dy.$$

Theorem 2.2.3. *Let $m = 1$ and $(f_0, f_1) \in H^1 \times L^2(\mathbf{R}; TN)$. Then there exists a unique weak solution f of class H^1 to the Cauchy problem (2.8) and (2.9) such that the energy inequality holds.*

Proof. (1) Existence. We can approximate the given data (f_0, f_1) by smooth data $(f_0^i, f_1^i) \in C^\infty(\mathbf{R}; TN)$ such that $(f_0^i, f_1^i) \to (f_0, f_1)$ in $H^1 \times L^2$ as $i \to \infty$, where f^i is the unique global smooth solution of (2.8) with data (f_0^i, f_1^i), the existence of which is ensured by Theorem 2.2.1. By the energy inequality, we may assume that $f^i \to f$ weakly in H^1_{loc} and locally uniformly as $i \to \infty$, and that

$$Flux(f; M(z)) + E(f(t)) \leq \int_{\mathbf{R}} (|\nabla f_0|^2 + |f_1|^2) dx = 2E(f(0))$$

for any $t \in \mathbf{R}$ and $z = (x, t) \in \mathbf{R} \times \mathbf{R}$.

Suppose that TN is parallelizable. Let $\bar{e}_1, \cdots, \bar{e}_n$ be a smooth orthonormal frame. We then obtain frames $e_j = \bar{e}_j \circ f$ or $e^i = \bar{e}_j \circ f^i$, $1 \le j \le n$, for the pull-back bundles $f^{-1}TN$ or ${f^i}^{-1}TN$, by composing with f or f^i. Then by (2.19) and the local uniform convergence of $e^i_j \mapsto e_j$, we get

$$< f^i_{\xi\eta}, e_j > = < f^i_{\xi\eta}, e_j - e^i_j > + < f^i_{\xi\eta}, e^i_j > = < f^i_{\xi\eta}, e_j - e^i_j >$$
$$= < A(f^i)(\partial_\xi f^i, \partial_\eta f^i), e^i_j - e_j > \to 0$$

in $\mathcal{D}'$ as $i \to \infty$. We also have

$$< f^i_{\xi\eta}, e_j > = \partial_\xi < f^i_\eta, e_j > - < f^i_\eta, \partial_\xi e_j >$$
$$\to \partial_\xi < f_\eta, e_j > - < f_\eta, \partial_\xi e_j > = < f_{\xi\eta}, e_j >$$

in $\mathcal{D}'$; thus, f solves (2.19) weakly.
(2) Uniqueness. Suppose that f and g are two weak finite energy solutions of (2.19), i.e., $Df, Dg \in L^\infty L^2(\mathbf{R} \times \mathbf{R})$ with $(f, f_t)_{t=0} = (g, g_t)_{t=0} = (f_0, f_1) \in H^1 \times L^2(\mathbf{R}; TN)$. It is sufficient to show that f and g coincide on any sufficiently small truncated cone $K^T_0(z_0)$.

By Sobolev's embedding $H^1(\mathbf{R}) \hookrightarrow C^{1/2}(\mathbf{R})$, $f(t)$ and $g(t)$ are uniformly Hölder continuous for $1 \le t \le T$, and since $\partial_t f, \partial_t g \in L^2$ for almost every x, the maps $t \mapsto f(t,s)$, $t \mapsto g(t,x)$ are also Hölder continuous. Then for $\delta > 0$ we can select T sufficiently small and $z_0 = (t_0, x_0)$ with $t_0 = 2T$ such that there is $p \in N$ such that $f(z), g(z) \in B_\delta(p)$ for all $z \in K^T_0(z_0)$. Select δ sufficiently small so that $|f(z) - g(z)| \le 2|\pi_{T_p(N)}(f(z) - g(z))|$ for all $z \in K^T_0(z_0)$, where $\pi_{T_p(N)} : \mathbf{R}^k \to T_pN$ is the orthonormal projection.

Let $h = f - g$. Using equation (2.19) for g and f, and multiplying by e^f_j, we derive

$$\partial_\xi < h_\eta, e^f_j > = < h_{\xi\eta}, e^f_j > + < h_\eta, \partial_\xi e^f_j > = - < g_{\xi\eta}, e^f_j > + < h_\eta, \partial_\xi e^f_j >,$$

because $< f_{\xi\eta}, e^f_j > = 0$. We also get

$$< g_{\xi\eta}, e^f_j > = < g_{\xi\eta}, e^f_j - e^g_j > = | < A(g)(f_\xi, g_\eta), e^f_j - e^g_j > | \le C|Dg|^2|f - g|$$

and

$$< h_\eta, \partial_\xi e^f_j > = < h_\eta, d\bar{e}_j(f)\partial_\xi f > = \partial_\eta < h, \partial_\xi e^f_j >$$
$$- < h, d^2\bar{e}_j(f)(\partial_\xi f, \partial_\eta f) > .$$

Thus we obtain

$$|\partial_\xi < h_\eta, e^f_j > - \partial_\eta < h, \partial_\xi e^f_j > | \le C|h|(|Df|^2 + |Dg|^2), \tag{2.20}$$

and likewise with ξ and η interchanged. Select $\bar{z} \in K_0^T(z_0)$ with $|h(\bar{z})| = ||h||_{L^\infty(K_0^T(z_0))}$, and note that $|h(\bar{z})| \leq C \sup_j | < h(\bar{z}), e_j^f > |$. For each j we may integrate over the right lateral boundary $\Lambda_r = \{(\eta, \bar{x} + \bar{t} - \eta) : 0 \leq \eta \leq \bar{t}\}$ of $K(\bar{z})$ to deduce

$$| < h(\bar{z}), e_j^f > | = \left| \int_{\Lambda_r} [< h_\eta, e_j^f > + < h, \partial_\eta e_j^f >] d\eta \right|.$$

Integrating (2.20) over $K(\bar{z})$ and using the energy inequality, we also have

$$\left| \int_{\Lambda_r} < h_\eta, e_j^f > d\eta - \int_{\Lambda_l} < h, \partial_\xi e_j^f > dx \right| \leq C \int_{K(\bar{z})} |h|(|Df|^2 + |Dg|^2) dz$$
$$\leq C ||h||_{L^\infty(K(\bar{z}))} \int_{K(\bar{z})} (|Df|^2 + |Dg|^2) dz \leq CT ||h||_{L^\infty(K_0^T(\bar{z}_0))},$$

where $\Lambda_l = \{(\xi, \bar{x} - \bar{t} + \xi) : 0 \leq \xi \leq \bar{t}\}$ is the left lateral boundary of $K(\bar{z})$. We estimate by the energy inequality again:

$$\left| \int_{\Lambda_l} < h, \partial_\xi e_j^f > d\xi + \int_{\Lambda_r} < h, \partial_\eta e_j^f > d\eta \right| \leq C \int_{\Lambda_l} |h||\partial_\xi f| d\xi + C \int_{\Lambda_r} |h||\partial_\eta f| d\eta$$
$$\leq C ||h||_{L^\infty(K(\bar{z}))} (\bar{t}\, Flux(f; M(\bar{z})))^{1/2} \leq C ||h||_{L^\infty(K_0^T(\bar{z}_0))} (TE(f(0)))^{1/2}.$$

Consequently, we obtain

$$||h||_{L^\infty(K^T(z_0))} \leq C \sup_j \Big[\left| \int_{\Lambda_r} < h_\eta, e_j^f > d\eta - \int_{\Lambda_l} < h, \partial_\xi e_j^f > d\xi \right|$$
$$+ \left| \int_{\Lambda_l} < h, \partial_\xi e_j^f > d\xi + \int_{\Lambda_r} < h, \partial_\eta e_j^f > d\eta \right| \Big]$$
$$\leq C(TE(f(0)) + (TE(f(0)))^{1/2}) ||h||_{L^\infty(K^T(\bar{z}_0))},$$

and we can conclude the uniqueness for sufficiently small $T > 0$. □

2.2.2 Brief Analytic Null Form Structure

Wave maps systems also possess the special analytic 'null form' structure, which is best interpreted when the target is a sphere. The non-linearity in the equation is a Lorentz gradient:

$$\Box f = (|f_t|^2 - |\nabla f|^2) f.$$

For simplicity, we consider solutions $f : \mathbf{R}^{1+m} \to \mathbf{R}$ of the equation

$$\Box f = |f_t|^2 - |\nabla f|^2 \text{ on } \mathbf{R} \times \mathbf{R}^m \tag{2.21}$$

with initial data $f|_{t=0} = 0$, $f_t|_{t=0} = f_1 \in H^{s-1}(\mathbf{R}^m)$.

Setting $g = e^f$, we calculate

$$\Box g = e^f(\Box f + |\nabla f|^2 - |f_t|^2) = 0$$

with $g|_{t=0} = 1$, $g_t|_{t=0} = f_1 \in H^{s-1}(\mathbf{R}^m)$.

By the dependence of the solution g on its data in $H^s \times H^{s-1}(\mathbf{R}^m)$, we have $g \in C^0(\mathbf{R}; H^s(\mathbf{R}^m))$. Also, a condition for g to arise as $g = e^f$ from a (local) solution f to (2.21) is $g > 0$ (for small time). This requires the embedding $H^s(R^m) \hookrightarrow L^\infty(\mathbf{R}^m)$ for $s > m/2$. Klainerman and Machedon [223] obtained the following theorem, which agrees with the above example.

Theorem 2.2.4. *For data $(f_0, f_1) \in H^1 \times H^{s-1}(\mathbf{R}^3; TN)$ with $s > 3/2$, the initial data problem for (2.8) and (2.9) is locally well-posed.*

The main tools for proving this result are special 'null form' estimates that hold for quadratic expressions $Q(Df, Df)$ like $(\partial^a f, \partial_a f)$ and involve the space-time gradient of a solution f to the wave equation, where the symbol of f vanishes on the null cone

$$\{(\tau, \eta) \in R \times R^m : \tau^2 = |\eta|^2\}.$$

Deducing these estimates is very technical (cf. [223]) in the general case. However, if $m = 1$, the computation is clear.

Let $f : \mathbf{R} \times \mathbf{R} \to N \hookrightarrow \mathbf{R}^k$ be a wave map of class H^2 such that

$$\Box f = (\partial_x + \partial_t)(\partial_x - \partial_t) f = (\partial_x - \partial_t)(\partial_x + \partial_t) f \perp T_f N.$$

Set $W_\pm = (\partial_x \pm \partial_t) f$; then $W_\pm$ satisfy

$$(\partial_x \pm \partial_t) W_\mp \perp T_f N.$$

Multiplying by $W_\mp$, we have the energy momentum conservation

$$(\partial_x \pm \partial_t)|W_\mp|^2 = 0. \tag{2.22}$$

Put

$$X(t) = \int_{x \geq y} \int W_-^2(t,x) W_+^2(t,y) dx dy,$$

which is bounded by the energy:

$$0 \leq X(t) \leq \int_{\mathbf{R}} W_-^2(t,x) dx \int_R W_+^2(t,x) dx \leq 4E(f(t))^2.$$

Apply (2.22) to calculate

$$\begin{aligned}\frac{d}{dt}X(t) &= \int_{x\geq y}\int \partial_t W_-^2(t,x)W_+^2(t,y)dxdy + \int_{x\geq y}\int W_-^2(t,x)\partial_t W_+^2(t,y)dxdy \\ &= -\int_{-\infty}^{\infty}\left[\int_y^{\infty} \partial_x W_-^2(t,x)dx\right] W_+^2(t,y)dy + \int_{-\infty}^{\infty}\left[\int_{-\infty}^{x} \partial_y W_+^2(t,y)dy\right] W_-^2(t,x)dx \\ &= 2\int_{\mathbf{R}} W_-^2(t,y)W_+^2(t,y)dy.\end{aligned}$$

It follows that for a wave map $f : \mathbf{R}\times\mathbf{R} \to N \hookrightarrow R^k$ of class H^2 the following space-time integral bound holds:

$$\int_{\mathbf{R}}\int_{\mathbf{R}} W_-^2(t,x)W_+^2(t,x)dx\,dt \leq 2E(f(0)).$$

Notice that in terms of characteristic coordinates $\xi = t + x$, $\eta = t - x$, the wave system is

$$-f_{\xi\eta} = A(f)(\partial_\xi f, \partial_\eta f) = A(f)(W_+, W_-).$$

Consequently, the above equation provides an L^2-bound for the right-hand side of the wave map equation in space-time.

The Klainerman-Machedon's theorem emphasizes the importance of the critical case $s = m/2$, and particularly for $s = 1$ and $m = 2$. Progress in this direction can be achieved by considering the algebraic structure of the wave map system and the results are described as follows (cf. [332, 333]).

1. *Suppose that $N = G/H$ is homogeneous, where G is a Lie group and H is a properly dis-continuous subgroup of G. Then for any $(f_0, f_1) \in H^1 \times L^2(\mathbf{R}^m;\, TN)$ there exists a global weak solution of (2.8) and (2.9) of class H^1.*
2. *Set $m = 2$. Assume that $\{f^i\}$ is a sequence of wave maps such that $f^i \to f$ in L^2 and $Df^i \to Df$ weakly in L^2, locally on $\mathbf{R}\times\mathbf{R}^2$, as $i \to \infty$. Then f is a (weak) wave map.*
3. *For any $(f_0, f_1) \in H^1 \times L^2(\mathbf{R}^2;\, TN)$, there exists a global weak solution to the Cauchy problem (2.8) and (2.9).*
4. *For any $\epsilon > 0$, any $(f_0, f_1) \in H^2 \times H^1(\mathbf{R}^2;\, TN)$, there exists a global unique H^2-solution f of the initial value problem*

$$\Box f - \epsilon\Delta f_t \perp T_f N,\ (f_0, f_t)|_{t=0} = (f_0, f_1),$$

such that the energy identity

$$\|Df(T)\|_{L^2}^2 + 2\epsilon\,\|\nabla f_t\|_{L^2([0,T]\times\mathbf{R}^2)}^2 = \|Df(0)\|_{L^2}^2$$

holds for any T. If $(f_0, f_1) \in H^s \times H^{s-1}$ for some $s \geq 2$, then f is also of class H^s. In particular, f is smooth if $(f_0, f_1) \in C^\infty(\mathbf{R}^2;\, TN)$.

2.2.3 Singularities

In higher space dimensions $m \geq 3$, solutions to the Cauchy problem might develop singularities in finite time. In fact, we can show smooth initial data that lead to self-similar blow-up in finite time and non-uniqueness of weak finite energy solutions. On the contrary, we can also exhibit situations in $m = 2$ no nontrivial self-similar solutions exist (see below). The question "For $m = 2$ do smooth solutions become singular in finite time?" is one of the open problems in the area these days.

The easiest way to generate initial data that lead to finite time singularities is to show the existence of self-similar solutions

$$f(t, x) = g(x/|t|)$$

to (2.8) with non-constant smooth Cauchy data

$$f_0 = g, \ \ f_1 = x \cdot \nabla g,$$

at $t = -1$. If g exists and is regular on a ball $B_{r_*}(0, \mathbf{R}^m)$ for $r_* > 1$, then g yields a self-similar solution on the truncated backward light cone $K^0_{-1}((\epsilon, 0))$ that is smooth on the base of the cone, but suffers a blow-up in the derivative at the origin (0,0).

Let $\varsigma = \sqrt{t^2 - |x|^2}$, $\xi = \frac{x}{|t|}$ be the similar coordinates in the backward light cone from 0, and set $|x| = r$, $|\xi| = \rho, x = rw$, and $\xi = \rho w$ with $w \in S^{m-1}$. We can rewrite the Minkowski metric $ds^2 = -dt^2 + dr^2 + r^2 dw^2$ as

$$ds^2 = -d\varsigma + \varsigma^2 \left(\frac{d\rho^2}{(1-\rho^2)^2} + \frac{\rho^2}{1-\rho^2} dw^2 \right).$$

Thus f is stationary for the Lagrangian $\mathcal{L} = \frac{1}{2}\partial^\alpha f \partial_\alpha f = \frac{1}{2}(|\nabla f|^2 - |f_t|^2)$ if and only if $g(\xi) = g(\rho, w)$ is stationary for

$$\frac{1}{2} \int \left\{ (1-\rho^2)^2 |g_\rho|^2 + \frac{1-\rho^2}{\rho^2} |g_w|^2 \right\} \frac{\rho^{m-1}}{(1-\rho^2)^{\frac{m+1}{2}}} d\rho \, dw$$

at $\varsigma = 1$. That means, f solves (2.8) if and only if g solves

$$-g_{\rho\rho} - \left(\frac{m-1}{\rho} + \frac{(m-3)\rho}{1-\rho^2} \right) g_\rho + \frac{1}{\rho^2(1-\rho^2)} \Delta_w g \perp T_g N. \tag{2.23}$$

Notice that (2.23) is an elliptic harmonic map equation on the unit m-ball B with the hyperbolic metric

$$\frac{d\rho^2}{(1-\rho^2)^2} + \frac{\rho^2}{1-\rho^2} dw^2. \tag{2.24}$$

We search for solutions g of (2.23) that extend smoothly to the boundary ($\rho = 1$) of B and thus can be continued smoothly to all of $\mathbf{R}^m$. Because the information propagates with speed ≤ 1, the unique solution of (2.8), (2.9) with initial data

$$f_0 = g,\ f_1 = x \circ \nabla g,\ at\ t = -1$$

will agree with $g(x/|t|)$ inside the backward light cone $|x| \leq -t$ and if $g \neq$ constant on B, we have blow-up at $t = 0$.

When $m = 2$, we can show that there are no self-similar solutions to the wave map system (2.8). For $m = 2$ the self-similar equation (2.23) yields

$$-(\rho\sqrt{1-\rho^2}g_\rho)_\rho + \frac{1}{\rho\sqrt{1-\rho^2}}\Delta_w g \perp T_g N.$$

Multiplying by $\rho\sqrt{1-\rho^2}g_\rho$ and integrating with respect to $w \in S^1$, we get

$$\frac{d}{d\rho}\left(\int_{S^1}\rho^2(1-\rho^2)|g_\rho|^2 dw - \int_{S^1}|g_w|^2 dw\right) = 0,$$

whence

$$\int_{S^1}\rho^2(1-\rho^2)|g_\rho|^2 dw - \int_{S^1}|g_w|^2 dw = C.$$

Examining at $\rho = 0$, we conclude that $C = 0$. Hence, for $\rho = 1$ we have $g_w = 0$, i.e., $g(1,\cdot) = \text{const}$.

We note that, by the Riemann mapping theorem, the hyperbolic metric (2.24) on the unit ball $B = B_1(0, \mathbf{R}^2)$ is locally conformal to the standard metric. Indeed, define

$$\sigma(\rho) = exp\left(-\int_\rho^1 \frac{d\rho}{\rho\sqrt{1-\rho^2}}\right)$$

and note that the metric

$$\begin{aligned} d\sigma^2 + \sigma^2 dw^2 &= \sigma^2\left(\frac{d\rho^2}{\rho^2(1-\rho^2)} + dw^2\right) \\ &= \left(\frac{\sigma}{\rho}\right)^2(1-\rho^2)\left(\frac{d\rho^2}{\rho^2(1-\rho^2)} + \frac{\rho^2}{\rho^2(1-\rho^2)}dw^2\right) \end{aligned}$$

is conformal to (2.24) on B. This means that the map $\psi : (\rho, w) \mapsto (\sigma, w)$ is a conformal diffeomorphism from B, endowed with the hyperbolic metric (2.24) to B with the standard metric.

Due to the conformal invariance of the Dirichlet integral, and thus of the harmonic map equation (2.8) in $m = 2$, g induces a harmonic map $\tilde{g} = g \circ \psi^{-1} \in H^1 \cap C^0(\bar{B}, N)$ on the standard ball with $\tilde{g}|_{\partial B} = \text{const}$. Applying Lemaire's uniqueness theorem [246], we have $\tilde{g} = 0$ and obtain the following theorem.

Theorem 2.2.5. *Set $m = 2$, and let $f(t, x) = g(x/|t|)$ be a map which solves (2.8) for $|x| \le |t|$. If g extends to a smooth map on a neighborhood of $\overline{B_1(0)}$, then $g = \text{const}$.*

When $m \ge 3$, we can construct self-similar solutions to (2.8) as follows. If the target is a surface of revolution N with the metric $ds^2 = dh^2 + k^2(h)dw^2$ in spherical coordinates $h > 0$, $w \in S^{m-1}$, we can seek solutions to (2.8) of the special form $f(t, rw) = h(t, r)w$, where $x = xw \in \mathbf{R}^m$ is spherical coordinates. Furthermore, we let the ansatz $f(t, x) = g(x/|t|)$, i.e., $h(tr) = \phi(r/|t|)$, $g(\xi) = \phi(\rho)w$. The action integral takes the form

$$\frac{1}{2}\int \left\{(1-\rho^2)^2|\phi_\rho|^2 + \frac{1-\rho^2}{\rho^2}(m-1)|k^2(\phi)|^2\right\} \frac{\rho^{m-1}}{(1-\rho^2)^{\frac{m+1}{2}}}d\rho\, dw,$$

and (2.23) becomes

$$-\phi_{\rho\rho} - \left(\frac{m-1}{\rho} + \frac{(m-3)\rho}{1-\rho^2}\right)\phi_\rho + \frac{(m-1)u(\phi)}{\rho^2(1-\rho^2)} = 0, \tag{2.25}$$

where $u(\phi) = k(\phi)k'(\phi)$. For special target metrics k, (2.25) admits non-constant solutions ϕ for $0 < \rho < 1$ that extend smoothly to all of $\mathbf{R}_+$. We explain this by an example where (2.25) can be solved explicitly. Set

$$k^2(\phi) = \phi^2 - \frac{1}{2}\phi^4 \text{ for } 0 < \phi < \phi_0,$$

where $\phi_0 > 0$ is a fixed number with $1 < \phi_0^2 < 2$ and extend k smoothly to $\mathbf{R}_+$. Then

$$u(\phi) = k(\phi)k'(\phi) = \frac{1}{2}(k^2(\phi))' = \phi - \phi^3$$

for $0 < \phi < \phi_0$ and the linear function

$$\phi(\rho) = \sqrt{\frac{2}{m-1}}\rho \equiv c_*\rho$$

solves (2.25) for $0 < \rho < c_*^{-1}\phi_0 \equiv \rho_0 > 1$, since $m \ge 3$. Observe that for k as above, the radius of convexity of N about 0 is $\phi_* = 1$, which is larger than c_* for $m \ge 4$ and equals c_* for $m = 3$. By appropriately changing the metric $k(\phi)$ on N for $\phi > c_*$, and by changing the initial data for h outside the unit ball, we can construct solutions to (2.25) with smooth finite initial energy that blow up in finite time. Remark that the target manifold is convex for $m \geq 4$, and slightly fails to be convex for $m = 3$. In three space dimensions blow-up may happen for more general metrics on the target surface as a more detailed discussion shows.

Theorem 2.2.6. *Let $k \in C^\infty$ be such that $k(0) = 0$, $k'(0) = 1$, and assume that k' has a smaller positive zero ϕ_* and $k''(\phi_*) \neq 0$. Then there is a class of regular initial data for which the corresponding Cauchy problem for equivariant harmonic maps from $\mathbf{R}^{1+3}$ to N has a solution that blows up in finite time.*

In higher space dimensions $m \geq 4$, there are classes of target manifolds N that admit solutions with self-similar blow up. The conditions involved are naturally analytical and have no obvious geometric interpretations. For instance, for $m = 5$ the condition for self-similar blow up permits the target manifold to be convex, and for $m = 7$ the target manifold is allowed to have negative sectional curvature.

2.2.4 Non-unique Weak Solutions

If the target manifold N satisfies the assumptions of Theorem 2.2.6, we can apply the self-similar solution to construct non-unique weak solutions. This is done by showing that the solution ϕ to (2.24) on $[0, \infty)$ such that $\phi_1 = \phi_*$ has the asymptotic expansion

$$\phi(\rho) = \alpha + \frac{\beta}{\rho} + \frac{\gamma}{\rho^2} + O(\frac{1}{\rho^3}),$$

$$\phi'(\rho) = -\frac{\beta}{\rho^2} + O(\frac{1}{\rho^3}),$$

for $\rho \to \infty$. The corresponding function $h(t,r)w = \phi(r/t)w$ is a weak solution of (2.8) on $\mathbf{R}_+ \times \mathbf{R}^m$, i.e., h is a weak solution of

$$h_{tt} - h_{rr} - \frac{2}{r}h_r + \frac{2u(h)}{r^2} = 0, \tag{2.26}$$

with local finite-energy initial data at $t = 0$ such that

$$h(0,r) = h_0(r) = \alpha = \lim_{t \searrow 0} \phi(\frac{r}{t}), \quad r \neq 0$$

$$h_t(0,r) = h_1(r) = \frac{\beta}{r} = \lim_{t \searrow 0} \frac{d}{dt}\phi(\frac{r}{t}), \quad r \neq 0. \tag{2.27}$$

This means, h fulfills

$$\int_0^T \int_0^\infty \left\{ -h_t\psi_t + h_r\psi_r + \frac{1}{r^2}\psi u(h) \right\} r^2 dr\, dt = \int_0^\infty \psi(0,r)\frac{\beta}{r} r^2\, dr \tag{2.28}$$

for any $\psi \in C^\infty([0,T] \times \mathbf{R}^3)$ such that $\psi(t,x) = \psi(t,r)$, $\psi(T,.) = 0$ and $supp\, \psi(t) \subset B_R(0)$ for some $R > 0$, and h satisfies the initial data (2.27) in the sense that

$$||h(t,r) - \alpha||_{H^1_{loc}(\mathbf{R}^3)} \to 0, \text{ as } t \to 0,$$

$$||h_t(t,r) - \frac{\beta}{r}||_{L^2(\mathbf{R}^3)} \to 0, \text{ as } t \to 0.$$

Next, we can define a new function

$$\hat{h}(t,r) = \begin{cases} \phi(r/t), & \text{if } r > t, \\ \phi_*, & \text{if } r \le t, \end{cases}$$

such that $\hat{h}(0,r) = h(0,r)$, $\hat{h}_t(0,r) = h_t(0,r)$, and h locally has finite energy $D\hat{h} \in L^\infty([0,1]; L^2(B_R(0)))$ for any $R > 0$. Moreover, $\hat{h}$ is a weak solution of (2.26) on $\mathbf{R}_+ \times \mathbf{R}^3$. To check that $\hat{h}$ solves (2.28), for any ψ we separate the integral into

$$\int_0^1 \int_0^\infty \left\{ -\hat{h}_t\psi_t + \hat{h}_r\psi_r + \frac{2}{r^2}\psi\, u(\hat{h}) \right\} r^2 dr\, dt - \int_0^\infty \psi(0,r)\frac{\beta}{r} r^2\, dr$$

$$= \left\{ \int_0^1 \int_t^\infty (\cdots) r^2 dr\, dt - \int_0^\infty \psi(0,r)\frac{\beta}{r} r^2 dr \right\} + \int_0^1 \int_0^t (\cdots) r^2\, dr\, dt = A + B.$$

The second integral $B = 0$, because $\hat{h}(t,r) = \phi_*$ for $r \le t$. Furthermore, the first integral reduces to the following boundary term, since $\hat{h} \equiv h$ for $r \ge t$ and h satisfies (2.28):

$$A = \frac{1}{\sqrt{2}} \int_0^1 (h_t(t,\, t) + h_r(t,\, t))\psi(t,\, t)\, t^2\, dt.$$

This equals to zero, because

$$h_t + h_r = -\frac{r}{t^2}\phi'(\frac{t}{r}) + \frac{1}{t}\phi'(\frac{t}{r}) = \frac{1}{t}(1 - \frac{r}{t})\phi'(\frac{r}{t}) = 0$$

for $r = t$.

The above discussion shows that there is a set of initial data of locally finite energy for (2.28) such that the Cauchy problem has more than one weak solution. However, these initial data are singular. In particular, when $N = S^3$ something stronger is true. The self-similar solution is

$$\phi(t,r) = 2\,\tan^{-1}(\frac{r}{-t}).$$

Because ϕ is the polar angle on S^3, $\phi = \pi$ and $\phi = -\pi$ correspond to the same point, i.e., the south pole. Therefore, the above solution, as a map into the sphere, is continuous as well as smooth across the line $t = 0$. Hence, we can pose the Cauchy problem (2.8) at $t = -1$ instead of $t = 0$, and we have non-uniqueness of a weak solution even for smooth data.

2.3 Equivariant Wave Maps

2.3.1 Equivariant Maps

Let $\mathbf{R}^{1+m}$ be a Minkowski space with spatial polar coordinates

$$(t, r, w) \in \mathbf{R} \times \mathbf{R}^+ \times S^{m-1}, \ r = |x|, \ w^i = x^i/r, \ i = 1, \cdots, m.$$

The metric on $\mathbf{R}^{1+m}$ has the form $-dt^2 + dr^2 + r^2 dw^2$ in the above coordinates, where dw^2 is the standard metric on $S^{m-1} \hookrightarrow \mathbf{R}^m$.

Let N be a smooth, n-dimensional rotationally symmetric, warped product manifold defined by $N = [0, R^*) \times_g S^{n-1}$, where $R^* \in \mathbf{R}^+ \cup \{+\infty\}$ and $g : \mathbf{R} \to \mathbf{R}$ is a smooth and odd function such that $g(0) = 0$, $g'(1) = 1$. On N we have the polar coordinates $(\phi, \chi) \in [0, R^*) \times S^{n-1}$. In these coordinates the metric of N takes the form $d\phi^2 + g^2(\phi) d\chi^2$ where $d\chi^2$ is the standard metric of $S^{n-1} \hookrightarrow \mathbf{R}^n$. Let $(f_1, \cdots, f_n)$ be the normal coordinates on N, where $f^i = \phi \cdot \chi^i$, $i = 1, \cdots, n$. Therefore, $(\phi, \chi) = (|f|, f/|f|)$. Hence, N can be identified with the ball $B_{R^*(0)}$ in $\mathbf{R}^n$.

Let M and N be rotationally symmetric manifolds, i.e., $SO(m)$ and $SO(n)$ act on M and N by isometries and let $f : M \to N$. Then f is *equivariant* if the orbit of any point $a \in M$ is mapped into the orbit of $f(a) \in N$ (or if f commutes with the actions on M and N).

For a map $f : \mathbf{R}^{1+m} \to N = [0, R^*) \times_g S^{n-1}$, the Cauchy problem is

$$\begin{aligned} \partial^\beta \partial_\beta f^a + \Gamma^a_{bc} \partial_\beta f^b \partial^\beta f^c &= 0, \\ f(0, x) = f_0, \ \partial_t f(0, x) &= f_1. \end{aligned} \tag{2.29}$$

Suppose that the initial data (f_0, f_1) are *equivariant* in the sense that there exist functions $\phi_0, \phi_1 : \mathbf{R} \to \mathbf{R}$ and a map $\chi : S^{m-1} \to S^{n-1}$ such that for $x = (r, w) \in \mathbf{R}^n$,

$$f_0^i = \phi_0(r) \cdot \chi^i(w), \ f_1^i = \phi_1(r) \cdot \chi^i(w), \quad i = 1, \cdots, n.$$

We observe that for f to be an equivariant solution of (2.29) the map χ has to be a *harmonic polynomial map*, i.e., the restriction of a map from $\mathbf{R}^m \to \mathbf{R}^n$ for which each component is a harmonic homogeneous polynomial of some degree $l > 0$. The solution f is then given by a radial function ϕ satisfying

$$f^i(t, x) = \phi(t, r) \cdot \chi_0^i(w).$$

When $m = n = 2$ and $l = 1$, (2.29) becomes

$$\phi_{tt} - \phi_{rr} - \frac{1}{r}\phi_r + \frac{1}{r^2}p(\phi) = 0, \tag{2.30}$$
$$\phi(0,r) = \phi_0(r),\ \phi_t(0,r) = \phi_1(r),$$

where $p(\phi) = g(\phi)g'(\phi)$. The energy of an equivariant map is equivalent to

$$E(\phi) = \int_0^\infty \left[\phi_t(t,r)^2 + \phi_r(t,r)^2 + \frac{g(\phi(t,r))^2}{r^2}\right] rdr;$$

and $\phi(t,0) = 0$ due to the finite energy requirement. Suppose that N satisfies

$$G(\phi) = \int_0^\phi g(s)ds \to \infty, \text{ as } \phi \to \infty, \tag{2.31}$$

then finite energy solutions are bounded pointwise by a constant depending on the energy of the initial data E_0. We require the following lemma to prove Theorem 2.3.1 obtained by Shatah and Struwe [332, 333].

(L1) *Under the above hypotheses, the smooth solutions of equation (2.30) have the property that* $|\phi(t,r)| \le C(E_0)$, *where* $C(s) \to 0$ *as* $s \to 0$.

Proof. For any solution $\phi(t,\ r)$ we get

$$G(\phi(t,\ r)) = \int_G \frac{\partial}{\partial r}G(\phi(t,r))dr = \int_0^r g(\phi)\partial_r\phi\, dr,$$

and thus

$$|G(\phi)| \le \left(\int_0^\infty \frac{g(\phi)^2}{r}dr\right)^{1/2} \left(\int_0^\infty \phi_r^2 rdr\right)^{1/2} \le C(E_0).$$

Hence, $|\phi|_{L^\infty} \le C(E_0)$, by the above inequality and (2.31). □

Theorem 2.3.1. *There exists a small* ϵ_0 *such that for any finite-energy initial data* (ϕ_0, ϕ_1) *with* $E(\phi) < \epsilon_0$, *(2.30) has a global finite energy solution* ϕ. *Furthermore, if the initial data are smooth, then the solution* ϕ *is regular, with* $\phi(t,0) = 0$.

Proof. We first have $|\phi| \le C(\epsilon_0)$ by (L1). Define the radial function $u : M \to \mathbf{R}$ by $f_i(t,x) = x^i u(t,r)$. Therefore, $\phi = ru$ and u satisfies:

$$u_{tt} - u_{rr} - \frac{3}{r}u_r = u^3 Y(ru), \tag{2.32}$$

$$u(0,r) = u_0 = \frac{\phi_0}{r},\ u_t(0,r) = u_1 = \frac{\phi_1}{r}$$

where Y is a smooth, even function. Since $\phi = ru$ is bounded, (2.32) is a critical wave equation when $m = 4$. In order to complete the proof, we only need to show that

$$||u^3 Y(ru)||_{\dot{B}^{1/2}_{q',2}(\mathbf{R}^{1+4})} \leq C(\epsilon_0)||u||^3_{\dot{B}^{1/2}_{q,2}(\mathbf{R}^{1+4})},$$

where $q' = 10/7$, using a Strichartz estimate for (2.32). We start with the bound

$$||u^3 Y(ru)||_{\dot{B}^{1/2}_{q',2}} \leq ||u^3||_{\dot{B}^{1/2}_{q',2}} + ||u^3(Y(ru) - Y(0))||_{\dot{B}^{1/2}_{q',2}}.$$

By Shatah and Struwe [332], we know that

$$||u^3||_{\dot{B}^{1/2}_{q',2}} \leq C||u||^3_{\dot{B}^{1/2}_{q',2}}.$$

We apply Sobolev's embedding to bound the second term

$$\begin{aligned} ||u^3(Y(ru) - Y(0))||_{\dot{B}^{1/2}_{q',2}} &\leq C||u^3(Y(ru) - Y(0))||_{\dot{W}^{1,s}} \leq C||u^4||_{L^s} + C||u^3 rDu||_{L^s} \\ &= A + B, \end{aligned}$$

where $1/s = 7/10 + 1/10 = 4/5$. The first term A can be estimated by Sobolev's embedding

$$A = C||u||^4_{L^5} \leq C||u||^4_{\dot{B}^{1/2}_{q',2}}.$$

The second term B can be estimated by observing that on the set $D(t)$, the Sobolev embedding and scaling yield that

$$|r^{7/10}u(t,r)| \leq C||u||^4_{\dot{B}^{1/2}_{q',2}(D(t))}.$$

Thus the spatial part of the norm in B can be bounded by

$$||u^3 rDu||_{L^s} \leq ||u||^{11/7}_{L^p}||u||^{10/7}_{\dot{B}^{1/2}_{q',2}}||Du||_{L^2},$$

where $p = 110/21$. By Sobolev's embedding, we obtain

$$||u||_{L^p(D(t))} \leq C||u||^{1-\alpha}_{L^4(D(t))}||u||^{\alpha}_{\dot{B}^{1/2}_{q',2}},$$

where $\alpha = 26/33$. Thus

$$||u^3 rDu||_{L^s(D(t))} \leq C||u||^{8/3}_{\dot{B}^{1/2}_{q',2}}||Du|^{4/3}_{L^2(D(t))},$$

and we derive

$$B \leq C(\epsilon_0)||u||^{8/3}_{\dot{B}^{1/2}_{q',2}}$$

by taking the L^s norm with respect to t. Hence, for a local solution u with sufficiently small energy bounded by ϵ_0, the Strichartz estimates imply that the solution is global and regular, if the data are smooth. □

We require the following lemma (see proof in [332]) to establish Theorem 2.3.2.

(L2) *If N is geodesically convex, i.e., $p(\phi) > 0$ for $\phi > 0$, then for any solution ϕ that is smooth away from the origin we have*

$$\int_0^{|t|} e(t,r) r dr \to 0 \text{ as } t \to 0.$$

Theorem 2.3.2. *If N is geodesically convex, then for smooth equivariant initial data equation (2.29) has a global regular solution.*

Proof. Let ϕ be a maximal smooth radial solution to (2.30). It follows from Theorem 2.3.1 and the energy inequality that the first singularity of ϕ can occur only at $r = 0$. Suppose that the first singularity of ϕ occurs at the origin. By (L2), the energy cannot concentrate at the origin, and thus, by Theorem 2.3.1 ϕ can be extended as a smooth solution beyond the origin, which contradicts the hypothesis that the origin is singular. Hence, the solution ϕ is globally regular. □

2.3.2 *Radial Wave Equation on* $\mathbf{R}^{1+2}$

On $\mathbf{R}^{1+2}$ we have polar coordinates $ds^2 = -dt^2 + dr^2 + r^2 d\theta^2$. Radially symmetric solutions of the linear non-homogeneous wave equation satisfy

$$f_{tt} - f_{rr} - \frac{1}{r} f_r = h, \tag{2.33}$$

where $f = f(t,r)$ and $h = h(t,r)$. To prove Theorem 2.3.3, we require the following lemmas. All the theorems and results were obtained in [332, 333].

(L3) *Solutions of (2.33) are given by*

$$f(t,r) = f_0(t,r) + \frac{1}{\sqrt{2}\pi} \int_K \sqrt{\frac{r'}{r}} J(\mu) h(t',r') dt' dr' \tag{2.34}$$

where

$$f_0 \text{ is a solution of the homogeneous wave equation,}$$
$$K = \{(r',t') : 0 \le t' \le t, max\{0, r-t+t'\} \le r' \le t-t'+r\},$$
$$\rho = \frac{r^2 + r'^2 - (t-t')^2}{2rr'},$$
$$J(\rho) = \int_{max\{-1,\rho\}}^{1} \frac{1}{\sqrt{1-\lambda^2}\sqrt{\lambda-\rho}} d\lambda.$$

Proof. If we express the fundamental solution

$$R(t,x) = \frac{1}{2\pi}\frac{1}{\sqrt{t^2-|x|^2}}\chi_{B_t(0)}$$

in polar coordinates, the solution f has the representation

$$f(t,r) = f_0(t,r) + \frac{1}{2\pi}\int_{|x-y|\le t} \frac{h(t',r')r'dr'd\theta'dt'}{\sqrt{(t-t')^2 - r^2 - r'^2 + 2rr'cos(\theta-\theta')}}.$$

Setting $\lambda = cos(\theta-\theta')$, we have that $d\lambda = \sqrt{1-\lambda^2}d\theta'$ and $max\{-1,\rho\} \le \lambda \le 1$. Replacing θ' by λ in the above formula, we derive (2.34). □

Remark that $\rho \in (-\infty, 1]$ on the region $K = \{(t',r') : \rho \in (-\infty,1], t' \ge 0\}$. The function J is well-defined on the cone $\rho = 1$ and has a logarithmic singularity on the straight line $\rho = -1$. For larger values of $|\rho|$ the function J behaves like $|\rho|^{1/2}$.

(L4) *The function* $J : (-\infty,1] \to \mathbf{R}^+$ *has the following properties:*

(a) $J(1)$ *is well-defined;*
(b) $J(\rho) \le C\, log\left(1 + \frac{1}{\sqrt{\rho+1}}\right)$ *for* $\rho \sim -1$;
(c) $J'(\rho) \le \frac{C}{|\rho+1|}$ *for* $\rho \sim -1$;
(d) $J(\rho) = \frac{\pi}{|\rho|^{1/2}} + O(\frac{1}{|\rho|^{3/2}})$ *for* $\rho \sim -\infty$;
(e) $J'(\rho) = \frac{\pi}{2|\rho|^{3/2}} + O(\frac{1}{|\rho|^{5/2}})$ *for* $\rho \sim -\infty$.

Proof. If $\rho \to 1$,

$$J(\rho) \to \frac{1}{\sqrt{2}}\int_\rho^1 \frac{dz}{\sqrt{(1-z^2)(z-\rho)}} \to \frac{\pi}{\sqrt{2}}.$$

If $\rho \to -1$, rewrite

$$\frac{1}{\sqrt{1-z^2}\sqrt{z-\rho}} = \frac{1}{\sqrt{1-z}}\frac{1}{\sqrt{(z+\frac{1-\rho}{2})^2 - (\frac{1+\rho}{2})^2}}$$

and integrate by parts to derive

$$\int_\rho^0 \frac{dz}{\sqrt{1-z^2}\sqrt{z-\rho}} = \frac{1}{\sqrt{1-z}} log\Big|z + \frac{1-\rho}{2} + \sqrt{(z+1)(z-\rho)}\Big|_\rho^0$$
$$- \int_\rho^0 \frac{log|z + \frac{1-\rho}{2} + \sqrt{(z+1)(z-\rho)}|dz}{2(1-z)^{3/2}}.$$

By the above formula, we can obtain (b) and (c).

For large values of $|\rho|$, (d) and (e) follow from

$$J(\rho) = \int_{-1}^1 \frac{dz}{\sqrt{(1-z^2)(z-\rho)}} = \frac{\pi}{|\rho|^{1/2}} + \frac{c_1}{|\rho|^{3/2}} + O(\frac{1}{|\rho|^{5/2}}). \qquad \square$$

The representation formula will be used to deduce singular estimates which have a derivative gain along the characteristic $t - r =$ const. So we utilize characteristic coordinates $\xi = t - r$, $\eta = t + r$, and express $\partial_\eta f$ as

$$\partial_\eta f = \partial_\eta f_0 \frac{1}{r^{5/2}} \int_K \Lambda h(r', t')dr'dt' + \frac{J(1)}{\pi\sqrt{2}} \int_{\eta'=\eta} \sqrt{\frac{r'}{r}} h(r', t')d\xi', \qquad (2.35)$$

where

$$\Lambda = \frac{1}{2\sqrt{2}\pi} \frac{1}{r'^{1/2}} \Big\{\big[(r-t+t')^2 - r'^2\big] J' - rr'J\Big\},$$
$$= \frac{1}{2\sqrt{2}\pi} \frac{1}{r'^{1/2}} \Big[(\xi' - \xi)(\eta' - \xi)J' - rr'J\Big].$$

Let $K_1 = K \cap \{(t', r') : -1 \le \rho \le 1\}$, $K_2 = K \cap \{(t', r') : \rho \le -1\}$.

(L5) *The kernel* Λ *is bounded by*

$$|\Lambda| \le Cr\sqrt{r'} log\left(2 + \frac{r}{|\eta' - \xi|}\right) \qquad (2.36)$$

on K_1 *and by*

$$|\Lambda| \le \frac{Cr^{5/2}r'}{|\eta - \eta'||\xi - \eta'|^{1/2}(r + r')^{1/2}} \qquad (2.37)$$

on K_2. (See the proof in [332, 333].)

Applying the bounds deduced in the above lemmas we can establish the regularity of radial wave maps from $\mathbf{R}^{1+2}$ into geodesically convex target manifolds with smooth initial data, governed by

$$f_{tt} - f_{rr} - \frac{1}{r} f_r = A(f)(f_\xi, f_\eta) \perp T_f N, \tag{2.38}$$

$$f(0,r) = f_0(r),\ \partial_t f(0,r) = f_1(r).$$

For simplicity, we assume that the image of $f : \mathbf{R}^{1+2} \to N \hookrightarrow \mathbf{R}^k$ lies in a geodesically convex region of a compact target manifold N.

Theorem 2.3.3. *Under the above assumption, there exists $\epsilon_0 > 0$ such that if $||Df(0)||^2_{L^2} < \epsilon_0$, then the solution f of equation (2.38) is globally smooth.*

We can apply the previous three lemmas to prove this theorem; the proof is lengthy and technical (cf. [332, 333]).

2.4 Global Regularity (1): Maps into Spheres in High Dimensions

2.4.1 Main Result

Let $m \geq 1$, $n \geq 1$ be fixed integers throughout this and next sections. In this section, we present Tao's work [360] showing that the wave maps to a sphere S^n are globally smooth if the initial data are smooth and have small norm in the critical Sobolev space $\dot{H}^m$, in the high dimensions $m \geq 5$. In next section we present the result in [361] asserting that the wave maps to a sphere are globally smooth in the similar situation, in the low dimensions $m = 2, 3, 4$, which is much harder than the first case. Following Tao's work, a number of mathematicians generalized the result to larger classes of target manifolds. This includes the work of Klainerman-Rodnianskii [226] ($m \geq 5$), Nahmod-Stefanov-Uhlenbeck [275] (see Sect. 2.6), Shatah-Struwe [333] ($m \geq 4$), Krieger [233] on maps into hyperbolic space ($m = 3$), and Tataru [367] on maps into Riemannian manifolds for low and high dimensions (see Sect. 2.7).

Let $\mathbf{R}^{1+m}$ be $m + 1$ dimensional Minkowski space with the metric $g = diag(-1, 1, \cdots, 1)$, and let S^n be the unit sphere in the Euclidean space $\mathbf{R}^{n+1}$. We denote by $\Box = \partial_\alpha \partial^\alpha = \Delta - \partial_t^2$ for the wave operator, and use $\dot{f}$ for $\partial_t f$. A wave map $f : \mathbf{R}^{1+m} \to S^n$ satisfies the equation

$$\partial_\alpha \partial^\alpha f = -f \partial_\alpha f^* \partial^\alpha f \tag{2.39}$$

in the sense of distributions, where f^* is the adjoint of f. For (2.39) to make sense, we require that f lies in $C_t^1 L_x^2 \cap C_t^0 H_x^1$ (it is easy to get regularity in the higher dimensional case). For any time t, we denote by $f[t] := (f(t), \dot{f}(t))$ the position and velocity of f at time t, and also denote by $f[0] = (f(0), \dot{f}(0))$ the initial datum of f. Suppose that $f[0]$ lies on the sphere, i.e., the initial datum $f[0]$ satisfies the following conditions

$$f^*(0)f(0) = 1,\ f^*(0)\dot{f}(0) = 0. \tag{2.40}$$

It is clear that relations (2.40) are preserved in time for smooth solutions (by Gronwall's inequality). The following main theorem and result were obtained by Tao [360].

Theorem 2.4.1. *Let $m \geq 5$ and $s > m/2$, and assume that $f[0] = (f(0), \dot{f}(0)) \in H^s \times H^{s-1}$ and has sufficiently small $\dot{H}^{m/2} \times \dot{H}^{m/2-1}$ norm. Then the solution for (2.39) with initial datum f[0] can be continued in $H^s \times H^{s-1}$ globally in time. In particular, smooth solutions remain smooth provided the initial data have small $\dot{H}^{m/2} \times \dot{H}^{m/2-1}$ norm. Moreover, if $|s - m/2| < 1/2$, we have the global bounds*

$$\|f[t]\|_{L_t^\infty(\dot{H}_x^s \times \dot{H}_x^{s-1})} \preceq \|f[0]\|_{\dot{H}_x^s \times \dot{H}_x^{s-1}}. \tag{2.41}$$

The techniques for proving Theorem 2.4.1 are Littlewood-Paley decomposition, Strichartz estimates (cf. [363]), some geometric equations (e.g. $f^*\partial_\alpha f = 0$), and a coordinate frame constructed by approximate parallel transport; this renormalization is important in order to remove the logarithmic divergence. The proof of the theorem relies on the geometric structure of the sphere, and it is much easier for the high dimension case $m \geq 5$ than the low dimension case, due to the strong decay of solutions to the wave equation (t^{-2} or better) and the rarity of parallel interactions.

2.4.2 Littlewood-Paley Projections and Strichartz Estimates

Let $f(t, x)$ be a function on $\mathbf{R}^{1+m}$. The *spatial* Fourier transform $\hat{f}(t, \xi)$ is defined by

$$\hat{f}(t, x) = \int_{\mathbf{R}^m} e^{-2\pi i x \cdot \xi} f(t, x) dx.$$

Let $\mu(\xi)$ be a non-negative radial bump function supported on $|\xi| \leq 2$ which equals 1 on the ball $|\xi| \leq 1$. For each k, we define the Littlewood-Paley operators $P_{\leq k} = P_{<k+1}$ onto the frequency ball $|\xi| \preceq 2^k$ (means $\xi \leq \text{const} \cdot 2^k$) by

$$\widehat{P_{\leq k}(f)}(\xi) = \mu(2^{-k}\xi)\hat{f}(t, \xi),$$

and the projection operators P_k to the frequency annulus $|\xi| \sim 2^k$ by $P_k = P_{\leq k} - P_{<k}$. We also define $P_{k_1 \leq \cdot \leq k_2} = P_{\leq k_2} - P_{<k_1}$, etc. Note that $P_k \to I$ in L^2 as $k \to \infty$, while $P_k \to 0$ in L^2 as $k \to -\infty$.

A pair of (q, r) of exponents are *admissible* if $2 \leq q, r \leq \infty$ and $\frac{1}{q} + \frac{m-1}{2r} \leq \frac{m-1}{4}$. For any integer $k \in \mathbf{Z}$, the *($\dot{H}^{m/2}$-normalized) Strichartz space $S_k(\mathbf{R}^{1+m})$ at frequency 2^k* is defined to be the space of functions on space-time whose norm is

$$\|f\|_{S_k} = \sup_{(q,r)\in F} 2^{k(\frac{1}{q}+\frac{m}{r})}(\|f\|_{L_t^q L_x^r} + 2^{-k}\|\partial_t f\|_{L_t^q L_x^r}),$$

where $F = \{(q,r)|2 \le q,r \le \infty, \frac{1}{q} + \frac{m-1}{2r} \le \frac{m-1}{4}\}$ is the set of admissible Strichartz exponents. Likewise, define $S_k(I \times \mathbf{R}^m)$ for the time interval I. In general, the large values (resp. small) of r are good for low-frequency (resp. high-frequency) terms. In high dimensions $m \ge 5$ we have a large set of Strichartz estimates which suffice for our objectives.

We only use specific values of q and r in our discussion. We notice that the control of S_k norm provides the following estimates:

$$||f||_{L_t^2 L_x^{2(m-1)/(m-3)}} \le 2^{-\frac{k}{2}-\frac{mk}{2}+\frac{mk}{m-1}} ||f||_{S_k}, \tag{2.42}$$

$$||f||_{L_t^2 L_x^4} \le 2^{-\frac{k}{2}-\frac{mk}{4}} ||f||_{S_k}, \tag{2.43}$$

$$||f||_{L_t^2 L_x^{m-1}} \le 2^{-\frac{k}{2}-\frac{mk}{m-1}} ||f||_{S_k}, \tag{2.44}$$

$$||f||_{L_t^2 L_x^\infty} \le 2^{-\frac{k}{2}} ||f||_{S_k}, \tag{2.45}$$

$$||f||_{L_t^4 L_x^{2(m-1)}} \le 2^{-\frac{k}{4}-\frac{mk}{2(m-1)}} ||f||_{S_k}, \tag{2.46}$$

$$||f||_{L_t^\infty L_x^\infty} \le ||f||_{S_k}, \tag{2.47}$$

$$||f||_{L_t^\infty L_x^2} \le 2^{-\frac{mk}{2}} ||f||_{S_k}. \tag{2.48}$$

Remark that in (2.43) and (2.44) we use the assumption $m \ge 5$. Similarly, we can also estimate the time derivative f_t in the above norms by playing the power of 2^k.

Assume that f is a smooth function and ψ is a rough function. The non-linearity is cubic in f and ψ with two derivatives somewhere. In order to estimate a term like $f \nabla f \nabla \psi$ in $L_t^1 L_x^2$, we usually estimate $\nabla \psi$ using (2.48) and f, and ∇f using (2.45); this works as long as the f term has equal or higher frequency than ∇f. The estimate (2.43) is helpful for achieving $L_t^1 L_x^2$ control on terms like $\nabla \psi \nabla \psi f$ which are quadratic in the high frequencies, and (2.42), (2.44) are helpful for handling terms such as $f \nabla^2 f \psi$. Furthermore, (2.42) and (2.46) can control the term $\nabla f \nabla f \psi$. There is some flexibility in our choice of exponents, especially in higher dimensions (the endpoint (2.42) can be avoided if $m \ge 6$). Then we have the following Strichartz estimates by Keel and Tao (cf. [216]).

Proposition 2.4.2 (Strichartz estimates). *Let k be an integer, and f be any function on $\mathbf{R}^{1+m}$ with spatial Fourier support in the annulus $|\xi| \sim 2^k$. Then*

$$||f||_{S_k} \preceq ||f[0]||_{\dot{H}_x^{m/2} \times \dot{H}_x^{m/2-1}} + 2^{-k+\frac{mk}{2}} ||\Box f||_{L_t^1 L_x^2}. \tag{2.49}$$

Likewise, f can be any function on $I \times \mathbf{R}^m$ for an interval I containing the origin.

If f, ψ are two functions, and ψ is much rougher (i.e. higher frequency) than f, then $(\nabla f)\psi$ is very small compared to $f \nabla \psi$. Because of this, one is able to ignore terms in which derivatives fail to fall on rough functions, and land instead on smooth ones. (In fact, these terms can be dealt with by Strichartz estimates.) Therefore, one has $\nabla(f\psi) \approx f \nabla \psi$ (i.e., f is approximately constant compared to ψ).

Suppose that the wave map f takes the form $f = \tilde{f} + \epsilon\psi$, where $\tilde{f}$ is a smooth wave map, $0 < \epsilon \ll 1$ and ψ is a $H^{m/2}$ function which is much rougher that $\tilde{f}$ (i.e., f is a small rough perturbation of a smooth wave map). If one neglects terms which are quadratic or better in ϵ, or which cannot differentiate the rough function ψ, one has the linearized equation

$$\partial_\alpha \partial^\alpha \psi = -2\tilde{f}\partial_\alpha \tilde{f}^* \partial^\alpha \psi \tag{2.50}$$

for ψ. Moreover, since $\tilde{f}$ and $\tilde{f} + \epsilon\psi$ lie on the sphere, one has that

$$\tilde{f}^*\psi = 0, \quad \tilde{f}^*\partial_\alpha\psi = 0 \tag{2.51}$$

(again neglecting the terms quadratic in ϵ, and terms where the derivative fails to land on ψ).

In order to keep the H^s norm of $\tilde{f} + \epsilon\psi$ from blowing up, we require to prevent the $H^{m/2}$ norm from being transferred from $\tilde{f}$ to $\epsilon\psi$. Especially, we need $L_t^\infty H_x^{m/2}$ bounds on ψ which are independent of ϵ. One also would like the corresponding Strichartz estimates for ψ, in order to handle the error terms that we neglected.

Though it is linear, equation (2.50) is not well-behaved without obvious cancellation structure (beyond the null form, which is not helpful in the high dimensional case). In order to iterate away the first order terms on the right-hand side of (2.50) one would like $\tilde{f}\partial_\alpha \tilde{f}^*$ in $L_t^1 L_x^\infty$. This might be doable if one had the Strichartz estimate $\nabla^{1/2}\tilde{f} \in L_t^2 L_x^\infty$; however, this estimate just barely fails to hold due to a logarithmic divergence in the frequencies. But, if one could ensure that the derivative in $\tilde{f}\partial_\alpha \tilde{f}^*$ always falls on a low-frequency component of $\tilde{f}$ and not on a high frequency component then one would have a chance of iterating away the non-linearity. This will be achieved by a renormalization using a coordinate frame adapted to f.

Applying (2.51), one can rewrite (2.50) in a parallel transport form:

$$\partial_\alpha \partial^\alpha \psi = 2A_\alpha \partial^\alpha \psi, \tag{2.52}$$

where A_α is the anti-symmetric matrix

$$A_\alpha = \partial_\alpha \tilde{f}\tilde{f}^* - \tilde{f}\partial_\alpha \tilde{f}^*. \tag{2.53}$$

Remark that (2.52) can have more cancellation than (2.50), since A_α is anti-symmetric. To solve (2.52), one considers the similar ODE

$$\ddot{\psi} = 2A_0\dot{\psi} \tag{2.54}$$

The matrix A_0 is anti-symmetric. Let $U(t)$ be the matrix-valued function solving the ODE $\dot{U}(t) = A_0 U(t)$ with $U(0)$ equal to the identity matrix. Then $\frac{d}{dt}(UU^*) = 0$ and U stays orthogonal for all time. Actually, one can see U as the parallel transport of the identity matrix along the path of $\tilde{f}$. Moreover, since $\tilde{f}$ is smooth, one deduces that U is also smooth, and is in fact much smoother than ψ. One can then use the linear change of variable $\psi = Uw$, and neglect terms which can not differentiate the rough function w, and then rewrite (2.54) as the trivial equation $\ddot{w} = 0$.

The ODE (2.54) suggests that (2.53) might be simplified by applying some orthogonal matrix U to the wave ψ, or by looking at ψ in a carefully selected coordinate frame (this works for harmonic maps in [177]). One would like U to be carried by parallel transport by $\tilde{f}$ in all directions, and would like U to solve the PDE

$$\partial_\alpha U = A_\alpha U \tag{2.55}$$

for each α. If one makes the unsuitable hypothesis that U obeys (2.55) exactly for all α, one can substitute $\psi = Uw$ as above and neglect all terms which can not differentiate the rough function w to convert (2.52) to the free wave equation $\partial_\alpha \partial^\alpha w = 0$, which one understands how to solve.

However, the PDE system (2.55) is over-determined, and has no solution (because the parallel transport connection induced by $\tilde{f}$ has a small but non-zero curvature). However, one can use Littlewood-Paley theory to construct an approximate solution U to (2.55). More precisely, one applies the Littelwood-Paley decomposition $\tilde{f} = f_{-M} + \sum_{-M<k} f_k$, where M is a large number, f_{-M} is the portion of f corresponding to frequencies $|\xi| \preceq 2^{-M}$, and f_k is the portion of f corresponding to frequencies $|\xi| \sim 2^k$. Thus one defines $U = U_{-M} + \sum_{-M<k} U_k$, where U_{-M} is the identity matrix, and the U_k are defined recursively by

$$U_k = (f_k f_{<k}^* - f_{<k} f_k^*) U_{<k} \tag{2.56}$$

where

$$f_{<k} = f_{-M} + \sum_{-M<k'<k} f_k, \quad U_{<k} = U_{-M} + \sum_{-M<k'<k} U_k.$$

It turns out that the matrix U is approximately orthogonal and approximately satisfies (2.55), provided the $\dot{H}^{m/2}$ norm of U is sufficiently small and M is sufficiently large. Since f_k is a rougher function than $f_{<k}$, one can ignore terms where the derivative falls on $f_{<k}$ instead of f_k. Likewise, for U_k and $U_{<k}$. Hence, one can differentiate (2.56) to get

$$\partial_\alpha U_k \approx (\partial_\alpha f_k f_{<k}^* - f_{<k} \partial_\alpha f_k^*) U_{<k} \tag{2.57}$$

and (2.55) follows by adding the telescoping series (keep ignoring the same type of terms as previously). The approximate orthogonality of U is based on the fact from (2.56) that $U_k^* U_{<k} + U_{<k}^* U_k = 0$. Summing this with respect to k and telescoping, one deduces

$$U^*U = I + \sum_{k>-M} U_k^* U_k.$$

The sum in which now on the right-hand side turns out to be negligible if we assume $\tilde{f}$ is small in $\dot{H}^{m/2}$, because this implies, by the Sobolev embedding, that the L^∞ norms of the f_k (and thus U_k) are small in l^2. (A analogous discussion can be used to dispose of the error terms which were ignored in (2.57).) If one converts (2.52) using $\psi = Uw$ as before, one obtains a non-linear wave equation for w, but the terms in the non-linearity either contain expressions like $\sum_k U_k U_k^*$ which are quadratic in the frequency parameter k, or have all derivatives falling on smooth functions rather than rough ones. Both types of terms can be handled by Strichartz estimates.

2.4.3 *Main Proposition and Linearization*

We now fix σ to be a constant depending only on m such that $0 < \sigma < 1/2$. We also fix $0 < \epsilon \ll 1$ to be a small constant depending only on m, n, σ.

Definition 2.4.3. A *frequency envelope* is a sequence $c = \{c_k\}_{k \in \mathbf{Z}}$ of positive reals which satisfies the l^2 bound

$$||c||_{l^2} \preceq \epsilon \tag{2.58}$$

and the local constancy condition

$$2^{-\sigma|k-k'|} c_{k'} \preceq c_k \preceq 2^{\sigma|k-k'|} c_{k'} \tag{2.59}$$
$$(\text{i.e., } \chi^{(0)}_{k=k'} c_{k'} \preceq c_k \preceq \chi^{-(0)}_{k=k'} c_{k'} \text{ in Sect. 2.5}),$$

for all $k, k' \in \mathbf{Z}$. In particular, we have $c_k \sim c_{k'}$ as $k = k' + O(1)$. If c is a frequency envelope and (f, g) is a pair of functions on $\mathbf{R}^m$, we say that (f, g) lies underneath the envelope c if

$$||P_k f||_{\dot{H}^{m/2}} + ||P_k g||_{\dot{H}^{m/2-1}} \preceq c_k \tag{2.60}$$

for all $k \in \mathbf{Z}$.

Remark that if (f, g) lies underneath an envelope c, then

$$||(f, g)||_{\dot{H}^{m/2} \times \dot{H}^{m/2-1}} \preceq \epsilon. \tag{2.61}$$

Conversely, if (2.61) holds then there exists an envelope c such that f lies underneath, for example we can take

$$c_k = \sum_{k' \in \mathbf{Z}} 2^{-\sigma|k-k'|} (||P_{k'} f||_{\dot{H}^{m/2}} + ||P_{k'} g||_{\dot{H}^{m/2-1}}). \tag{2.62}$$

Main Proposition 2.4.4. *Let $0 < T < \infty$, c be a frequency envelope, and f be a wave map on $[0, T] \times \mathbf{R}^m$ such that $f[0]$ lies underneath the envelope c. If ϵ is sufficiently small, then*

$$||P_k f||_{S_k([0,T]\times\mathbf{R}^m)} \leq C_0 c_k \tag{2.63}$$

for all $k \in \mathbf{Z}$, where $C_0 \gg 1$ is an absolute constant depending only on m, n (i.e. independent of T, f, c, ϵ). It follows from (2.63), (2.48) and Definition 2.4.3 that $f[t]$ lies underneath the envelope $C_0 c$ for all $t \in [0, T]$.

By a slight modification we can rewrite (2.63) as

$$||P_k f||_{S_k([0,T]\times\mathbf{R}^m)} \leq 2C_0 c_k.$$

Since the differential operator $2^{-k}\nabla$ is bounded on frequencies $|\xi| \sim 2^k$, this inequality yields

$$||\nabla^j P_k f||_{S_k([0,T]\times\mathbf{R}^m)} \lesssim 2^{jk} 2C_0 c_k. \tag{2.64}$$

for all $k \in \mathbf{Z}$ and all j, with the implicit constant depending on j.

Let us verify (2.63). By scale-invariance (scaling T, c and f), it suffices to show that

$$||\psi||_{S_0} \leq C_0 c_0 \tag{2.65}$$

where we define $\psi = P_0 f$. By applying P_0 to (2.39) we obtain

$$\Box\psi = -P_0(f\partial_\alpha f^* \partial^\alpha f). \tag{2.66}$$

We want to transform this non-linear equation into the linearized equation (2.50), modulo acceptable errors.

A function E on $[0, T] \times \mathbf{R}^m$ has an *acceptable error* if

$$||E||_{L^1_t L^2_x} \preceq C_0^3 \epsilon c_0,$$

where we denote this by $E = \mathit{error}$.

In fact, we would like the whole non-linearity in (2.66) to be an acceptable error, because then we could use Proposition 2.4.2 to derive (2.65). Though we cannot quite to do this, we can show that almost all the non-linearity is an acceptable error, and the remaining term can be renormalized by an appropriate change of coordinates to be an acceptable error.

Let $\tilde{f} = P_{\leq -10} f$ be the regularization of f. Because f lies on the sphere, we get

$$||\tilde{f}||_{L^\infty_t L^\infty_x} \preceq ||f||_{L^\infty_t L^\infty_x} = 1 \tag{2.67}$$

By (2.45), (2.64) and the triangle inequality, we obtain

$$||\nabla \tilde{f}||_{L^2_t L^\infty_x} \preceq C_0 \epsilon. \tag{2.68}$$

Proposition 2.4.5 ([360]).

(1) *We have*

$$P_0(f\partial_\alpha f^*\partial^\alpha f) = 2\tilde{f}\partial_\alpha \tilde{f}^*\partial^\alpha \psi + error. \tag{2.69}$$

In particular, it follows from (2.66) that

$$\Box\psi = -2\tilde{f}\partial_\alpha \tilde{f}^*\partial^\alpha \psi + error. \tag{2.70}$$

(2) *We have the estimate* $||\tilde{f}^*\partial^\alpha\psi||_{L_t^2 L_x^2} \preceq C_0^2\epsilon c_0$.

As a corollary of Proposition 2.4.5 (2), (2.68) and Hölder, we get $\partial_\alpha \tilde{f}\tilde{f}^*\partial^\alpha\psi =$ *error*. Therefore, by (2.70), we have the analogue

$$\Box\psi = 2A_\alpha\partial^\alpha\psi + error \tag{2.71}$$

of (2.52), where $A_\alpha = \partial_\alpha \tilde{f}\tilde{f}^* - \tilde{f}\partial_\alpha \tilde{f}^*$ is an anti-symmetric $n \times n$ matrix.

2.4.4 Approximate Parallel Transport and Summary

2.4.4.1 Approximate Parallel Transport

We now construct a matrix field U which is approximately orthogonal and which will re-normalize (2.71) into a better form, namely, $\Box w = error$.

We use the process described earlier. Let M be a large integer (depending on T) to be selected later and define the real $(n \times n)$-matrix valued field U by

$$U = I + \sum_{-M<k\leq -10} U_k,$$

where I is the identity matrix and U_k are defined successively by

$$U_k = ((P_k f)(P_{<k} f^*) - (P_{<k} f)(P_k f^*))U_{<k}, \tag{2.72}$$

and

$$U_{<k} = I + \sum_{-M<k'<k} U_{k'}.$$

Proposition 2.4.6. *Suppose that ϵ is sufficiently small depending on C_0, and M is sufficiently large depending on T, C_0, ϵ. Then we have the almost orthogonality property*

$$||U^*U - I||_{L_t^\infty L_x^\infty},\ ||\partial_t(U^*U - I)||_{L_t^\infty L_x^\infty} \preceq C_0^2\epsilon. \tag{2.73}$$

In particular, if ϵ is sufficiently small depending on C_0, then U is invertible, and

$$||U||_{L_t^\infty L_x^\infty}, ||U^{-1}||_{L_t^\infty L_x^\infty} \preceq 1. \tag{2.74}$$

We also have the approximate parallel transport property

$$||\partial_\alpha U - A_\alpha U||_{L_t^1 L_x^\infty} \preceq C_0^2 \epsilon, \tag{2.75}$$

and bounds (required to control error terms)

$$||\partial_\alpha U||_{L_t^\infty L_x^\infty} \preceq C_0^2 \epsilon, \tag{2.76}$$

$$||\partial_\alpha U||_{L_t^2 L_x^\infty} \preceq C_0^2 \epsilon, \tag{2.77}$$

$$||\Box U||_{L_t^2 L_x^{m-1}} \preceq C_0^2 \epsilon, \tag{2.78}$$

for all α. (See the proof in [360].)

2.4.4.2 Summary

By applying Proposition 2.4.6, we can prove (2.65), and hence conclude Proposition 2.4.4. Since U is invertible, we can write $\psi = Uw$ for some w, which is smooth by our hypotheses. By (2.74), (2.76) and the Leibniz's rule in the time variable we get

$$||\psi||_{S_0} \preceq ||w||_{S_0}$$

and it is sufficient to show that

$$||w|| \ll C_0 c_0. \tag{2.79}$$

We expand (2.71) by Leibniz's rule as

$$U\Box w + 2\partial_\alpha U \partial^\alpha w + (\Box U)w = 2A_\alpha U \partial^\alpha w + 2A_\alpha(\partial^\alpha U)w + error.$$

By (2.74), $U^{-1}error = error$, and we can rewrite the above equation as

$$\Box w = -2U^{-1}(\partial_\alpha U - A_\alpha U)\partial^\alpha w + 2U^{-1}A_\alpha(\partial^\alpha U)U^{-1}\psi - U^{-1}(\Box U)U^{-1}\psi + error. \tag{2.80}$$

We now claim that all the terms on the right-hand side are of the form *error*. In order to control the first term, it suffices, by (2.74) and (2.75) (if ϵ is sufficiently small depending on C_0), to show that

$$||\partial^\alpha w||_{L_t^\infty L_x^2} \preceq C_0 c_0.$$

Since $\psi = Uw$ and $\partial^\alpha \psi = U\partial^\alpha w + (\partial^\alpha U)w$, we obtain

$$\partial^\alpha w = U^{-1}\partial^\alpha \psi + U^{-1}(\partial^\alpha U)U^{-1}\psi.$$

Hence, the claim follows from (2.74), (2.48) and (2.64) for the first term, and (2.74), (2.76), (2.48) and (2.64) for the second term.

To deal with the second term in (2.80), we note that (2.67) and (2.68) yield

$$||A_\alpha||_{L_t^2 L_x^\infty} \preceq C_0 \epsilon,$$

and (2.45) and (2.48) yield

$$||\psi||_{L_t^\infty L_x^2} \preceq C_0 c_0.$$

Hence, the claim follows from (2.74) and (2.77).

To deal with the third term in (2.80), we note that (2.42) and (2.64) yield

$$||\psi||_{L_t^2 L^{2(m-1)/(m-3)}} \preceq C_0 c_0.$$

Then the claim follows from (2.78) and (2.74). Thus, we have shown that $\Box w =$ *error*, that is

$$||\Box w||_{L_t^1 L_x^2} \preceq C_0^3 \epsilon c_0.$$

By (2.74) and the hypothesis on $\psi[0]$, we also have

$$||w[0]||_{L_x^2} \preceq c_0.$$

So far, we can obtain (2.79) from Proposition 2.4.2, but w is not quite supported on the frequency annulus $|\xi| \sim 1$ (we have Fourier support control on U, but not on U^{-1}). But, we can apply $P_{-10<.<10}$ to the previous estimates and apply Proposition 2.4.2 to assert that

$$||P_{-10<.<10} w||_{S_0} \preceq C_0 c_0 \tag{2.81}$$

(if C_0 is sufficiently large depending only on m, n, σ, and ϵ is sufficiently small depending on C_0).

We start with the identity

$$w = U^*\psi - (U^*U - I)w$$

to pass from (2.81) to (2.79). Examining the Fourier support of U and ψ, we see that $U^*\psi$ has Fourier support in the annulus $2^{-5} \leq |\xi| \leq 2^5$, which implies that

$$(1 - P_{-10<.<10})w = -(1 - P_{-10<.<10})(U^*U - I)w.$$

Applying S_0 norms of both sides, we observe that

$$||(1 - P_{-10<.<10})w||_{S_0} \preceq C_0^2 \epsilon^2 ||w||_{S_0}.$$

It follows from (2.73) and Hölder that

$$||w - P_{-10<.<10}w||_{S_0} \preceq ||(U^*U - I)w||_{S_0}.$$

Then (2.81) implies (2.79), provided that C_0 is sufficiently large and ϵ is sufficiently small. This completes the proof of Proposition 2.4.4 and hence of Theorem 2.4.1.

2.5 Global Regularity (2): Maps into Spheres in Low Dimensions

We provide an outline of Tao's work [361] (109 pages), which is rather lengthy and technical. Since the global regularity of wave maps into spheres in low dimensions is much harder to establish than that of maps into spheres in high dimensions discussed in the last section, it is impossible to give here all the details. However, we try to present the crucial ingredients of the topics.

2.5.1 Main Result

Following the notations from last section, let $\mathbf{R}^{1+m}$ be $(m+1)$-dimensional Minkowski space with the metric $g = diag(-1, 1, \cdots, 1)$, and let S^n be the unit sphere in the Euclidean space $\mathbf{R}^{n+1}$. A (classical) wave map $f : \mathbf{R}^{1+m} \to S^n$ is a function defined on an open set in $\mathbf{R}^{1+m}$ with values on S^n which is smooth, is constant outside of a finite union of light cones, and satisfies

$$\Box f = -f \partial_\alpha f^* \partial^\alpha f \tag{2.82}$$

in the sense of distributions, where f^* is the adjoint of f. Recall that $f[t] := (f(t), \dot{f}(t))$ is the position and velocity of f at time t, $f[0] = (f(0), \dot{f}(0))$ is the initial datum of f. In order to let $f[0]$ be the initial datum for a (classical) wave map, $f[0]$ has to be smooth, constant outside of a compact set, and obey the following conditions:

$$f^*(0)f(0) = 1, \; f^*(0)\dot{f}(0) = 0.$$

Let $\dot{H}^s = \sqrt{-\Delta}^{-s} L^2(\mathbf{R}^m)$ be the usual homogeneous Sobolev spaces. The following main theorem and results were obtained by Tao [361].

Theorem 2.5.1. *Suppose that f[0] is a classical initial datum which has sufficiently small $\dot{H}^{m/2} \times \dot{H}^{m/2-1}$ norm for $m \geq 2$. Then f can be extended to a classical wave map globally in time. Moreover, if f is sufficiently close to m/2, we have the global bounds*

$$||f[t]||_{L_t^\infty(\dot{H}_x^s \times \dot{H}_x^{s-1})} \preceq ||f[0]||_{\dot{H}_x^s \times \dot{H}_x^{s-1}}. \tag{2.83}$$

The above theorem was proved in the high dimensions $m \geq 5$ in last section. In low dimensions $m = 2, 3, 4$, we will continue to use the Littlewood-Paley theory and the adapted coordinate frame constructed by parallel transport, but will discard the Strichartz estimates because the range of these estimates is too restrictive to be used, especially for $m = 2$. Instead, we will adapt the more complicated spaces $\dot{X}^{s,b}$ and estimates developed by Tataru [365], as substitutes for the Strichartz estimates. This will lengthen the discussion, although the overall strategy is almost the same. The main difficulty is that multiplication by $L_t^\infty L_x^\infty$ functions is not well-behaved on $\dot{X}^{s,b}$ spaces, and so we will need to substitute $L_t^\infty L_x^\infty$ with a more complicated Banach algebra.

In the preceding section, the non-linearity was settled and placed (after localizing in frequency and exchanging to the adapted coordinate frame) in the space $L_t^1 \dot{H}_x^{m/2-1}$. If $m \geq 5$, this is easy to achieve, since we can apply $L_t^2 L_x^4$ and $L_t^2 L_x^\infty$ Strichartz estimates. For $m = 4$ we lose the $L_t^2 L_x^4$ estimate but we could use $\dot{X}^{s,b}$ and null form estimates (which replace $f_{,\alpha}^* f^{,\alpha}$ in a space like $L_t^2 L_x^2$) as a substitute, so that we could continue to place the non-linearity in good spaces, such as $L_t^1 \dot{H}_x^{m/2-1}$. For $m = 3$ we (barely) lose the $L_t^2 L_x^\infty$ estimate, though this could be compensated by the L^p null form estimates in [362, 409] for some $p < 2$. But, in the energy-critical case $m = 2$, the only Strichartz estimate available is $L_t^4 L_x^\infty$, and it seems that even the best possible L^p null form estimates are not strong enough to place the non-linearity in a space like $L_t^1 L_x^2$, even after applying the adapted coordinate frame and $\dot{X}^{s,b}$ type spaces.

Due to the above reason, we can only place a small portion of the non-linearity in $L_t^1 L_x^2$. Following [365], we will place the other portions of the non-linearity either in an $\dot{X}^{s,b}$ type space, or in $L_t^1 L_x^2$ spaces corresponding to the null frames. In order to control the non-linearity, we will use null-norm estimates and the decomposition (cf. [365]) of free solutions as a superposition of traveling waves, each of which is in $L_t^2 L_x^\infty$ via a certain null frame. This decomposition, combined with the $L_t^2 L_x^2$ control coming from $\dot{X}^{s,b}$ estimates, is important for recovering the $L_t^1 L_x^2$ type control of the non-linearity. The high dimensional case in the last section does not require the null structure. However, for the low dimensional case, we need to exploit the identity

$$2 f_{,\alpha} \psi^{,\alpha} = \Box(f\psi) - f\Box\psi - \Box f \psi, \tag{2.84}$$

(see [363, 366]). This identity is helpful when $f, \psi, f\psi$ are relatively close to the light cone in frequency space, and less helpful, when one is far away from the light cone. In order to prove Theorem 2.5.1, we require the following proposition.

Proposition 2.5.2. *Let $T_0 > 0$ and c be a frequency envelope, and assume that f is a wave map on $[-T_0, T_0] \times \mathbf{R}^m$ such that $f[0]$ lies underneath ϵc. Then $f[t]$ lies underneath Cc for all $t \in [-T_0, T_0]$, where C is an absolute constant depending only on m, n.*

Proposition 2.5.2 implies Theorem 2.5.1 as follows: By the regularity theorem in [227, 366], it is sufficient to show that (2.83) holds in $[-T_0, T_0]$ for all wave maps f on $[-T_0, T_0] \times \mathbf{R}^m$ whose initial data $f[0]$ have $\dot{H}^{m/2} \times \dot{H}^{m/2-1}$ norm $\ll \epsilon^2$. If f is such a wave map, we can define the envelope c by

$$c_k = \epsilon^{-1} \sum_{k' \in Z} 2^{-\sigma_0|k-k'|} ||f_{k'}[0]||_{\dot{H}^{m/2} \times \dot{H}^{m/2-1}}. \tag{2.85}$$

It is easy to verify that (2.58)–(2.60) hold, so that $f[0]$ lies underneath ϵc. By Young's inequality on l^2, we also have

$$\Big(\sum_k (2^{\sigma k} c_k)^2\Big)^{1/2} \sim ||f[0]||_{\dot{H}^{m/2+\sigma} \times \dot{H}^{m/2-1+\sigma}}$$

for all $|\sigma| \leq \delta_0$. It follows from Proposition 2.5.2 that $f[t]$ lies underneath Cc for $t \in [-T_0, T_0]$, which implies from (2.59) that

$$||f[t]||_{\dot{H}^{m/2+\sigma} \times \dot{H}^{m/2-1+\sigma}} \precsim \Big(\sum_k (2^{\sigma k} c_k)^2\Big)^{1/2}.$$

This yields (2.83), and completes the proof of Theorem 2.5.1. □

2.5.2 $\dot{X}^{s,b}$ Type Spaces

We first introduce a few notations. Let $0 < \delta_0 \ll \delta_1 \ll \delta_2 \ll \delta_3 \ll \delta_4 \ll 1$. We choose $0 < \delta_4 \ll 1$ to be a small absolute constant depending on m ($\delta_4 = \frac{1}{1{,}000m}, \frac{1}{100m}$, etc.), and then put $\delta_i = \delta_{i+1}^{10}$ for $i = 3, 2, 1, 0$. We assume that δ_4 is sufficiently small. Let j and k be integers and $i = 0, 1, 2, 3, 4$. Designate by $\chi^{(i)}_{j \leq k}$ or $\chi^{(i)}_{k \geq j}$ a quantity of the form min $(1, 2^{\delta(j-k)})$, where $\delta > C^{-1}\delta_i^2$ for some absolute constant $C > 0$ depending on m. Designate by $\chi^{(i)}_{j=k}$ a quantity of the form $2^{-\delta|j-k|}$ with the same assumption on δ. So $\chi^{(i)}_{j \leq k}$ is small if $j \leq k + O(1)$, and $\chi^{(i)}_{j=k}$ is small if $j = k + O(1)$. Likewise, designate by $\chi^{-(i)}_{j \leq k} = \chi^{-(i)}_{k \geq j}$ a quantity of the form max $(1, 2^{\delta(j-k)})$ where $\delta < C\delta_i^{1/2}$ for some constant $C > 0$ depending only on m, and also designate by $\chi^{-(i)}_{j \leq k}$ the quantity of the form $2^{\delta(j-k)}$ with the same assumption on δ. So the $\chi^{(i)}$ represent exponential gains in our estimates, while $\chi^{-(i)}$ represent various losses. Notice that a $\chi^{(i)}$ dominates a corresponding $\chi^{-(j)}$ loss if $i > j$.

In order to prove Proposition 2.5.2, we require to introduce Banach spaces $S(c)$, S_k, N_k to iterate in. We first need to study $\dot{X}^{s,b}$ type spaces to define these three spaces.

If $X(\mathbf{R}^{1+m})$ is a Banach space of functions on $\mathbf{R}^{1+m}$ and $T \geq 0$, we define $X([-T,T] \times \mathbf{R}^m)$ to consist of the restrictions of functions in X to $[-T,T] \times \mathbf{R}^m$, and equip it with the norm

$$||f||_{X([-T,T]\times\mathbf{R}^m)} = \inf\{||F||_{X(\mathbf{R}^{1+m})} : F|_{[-T,T]\times\mathbf{R}^m} = f\}.$$

We now work globally in the space-time $\mathbf{R}^{1+m}$, and use the Fourier transform

$$\mathcal{F}f(\tau,\xi) = \int\int e^{-2\pi i(x\cdot\xi+t\tau)} f(t,x)dt\,dx.$$

We define the non-negative quantities D_0, D_+, D_- on the frequency space $\{(\tau,\xi) : \tau \in \mathbf{R}, \xi \in \mathbf{R}^m\}$ by

$$D_0 = |\xi|,\ D_+ = |\xi| + |\tau|,\ D_- = ||\tau| - |\xi||.$$

Note that $D_+ \sim D_0 + D_-$. We also define the direction $\Theta \in S^{m-1}$ by $\Theta = \tau\xi/|\tau\xi|$ if $\tau\xi \neq 0$, and $\Theta = 0$ otherwise. We regard D_0 as the *frequency*, D_- as the *modulation*, and Θ as the *direction*. We often use D_0, D_-, D_+ and Θ to define various regions in frequency space, for example, $\{D_0 \sim 2^k, D_- \sim 2^j\}$ denotes the frequency region $\{(\tau,\xi) : |\xi| \sim 2^k, ||\xi| - |\tau|| \sim 2^j\}$. Most activities will happen in the region $D_0 \sim D_+$ and we can think of these two as being equivalent. Usually, the region $D_+ \gg D_0$ will generate numerous minor sub-cases in the estimates, which often can be disposed of quickly.

We can define the Littlewood-Paley projection adapted to the light cone using the space-time Fourier transform. For any integer j, we define the projection operator $Q_{\leq j} = Q_{<j+1}$ by

$$\mathcal{F}(Q_{\leq j}f)(\tau,\xi) = \mu_0(2^{-j}D_-)\mathcal{F}(\tau,\xi).$$

Likewise, define $Q_j, Q_{\geq j}, Q_{j_1\leq\cdot\leq j_2}$, etc., similarly to the corresponding P multipliers. For instance, Q_j has symbol $\mu(2^{-j}D_-)$. Note that we have the decomposition $f = \sum_j Q_j f$ for Schwartz functions f. For any sign $\pm$, define the operator $Q^\pm_{\leq j}$ to be the restriction of $Q_{\leq j}$ to the frequency region $\pm\tau \geq 0$, so $Q_{\leq j} = Q^+_{\leq j} + Q^-_{\leq j}$ and

$$\mathcal{F}(Q^\pm_{\leq j}f)(\tau,\xi) = \chi_{[0,\infty)}(\pm\tau)\mu_0(2^{-j}(\pm\tau - |\xi|))\mathcal{F}f(\tau,\xi).$$

Likewise, define $Q^\pm_j$, $Q^\pm_{\geq j}$, $Q^\pm_{j_1\leq\cdot\leq j_2}$, etc. Note that the sum of $Q^\pm_{\leq j} + Q^\pm_{>j}$ is not the identity, but rather the Riesz projection to the half-space $\{\pm\tau \geq 0\}$. Observe that $P_k, Q_j, Q^\pm_j, Q_{\leq j}$, etc., are all space-time Fourier multipliers and commute with each other and with constant-coefficient differential operators. The operator $P_k Q_j$ is the projection onto the functions with spatial frequency 2^k and modulation 2^j.

Definition 2.5.3. A space-time Fourier multiplier is called *disposable* if its (distributional) convolution kernel is given by a measure with total mass $O(1)$.

A disposable multiplier is bounded on all space-time translation-invariant Banach spaces thanks to Minkowski's inequality, which means that these operators can be discarded whenever we wish. Also, the composition of two disposable multiplier is also multiplier. The Littlewood-Paley operators P_k, $P_{\leq k}$, $P_{\geq k}$, etc., are disposable. However, the operators Q_j, $Q_j^{\pm}$, $Q_{\leq j}$, etc., are not disposable, but we have the following facts [361]:

1. If j and k are integers such that $j \geq k + O(1)$, then $P_{\leq k} Q_{\leq j}$, $P_k Q_{\leq j}$, $P_{\leq k} Q_j$, $P_k Q_{\leq j}$ are disposable multipliers. Moreover, if we assume $|k - j| > 10$, then $P_k Q_j^{\pm}$, $P_{\leq k} Q_{\leq j}^{\pm}$, $P_k Q_{\geq j}^{\pm}$ are disposable for either choice of sign $\pm$.
2. The operators Q_j, $Q_{\leq j}$, $Q_{\geq j}$, $Q_{j_1 \leq \cdot \leq j_2}$, etc., are bounded on the spaces $L_t^p L_x^2$ for all $1 \leq p \leq \infty$.
3. (Improvement to Bernstein's Inequality): If f has Fourier support in the region $D_0 \sim 2^k$, $D_- \sim 2^j$, then

$$||f||_{L_t^2 L_x^\infty} \precsim \chi_{j \geq k}^{(4)} 2^{mk/2} ||f||_{L_t^2 L_x^2}.$$

Definition 2.5.4. For any $s, b \in \mathbf{R}$ and $1 \leq q \leq \infty$, $\dot{X}_k^{s,b,q}$ is the completion of the space of all Schwartz functions with Fourier support in $2^{k-5} \leq D_0 \leq 2^{k+5}$ with respect to the norm

$$||f||_{\dot{X}_k^{s,b,q}} = 2^{sk} \left[\left(\sum_j 2^{bj} ||Q_j f||_{L_{t,x}^2} \right)^q \right]^{1/q} \quad (< \infty),$$

with the usual supremum convention if $q = \infty$.

For a first approximation, we consider f_k in spaces of strength comparable to $\nabla_{x,t}^{-1} \dot{X}_k^{m/2-1,1/2,q}$ for certain q, while $\Box f_k$ belongs to spaces of strength comparable to $\dot{X}_k^{m/2-1,1/2,q}$. The space $\dot{X}_k^{s,b,1}$ is an atomic space whose atoms are functions with space-time Fourier support in the region $\{2^{k-5} \leq D_0 \leq 2^{k+5}, D_- \sim 2^j\}$ and have an $L_t^2 L_x^2$ norm of order $O(2^{-sk} 2^{-j/2})$ for certain integer j. The proof of multi-linear estimates on these spaces then reduces to proving the estimates on atoms. However, only a portion of a wave map f can be placed into such a good space.

We notice the Plancherel duality

$$\sup\{| < f, \psi > | : ||f||_{\dot{X}_k^{s,b,1}} \leq 1\} \sim 2^{-k(s+s')} ||f||_{\dot{X}_k^{s',-b,\infty}} \tag{2.86}$$

for all k, s, s', b and Schwartz $\psi \in \dot{X}_k^{s',-b,\infty}$, where

$$< f, \psi > = \int\int f \bar{\psi} \, dx \, dt$$

is the standard inner product.

2.5.3 Null Frames

We discuss the null frame spaces $NKA[\kappa]$ and $S[\kappa, \kappa]$, which will play an important role in the construction of N_k and S_k.

We define a *spherical cap* to be any subset κ of S^{m-1} of the form

$$\kappa = \{\omega \in S^{m-1} : |\omega - \omega_k| < r_k\}$$

for some $\omega_k \in S^{m-1}$ and $0 < r_k < 2$. ω_k and r_k are called the center and radius of κ. If κ is a cap and $C > 0$, C_κ denotes the cap with the same center but C times the radius. If $\pm$ is a sign, $\pm\kappa$ mean the cap with the same radius and center $\pm\omega_\kappa$.

For any direction $\omega \in S^{m-1}$, we define the null direction $\theta_\omega = \frac{1}{\sqrt{2}}(1, \omega)$, and the null plane $NP(\omega)$ by

$$NP(\omega) = \{(t, x) \in \mathbf{R}^{1+m} : (t, x) \cdot \theta_\omega = 0\},$$

where $(t, x) \cdot (t', x') = tt' + x \cdot x'$ is the standard Euclidean inner product.

We can parametrize physical space $\mathbf{R}^{1+m}$ by the null coordinate $(t_\omega, x_\omega) \in \mathbf{R} \times NP(\omega)$, defined by

$$t_\omega = (t, x) \cdot \theta_\omega, \quad x_\omega = (t, x) - t_\omega \theta_\omega.$$

Likewise, we can parameterize frequency space by $(\tau_\omega, \xi_\omega) \in \mathbf{R} \times NP(\omega)$, defined by

$$\tau_\omega = (\tau, \xi) \cdot \theta_\omega, \quad \xi_\omega = (\tau, \xi) - \tau_\omega \theta_\omega.$$

We can then define Lebesgue space $L^q_{t_\omega} L^r_{x_\omega}$, $L^q_{\tau_\omega} L^r_{\xi_\omega}$, in the usual manner. The trivial identity

$$||f||_{L^2_t L^2_x} \sim ||f||_{L^2_{t_\omega} L^2_{x_\omega}}$$

is important for connecting the null frame estimates to the Euclidean frame estimates.

The null plane $NP(\omega)$ contains the null direction $(1, -\omega)$, but it does not produce nice energy estimates. For example, the control of $\Box f$ in $L^1_{t_\omega} L^2_{x_\omega}$ does not give a good control of f. But, if we know that f has Fourier support in a sector $\{\Theta \in \kappa\}$ and ω is outside of 2κ, then we can recover nice energy estimates in this coordinate system (loosing a factor of $\frac{1}{dist(\omega,\kappa)}$). Thus we are motivated to introduce a Banach space $NFA[\kappa]$ of null frame atoms oriented away from κ.

Definition 2.5.5. For any κ, let $NFA[\kappa]$ be the atomic Banach space whose atoms are functions F with

$$||F||_{L^1_{t_\omega} L^2_{x_\omega}} \leq dist(\omega, \kappa)$$

for some $\omega \notin 2\kappa$.

The space $NFA[\kappa]$ is a strange building block for our non-linearity space N_k, which was used in [363]. Because $NFA[\kappa]$ is defined using Lebesque spaces, we have the following estimates:

$$||f\psi||_{NFA[\kappa]} \preceq ||f||_{L^\infty_t L^\infty_x} ||\psi||_{NFA[\kappa]} \tag{2.87}$$

for all Schwartz f, ψ. We also note the nesting inequality

$$||F||_{NFA[\kappa]} \leq ||F||_{NFA[\kappa']}, \tag{2.88}$$

if $\kappa' \subset \kappa$.

Define $NFA^*[\kappa]$ to be the space of functions f whose norm

$$||f||_{NFA^*[\kappa]} = \sup_{\omega \notin 2\kappa} dist(\omega, \kappa) ||f||_{L^\infty_{t_\omega} L^2_{x_\omega}}$$

is finite. Notice that $NFA^*[\kappa]$ is the dual of $NFA[\kappa]$.

We next construct a Banach space $S[\kappa, \kappa]$ which in some sense is a dual space to $NFA[\kappa]$, and will be a crucial part of the construction of the spaces S_k and $S(c)$. We define the plane wave space $PW[\kappa]$ to be the atomic Banach space whose atoms are functions f with

$$||f||_{L^2_{t_\omega} L^\infty_{x_\omega}} \leq 1$$

for some $\omega \in \kappa$. Now we define $S[\kappa, \kappa]$ by means of the norm

$$||f||_{S[\kappa,\kappa]} = 2^{mk/2} ||f||_{NFA^*[\kappa]} + |\kappa|^{-1/2} 2^{k/2} ||f||_{PW[\kappa]} + 2^{mk/2} ||f||_{L^\infty_t L^2_x}. \tag{2.89}$$

2.5.4 Construction of S_k, $S(c)$ and N_k

In order to construct the spaces N_k, S_k, $S(c)$ required for Theorem 2.5.9, we need to define the frequency localized variants $N[k]$ and $S[k]$ as follows.

Definition 2.5.6. For each k, $S[k]$ is the completion of the space of Schwartz functions with Fourier support in the region $\{\xi \in R^m | 2^{k-3} \leq D_0 \leq 2^{k+3}\}$ with respect to the norm

$$||f||_{S[k]} = ||\nabla_{x,t} f||_{L_t^\infty \dot{H}_x^{m/2-1}} + ||\nabla_{x,t} f||_{\dot{X}_k^{m/2-1,1/2,\infty}}$$

$$+ \sup_{\pm} \sup_{l>10} \left(\sum_{\kappa \in K_l} ||P_{k,\pm\kappa} Q^{\pm}_{<k-2l} f||^2_{S[\kappa,\kappa]} \right)^{1/2}. \tag{2.90}$$

In fact, if $f \in \dot{X}_k^{m/2,1/2,1}$, then $||f||_{S[k]} \preceq ||2^{-k}\nabla_{x,t} f||_{\dot{X}_k^{m/2,1/2,1}}$. Thus the space $S[k]$ contains $\dot{X}_k^{m/2,1/2,1}$.

We define the full space $S(c)(\mathbf{R}^{1+m})$ to be the closure of the space of functions f such that $f - e$ is a Schwartz function for some constant e (then f is asymptotically constant) with respect to the norm

$$||f||_{S(c)(\mathbf{R}^{1+m})} = ||f||_{L_t^\infty L_x^\infty} + \sup_k c_k^{-1} ||f_k||_{S[k]}. \tag{2.91}$$

We can form the restricted space $S(c) = S(c)([-T,T] \times \mathbf{R}^m)$. We then define the space $S_k(\mathbf{R}^{1+m})$ to be the subspace of $S(c)(\mathbf{R}^{1+m})$ given by the norm

$$||f||_{S_k(\mathbf{R}^{1+m})} = \sup_{k'} 2^{\delta_1|k-k'|} ||f_k||_{S[k]}, \tag{2.92}$$

and define $S_k = S_k([-T,T]\times\mathbf{R}^m)$. So S_k is a larger variant of $S[k]$ which allows for some leakage out of the frequency region $D_0 \sim 2^k$. Most estimates in this section will contain a decay of $\chi^{(2)}$ or stronger, which is more than enough to overcome the leakage $2^{\delta_1|k-k'|} = \chi^{-(1)}_{k=k'}$. We next define $N[k]$ and N_k.

Definition 2.5.7. Let k be an integer, and let F be a Schwartz function with Fourier support in the region $2^{k-4} \le D_0 \le 2^{k+4}$. F is called an $L_t^1 \dot{H}_x^{m/2-1}$-atom at frequency 2^k if

$$||F||_{L_t^1 L_x^2} \le 2^{-(m/2-1)k}. \tag{2.93}$$

If $j \in \mathbf{Z}$, we say that F is a $\dot{X}^{m/2-1,-1/2,1}$-atom with frequency 2^k and modulation 2^j if F has Fourier support in the region $2^{k-4} \le D_0 \le 2^{k+4}$, $2^{j-5} \le D_- \le 2^{j+5}$, and

$$||F||_{L_t^2 L_x^2} \le 2^{j/2} 2^{-(m/2-1)k}. \tag{2.94}$$

If $l > 0$ is a real number and $\pm$ is a sign, F is called a $\pm$ null frame atom with frequency 2^k and angle 2^{-l} if there exists a decomposition $F = \sum_{k \in K_l} F_k$ such that each F_k has Fourier support in the region $\{(\tau,\xi) | \pm\tau > 0,\ D_- \le 2^{k-2l-50},\ 2^{k-4} \le D_0 \le 2^{k+4},\ \Theta \in 1/2k\}$ and

$$\Big(\sum_{k \in K_l} ||F_k||^2_{NFA[k]}\Big)^{1/2} \le 2^{-(m/2-1)k}. \tag{2.95}$$

Let $N[k]$ be the atomic Banach space generated by the $N[k]$ atoms.

Definition 2.5.8. Let F be a Schwartz function and k be an integer. We say that F is an N_k atom if there exists a $k' \in \mathbf{Z}$ such that $2^{100m|k-k'|}F$ is an $N[k]$ atom. We define $N_k(\mathbf{R}^{1+m})$ to be the atomic Banach space generated by the N_k atoms. We define $N_k = N_k([-T,T] \times \mathbf{R}^m)$ to be the restriction of $N_k(\mathbf{R}^{1+m})$ to the slab $[-T,T] \times \mathbf{R}^m$.

N_k is a very slight enlargement of $N[k]$ which barely allow us for some frequency leakage out of the region $D_0 \sim 2^k$.

2.5.5 *Iteration Space and Key Estimates*

Since we have defined S_k, $S(c)$, N_k, we can discuss the proof of Proposition 2.5.2, which is based on the following informal procedures. (1) Bootstrap assumption: By a continuity argument, one assumes a priori that the f_k already lie in S_k, and then boostrap this to better control on f_k in S_k. (2) Control of f: Because f lies on the sphere, it is in $L_t^\infty L_x^\infty$. Combining this with the assumption $f_k \in S_k$, one has $f \in S(c)$. (3) Construction of a gauge: One constructs for each frequency 2^k a gauge transform $U_{\le k}$, which is a polynomial in f and f_k. Using the preceding control on f_k and f, one can deduce that $U_{\le k} \in S(c)$ for all k. (4) Control of renormalized non-linearity: One defines $w_k = U_{\le k-10} f_k$, and use the previous estimates to show that $\Box w_k \in N_k$. (5) Energy estimates: One has $w_k[0] \in \dot{H}^{m/2} \times \dot{H}^{m/2-1}$ as well. Using this and the above step, one has $w_k \in S_k$. (6) Inverting the gauge: One then shows that $U_{\le k-10}$ is invertible in $S(c)$, and use this and the previous step to show that $f_k = U_{\le k-10}^{-1} w_k$ is in S_k, hence close the bootstrap circle. (7) Conclusion: One lastly shows that the control of $f_k \in S_k$ implies that f lies underneath the envelope C_c.

In order to make the above procedures work, we require a few estimates relating S_k, N_k, and $S(c)$, which will be listed in Theorem 2.5.9 [361]. One of the main tasks in proving Proposition 2.5.2 is to choose spaces $S(c)$, S_k, N_k which satisfy all these estimates.

In the high dimensional case $m \ge 5$, N_k was the energy method space $L_t^1 \dot{H}_x^{m/2-1}$ (localized to frequency 2^k), and the spaces S_k were Strichartz spaces. The relation with of the $S(c)$ norm in [360] was given by

$$||f||_{S(c)} \sim ||f||_{L_t^\infty L_x^\infty} + \sup_k c_k^{-1} ||f_k||_{S_k}.$$

The properties in Proposition 2.5.2 are easily verified (after some minor modification) using standard Strichartz estimates and Hölder's inequality when $m \ge 5$.

In the low dimensional case, we can not place the non-linearity in $L_t^1 \dot{H}_x^{m/2-1}$ even after re-normalization. Therefore, we must enlarge the space N_k, incorporating not only $L_t^1 \dot{H}_x^{m/2-1}$ but also $\dot{X}^{s,b}$ type norms, as well as some more complicated null frame spaces of Tataru [366]. Since we wish to prove energy estimates, this enlargement of N_k makes S_k more complicated; it shall involve $\dot{X}^{s,b}$ norms and null frame spaces. This also forces the space $S(c)$ to become a more sophisticated Banach algebra, similar to $\dot{X}^{m/2,1/2}$ but still controlling $L_t^\infty L_x^\infty$ and closed under multiplication. Luckily, examples of such algebras $S(c)$ exist (cf. [225, 363, 366]).

We summarize all the properties of the spaces $S(c)$, S_k, N_k that we require as follows.

Theorem 2.5.9 ([361]). *Let $T \geq 0$ and c be a frequency envelope. Then there exist Banach spaces $S(c) = S(c)([T,T] \times \mathbf{R}^m)$, $S_k = S_k(-T, T \times \mathbf{R}^m)$, and $N_k = N_k([-T,T] \times \mathbf{R}^m)$ for $k \in \mathbf{Z}$ of functions in $[-T,T] \times \mathbf{R}^m$ such that the following properties hold for all integers k, k_1, k_2, k_3:*

(1) *Quasi-continuity: Assume that f is a wave map on $[-T_0, T_0] \times \mathbf{R}^m$. Then the function*

$$a(T) = \max\left(1, c_k^{-1} \sup_k ||f_k|_{[-T,T]\times\mathbf{R}^m}||_{S_k([-T,T]\times\mathbf{R}^m)}\right) \tag{2.96}$$

defined on $[0, T_0]$ satisfies the quasi-continuity property

$$\limsup_{T'\to T} a(T') \preceq \liminf_{T'\to T} a(T')$$

for all $0 \leq T' \leq T_0$.

(2) *Invariance properties: The space S(c), S_k, N_k are invariant under spatial translations. For any $j \in \mathbf{Z}$, the scaling $f(t,x) \mapsto f(2^j t, 2^j x)$, $T \mapsto T/2^j$ maps $S(\{c_k\})$ to $S(\{c_{k-j}\})$ and S_k to S_{k+j}. Likewise, the scaling $F(t,x) \mapsto 2^{2j} F(2^j t, 2^j x)$, $T \mapsto T/2^j$ maps N_k to N_{k+j}.*

(3) *S(c) is a Banach algebra: The space S(c) is a Banach algebra with respect to pointwise multiplication, i.e., S(c) contains the identity 1 and satisfies the estimate*

$$||L(f, \psi)||_{S(c)} \preceq ||f||_{S(c)} ||\psi||_{S(c)} \tag{2.97}$$

for all $f, \psi \in S(c)$. Moreover,

$$||f||_{L_t^\infty L_x^\infty} \preceq ||f||_{S(c)}$$

for all $f \in S(c)$.

(4) *Frequency-localized algebra property: If $f, \psi \in S_k$, then*

$$||L(f, \psi)||_{S_k} \preceq ||f||_{S_k} ||\psi||_{S_k}. \tag{2.98}$$

Likewise, if $f \in S_k, \psi \in S(c)$ and ψ has Fourier support in $D_0 \preceq 2^k$, then

$$||L(f, \psi)||_{S_k} \preceq ||f||_{S_k} ||\psi||_{S(c)}.$$

(5) *S(c) insensitive to c: For all* $f \in S(c)$ *and* $C > 0$ *we have*

$$||f||_{S(Cc)} \sim ||f||_{S(c)}. \tag{2.99}$$

with the implicit constants depending at most polynomially on C.

(6) *S(c) is built up from* S_k*: Let f be a smooth function on* $[-T, T] \times \mathbf{R}^m$ *which is constant outside of a compact set. Assume that we have a decomposition* $f = \sum_k f^{(k)}$ *with each* $f^{(k)}$ *in* S_k*. Then we obtain*

$$||f||_{S(c)} \preceq ||f||_{L_t^\infty L_x^\infty} + \sup_k c_k^{-1} ||f^{(k)}||_{S_k}. \tag{2.100}$$

(7) N_k *contains* $L_t^1 \dot{H}_x^{m/2-1}$*: Let F be an* $L_t^1 L_x^2$ *function on* $[-T, T] \times \mathbf{R}^m$ *which has Fourier support on the region* $D_0 \sim 2^k$ *for some integer k. Then F is in* N_k *and*

$$||F||_{N_k} \preceq ||F||_{L_t^1 \dot{H}_x^{m/2-1}} \sim 2^{(m-1)k} ||F||_{L_t^1 L_x^2}. \tag{2.101}$$

(8) *Adjacent spaces* N_k *are equivalent: We have the compatibility property*

$$||F||_{N_{k_1}} \sim ||F||_{N_{k_2}}, \tag{2.102}$$

where $F \in N_k$ *and* $k_1 = k_2 + O(1)$.

(9) *Energy estimate: For any Schwartz function f on* $[-T, T] \times \mathbf{R}^m$ *with Fourier support in* $D_0 \sim 2^k$*, we have*

$$\begin{aligned} ||f||_{S_k} &\preceq ||\Box f||_{N_k} + ||f[0]||_{\dot{H}^{m/1} \times \dot{H}^{m/2-1}} \\ &\sim ||\Box f||_{N_k} + 2^{mk} ||f(0)||_{L^2} + 2^{(m-1)k} ||\partial_t f(0)||_{L^2}. \end{aligned} \tag{2.103}$$

(10) *Product estimates: We have*

$$||P_k L(f, F)||_{N_k} \preceq \chi^{(1)}_{k \geq k_2} ||f||_{S(c)} ||\psi||_{N_{k_2}} \tag{2.104}$$

for $f \in S(c)$ *and* $F \in N_{k_2}$*. Also,*

$$||P_k L(f, F)||_{N_k} \preceq \chi^{(1)}_{k \geq k_2} ||f||_{S_{k_2}} ||\psi||_{N_{k_2}} \tag{2.105}$$

for $f \in S_{k_2}$ *and* $F \in N_{k_2}$.

(11) *Null form estimates: We have*

$$||P_k L(f_{,\alpha}, \psi^{,\alpha})||_{N_k} \preceq \chi^{(1)}_{k = max(k_1, k_2)} ||f||_{S_{k_1}} ||\psi||_{S_{k_2}} \tag{2.106}$$

for $f \in S_{k_1}$ *and* $\psi \in S_{k_2}$.

(12) *Tri-linear estimate: We have*

$$||P_k L(f^{(1)}, f^{(2)}_{,\alpha}, f^{(3),\alpha})||_{N_k} \preceq \chi^{(1)}_{k=max(k_1,k_2,k_3)} \chi^{(1)}_{k \leq min(k_2,k_3)} \prod_{i=1}^{3} ||f^{(i)}||_{S_{k_i}},$$

for $f^{(i)} \in S_{k_i}$, $i = 1, 2, 3$.

(13) *Conclusion: For any* $f \in S_k$ *with Fourier support in* $D_0 \preceq 2^k$, *we obtain*

$$\sup_t ||f[t]||_{\dot{H}_x^{m/2} \times \dot{H}_x^{m/2-1}} \preceq 2^{mk/2} \sup_t ||f[t]||_{L_x^2 \times L_x^2} \preceq ||f||_{S_k}. \tag{2.107}$$

Remark that in all cases we have obtained the null structure of the non-linearity. In the low dimensions $m = 2, 3, 4$ this is crucial in order for the above non-linearities to be nice. In all of the above cases we have various exponential gains which allow us to sum with respect to the k_i indices.

At a first approximation, we consider these nice non-linearities as negligible errors. The purpose is to gauge transform (the Littlewood-Paley localized version of) (2.82), utilizing such geometric identities as $f^* f_{,\alpha} = 0$ as well as using a lemma (Leibnitz rule for P_k), namely, the identity

$$P_k(fg) = f P_k g + L(\nabla_x f, 2^{-k} g),$$

until all the non-linearities are negligible. In this situation, the Littlewood-Paley decomposition seems to play a key role, since it permits us to separate the core component of the non-linearity (which for wave maps is a connection term $A_{\alpha;\leq k}$ that has small curvature) from the remaining error terms which are nice non-linearities, and hence negligible. □

Let $S(c)$, S_k, N_k be defined as above. The next main proposition shows the "bootstrap" property of the S_k norms. All the arguments and results are based on [360, 361].

Proposition 2.5.10. *Let c be a frequency envelope,* $0 < T < \infty$, *and let f be a classical wave map on* $[T, -T] \times \mathbf{R}^m$, *extended to* $\mathbf{R}^{1+m}$ *by the free wave equation, such that* $f[0]$ *lies underneath* ϵc *and that*

$$||f_k||_{S_k} \preceq c_k$$

for all k. Then

$$||f_k||_{S_k} \leq c_k$$

for all k.

We now provide the continuity argument which derives Proposition 2.5.2 from Proposition 2.5.10. Let T_0, c, f be as in Proposition 2.5.2, and let $a(T)$ be the quantity in (2.96). By (2.103) and the assumption that f lies underneath ϵc, we have that $a(0) = 1$. By Proposition 2.5.10, we observe that if $0 < T \leq T_0$ obeys $a(T) \preceq 1$, then we can boostrap this bound to $a(T) = 1$. Combining this and (2.97), we see that the set $\{T \in [0, T_0] : a(T) = 1\}$ is both open and closed in $[0, T]$. Because this set contains the origin, we get $a(T_0) = 1$ and so, using (2.107), we conclude that $f[t]$ lies underneath Cc for all $0 \leq t \leq T_0$.

For the details of the proofs of Theorem 2.5.9 and Proposition 2.5.10 see [361].

2.6 Well-Posedness for Maps into Lie Groups in High Dimensions

Nahmod, Stefanov and Uhlenbeck [275] constructed a gauge theoretic change of variables for the wave maps from $\mathbf{R} \times \mathbf{R}^m$ into a compact Lie group or Riemannian symmetric space. They showed a new multiplication theorem for mixed Lebesgue-Besov spaces and proved the global well-posedness of a modified wave map system for $m \geq 4$ and small critical initial data. They obtained global existence and uniqueness for the Cauchy problem of wave maps into compact Lie groups and symmetric spaces with small initial data for $m \geq 4$. Their results generalized the results of Tao [360, 361] and Tataru [366]. The Shatah-Struwe's [333] methods using Lorentz spaces were stronger since they produced estimates for solutions with variable curvature (F. Planchon showed that the multiplication theorem for Besov spaces and L^∞ includes the variable curvature case).

2.6.1 *Formulation*

We consider the wave map equation as an equation given through covariant derivatives as follows: $f : \mathbf{R}^{1+m} \to N$ is a map from Minkowski space into a Riemannian manifold N and $df : T(\mathbf{R}^{1+m}) \to TN$ where $T(\mathbf{R}^{1+m}) = \mathbf{R} \times \mathbf{R}^m \times (\mathbf{R} \oplus \mathbf{R}^m)$. Let $f^{-1}\nabla$ be the pull-back of the Levi-Civita connection on N to $f^{-1}TN$ via the map f. Then the wave map equation reads

$$f^{-1}\nabla_0 \frac{\partial f}{\partial t} - \sum_{i=1}^{m} f^{-1}\nabla_i \frac{\partial f}{\partial x^i} = 0.$$

Since the Levi-Civita connection on N is torsion free, we have

$$f^{-1}\nabla_i \frac{\partial f}{\partial x^k} = f^{-1}\nabla_k \frac{\partial f}{\partial x^i} = 0,$$

for $i = 0, 1, \cdots, m$, $k = 1, \cdots, m$, and where we put $t = x^0$.

Suppose that the map f is topological trivial which will follow from curvature bounds given below. (The wave map fixes spatial infinity topologically, $f : \mathbf{R}^m \cup \{\infty\} \to N$.) Thus $f^{-1}TN$ is the trivial bundle $(\mathbf{R}^1 \times \mathbf{R}^m) \times \mathbf{R}^n$. Using the equation $[f^{-1}\nabla_i, f^{-1}\nabla_k] = R(f)(\frac{\partial f}{\partial x^i}, \frac{\partial f}{\partial x^k})$, we have control on the curvature of $f^{-1}\nabla$.

The following theorem verifies that there is a unique choice of coordinates for $f^{-1}TN$ under a smallness assumption on $f \in L_t^\infty \dot{W}_x^{1,m/2}$. Given a smooth map f with sufficient decay in asymptotics to a point at infinity, the initial coordinates can be formed by a partition of unity. The proposition needed is described in general, as we hope to find applications for this proposition in gauge theory. All the following results and theorems were obtained by Nahmod, Stefanov and Uhlenbeck [275].

Proposition 2.6.1. *Let $d + A$ be a smooth connection with compact structure group G over $\mathbf{R} \times \mathbf{R}^m$ or $I \times \mathbf{R}^m$. Suppose $A \sim 0$ at spatial infinity. Let $F_A = dA + [A, A]$ be the space-time curvature. Then there is a positive constant $\epsilon = \epsilon(m, G)$ such that if the mixed space-time norm $||F_A||_{L_t^\infty L_x^{m/2}} < \epsilon$, then there exists a unique smooth gauge change g, with $g \sim I$ at spatial infinity, such that if $\tilde{A} = gAg^{-1} - dgg^{-1}$, then*

(a) $||\tilde{A}||_{L_t^\infty \dot{W}_x^{1,m/2}} \le c(m, G)||F_A||_{L_t^\infty L_x^{m/2}}$,

(b) $\sum_{i=1}^m \frac{\partial}{\partial x^i}\tilde{A}_i = 0$.

See the proof in [275]. This proposition still holds if $A \in L_t^\infty \dot{W}_x^{1,m/2}$ and $F_A \in L_t^\infty L_x^{m/2}$.

We provide a coordinate invariant expression of the wave equations. Let $D = f^{-1}\nabla = d + a$, and the curvature of d be $F_A = (R \circ f)(df, df)$. The term $R \circ f$ is not explicit except for the case of a Lie group or symmetric space. Let $b = df$. Thus the equations are expressed by

$$D_0 b_0 - \sum_{i=1}^m D_i b_i = 0.$$

Since the Levi-Civita connection on N has no torsion, we have

$$D_k b_i = D_i b_k, \text{ for } k = 0, 1, \cdots, m, \ i = 1, 2, \cdots, m.$$

This is a non-linear first-order hyperbolic system. Using the standard method, we transform it to a single equation by utilizing Hodge theory [90]. Let d be the exterior differentiation and $d* = div_{(time, space)}$ be its dual, the time-space divergence calculated in the Lorentz metric.

Proposition 2.6.2. *Let $b = d\phi + d^*\psi$. The wave map equations can be reformulated as follows:*

$$(1)\ \Box\phi + (a, b) = 0, \ (2)\ \ \Box\psi + a \wedge b = 0, \ (3)\ \ b = d\phi + d * \psi$$

$$(4)\ da + [a, a] = R(x)[b, b], \ (5)\ \sum_{i=1}^m \frac{\partial}{\partial x^i} a_i = 0, \tag{2.108}$$

where R(x) is the Riemannian curvature of N evaluated at f(x). The initial data on ϕ and ψ are

$$\phi(0,x) = 0, \ \psi(0,x) = 0,$$
$$\frac{\partial \phi}{\partial t}(0,x) = b_0(0,x), \ \frac{\partial \psi}{\partial t_{0,i}}(0,x) = b_i(0,x),$$
$$\frac{\partial \psi_{i,k}}{\partial t}(0,x) = 0, \ i,k \neq 0.$$

Proof. Set $b = \Box q$ with $q(0,x) = \frac{\partial q}{\partial t}(0,x) = 0$, where $\Box = dd * + d * d$. Let

$$\phi = d * q = \frac{\partial}{\partial t} b_0 - \sum_{i=1}^{m} \frac{\partial}{\partial x_i} b_i$$

and

$$\psi = dq = \frac{\partial b_i}{\partial t} - \frac{\partial b_0}{\partial x_i}, \ \frac{\partial b_i}{\partial x^k} - \frac{\partial b_k}{\partial x^i}.$$

Thus $\Box\phi = d * b$ and $\Box\psi = d \wedge b$. Therefore, $b = d\phi + div_{(time,space)}\psi$. Observe that $d\psi = 0$.

The initial data obviously contain $\phi(0,x) = 0$, $\psi(0,x) = 0$. Then

$$b_0 = \frac{\partial \phi}{\partial t} - \sum_{i=1}^{m} \frac{\partial \psi_{i,0}}{\partial x^i} \Rightarrow b_0(0,x) = \frac{\partial \phi(0,x)}{\partial t}.$$

Similarly,

$$b_i = \frac{\partial \phi}{\partial x^i} - \frac{\partial \psi_{0,i}}{\partial t} + \sum_{k=1}^{m} \frac{\partial \psi_{k,i}}{\partial x_k} \Rightarrow b_i(0,x) = \frac{\partial \psi_{0,i}(0,x)}{\partial t}.$$

Observe also that

$$\frac{\partial \psi_{i,k}}{\partial t} = \frac{\partial \psi_{0,k}}{\partial x^i} + \frac{\partial \psi_{i,0}}{\partial x^k} \Rightarrow \frac{\partial \psi_{i,k}}{\partial t}(0,x) = 0.$$

The equation (4) of (2.108) is not determined by the rest of the data, because the curvature depends on the original map (and the gauge change). We don't have a general formula in our hands. This would not preclude a priori estimates. But the estimates for our global existence theorem for wave maps are done in Besov spaces (inferior to the Lorentz spaces). The identity $da + [a,a] = R(x)[b,b]$ behaves badly (for bounded $R(x)$) in this text. Therefore, we have to restrict the manifold N to a group or a Riemannian symmetric space.

Proposition 2.6.3. *Let G be a compact Lie group. If $N = G$ or $N = H/G$, then (4) in (2.108) becomes*

$$(4)' \; da + [a, a] + [b, b] = 0.$$

Furthermore, the original map $f : \mathbf{R}^1 \times \mathbf{R}^m \to G$ (or H/G) can be reconstructed by using the fact that $d + a + b$ and $d + a - b$ are flat connections. Namely, let $(d + a + b)g^+ = 0$ and $(d + a - b)g^- = 0$. Then the original map can be expressed as $f = g^+ \cdot g^-$.

Proof. Recall that $T^*G = \mathcal{G}$ (the Lie algebra of G), that the curvature is expressed in terms of the Lie commutator $[\cdot, \cdot]$, and that the structure group is a specialization of the orthogonal group. One can consider the symmetric space N as an Ad orbit in the (possibly) non-compact group H, i.e., $N = AdH(\hat{i})$ and G is the (compact) isotropy subgroup of $\hat{i}$. For H^n, H is the Lorentz group $O(1, n)$ and G is the Euclidean group $O(n)$. Set $\hat{i} = diag(1, -1, \cdots, -1)$. Thus one can write b (with off-diagonal vectors) as

$$b_i = \begin{bmatrix} 0 & \cdot & v_i & \cdot \\ \cdot & 0 & \cdots & 0 \\ -v_i^* & 0 & \cdots & 0 \\ \cdot & 0 & \cdots & 0 \end{bmatrix}$$

and the compact structure group $O(n)$ is described on the diagonal. This construction doesn't work for non-compact groups like $O(1, n)$ because they do not have a bi-invariant Riemannian metric. □

Assume that $N = G$ or $N = H/G$. Then a subset of the gauged wave map equations (1)–(5) in (2.108) (GMW) has a structure of a non-linear wave system of integral differential equations.

$$(1) \quad \Box\phi + (a, b) = 0, \;\; (2) \quad \Box\psi + a \wedge b = 0, \;\; (3) \quad b = d\phi + d * \psi,$$

$$(4) \quad \Delta a_i + \sum_{k=1}^{m} \frac{\partial}{\partial x^k}[a_k, a_i] + \frac{\partial}{\partial x^k}[b_k, b_i] = 0, \; i = 0, 1, \cdots, m.$$

Since $\sum_{k=1}^{m} \frac{\partial}{\partial x^k} a_k = 0$, one can have the last equation in terms of divergence. This system is called the *modified wave map* (MWM) system.

2.6.2 Multiplication Estimates

To prove that the Cauchy problem for the MWM system has a unique global solution in $L^\infty(\mathbf{R}; \dot{H}_x^{m/2})$ provided the initial data has sufficiently small $\dot{H}_x^{m/2} \times \dot{H}_x^{m/2-1}$ norm, in an indirect way, we need to recall the following facts from Littlewood-Paley theory.

Let $f(t,x)$ be a function on $\mathbf{R} \times \mathbf{R}^m$. The *spatial* Fourier transform $\hat{f}(t,\xi)$ is defined by

$$\hat{f}(t,x) = \int_{\mathbf{R}^m} e^{-2\pi i x\cdot\xi} f(t,x)dx.$$

The Littlewood-Paley projection operators P_k and Q_k are defined as follows. Let $\mu(\xi)$ be a non-negative radial bump function supported on the ball $|\xi| \leq 2$ and equal to 1 on the ball $|\xi| \leq 1$. Then for each k we define $P_k(\phi)$ the projection onto the frequency ball $|\xi| \preceq 2^k$ (means $|\xi| \leq const \cdot 2^k$) by

$$\widehat{P_k(f)}(\xi) = \mu(2^{-k}\xi)\hat{f}(t,\xi).$$

Note that $P_k \to I$ in L^2 as $k \to \infty$, and $P_k \to 0$ in L^2 as $k \to -\infty$.

The operator $Q_k = P_k - P_{k-1}$ is the projection onto the frequency annulus $|\xi| \sim 2^k$. If we set $\psi(\xi) = \mu(\xi) - \mu(2\xi)$, then ψ is supported in the annulus $\frac{1}{2} \leq |\xi| \leq 2$, for all $\xi \neq 0$, $\sum_{k\in\mathbf{Z}} \psi(2^{-k}\xi) = 1$ and $\widehat{Q_k(f)}(t,\xi) = \psi(2^{-k}\xi)\hat{f}(t,\xi)$. The Littlewood-Paley projections are bounded operators in all the Lebesque spaces and commute with any constant-coefficient differential operator. Observe that Q_k is given by a convolution kernel whose L^p-norm equals $2^{(km)(1-1/p)}$ for all $1 \leq p \leq \infty$. In particular, its L^1-norm is 1 for all $k \in \mathbf{Z}$.

Let $i = 0$ or $i = 1$ and $k \in \mathbf{Z}$. As in Tao's work [360, 361] (see the previous two sections) and Klainerman's work [226], the *Strichartz space* $S_k^{(-i)}(\mathbf{R}\times \mathbf{R}^m)$ *at frequency* 2^k is defined to be the space of functions whose space-time norm is

$$||f||_{S_k^{(-i)}} = \sup_{\mathrm{q,r}\in\mathrm{F}} 2^{k(\frac{1}{q}+\frac{m}{r}-i)}(||f||_{L_t^q L_x^r} + 2^{-k}||\partial_t f||_{L_t^q L_x^r}),$$

where $F = \{(q,r) | 2 \leq q,r \leq \infty, \frac{1}{q} + \frac{m-1}{2r} \leq \frac{m-1}{4}\}$ is the set of admissible Strichartz exponents. Notice that if $i = 0$ the preceding spaces are $\dot{H}^{m/2}$-normalized and correspond to Tao's spaces S_k of Sect. 2.5. Notice also that for $m \geq 4$, specific values of (q,r) are required. Lastly, we have the following estimates given by the control of the $S_k^{(-i)}$:

$$||Q_k(f)||_{L_t^2 L_x^{\frac{2(m-1)}{(m-3)}}} + 2^{-k}||\partial_t Q_k(f)||_{L_t^x L_x^{\frac{2(m-1)}{(m-3)}}} \leq 2^{k(i+\frac{m}{m-1}-\frac{m+1}{2})} ||Q_k(f)||_{S_k^{(-i)}}, \tag{2.109}$$

$$||Q_k f||_{L_t^q L_x^r} + 2^{-k}||\partial_t Q_k(f)||_{L_t^\infty L_x^r} \leq 2^{k(i-\frac{m}{2})} ||Q_k(f)||_{S_k^{(-i)}}, \tag{2.110}$$

$$||Q_k f||_{L_t^2 L_x^\infty} + 2^{-k}||\partial_t Q_k(f)||_{L_t^2 L_x^\infty} \leq 2^{k(i-\frac{1}{2})} ||Q_k(f)||_{S_k^{(-i)}}. \tag{2.111}$$

It follows from Keel and Tao [216, 217] that we have the following Strichartz estimates. In the particular case $i = 0$, one recovers Proposition 2.4.2.

Proposition 2.6.4 (Strichartz estimates). *Let k be an integer and let f be any function on $\mathbf{R} \times \mathbf{R}^m$ with spatial Fourier support in the annulus $|\xi| \sim 2^k$. Then*

$$||f||_{S_k^{(-i)}} \preceq ||f(0,\cdot)||_{\dot{H}_x^{m/2-i}} + ||\partial_t f(0,\cdot)||_{\dot{H}_x^{m/2-(i+1)}} + 2^{k(m/2-(i+1))}||\Box f||_{L_t^1 L_x^2}.$$

Definition 2.6.5. Let $S^{(-i)}$ be the space of functions on $\mathbf{R} \times \mathbf{R}^m$ whose norm is defined by

$$||f||_{S^{(-i)}} = (\sum_{k \in \mathbf{Z}} ||Q_k(f)||^2_{S_k^{(-i)}})^{1/2}.$$

A pair (q, r) is called *sharp admissible* if $2 \leq q, r \leq \infty$ and $\frac{1}{q} + \frac{m-1}{2r} = \frac{m-1}{4}$. Note that if $m \geq 4$ and (q, r) is sharp admissible, then $s = 1/q + m/r - 1 > 0$. In particular, $q \geq 2$ and $2 \leq r \leq \frac{2(m-1)}{m-3}$.

Lemma 2.6.6. *For any $i \geq 0$ we have*

$$\sup_{(q,r)-\text{admissible}} 2^{k(1/q+m/r-i)}||Q_k(f)||_{L_t^q L_x^r} = \sup_{(q,r)-\text{sharp admissible}} 2^{k(1/q+m/r-i)}||Q_k(f)||_{L_t^q L_x^r}.$$

That is,

$$||f||_{S^{(-i)}} = \left(\sum_{k \in \mathbf{Z}} \left| \sup_{(q,r)-\text{sharp admissible}} 2^{k(1/q+m/r-i)}(||Q_k(f)||_{L_t^q L_x^r} + 2^{-k}||\partial_t Q_k(f)||_{L_t^q L_x^r}) \right|^2 \right)^{1/2}.$$

(See the proof in [275].)

Let $|\nabla|^{-1} = \nabla \Delta^{-1}$ be the pseudo-differential operator given by

$$\widehat{|\nabla|^{-1} f}(t, \xi) = \frac{1}{|\xi|} \hat{f}(t, \xi).$$

Definition 2.6.7. Let B_p be the Banach space of functions on $\mathbf{R} \times \mathbf{R}^m$ whose norm is defined by

$$||f||_{B_p} = \left(\sum_{k \in \mathbf{Z}} ||Q_k(f)||^p_{L_t^1 L_x^\infty} \right)^{1/p}$$

for $1 \leq p < \infty$, suitably modified with the l^∞-norm when $p = \infty$. (Note that $l^p \subset l^q$ implies that $B_p \subset B_q$ for $1 \leq p < q \leq \infty$.)

In order to discuss the multiplication estimate, we introduce a few sets. Let C, D, D_t, E be the following sets of pairs (q, ν) with $q, \nu \geq 1$:

$$C = \{(q, \nu) \in F | \frac{1}{q} + \frac{m}{\nu} \leq 1^-\},$$

$$D = \{(q, \nu) | \frac{1}{2q} + \frac{m-1}{4\nu} \leq (\frac{m-1}{4})^-\},$$

$$D_t = \{(q, \nu) | q \geq 2, \frac{1}{2q} + \frac{m-1}{4\nu} \leq (\frac{m-1}{4})^-\},$$

$$E = \{(q, \nu) | \frac{1}{q} = \frac{1}{q_1} + \frac{1}{q_2}; \frac{1}{\nu} = \frac{1}{\nu_1} + \frac{1}{\nu_2}, (q_1, \nu_1) \in F, (q_2, \nu_2) \in C\}$$

where F is as before, the set of all (wave) admissible pairs. Lastly, let $\mathcal{G} = D \cap E$ and $\mathcal{G}_t = D_t \cap E$. Remark that $F \subset \mathcal{G}$ and $F \subset \mathcal{G}_t$ because $F \subset D$, $F \subset D_t$ and $F \subset E$. We refer to the pairs in $\mathcal{G}$ as the set of good pairs of frequency localized wave products.

We define $S_+^{(-1)}$ to be the space of functions on $\mathbf{R} \times \mathbf{R}^m$ whose norm is

$$||f||_{S_+^{(-1)}} = \sum_{k \in \mathbf{Z}} ||Q_k(f)||_{S_+^{(-1)}},$$

where

$$||f||_{S_{k+}^{(-1)}} = \sup_{(q,\nu) \in \mathcal{G}} 2^{k(1/q+m/\nu-1)} ||f||_{L_t^q L_x^\nu} + \sup_{(q,\nu) \in \mathcal{G}_t} 2^{k(1/q+m/\nu-1)} 2^{-k} ||\partial_t f||_{L_t^q L_x^\nu}.$$

(L1) *We have the following embeddings*

$$S_+^{(-1)} \hookrightarrow S^{-1}, \ S_+^{(-1)} \hookrightarrow B_1,$$

$$S_+^{(-1)} \hookrightarrow L_t^q \dot{B}_{\hat{\nu},2}^s, \ \textit{for}\, q \geq 2, \hat{\nu} \geq 2, \ s = \frac{1}{q} + \frac{m}{\hat{\nu}} - 1.$$

Proof. This follows from the definition of $\mathcal{G}$, Lemma 2.6.6, the embeddings $l^\nu \subset l^q$ for $\nu < q$, and

$$||f||_{L_t^q \dot{B}_{\hat{\nu},2}^s} \preceq \left(\sum_{k \in \mathbf{Z}} 2^{2ks} ||Q_k(f)||_{L_t^q L_x^{\hat{\nu}}}^2 \right)^{1/2}.$$

Remark that $(q, \hat{\nu}) \in \mathcal{G}$ for $q, \hat{\nu} \geq 2$. □

Proposition 2.6.8 (Multiplication Estimate). *The operator* $|\nabla|^{-1}$ *acts from* $S^{(-1)} \times S^{(-1)}$ *into* $S_+^{(-1)}$ *as*

$$|\nabla|^{-1} : S^{(-1)} \times S^{(-1)} \to S_+^{(-1)}.$$

Proof. Apply (L1) and see the lengthy and technical proof in [275].

2.6.3 Modified Wave System

We show that the Cauchy problem for the MWM system has a unique global solution in $L^\infty(\mathbf{R}; \dot{H}_x^{m/2})$ provided the initial data have sufficiently small $\dot{H}_x^{m/2} \times \dot{H}_x^{m/2-1}$ norm.

Let $B(a,b)$ be the quadratic form equal to any finite linear combination of functions $a \in S_{+}^{(-1)}$ and $b \in S^{(-1)}$ of the form $\sum_{k,l} c_{kl} a_k b_l$, where $a_k \in S_{+}^{(-1)}$, $b_l \in S^{(-1)}$ and $c_{kl} \in \mathbf{C}$. Based on reduction as above, we consider the following system of coupled wave equations in $\mathbf{R}^{m+1}$, $m \geq 4$:

$$\Box v = B(a,b),\ v(x,0) = f(x),\ v_t(x,0) = g(x). \tag{2.112}$$

(L2) *If $a \in S_{+}^{(-1)}$ and $b \in S^{(-1)}$, then the solution to the MWM system (2.112) with initial data $(f,g) \in \dot{H}_x^{m/2} \times \dot{H}_x^{m/2-1}$ satisfies*

$$||v||_S \preceq ||f||_{\dot{H}^{m/2}} + ||g||_{\dot{H}^{m/2-1}} + ||a||_{S_{+}^{(-1)}} ||b||_{S^{(-1)}}.$$

Theorem 2.6.9 (Existence [275]). *There exists $\epsilon > 0$ such that if the initial data satisfy*

$$||(f,g)||_{\dot{H}^{m/2} \times \dot{H}^{m/2-1}} < \epsilon,$$

then the system (2.112) has a unique global solution $v \in S$. In particular, the solution v lies in $L^{\infty}(\mathbf{R}; \dot{H}_x^{m/2}) \cap L^2(\mathbf{R}; \dot{B}^1_{2m,2})$ and $W^{1,\infty}(\mathbf{R}; \dot{H}^{m/2-1}) \cap W^{1,2}(\mathbf{R}; \dot{B}^0_{2m,2})$. Furthermore, stability holds, i.e.,

$$\operatorname*{ess\,sup}_t ||v_1 - v_2||_{\dot{H}^{m/2}} \preceq ||(f_1, g_1) - (f_2, g_2)||_{\dot{H}^{m/2} \times \dot{H}^{m/2-1}}$$

provided the right-hand side is sufficiently small.

Proof. The result follows by Picard's iteration based on the a priori estimates and the smallness of the data. Assume $||(f,g)||_{\dot{H}^{m/2} \times \dot{H}^{m/2-1}} = \delta$ and let v_0 be the solution to

$$\Box v_0 = 0,\ v_0(0,\cdot) = f,\ \partial_t v_0(0,\cdot) = g. \tag{2.113}$$

By the Strichartz estimates, we have

$$||v_0||_S \leq c_1 ||(f,g)||_{\dot{H}^{m/2} \times \dot{H}^{m/2-1}} = c_1 \delta.$$

Note that $v_0 = (\phi_0, \psi_0)$ yields $b_0 = d\phi_0 + div_{(space-time)} \psi_0$ with $||b_0||_{S^{(-1)}} \leq c_2 ||v_0||_S \leq c_3 \delta$. Then the multiplication estimates enable us to use a fixed-point argument to generate a_0 from b_0 by solving the equation

$$a_0 = |\nabla|^{-1}(a_0, a_0) + |\nabla|^{-1}[b_0, b_0].$$

Furthermore, the resulting a_0 satisfies

$$||a_0||_{S_{+}^{(-1)}} \leq c_4 ||b_0||^2_{S^{(-1)}} \leq c_5 \delta^2.$$

Let v_1 be the solution of $\Box v_1 = B(a_0, b_0)$, $v_1(0,\cdot) = f$, $\partial_t v_1(0,\cdot) = g$. It follows from the a priori estimate that

$$||v_1||_S \leq c_0(\delta + ||a_0||_{S_+^{(-1)}}||b_0||_{S^{(-1)}}) \leq 2c_0\delta$$

if δ is sufficiently small.

We then use induction to show that for any $i \geq 0, ||b_i||_S \leq 2c_2c_0\delta$, $||a_i||_S \leq c_5\delta^2$ and so $||v_{i+1}||_S \leq 2c_0\delta$ if $\delta > 0$ is sufficiently small (independently of i), where v_{i+1} is the solution to

$$\Box v_{i+1} = B(a_i, b_i),\ v_{i+1}(x,0) = f,\ \partial_t v_{i+1}(x,0) = g. \tag{2.114}$$

Remark that we have

$$||v_{i+1}||_S \leq c_0(||(f,g)||_{\dot{H}^{m/2}\times\dot{H}^{m/2-1}} + ||a_i||_{S_+^{(-1)}}||b_i||_{S^{(-1)}}),$$

thanks again to the a priori estimates.

Finally, for the differences

$$\Box(v_{i+2} - v_{i+1}) = B(a_{i+1}, b_{i+1}) - B(a_i, b_i) = B(a_{i+1} - a_i, b_{i+1}) + B(a_i, b_{i+1} - b_i).$$
$$v_{i+1}(0,\cdot) = 0,\ \partial_t v_{i+1}(0,\cdot) = 0.$$

Since

$$\begin{aligned} a_{i+1} - a_i &= |\nabla|^{-1}[a_{i+1} - a_i, a_{i+1}] + |\nabla|^{-1}[a_i, a_{i+1} - a_i] \\ &+ |\nabla|^{-1}[b_{i+1} - b_i, b_{i+1}] + |\nabla|^{-1}[b_i, b_{i+1} - b_i], \\ &||b_{i+1} - b_i||_{S^{(-1)}} \leq c_2||v_{i+1} - v_i||_S, \end{aligned}$$

and

$$||a_i||_{S^{(-1)}},\ ||a_{i+1}||_{S^{(-1)}},\ ||b_i||_{S^{(-1)}}||,\ |b_{i+1}||_{S^{(-1)}} \leq c\delta,$$

we obtain

$$\begin{aligned} ||a_{i+1} - a_i||_{S_+^{(-1)}} &\leq c||a_{i+1}||_{S^{(-1)|}}||a_{i+1} - a_i||_{S^{(-1)}} + ||a_i||_{S^{(-1)|}}||a_{i+1} - a_i||_{S^{(-1)}} \\ &+ ||b_{i+1}||_{S^{(-1)}}||b_{i+1} - b_i||_{S^{(-1)}} + ||b_i||_{S^{(-1)}}||b_{i+1} - b_i||_{S^{(-1)}} \\ &\leq c\delta||a_{i+1} - a_i||_{S^{(-1)|}} + c'\delta||b_{i+1} - b_i||_{S^{(-1)}}. \end{aligned}$$

Thus,

$$||a_{i+1} - a_i||_{S_+^{(-1)|}} \leq c\delta||b_{i+1} - b_i||_{S^{(-1)}} \leq c\delta||v_{i+1} - v_i||_S.$$

We then have

$$||v_{i+2} - v_{i+1}||_S \leq c(||a_{i+1} - a_i||_{S_+^{(-1)}}||b_{i+1}||_{S^{(-1)}}$$
$$+ ||a_{i+1}||_{S^{(-1)}}||b_{i+1} - b_i||_{S^{(-1)}}) \leq c\delta^2||v_{i+1} - v_i||_S.$$

Finally, we obtain the inequality

$$||v_{i+2} - v_{i+1}||_S \leq \frac{1}{2}||v_{i+1} - v_i||_S$$

by choosing δ sufficiently small. Therefore, v_i is Cauchy in S, which yields the existence and uniqueness. For the stability, we follow arguments similar to those in the proof of being Cauchy. □

Theorem 2.6.10 (Uniqueness [275]). *Assume that* (v_1, a_1) *and* (v_2, a_2) *are two solutions to*

$$\Box v + B(a, dv) = 0, \quad \Delta a + div\, B(a, a) + div\, B(dv, dv) = 0$$

such that $dv_i = b_i$ *are small in* $L_t^\infty L_x^m$ *for* $i = 1, 2$. *Assume further that* $dv_i = b_i \in L_t^2 L_x^{2m}$ *for* $i = 1, 2$ *and* $a_1 = a_1(v_1) \in L_t^1 L_x^\infty$. *Then* $v_1 = v_2$.

Proof. Let $\delta v = v_1 - v_2$, $\delta a = a_1 - a_2$, $\delta b = dv_1 - dv_2$. We apply the Shatah-Struwe technique for uniqueness [332, 333]. We have

$$\Box \delta v = B(a_1, \delta b) + B(\delta a, b_2),$$
$$\int < \frac{\partial}{\partial t}\delta v, \Box \delta v > dx = \frac{1}{2}\frac{\partial}{\partial t}E_2,$$

where

$$E_2 = \int \Big(|d(\delta v)|^2 + |\frac{(\delta v)}{\partial t}|^2\Big)dx = \int |\delta b|^2 dx.$$

Thus

$$\frac{1}{2}\frac{\partial}{\partial t}E_2 \leq ||a_1||_{t,L_x^\infty}E_2(t) + E(t)||\delta a||_{t,L_x^{2m/m-1}}||b||_{t,L_x^{2m}}.$$

Integrating with respect to t, we get

$$E_2(t) \leq \max_{\tau\leq t} E_2(\tau)\int_0^\tau ||a_1||_{t,L_x^\infty}dt + \max_{\tau\leq t} E(\tau)||\delta a||_{L^2_{(0,t)}L_x^{2m/m-1}}||d||_{L^2_{(0,t)}L_x^{2m}}, \tag{2.115}$$

where $L^2_{(0,t)}$ is the L^2 norm on the time interval $(0,\ t)$. We have

$$\Delta \delta a + div\, B(\delta a, a_1 + a_2) + div\, B(\delta b, b_1, b_2) = 0.$$

Therefore,

$$||\delta a||_{t,W_x^{1,2m/m+1}} \le ||\delta a||_{t,L_x^{2m/m-1}}(||a_1||_{(t,L_x^m)} + ||a_2||_{(t,L_x^m)})$$
$$+ ||\delta b_1||_{(t,L_x^2)}(||b_1||_{(t,L_x^{2m})} + ||b_2||_{(t,L_x^{2m})}). \qquad (2.116)$$

Moreover, $||a_i||_{W^{1,m/2}} \le ||a_i||_{L^m}^2 + ||b_i||_{L^m}^2$ and $||b_i||_{L^m}^2$ is small for each t. Using the Sobolev embedding we have $||a_i||_{L^m} \le c(m)||a_i||_{\dot{W}^{1,m/2}}$, so we conclude that $||a_i||_{L^m}$ is small for each fixed t.

By (2.116), we obtain

$$||\delta a||_{t,L_x^{2m/m-1}} \le \hat{c}(m)\,||\delta a||_{t,L_x^{2m/m+1}} \le \hat{c}(m)E(t)(||b_1||_{t,L_x^{2m}} + ||b_2||_{t,L_x^{2m}}).$$

We have, upon integrating for $\tau \le t$, that

$$||\delta a||_{L^2_{(0,t)},L_x^{2m/m-1}} \le \hat{c}(m)\max E_{\tau\le t}(\tau)(||b_1||_{L^2_{(0,t)},L_x^{2m}} + ||b_2||_{L^2_{(0,t)},L_x^{2m}}).$$

Substituting this estimate into (2.115), we deduce

$$E_2(t) \le \max_{\tau\le t} E_2(\tau)(||a_1||_{L^1_{(0,t)}L_x^\infty} + \hat{c}(m)(||b_1||_{L^2_{(0,t)}L_x^{2m}}||b_2||_{L^2_{(0,t)}L_x^{2m}})^2).$$

It follows that $E(t) = 0$ since $E(0) = 0$. □

If we differentiate (2.112) and the resulting non-linearity has the same bilinear structure for which we can apply the multiplication estimates, then we can obtain the following higher regularity theorem [275]:

Assume the initial datum (f, g) to (2.112) is in $H^{m/2+1} \times H^{m/2}$ and has sufficiently small $\dot{H}^{m/2} \times \dot{H}^{m/2-1}$ norm. Then the solution v to (2.112) with initial datum (f, g) can be continued in $H^{m/2+1} \times H^{m/2}$ globally in time. Moreover, we have the global bounds

$$||v||_{L_t^\infty(\mathbf{R};\,\dot{H}_x^{m/2+1})} \preceq ||(f,g)||_{\dot{H}_x^{m/2+1}\times\dot{H}_x^{m/2}}.$$

2.7 Global Well-Posedness: Maps into Riemannian Manifolds

We now consider the wave maps into a Riemannian manifold which is isometrically embedded in $\mathbf{R}^k$. The main result obtained by Tataru [366] (85 pages) asserts that the corresponding Cauchy problem is globally well-posed for initial data which are small in the critical Sobolev spaces. This result generalized the main theorems of

Sects. 2.4 and 2.5 obtained by Tao [360, 361] and other authors. Krieger, Schlag and Tataru [236] also studied renormalization and blow up for charge-one equivariant critical wave maps.

2.7.1 Main Results

Let $f = (f^1, \cdots, f^n) : \mathbf{R}^{1+m} \to N \subset \mathbf{R}^k$ be a wave map into a Riemannian manifolds N isometrically embedded in $\mathbf{R}^k$ with second fundamental form A.

Definition 2.7.1. A Riemannian manifold N is *uniformly isometrically embedded* into $\mathbf{R}^k$ if there is $r > 0$ such that (i) for each $y_0 \in N$ the intersection $N \cap B(y, r)$ is the graph of a smooth function $y^1 = f_{y_0}(y')$ in an appropriate orthonormal frame (y_0, y'); (ii) the derivatives of the function f_{y_0} are bounded uniformly in $y_0 \in N$.

We always suppose that N is isometrically embedded into $\mathbf{R}^k$, which is similar to the assumption in the work of Eell and Sampson [129] on harmonic maps. If N is compact, Nash's embedding theorem guarantees that such an embedding always exists. In general, a uniform isometric embedding exists if the curvature tensor and its covariant derivatives are uniformly bounded on N.

Recall from (2.8) and (2.9) that f satisfies

$$\begin{cases} \Box f^k = (-)A^k_{ij}(f)(\partial^\alpha f^i, \partial_\alpha f^j) & \text{in } \mathbf{R} \times \mathbf{R}^m, \\ f(0,x) = f_0(x),\ \partial_t f(0,x) = f_1(x) & \text{in } \mathbf{R}^m, \end{cases} \tag{2.117}$$

such that the initial data $f_0(x) \in N$ and $f_1(x) \in T_{f_0(x)}(N)$. (Note that there is a $+$ or $-$ sign convention in the wave operator $\Box = (-)(-\frac{\partial^2}{\partial t^2} + \Delta)$.) Let $f[t] = (f(t), \partial_t f(t))$. One can ask a question whether the Cauchy problems is locally well-posed for initial data in Sobolev spaces, $f[0] \in H^s \times H^{s-1}$.

The wave maps equation is invariant with respect to the dimensionless scaling $f(t,x) \mapsto f(\lambda t, \lambda x)$, $\lambda \in \mathbf{R}$. The scale invariant data space corresponds to $s = m/2$. In order to have a scale invariant problem, we need to use the homogeneous Sobolev spaces, $f[0] \in \dot{H}^{m/2} \times \dot{H}^{m/2-1}$. Since it remains unchanged under scaling, the size of the initial data is important. We show here that for initial data being small in the preceding space the Cauchy problem for the wave map equation is globally well-posed.

Since we assume that $N \subset \mathbf{R}^k$, we consider the N-valued functions as $\mathbf{R}^k$ valued functions which take values in N a.e. For test functions $\phi \in D(\mathbf{R}^m, \mathbf{R}^k)$, we use the Fourier transform to define $||\phi||_{\dot{H}^s} = |||\xi|^s \hat{f_1}||_{L^2}$. Thus we define $\dot{H}^s$ as the completion of D with respect to this norm. If $s < m/2$, this is a space of distributions with locally integrable transform. But, for $s = m/2$ the space $\dot{H}^{m/2}$ is not a space of distributions. However, it can be identified as a subspace of a quotient space, $\dot{H}^{m/2} \subset BMO \subset D'(\mathbf{R}^m)/const$, where *const* denotes the space of constant functions. Because the wave maps equation is non-linear, the constants are crucial.

When we add them back in, we change a norm into a semi-norm that vanishes on a one-dimensional subspace. It can not be used well to compare different solutions, while the $\dot{H}^{m/2} \times \dot{H}^{m/2-1}$ norm gives a good measurement for the size of a solution.

All the following theorems and results were obtained by Tataru [367].

Main Theorem 2.7.2. *Suppose the manifold N is uniformly isometrically embedded into $\mathbf{R}^k$. Then the wave maps equation (2.117) is well-posed for small initial data in $\dot{H}^{m/2} \times \dot{H}^{m/2-1}$ as follows.*

(1) *If the initial data is small, then there is a global smooth solution f which for $s \geq m/2$ obeys the global bounds*

$$||f[t]||_{L^\infty(\dot{H}^s \times \dot{H}^{s-1})} \preceq ||f[0]||_{\dot{H}^s \times \dot{H}^{s-1}} \tag{2.118}$$

in the sense that the left-hand side is finite and equality holds if the right-hand side is finite (smooth solutions).

(2) *For each small initial data set in $\dot{H}^{m/2} \times \dot{H}^{m/2-1}$ there is a solution f such that*

$$||f[t]||_{L^\infty(\dot{H}^{m/2} \times \dot{H}^{m/2-1})} \preceq ||f[0]||_{\dot{H}^{m/2} \times \dot{H}^{m/2-1}} \tag{2.119}$$

which is the unique limit of smooth solutions in the $[L^\infty(H^{m/2} \times H^{m/2-1})]_{loc}$ topology (rough solutions as limits of smooth solutions).

(3) *If $f^{(1)}, f^{(2)}$ are as in (1), then for $s < m/2$ and close to it we obtain*

$$||f^{(1)}[t] - f^{(2)}[t]||_{L^\infty(\dot{H}^s \times \dot{H}^{s-1})} \preceq ||f[0]||_{\dot{H}^s \times \dot{H}^{s-1}} \tag{2.120}$$

(weak stability).

(4) *The solution f depends continuously on the initial data, in the sense that for $s < m/2$ and close to it*

$$f_n[0] - f[0] \to 0 \text{ in } \dot{H}^{m/2} \times \dot{H}^{m/2-1} \cap \dot{H}^s \times \dot{H}^{s-1} \tag{2.121}$$

implies

$$f_n - f_0 \to 0 \text{ in } L^\infty(\dot{H}^{m/2} \times \dot{H}^{m/2-1}) \cap L^\infty(\dot{H}^s \times \dot{H}^{s-1}) \tag{2.122}$$

(continuous dependence).

In order to state a local version of the result, let

$$B_\sigma = \{|x| < \sigma\}, \ Q_\sigma = \{|t| + |x| \leq \sigma\}.$$

For functions in B_σ and Q_σ we define their Sobolev norms as the norms of their extensions to $\mathbf{R}^m$ and $\mathbf{R} \times \mathbf{R}^m$, respectively.

Theorem 2.7.3. *Consider the wave maps equation (2.117) with initial datum* $f[0]$ *in* B_σ *which is small in* $(\dot{H}^{m/2} \times \dot{H}^{m/2-1})(B_\sigma)$.

(1) *If the initial datum is small, then there is a global smooth solution f in* Q_σ *which for* $s \geq m/2$ *obeys the global bounds*

$$||f[t]||_{[L^\infty(\dot{H}^s \times \dot{H}^{s-1})](Q_\sigma)} \preceq ||f[0]||_{[\dot{H}^s \times \dot{H}^{s-1}](B_\sigma)}. \tag{2.123}$$

(2) *For each small initial datum there is a solution f such that*

$$||f[t]||_{[L^\infty(\dot{H}^{m/2} \times \dot{H}^{m/2-1}](Q_\sigma)} \preceq ||f[0]||_{[\dot{H}^{m/2} \times \dot{H}^{m/2-1}](B_\sigma)}, \tag{2.124}$$

which is the unique limit of smooth solutions in the $L^\infty(H^{m/2} \times H^{m/2-1})$ *topology.*

(3) *The solution f depends continuously on the initial data, in the sense that*

$$f_n[0] - f[0] \to 0 \; in \; (\dot{H}^{m/2} \times \dot{H}^{m/2-1})(B_\sigma) \tag{2.125}$$

implies

$$f_n - f_0 \to 0 \; in \; L^\infty(\dot{H}^{m/2} \times \dot{H}^{m/2-1})(Q_\sigma). \tag{2.126}$$

The local well-posedness problem for wave maps has recently been studied extensively. The method of energy estimates is a natural approach. However, it only provides results when s is at least one unit above scaling. Some improvement can be obtained by applying the Strichartz estimates for wave maps, which works for the generic equation of the form $\Box f = A(f)(\nabla f^2)$. However, the wave map equation does not behave like the above equation. Instead, its quadratic non-linearity $Q_0(f, f) = \partial^\alpha f \partial_\alpha f$ exhibits a cancellation property called the null condition, as in the work of Klainerman and Machedon [222] ($m \geq 3$) and Klainerman and Selberg [227] ($m = 2$), who verified local well-posedness for all s above scaling, $s > m/2$. A key step in their approach is to use the $X^{s,b}$ spaces for bilinear estimates, which are suitable for the wave operator.

The scaling problem becomes more intriguing. If one attempts to use homogeneous versions of $X^{s,b}$ spaces, then one runs into trouble due to logarithmic divergences. The first scale invariant result was obtained by Tataru (for $m \geq 4$), and then for more difficult cases for $m = 2, 3$. These results for initial data are not in the scale invariant Sobolev space, but instead in the slightly smaller Besov space $\dot{B}^{m/2}_{2,1} \times \dot{B}^{m/2-1}_{2,1}$. In the high-dimensional case, this is accomplished by a mix of homogeneous $X^{s,b}$ spaces and Stricharts type norms for the bilinear estimates. In the low-dimensional case, the construction of the function spaces for the solution becomes much more intricate and involves the use of energy spaces with respect to rotating null frames.

As described in Sects. 2.4 and 2.5, Tao [360, 361] considered the case when the target manifolds are spheres in high dimensions $m \geq 5$ and low dimensions $m = 2, 3, 4$, and proved that if the initial data are smooth and small in $\dot{H}^{m/2} \times \dot{H}^{m/2-1}$, then there is a global smooth solution (Theorem 2.5.1). Tao's function spaces are based on Tataru's work [363, 365]. He made some modifications in order to obtain a key algebra property. He introduced a re-normalization argument such that he was still able to use linear estimates for the wave equation. This eliminates the nonlinear part of the equation, leaving only a semi-linear part which can be appropriately estimated in a semi-linear fashion. He used Helein's [178] ideas on harmonic maps, however, the adaptation to wave equation was more involved.

In this section we focus more on the two-dimensional case since it can be simplified relative to the high dimensional case [364]. We not only present the behavior of smooth solutions, but we also discuss rough solutions depending continuously on the initial data. Since Tataru's [367] original paper is very lengthy and technical, we are only able to present the essence of the proof of the main theorem.

2.7.2 Auxiliary Lemmas

We present the following auxiliary lemmas (see proofs in [367]) which are required for proving the Main Theorem 2.7.2. Remark that if $N = \mathbf{R}^k$ then each initial datum $(f_0, f_1) \in \dot{H}^{m/2} \times \dot{H}^{m/2-1}$ is the trace as $h \to \infty$ of the $\mathcal{H}$ function $\tilde{f}(h) = P_{<h} f$. If $N \subset \mathbf{R}^k$ then this construction fails since there is no guarantee that the regularized functions f_h have values in N. We will project these regularized functions to N to remedy this, which is the purpose of the following lemma.

(L1) *Let $(f_0, f_1) \in \dot{H}^{m/2} \times \dot{H}^{m/2-1}(\mathbf{R}^m, \mathbf{R}^k)$ be a small initial data set for the wave maps equation. Then there exists a smooth function $\tilde{f} : \mathbf{R} \times \mathbf{R}^m \to N \subset \mathbf{R}^k$ obeying*

$$|\partial_h^j \partial_x^\beta \tilde{f}_0(h, x)| \leq 2^{|\beta| h} c_{j,\beta}, \; |\beta| + j \geq 1, \tag{2.127}$$

$$|\partial_h^j \partial_x^\beta \tilde{f}_1(h, x)| \leq 2^{(|\beta|-1)h} c_{j,\beta}, \; |\beta| + j \geq 0, \tag{2.128}$$

such that

(i) *For any non-negative multi-index β we have*

$$\int_{\mathbf{R}^m} \left[2^{(m/2-|\beta|)h} |\partial^\beta \partial_h \tilde{f}_0|^2 + 2^{(m/2-|\beta|-1)h} |\partial^\beta \partial_h \tilde{f}_1|^2 \right] dx$$
$$\leq c_\beta ||P_h(f_0, f_1)||^2_{\dot{H}^{m/2} \times \dot{H}^{m/2-1}(\mathbf{R}^m, \mathbf{R}^k)}.$$

In particular, if $m = 2$, then

$$2^h ||\partial^h \tilde{f}_1||^2_{\partial L^2 + L^1} \preceq ||P_h(f_0, f_1)||_{\dot{H}^{m/2} \times \dot{H}^{m/2-1}(\mathbf{R}^m, \mathbf{R}^k)},$$

where $P_h = P_{\leq h} - P_{<h}$ is the Littlewood-Paley projection.

(ii) $(\tilde{f}_1(h), \tilde{f}_1(h))$ *are uniformly small in* $\dot{H}^{m/2} \times \dot{H}^{m/2-1}(\mathbf{R}^m, \mathbf{R}^k)$ *and*

$$\lim_{h\to\infty} (\tilde{f}_0(h), \tilde{f}_1(h)) = (f_0, f_1) \text{ in } \dot{H}^{m/2} \times \dot{H}^{m/2-1}.$$

(L2) *Let* B *be a ball in* $\mathbf{R}^m$. *Assume that* N *is uniformly isometrically embedded into* $\mathbf{R}^k$. *Let* f *be an initial data set which is small in* $\dot{H}^{m/2} \times \dot{H}^{m/2-1}(B, \mathbf{R}^k)$. *Then* f *is in* $\dot{H}^{m/2} \times \dot{H}^{m/2-1}(B, N)$ *and*

$$||f||_{\dot{H}^{m/2}\times\dot{H}^{m/2-1}(B,N)} \preceq ||f||_{\dot{H}^{m/2}\times\dot{H}^{m/2-1}(B,\mathbf{R}^k)}$$

The extension of f *in* $\dot{H}^{m/2} \times \dot{H}^{m/2-1}(\mathbf{R}^m, N)$ *can be taken to be constant outside* $2B$. *Moreover, this extension is continuous in all better topologies* $\dot{H}^s \times \dot{H}^{s-1}$, $s > m/2$.

(L3) (i) *For each initial datum* $f[0] \in H^k_{(1,\infty)} \times H^{k-1}_{(0,\infty)}$, *there exists some* $T_0 > 0$ *depending only on the size of f[0], and a solution* $f \in C(-T_0, T_0; H^k_{(0,\infty)} \times H^{k-1}_{(0,\infty)})$. *Moreover, the solution depends smoothly on the initial datum in the above topologies.*

(ii) *If the initial datum* $f[0]$ *is sufficiently small in* $H^k_{(0,\infty)} \times H^{k-1}_{(0,\infty)}$, *then we can take* $T_0 \geq T$.

(iii) *If the initial datum is smooth, then the solution is smooth up to time* T_0.

(L4) *Suppose that for some initial datum* $f[0] \in H^k_{(1,\infty)} \times H^{k-1}_{(0,\infty)}$, *there is a solution* $f \in C(-T, T; H^k_{(1,\infty)} \times H^{k-1}_{(0,\infty)})$.

(i) *If* $f[0] \in S[0]$, *then* $f \in S$ *and* $||f||_S \preceq A||f[0]||_{S[0]}$, *where A depends only on the norm of f in* $C(-T, T; H^k_{(0,\infty)} \times H^{k-1}_{(0,\infty)})$ *and on the* $S[0]$ *norm of* $f[0]$.

(ii) *Let* c *be an admissible frequency envelope. If the initial datum* $f[0]$ *is in* $S_c[0]$, *then* $f \in S_c$ *and* $||f||_{S_c} \preceq B||f[0]||_{S_c[0]}$, *where B depends only on the* $C(-T, T; S[0] \cap H^k_{(0,\infty)} \times H^{k-1}_{(0,\infty)})$ *norm of* f.

(L5) *Let* $f \in C(-T, T; H^k_{(1,\infty)} \times H^{k-1}_{(0,\infty)})$ *be a solution of*

$$\Box f = A(f)(\nabla f)^2.$$

(i) *For any initial datum* $\tilde{f}[0]$ *which is close to* f *in* $S[0] \cap H^k_{(0,\infty)} \times H^{k-1}_{(0,\infty)}$ *there is a unique solution* $\tilde{f}$ *which satisfies*

$$\tilde{f} - f \in S \cap C(-T, T; H^k_{(0,\infty)} \times H^{k-1}_{(0,\infty)});$$

f depends smoothly on the initial datum in the above topology.

(ii) *If c is an admissible frequency envelope for the linearized equation, then for each initial datum* $\psi[0] \in S[0] \cap H^k_{(0,\infty)} \times H^{k-1}_{(0,\infty)}$ *there is a solution* $\psi \in C(-T, T; H^k_{(0,\infty)} \times H^{k-1}_{(0,\infty)})$ *for the linearized equation which fulfills*

$$||\psi||_{S_c} \leq B||\psi[0]||_{S_c[0]},$$

where B depends only on the $C(-T, T; H^k_{(1,\infty)} \times H^{k-1}_{(0,\infty)})$ *norm of* f.

In the next lemma, we show how to bootstrap the regularity of smooth wave maps in S and S_c. Because we do not know that the solutions are global, the results hold on a fixed time interval $[T, -T]$. But they hold uniformly with respect to T, by rescaling.

(L6) *There is an* $\epsilon > 0$ *such that for* $f \in S$ *a smooth solution to the wave maps equation in* $[-T, T]$ *we have*

$$||f[0]||_{S[0]} \leq \epsilon^2, \; ||f||_S \leq 2\epsilon \Rightarrow ||f||_S \leq \epsilon.$$

Moreover, if $f \in S_c$ *for some admissible frequency envelope c, then*

$$||f||_S \ll \epsilon, \; ||f||_{S_c} \leq 2M, \; ||f[0]||_{S_c[0]} \leq \epsilon M \Rightarrow ||f||_{S_c} \leq M.$$

If we differentiate (2.117), we obtain the linearized equations

$$\Box\psi^l = -(\partial_n A^l_{ij})(f)\psi^n \partial^\alpha f^i \partial_\alpha f^j - 2A^l_{ij}(f)\partial^\alpha f^i \partial_\alpha \psi^j. \tag{2.129}$$

The function ψ must obey the compatibility condition

$$\psi(t, x) \in T_{f(t,x)}N.$$

The behavior of these equations is the key to compare different solutions of the wave maps equations.

(L7) *There is an* $\epsilon > 0$ *such that for any smooth solution* f *to the wave maps equation in* $[-T, T]$ *with uniformly bounded derivatives and which obeys* $||f||_S \preceq \epsilon^2$, *the linearized equation (2.129) is well-posed for initial data in* $S_c[0]$ *which satisfy the above compatibility condition, and*

$$||\psi||_{S_c} \preceq ||\psi[0]||_{S_c[0]}$$

Here c is any admissible frequency envelope for the linearized equation.

2.7.3 *Proof of Main Theorem*

We prove the Main Theorem 2.7.2 using the previous auxiliary lemmas as tools in the following steps.

(1) Continuity: For $T > 0$ and an initial datum set $f[0]$ with $||f[0]||_{S[0]} \leq \epsilon$, we consider the approximations $f^h[0]$ furnished by (L1) (see the proof of Lemma 3.9 in [366]). These are smooth functions, depending smoothly on h. Moreover, they satisfy the uniform bounds and the following h-dependent bounds:

$$||f^h[0]||_{S[0]} \leq \epsilon^2, \; ||f^h[0]||_{S_m[0]} \preceq 2^{hm}, \; m > 0.$$

By the Sobolev embedding theorem, we have the uniform bounds

$$||\partial^\beta f^h[0]||_{W^{1,\infty}\times L^\infty} \preceq c_\beta 2^{h(1+|\beta|)}$$

It follows from (L3) (ii) that for small enough h the wave map with initial datum $f^h[0]$ has a unique small smooth solution f^h in $C(-T, T; H^k_{(1,\infty)} \times H^{k-1}_{(0,\infty)})$. By (L4) (i) this solution satisfies

$$||f^h||_S \leq \epsilon. \tag{2.130}$$

Let us show that smooth solutions satisfying (2.130) exist for all $h \in \mathbf{R}$. To prove this we utilize a continuity argument. Let $A \subset (-\infty, \infty)$ be the set of those h for which there is a smooth solution f^h in $[-T, T]$ satisfying (2.130). As the previous discussion shows, A is non-empty. If we show that A is both open and closed, then $A = (-\infty, \infty)$.

(a) A is open: Let $h_0 \in A$. Thus the wave map equation with initial datum $f^{h_0}[0]$ has a smooth solution

$$f^{h_0} \in S \cap C(-T, T; H^k_{(1,\infty)} \times H^{k-1}_{(0,\infty)}).$$

Because $f^h[0]$ depends smoothly on h in $S[0] \cap H^k_{(0,\infty)} \times H^{k-1}_{(0,\infty)}$, by (L5) (i), for h close to h_0 there is also a smooth solution

$$f^h \in S \cap C(-T, T; H^k_{(1,\infty)} \times H^{k-1}_{(0,\infty)}),$$

depending smoothly on h in the $S \cap C(-T, T; H^k_{(0,\infty)} \times H^{k-1}_{(0,\infty)})$ topology. Therefore, for h close to h_0 we have $||f^h||_S \leq 2\epsilon$. Then (L6) shows that (2.130) must hold.

(b) A is closed: Let $h_j \in A$, $h_j \to h$. Then the corresponding solutions f^{h_j} are uniformly bounded in S. Moreover, their initial data are uniformly bounded in $S_N[0]$ for large N. By (L4) (ii) the solutions f^{h_j} are in S_N for all N. Then the solutions f^{h_j} are uniformly bounded in S_N by (L6). It follow that their derivatives are uniformly bounded. Since $f^{h_j} \to f^h[0]$ uniformly, by Arzelà-Ascoli's Theorem we have a subsequence which converges uniformly on compact sets, together with all its derivatives. The limit is in $C(-T, T; H^k_{(1,\infty)} \times H^{k-1}_{(0,\infty)})$ for some large k. Therefore, it solves the wave map equations in $[-T, T]$ and has initial datum $f^h[0]$. Then this solution is also in S, and the previous discussion shows that f^h is the strong limit of f^{h_j} in S.

(2) Construction of rough solutions: If the initial datum $f[0]$ is small in $S[0]$, the above step shows that the approximate solutions f^h with initial datum $f^h[0]$ are uniformly small in S. Since $f[0] \in S[0]$, there exists an admissible frequency envelope $c \in l^2$ such that $f[0]$ is small in $S_c[0]$, thus by (L4) (ii) we have $f^h \in S_c$. We can bootstrap this with (L6) to deduce that f^h are uniformly small in S_c.

The approximate solutions f^h also depend smoothly on h in the $C(-T, T; H^k_{(1,\infty)} \times H^{k-1}_{(0,\infty)})$ topology. Thus the functions $\frac{d}{dh} f^h$ are also in $S \cap C(-T, T; H^k_{(1,\infty)} \times H^{k-1}_{(0,\infty)})$ and solve the linear equation. For small positive δ their initial datum $\frac{d}{dh} f^h[0]$ obeys

$$\left\| \frac{d}{dh} f^h[0] \right\|_{S^{-\delta}[0]} \preceq 2^{-\delta h}.$$

It follows from L(5) (ii) that the functions $\frac{d}{dh} f^h$ are in $S_{-\delta[0]}$. We can use (L7) to boostrap their regularity and obtain

$$\left\| \frac{d}{dh} f^h \right\|_{S_{-\delta}} \preceq 2^{-\delta h}.$$

Integrating this, there exists a limit function f such that

$$\left\| f - f^h \right\|_{S_{-\delta}} \preceq 2^{-\delta h}.$$

It follows that

$$f^h \to f \text{ in } S,$$

since the f^h are uniformly bounded in S_c. It is easy to see, passing to the limit, that f solves the wave maps equation in the sense of distributions.

(3) Regularity of solutions for the wave maps: Let $f[0]$ be small in S and $f \in S$ be the solution constructed as above. Assume that $f[0] \in S_c[0]$ for some admissible frequency envelope c. Then the approximate initial data $f^h[0]$ are uniformly bounded in $S_c[0]$, $||f^h[0]||_{S_c[0]} \preceq ||f[0]||_{S_c[0]}$. By (L4) (ii), it follows that $f^h \in S_c$. We can bootstrap this with (L6) and show that f^h are uniformly bounded in S_c:

$$||f^h||_{S_c} \preceq ||f^h[0]||_{S_c[0]} \preceq ||f[0]||_{S_c[0]}.$$

Passing to the limit we arrive at $||f||_{S_c} \preceq ||f[0]||_{S_c[0]}$.

(4) Smooth solutions to the wave maps: Let $f[0]$ be smooth and small in $S[0]$. Given $\sigma > 0$ we use (L2) to construct a smooth initial datum $f^\sigma[0]$ which is small in $S[0]$, is constant outside of $B_{2\sigma}$ and coincides with $f[0]$ in $B(0, \sigma)$. Then $f^\sigma[0] \in S_c[0]$ for all admissible frequency envelopes c. It follows that the corresponding solution f^σ is in S_c. Because its initial datum is smooth, the solution f^σ must be smooth. Due to the finite speed of propagation, these solutions must agree on the common cone of influence. As $\sigma \to 0$, we obtain a global smooth solution for the initial datum $f[0]$.

(5) Weak stability estimates: Let $f^{(1)}[0]$ and $f^{(2)}[0]$ be two smooth initial data which are small in $S[0]$ and such that $f^{(1)}[0] - f^{(2)}[0] \in S_{-\delta}[0]$. The following proposition is given in [367]: *Assume that N is uniformly isometrically embedded in $\mathbf{R}^k$. Let $f^{(1)}$ and $f^{(2)}$ be two initial data which are small in $\dot{H}^{m/2} \times \dot{H}^{m/2-1}(\mathbf{R}^m, N)$. Then*

$$d_N(f^{(1)}, f^{(2)})_{\dot{H}^s \times \dot{H}^{s-1}} \approx ||f^{(1)} - f^{(2)}||_{\dot{H}^s \times \dot{H}^{s-1}}.$$

We construct a smooth one-parameter family $f^{(\alpha)}[0], \alpha \in [0, 1]$, of initial data which combines them, such that $f^{(\alpha)}$ are uniformly small in $S[0]$ and

$$\int_0^1 \left|\left|\frac{d}{d\alpha} f^{(\alpha)}[0]\right|\right|_{S_{-\delta}[0]} \preceq \left|\left|f^{(1)}[0] - f^{(2)}[0]\right|\right|_{S_{-\delta}[0]}.$$

The corresponding solutions are smooth and small in S, depending smoothly on $\alpha \in (0, 1)$ in the S topology.

The solutions $\frac{d}{d\alpha} f^{(\alpha)}$ to the linearized equation are also smooth. By (L5) (ii) they lie in $S_{-\delta}$ and by (L7) we have the estimate

$$\left|\left|\frac{d}{d\alpha} f^{(\alpha)}[0]\right|\right|_{S_{-\delta}} \preceq \left|\left|\frac{d}{d\alpha} f^{(\alpha)}\right|\right|_{S_{-\delta}[0]}.$$

Integrating this we conclude that

$$\left|\left|f^{(1)} - f^{(2)}\right|\right|_{S_{-\delta}} \preceq \left|\left|f^{(1)}[0] - f^{(2)}[0]\right|\right|_{S_{-\delta}[0]}.$$

We have shown this for smooth solutions, but rough solutions are constructed as limits of smooth solutions in $S \cap S_{-\delta}$, thus the above inequality can be transferred to rough solutions.

(6) Global continuous dependence: Let $f^{(k)}[0]$ be a sequence of initial data which is small in $S[0]$ and such that $f^{(k)}[0] - f[0] \to 0 \in S[0] \cap S_{-\delta}[0]$. Then it is equibounded in the sense that

$$\lim_{j\to\infty} \sup_k ||P_{>j} f^{(k)}[0]||_{S[0]} = 0.$$

Thus we can select a slowly decreasing sequence a_j (i.e., $i < j$ implies $a_i > a_j > 2^{-\delta|i-j|}a_i$ for a sufficiently small positive constant δ) such that

$$||P_{>j} f^{(k)}[0]||_{S[0]} \leq a_j, \quad \lim_{j\to\infty} a_j = 0. \tag{2.131}$$

If $f[0] \in S[0]$, then $f[0] \in S_c[0]$ for some admissible frequency envelope $c \in l^2$. If (2.131) holds, then c can be chosen satisfying $\sum_{i>j} c_i^2 \preceq a_j^2$. Since

$$||f^{(k)}||_{S_c} \preceq ||f^{(k)}[0]||_{S_c},$$

we also have

$$||P_{>j} f^{(k)}[0]||_S \leq a_j, \tag{2.132}$$

which holds uniformly for all solutions $f^{(k)}$. On the other hand, the weak stability result shows that $f^{(k)}[0] - f[0] \to 0 \in S_{-\delta}$. Combining this with the equiboundedness in (2.132), we conclude that $f^{(k)} \to f$ in S.

(7) Local continuous dependence: Let $\sigma > 0$ and $f^{(k)}[0]$, $f[0]$ be small initial data in $\dot{H}^{m/2} \times \dot{H}^{m/2-1}(B_\sigma)$ such that

$$\lim_{k\to\infty} f^{(k)}[0] = f[0] \text{ in } \dot{H}^{m/2} \times \dot{H}^{m/2-1}(B_\sigma).$$

We can extend them to functions in $S = \dot{H}^{m/2} \times \dot{H}^{m/2-1}(B_\sigma)$ so that this property still holds. Moreover, by (L2) we can also arrange that the extensions are initial data which are equal and constant outside of $B_{2\sigma}$. For the extension we use the global stability result. Then we apply the finite speed of propagation to conclude that the initial solutions fulfill

$$||\nabla(f^{(k)} - f)||_{L^\infty \dot{H}^{m/2-1}(\{|t|+|x|\leq\sigma\})} \to 0.$$

This completes the proof of the main theorem. □

2.8 Transversal Wave Maps

Transversal wave maps are transversally harmonic maps on Minkowski spaces. The equations of transversal wave maps form a second-order hyperbolic system of PDEs on the transverse manifolds, are different from the equations for transversally harmonic maps which form a second-order elliptic system of PDEs on the transverse manifolds. Our treatment of transversal wave maps is based on Chiang and Wolak [87].

2.8.1 Definitions and Examples

We follow the notions in Sect. 1.9.1. Let $\mathcal{H}$ be a foliation on a n-manifold (M, g) for the moment such that $\mathcal{H}$ is defined by a cocycle $\mathcal{U} = \{U_i, f_i, g_{ij}\}_{i\in I}$ modeled on a q-manifold N. Recall that the foliation $\mathcal{H}$ is said to be transversely semi-Riemannian (resp. Minkowskian, Lorentzian) if its normal bundle admits a semi-Riemannian (resp. Minkowskian, Lorentzian) metric h such that for any vector field X tangent to the leaves of $\mathcal{H}$ we have $L_X h = 0$. This condition is equivalent to the existence of an $\mathcal{H}_{\mathcal{U}}$-invariant semi-Riemannian (resp. Minkowskian, Lorentzian) metric $\bar{h}$ on the transverse manifold $N_{\mathcal{U}}$, cf. [230].

Let $\mathbf{R}^{1+m}$ be the $m+1$ dimensional Minkowski space with the metric $(\eta_{ab}) = diag(-1, 1, \cdots, 1)$ and the coordinates $x^0 = t, x^1, x^2, \cdots, x^m$, foliated by planes parallel to $\{0\} \times \mathbf{R}^p \subset \mathbf{R} \times \mathbf{R}^m$, $(p+q=m)$. Then $(\mathbf{R}^{1+m}, \mathcal{H}^p)$ is a transversally Minkowski foliation defined by the global submersion $\iota \times \phi\colon \mathbf{R} \times \mathbf{R}^m \to \mathbf{R} \times \mathbf{R}^q$; $\mathbf{R} \times \mathbf{R}^q$ can be considered as its complete transverse manifold.

Let $\mathcal{F}$ be a Riemannian foliation for a Riemannian metric g of an n-dimensional Riemannian manifold M which induces a Riemannian metric $\bar{g}$ on a q_1 $(p_1 + q_1 = n)$ dimensional transverse manifold $N_{\mathcal{U}} = \coprod_i \bar{U}_i$. Let $f : (\mathbf{R}^{1+m}, \mathcal{H}) \to (M, \mathcal{F})$ be a smooth foliated map from a foliated Minkowski space into a foliated Riemannian manifold. Denote $V_i = f^{-1}(U_i) \subset \mathbf{R}^{1+m}$ for each i. Let $\bar{V}_i$ be the quotient of V_i; then $\bar{V}_i$ is an open subset of $\mathbf{R}^{1+q}$ for each i. The map f induces a map $\bar{f} = \coprod_i \bar{f}_i : \coprod_i \bar{V}_i \to \coprod_i \bar{U}_i$ with $\bar{f}_i : \bar{V}_i \to \bar{U}_i$ such that the following diagram (for the sake of convenience, we drop the subscript i from $\bar{f}_i$ if there is no confusion)

$$\begin{array}{ccc} V_i \subset \mathbf{R}^{1+m} & \xrightarrow{f} & U_i \\ \Big\downarrow{\scriptstyle \iota\times\phi} & & \Big\downarrow{\scriptstyle \phi_1} \\ \bar{V}_i \subset \mathbf{R}^{1+q} & \xrightarrow{\bar{f}} & \bar{U}_i \end{array}$$

Diagram 2.8.1.

commutes, i.e., $\bar{f}_i \circ (\iota \times \phi) = \phi_1 \circ f_i$ for each i, where $\iota\times\phi : V_i \to \bar{V}_i$ is a submersion defined by the foliation $\mathcal{H}$ on an open subset V_i and $\phi_1 : U_i \to \bar{U}_i$ is a Riemannian submersion defining the foliation $\mathcal{F}$ on the open set U_i, $\iota(t) = t$. By taking a smaller V_i we can assume that $V_i = T_i \times W_i \subset \mathbf{R} \times \mathbf{R}^m$ and $\bar{V}_i = T_i \times \bar{W}_i \subset \mathbf{R} \times \mathbf{R}^q$, where T_i is an open subset of $\mathbf{R}$ and $\bar{W}_i$ is an open subset of $\mathbf{R}^q$. A transversal wave map $f : \mathbf{R}^{1+m} \to (M, \mathcal{F})$ is a transversally harmonic map with the transversal energy

$$E(\bar{f}) = \frac{1}{2}\int_{\coprod \bar{V}_i} (-|\bar{f}_t|^2 + |\nabla_x \bar{f}|^2) dt\, dx = \frac{1}{2}\int_{\coprod \bar{V}_i} \bar{g}_{rs}\left(-\bar{f}_t^r \bar{f}_t^s + \sum_{a=1}^{q} \bar{f}_a^r \bar{f}_a^s\right) dt\, dx_a. \tag{2.133}$$

The Euler-Lagrange equations describing the critical point of (2.132) are

$$\tau_{\Box}(\bar{f})^k = \Box \bar{f}^k + \bar{\Gamma}^k_{rs}(-\bar{f}^r_t \bar{f}^s_t + \bar{f}^r_a \bar{f}^s_a) = 0 \tag{2.134}$$

for each i, where $\Box = -\frac{\partial^2}{\partial t^2} + \Delta_x$ is the d'Alembertian and $\bar{\Gamma}^k_{rs}$ are the Christoffel symbols of each $\bar{U}_i$. Similarly to Sect. 1.9, we have the transversal wave field $(\tau_{\Box})_b(f)$ and the wave fields $\tau_{\Box}(\bar{f})$ of the induced maps $\bar{f}$, which are obtained by the local submersions defining the foliations $\mathcal{H}$ and $\mathcal{F}$ such that

$$d\phi_1 \circ (\tau_{\Box})_b(f)_x = \tau_{\Box}(\bar{f})_{(\iota\times\phi)(x)}.$$

Definition 2.8.1. The map f is a transversal wave map if and only if $\bar{f} : N_{\mathcal{V}} \to N_{\mathcal{U}}$ is a wave map (i.e., $\bar{f} : \bar{V}_i \to \bar{U}_i$ is a wave map locally for each i).

Since Diagram 2.8.1 commutes, the definition of a transversal wave map does not depend on the choice of local Riemannian submersions defining the Riemannian foliation.

Let $(M_1, \mathcal{F}_1, g_1)$ and $(M_2, \mathcal{F}_2, g_2)$ be two Riemannian manifolds with Riemannian foliations. Suppose that $h : (M_1, \mathcal{F}_1) \to (M_2, \mathcal{F}_2)$ is a smooth foliated leaf-preserving map, i.e. $dh(T\mathcal{F}_1) \subset T\mathcal{F}_2$. Let $U_i \subset M_i$ be open subsets and let $\phi_i : (U_i, g_i) \to (\bar{U}_i, \bar{g}_i)$ be Riemannian submersions on U_i which define locally the Riemannian foliations $\mathcal{F}_i$, $i = 1, 2$. Suppose that $h(U_1) \subset U_2$. There is a closed relationship between the transversally second fundamental form A of h and the second fundamental form A of the induced maps $\bar{h}$, obtained by using the local submersions defining the foliations $\mathcal{F}_1$ and $\mathcal{F}_2$. It follows from Sect. 1.9 and Diagram 1.9.1 that

$$d\phi_2 A_b(h)_x = A(\bar{h})_{\phi_1(x)} \tag{2.135}$$

holds for each of the foliation defining local submersions $\phi_i : U_i \to \bar{U}_i$ such that $h(U_1) \subset U_2$.

Definition 2.8.2. $h : (M_1, \mathcal{F}_1) \to (M_2, \mathcal{F}_2)$ is a *transversally totally geodesic* map if $A(\bar{h})_{\phi_1(x)} = \nabla d(\bar{h})_{\phi_1(x)} = 0$ in each $\bar{U}_1$, where ∇ is the connection on $T^*\bar{U}_1 \otimes \bar{h}^{-1}T\bar{U}_2$.

Proposition 2.8.3. *If* $f : (\mathbf{R}^{1+m}, \mathcal{H}) \to (M_1, \mathcal{F}_1)$ *is a transversal wave map and* $h : (M_1, \mathcal{F}_1) \to (M_2, \mathcal{F}_2)$ *is a transversally totally geodesic map, then* $h \circ f$ *is a transversal wave map.*

Proof. $f : (\mathbf{R}^{1+m}, \mathcal{H}) \to (M_1, \mathcal{F}_1)$ and $h : (M_1, \mathcal{F}_1) \to (M_2, \mathcal{F}_2)$ induce maps $\bar{f} : N_{\mathcal{V}} \to N_{\mathcal{U}}$ and $\bar{h} : N_{\mathcal{U}} \to N_{\mathcal{V}}$. By Eells and Sampson [129], we have

$$\tau_{\Box}(\bar{h} \circ \bar{f}) = d\bar{h} \circ \tau_{\Box}(\bar{f}) + trace\, \nabla d\bar{h}(d\bar{f}, d\bar{f}). \tag{2.136}$$

Since f is a transversal wave map and h is a transversally totally geodesic map, the result follows. We note that the transversally total geodesicity of h cannot be weakened to transversal harmonicity.

Example 1. Let $(M_1, \mathcal{F}_1)$ be a foliated submanifold of $(M_2, \mathcal{F}_2)$ such that the traces of leaves of $\mathcal{F}_2$ on M_1 are leaves of $\mathcal{F}_1$. This condition implies that for suitable choices of foliation cycles, the transverse manifold N_1 is a submanifold of the transverse manifold N_2. Are the transversal wave maps into $(M_1, \mathcal{F}_1)$ also transversal wave maps into $(M_2, \mathcal{F}_2)$? By Proposition 2.8.3, the answer is affirmative if $(M_1, \mathcal{F}_1)$ is a transversally totally geodesic foliated submanifold of $(M_2, \mathcal{F}_2)$, i.e. $N_1 = \coprod_i (\bar{U}_1)$ is a totally geodesic submanifold of N_2, that is, N_1 geodesics are also N_2 geodesics. If γ is a transversal geodesic of $(M_1, \mathcal{F}_1)$, i.e. $\bar{\gamma} = \phi \circ \gamma : \mathbf{R} \to U_1 \to \bar{U}_1$ is a N_1 geodesic, then $\bar{\gamma}$ is also a N_2 geodesic, $\bar{\gamma}$ has dimension one and has no curvature. For a map $v : \mathbf{R}^{1+m} \to \mathbf{R}$, let $u = \gamma \circ v : \mathbf{R}^{1+m} \to \mathbf{R} \to U_1$, which induces $\bar{u} = \bar{\gamma} \circ \bar{v} : \bar{V}_1 \to \mathbf{R} \to \bar{U}_1$. By (2.136) we have

$$\Box \bar{u} = \dot{\bar{\gamma}} \Box \bar{v} + trace\, \nabla \dot{\bar{\gamma}}(d\bar{v}, d\bar{v}), \tag{2.137}$$

where the second term vanishes since $\bar{\gamma}$ is a geodesic. Therefore, u is a transversal wave map iff $\bar{v}$ solves the homogeneous linear wave equation $\Box \bar{v} = 0$. Hence, by (2.137), with respect to the arc length parameterization, the transversal wave map equation into $\bar{\gamma}$ is equivalent to a linear wave equation. Then for any target foliated manifold $(M_2, \mathcal{F}_2)$ we can provide many transversal wave maps associated to the transversal geodesics of $(M_2, \mathcal{F}_2)$.

There are wave maps which are not transversal wave maps. We construct in Example 2 a wave map that is not a transversal wave map by using a warped product of two manifolds. By O'Neill [279], a warped product can be defined for semi-Riemannian manifolds (i.e., pseudo-Riemannian manifolds) or Riemannian manifolds. Let (B, g), (F, h) be semi-Riemannian manifolds or Riemannian manifolds and $\alpha : B \to \mathbf{R}$ be a smooth function. On the product manifold $B \times F$, consider the metric tensor $k = g \oplus e^{2\alpha} h$. Let ∇^g and ∇^h be the Levi-Civita connections on (B, g) and (F, h) respectively. The Levi-Civita connection ∇^k on $B \times F$ can be related to those on B and F as in Sect. 1.9.

Example 2. Let $f : B_1 \times F_1 \to B_2 \times F_2$ be a smooth map preserving the leaves of the foliations such that $f(t, x, y) = (f_1(t, x), f_2(t, x, y))$, where $B_1 = \mathbf{R} \times \mathbf{R} = \mathbf{R}^{1,1}$, $F_1 = \mathbf{R}$, $B_2 = F_2 = \mathbf{R}, \alpha_1(x) = 0, \alpha_2(x) = x, f_1(t, x) = t + 8x^2$, $f_2(t, x, y) = 4y$. Similarly to (1.58) and (1.59), we have

$$\tau_\Box(f) = \tau_\Box(f_1) + \tau_\Box(f_2|_{B_1}) + \tau_\Box(f_2|_{F_1}) - ||df_2||^2 (grad_{g_2} \alpha_2) \circ f_1 = 16 - 16 = 0$$

(the second and third terms vanish). But, $\tau_\Box(f_1) = 16 \neq 0$. Therefore, f is a wave map, but it is not a transversal wave map.

However, if we impose some additional assumptions, then we obtain the following result.

Theorem 2.8.4. *Let $f : (\mathbf{R}^{1+m}, \mathcal{H}) \to (M, \mathcal{F})$ be a leaf-preserving wave map from a foliated Minkowski space into a foliated Riemannian manifold. Suppose that all the leaves of the minimal foliation $\mathcal{H}$ are minimal, the foliation $\mathcal{F}$ is totally geodesic, and f is horizontal (i.e., $df(T\mathcal{H}^\perp) \subset T\mathcal{F}^\perp$). Then f is a transversal wave map.*

Proof. The wave map f induces maps $\bar{f} : \bar{V} \to \bar{U}$ as in Diagram 2.8.1 (we drop i for convenience). Because the properties of wave and transversal wave maps are local, we consider open subsets $V \subset \mathbf{R}^{1+m}$ and $U \subset M$ with Riemannian submersions $\iota \times \phi$ and ϕ_1 such that the foliations on V and U are fibres of $\iota \times \phi$ and ϕ_1, and $f(V) \subset U$. On one hand, by Eells and Sampson [129] we have

$$\begin{aligned}\tau_\Box(\bar{f} \circ (\iota \times \phi)) &= d\bar{f}(\tau_\Box(\iota \times \phi)) + trace\, \nabla d\bar{f}(d(\iota \times \phi), d(\iota \times \phi)) \\ &= trace_{T\mathcal{H}^\perp} \nabla d\bar{f}(d(\iota \times \phi), d(\iota \times \phi)) = \tau_\Box(\bar{f}),\end{aligned}$$

where the second equality holds since $\tau_\Box(t, \phi) = 0$ and the last equality follows from the fact that the trace over the tangent space to the leaves vanishes since the leaves of $\mathcal{H}$ are minimal.

On the other hand, we derive

$$\begin{aligned}\tau_\Box(\phi_1 \circ f) = d\phi_1(\tau_\Box(f)) + trace\, \nabla d\phi_1(df, df) = trace_{T\mathcal{H}} \nabla d\phi_1(df, df) \\ + trace_{T\mathcal{H}^\perp} \nabla d\phi_1(df, df) = 0,\end{aligned}$$

where the second equality holds since $\tau_\Box(f) = 0$, and the third equality holds since $\mathcal{F}$ is totally geodesic and $trace_{T\mathcal{H}^\perp} \nabla d\phi_1(df, df) = 0$, since ϕ_1 is a Riemannian submersion and the connection on $\bar{U}$ is projected from U. Hence, we can conclude the result by the commutative Diagram 2.8.1. □

The following example shows that there are also transversal wave maps which are not wave maps.

Example 3. Let (B_1, g_1), (B_2, g_2), (F_1, h_1) and (F_2, h_2) be Riemannian manifolds. Consider the foliations on the Riemannian manifolds $B_1 \times F_1$ and $B_2 \times F_2$ given by the projections on the first component $\pi_1 : B_1 \times F_1 \to B_1$ and $\pi_2 : B_2 \times F_2 \to B_2$, respectively. The π_1 and π_2 are Riemannian submersions, and the foliations defined by them are Riemannian. Let $h : B_1 \times F_1 \to B_2 \times F_2$ be a smooth map which preserves the leaves of the foliations. Then h must be of the form $h(x, y) = (h_1(x), h_2(x, y))$, $x \in B_1$, $y \in F_1$, where $h_1 : B_1 \to B_2$, $h_2 : B_1 \times F_1 \to F_2$ are smooth. For the product Riemannian metrics on $B_1 \times F_1$ and $B_2 \times F_2$, the connection of dh is equal to

$$\nabla d(h) = (\nabla d(h_1), \nabla d(h_2|_{B_1}) + \nabla d(h_2|_{F_1})), \tag{2.138}$$

where $\nabla d(h_1)$ is the connection derivative of dh_1 at x of the map $h_1 : B_1 \to B_2$, $\nabla d(h_2|_{B_1})$ is the connection derivative of dh_2 at x of the map $x \mapsto h_2(x, y)$ for fixed y, and $\nabla d(h_2|_{F_1})$ is the connection derivative of dh_2 at y of the map $y \mapsto h_2(x, y)$ for fixed x. On one hand, by (2.138) the property "totally geodesic" of $h = (h_1, h_2)$ is equivalent to h_1 is totally geodesic and $\nabla d(h_2|_{B_1}) + \nabla d(h_2|_{F_1}) = 0$, i.e., the vertical and horizontal contributions to the totally geodesic annihilate each other. On the other hand, if h_1 is totally geodesic and $h_2|_{B_1}$, $h_2|_{F_1}$ are totally geodesic for $x \in B_1, y \in F_1$, then h is totally geodesic. Therefore, it follows that there are maps h which are transversally totally geodesic, but not totally geodesic. Hence, by (2.138) there are transversal wave maps which are not wave maps.

2.8.2 Properties

Let $f : \mathbf{R}^{1+m} \to (M, \mathcal{F}, g)$ be a smooth foliated map from a foliated Minkowski space to a foliated Riemannian manifold with a transverse manifold $(N, \bar{g}) = (\coprod \bar{U}_i, \bar{g})$ which induces $\bar{f} : \bar{V}_i \to \bar{U}_i$ for each i. The *transversal stress energy* is defined by $T(\bar{f}) = e(\bar{f})\eta - \bar{f}^*\bar{g}$, where $e(\bar{f}) = \frac{1}{2}||d\bar{f}||^2$ is the energy density for each $\bar{f} : \bar{V}_i \to (\bar{U}_i, \bar{g}) \subset (N, \bar{g})$, $\eta = \begin{pmatrix} -1 & 0 \\ 0 & I \end{pmatrix}$, I is a $q \times q$ matrix. f satisfies the *transverse conservation law* if $div\, T(\bar{f}) = 0$ for each i.

Theorem 2.8.5. (1) *Let $f : \mathbf{R}^{1+m} \to (M, \mathcal{F})$ be a smooth foliated map from a foliated Minkowski space into a foliated Riemannian manifold with a transverse manifold $N = \coprod \bar{U}_i$ which induces $\bar{f} : \bar{V}_i \to \bar{U}_i$ for each i. Then*

$$div\, T(\bar{f})(X) = - < \tau_{\Box}(\bar{f}), d\bar{f}(X) >, \ \forall X \in T\bar{V}_i. \tag{2.139}$$

(2) *If $f : \mathbf{R}^{1+m} \to (M, \mathcal{F})$ is a transversal wave map, then f satisfies the transverse conservation law.*

Proof. (1) Let $x^0 = t, x^1, \cdots, x^q$ be the coordinates in $\bar{V}_i \subset \mathbf{R}^{1+q}$, and $e_0 = \frac{\partial}{\partial t}$, $e_1 = (1, 0, \cdots, 0), \cdots, e_q = (0, 0, \cdots, 1)$. For each $\bar{f} : \bar{V}_i \to \bar{U}_i$, we compute

$$\begin{aligned}
div\, T(\bar{f})(X) &= \nabla_{e_i} T(\bar{f})(e_i, X) = \nabla_{e_i} (\frac{1}{2}|d\bar{f}|^2 \begin{pmatrix} -1 & 0 \\ 0 & I \end{pmatrix} - \bar{f}^*\bar{g})(e_i, X) \\
&= \nabla_{e_i} \left(\frac{1}{2}|d\bar{f}|^2 \begin{pmatrix} -1 & 0 \\ 0 & I \end{pmatrix} (e_i, X) \right) - (\nabla_{e_i} \bar{f}^*\bar{g})(e_i, X) \\
&= \left[-(\nabla \frac{\partial \bar{f}}{\partial t}, \frac{\partial \bar{f}}{\partial t})(-1) + (\nabla \frac{\partial \bar{f}}{\partial x_i}, \frac{\partial \bar{f}}{\partial x_i})(I) \right] (e_i, X) \\
&\quad - \nabla_{e_i} (\bar{f}_* e_i, \bar{f}_* X) \\
&= \left[(\nabla \frac{\partial \bar{f}}{\partial t}, \frac{\partial \bar{f}}{\partial t})(e_i, X) + (\nabla \frac{\partial \bar{f}}{\partial x_i}, \frac{\partial \bar{f}}{\partial x_i})(e_i, X) \right] \\
&\quad - (\nabla_{e_i} \bar{f}_* e_i, \bar{f}_* X) - (\bar{f}_* e_i, \nabla_{e_i} \bar{f}_* X) \\
&= ((\nabla_X d\bar{f}) e_i, \bar{f}_* e_i) - (\tau_{\Box}(\bar{f}), \bar{f}_* X) - (\bar{f}_* e_i, \nabla_{e_i} \bar{f}_* X),
\end{aligned} \tag{2.140}$$

where the first term and the third term cancel out and $\nabla_{e_i} \bar{f}_* e_i = \tau_{\Box}(\bar{f})$.
(2) Since $f : \mathbf{R}^{1+m} \to (M, \mathcal{F})$ is a transversal wave map which induces $\bar{f} : \bar{V}_i \to \bar{U}_i$ such that $\tau_{\Box}(\bar{f}) = 0$ for each i, one has $div\, T(\bar{f}) = 0$ by (2.139). Therefore, f satisfies the transverse conservation law. □

Let $f : \mathbf{R}^{1+m} \to (M, \mathcal{F}, g)$ be a smooth foliated map from a foliated Minkowski space into a foliated Riemannian manifold with a transverse manifold $(N, \bar{g}) = (\coprod \bar{U}_i, \bar{g})$ which induces $\bar{f} : \bar{V}_i \to (\bar{U}_i, \bar{g}) \subset (N, \bar{g})$ for each i. f is *transversally pseudo-weakly conformal* if there is a smooth function $\mu : \bar{V}_i \to \mathbf{R}$ such that $\bar{f}^*(\bar{g}) = \mu \begin{pmatrix} -1 & 0 \\ 0 & I_{q\times q} \end{pmatrix}$ for each i. f is *transversally homothetic* if μ is constant for each i.

Proposition 2.8.6. *Let $f : \mathbf{R}^{1+m} \to (M, \mathcal{F})$ be a transversal wave and transversally pseudo-weakly conformal map which on the level of the transverse manifold N^q induces a map $\bar{f} : \bar{V}_i \to (\bar{U}_i, \bar{g})$ for each i. If $q \neq 3$, then f is transversally homothetic.*

Proof. Since f is transversal wave and transversally pseudo-conformal, f induces $\bar{f} : \bar{V}_i \to \bar{U}_i$ for each i and we have

$$T_{\bar{f}} = e(\bar{f})\eta - \bar{f}^*\bar{g} = \frac{q-3}{2}\mu\,\eta,\ \eta = \begin{pmatrix} -1 & 0 \\ 0 & I_{q\times q} \end{pmatrix}.$$

By Theorem 2.8.5, $div\, T_{\bar{f}} = 0$ for each i we have

$$\frac{q-3}{2}\mu_{,j}\,\eta_{ij} = 0\ (0 \leq i \leq q), \quad (q \neq 3)$$

which implies $d\mu = 0$. Therefore μ is constant for each i, and so f is transversally homothetic.

Chapter 3
Yang-Mills Fields

Yang-Mills fields are the critical points of the Yang-Mills functionals of connections whose curvature tensors are harmonic. They were first explored by several physicists in the 1950s. Yang-Mills theory had a profound impact on the developments of differential and algebraic geometry over the last quarter of the twentieth-century. Here is a brief online: (1) In the calculus of variations associated with the Yang-Mills functional, the emphasis was on differential geometric aspects, in the well-known results of Bourguignon and Lawson [38, 39], and on analytic aspects, in the famous results of Uhlenbeck [382, 383]. (2) Algebraic-geometric aspects, involving Ward's description of the Yang-Mills instantons in terms of holomorphic bundles over Penrose's twistor space, leading to the description of solutions via the ADHM construction [15]. The reader is referred to the overview of Yang-Mills theory by Donaldson and Kronheimer [102] and Donaldson [101] for details.

In the last two decades, Taubes [369] introduced a highly technical method to attack the questions in the calculus of variations. He also took the critical step of studying Yang-Mills instantons over general Riemannian four-dimensional manifolds [368] (which was different than the previous work focused on symmetric spaces or "self-dual" manifolds [16]). In both cases, one can have small, highly concentrated "bubble-like" instantons, related to the conformal invariance of Yang-Mills theory in four dimensions.

The instanton equations and the moduli spaces of their solutions were applied to solve problems in four-dimensional topology. Two main themes were first to show that certain interaction forms could not be realized by smooth four-manifolds and second, to define new invariants distinguishing smooth manifolds with same interaction forms. This development happened about the same time as, coming from a different direction, Freedman produced his theory of topological four-manifolds.

The Hitchin-Kobayshi conjecture (which was established in [99, 100, 388] by different methods) set out a general relation between Yang-Mills theory over complex Kähler manifolds and holomorphic bundles, specially Mumford's theory of "stability".

Y.-J. Chiang, *Developments of Harmonic Maps, Wave Maps and Yang-Mills Fields into Biharmonic Maps, Biwave Maps and Bi-Yang-Mills Fields*, Frontiers in Mathematics, DOI 10.1007/978-3-0348-0534-6_3, © Springer Basel 2013

In this chapter, we concentrate on the developments of the differential geometric and analytic aspects of Yang-Mills fields by Bourguignon and Lawson [38, 39], Uhlenbeck [382, 383], Price [304], Nakajima [276], Tian [374], etc. We show the interaction between Yang-Mills connections, which are critical points of Yang-Mills functionals associated to vector bundles, and minimal submanifolds, which have been investigated for years. We finally make a brief overview of Taubes' work [368–370]. Recently, Taubes has studied the Seiberg-Witten equations further in his papers [371–373].

3.1 Yang-Mills Fields: Differential Geometric Aspects

3.1.1 *Preliminaries*

Let M be a compact Riemannian manifold and P be a principal G-bundle over M, where G is a compact Lie group. On the space $\mathcal{C}_P$ of connections on G, we consider the Yang-Mills fields functional $\mathcal{YM}(D) = \frac{1}{2}\int_M ||R^D||^2 dv_M$, where R^D is the curvature of the connection $D \in \mathcal{C}_P$. The critical points of the smooth functional $\mathcal{YM} : \mathcal{C}_P \to \mathbf{R}$ are those connections whose curvature tensors are 'harmonic'. These critical points are called *Yang-Mills connections* and their associated curvature tensors are called *Yang-Mills fields.*

In general, let (E, h) be a real vector bundle of rank N with an inner product over an m-dimensional compact Riemannian manifold (M, g). Let $\mathcal{C}(E, h)$ be the space of all C^∞ connections of E satisfying the compatibility condition

$$X(s,t)_h = (D_X s, t)_h + (s, D_X t)_h, \quad s, t \in \Gamma(E),$$

where X is a vector field in $\mathcal{X}(M)$ (the set of all vector fields on M) and $\Gamma(E)$ is the space of all C^∞ sections of E. For $D \in \mathcal{C}(E, h)$, let R^D be its curvature tensor, defined by

$$R(X,Y)s = D_X(D_Y s) - D_Y(D_X s) - D_{[X,Y]}s,$$

for all $X, Y \in \mathcal{X}(M)$ and all $s \in \Gamma(E)$. Let $F = End(E, h)$ be the bundle of endomorphisms of E which are skew-symmetric with respect to the inner product h on E. We define the inner product $<,>$ on F by

$$<\phi, \psi> = \sum_{i=1}^{N} (\phi u_i, \psi u_i)_h, \quad \phi, \psi \in F_x,$$

where $\{u_i\}_{i=1}^N$ is an orthonormal basis of E_x with respect to h and $x \in M$. Let $\Omega^k(F) = \Gamma(\wedge^k T^*M) \otimes F$ be the space of F-valued k-forms on M, which admits the global inner product $(\cdot,\cdot)$ defined by

$$(\alpha,\beta)=\int_M <\alpha,\beta> dv_g,$$

where the pointwise inner product $<\alpha,\beta>$ is given by

$$<\alpha,\beta>=\sum_{i_1<\cdots<i_k} <\alpha(e_{i_1},\cdots,e_{i_k}),\beta(e_{i_1},\cdots,e_{i_k})>.$$

Here $\{e_i\}_{i=1}^m$ is a locally defined orthonormal frame on (M,g).

For every $D\in C(E,h)$, let $d^D:\Omega^k(F)\to\Omega^{k+1}(F)$ be the exterior differentiation with respect to D. Then the adjoint operator $\delta^D:\Omega^{k+1}(F)\to\Omega^k(F)$ is defined by

$$\delta^D\alpha=(-1)^{k+1}*d^D*\alpha,\quad \alpha\in\Omega^{k+1}(F),$$

where $*:\Omega^k(F)\to\Omega^{m-k}(F)$ is the extension of the usual Hodge star operator on (M,g). Then we have

$$(d^D\alpha,\beta)=(\alpha,\delta^D\beta),\ \alpha\in\Omega^k(F),\quad \beta\in\Omega^{k+1}(F).$$

Let

$$\mathcal{YM}(D)=\frac{1}{2}\int_M ||R^D||^2dv,\ D\in C(E,h) \tag{3.1}$$

be the Yang-Mills functional, where $||R^D||$ is the norm of $R^D\in\Omega^2(F)$ with respect to $<,>$. $D\in C(E,h)$ is a *Yang-Mills connection* if for any smooth one-parameter family D^t ($|t|<\epsilon$) with $D^0=D$, $\frac{d}{dt}\big|_{t=0}\mathcal{YM}(D^t)=0$. Its associated curvature R^D is called a *Yang-Mills field.*

Definition 3.1.1. Let $\alpha=\frac{d}{dt}\big|_{t=0}D^t\in\Omega^1(F)$. Then we have

$$\frac{d}{dt}\Big|_{t=0}\mathcal{YM}(D^t)=\int_M <\delta^D R^D,\alpha> dv, \tag{3.2}$$

where $R^D(\beta)\in\Omega^1(F)(\beta\in\Omega^1(F))$ is defined by

$$R^D(\beta)(X)=\sum_{j=1}^m[R^D(e_j,X),\beta(e_j)],\quad X\in\mathcal{X}(M).$$

Hence, R^D is a Yang-Mills field if and only if $\delta^D R^D=0$.

Remark. We may compute the first variation of the functional informally as follows:

$$\frac{d}{dt}\Big|_{t=0}\mathcal{YM}(D^t) = \frac{1}{2}\int_M < \frac{d}{dt}R^{D_t}\frac{d}{dt}\Big|_{t=0}D^t, R^{D_t} > dv + \frac{1}{2}\int_M < R^{D_t}, \frac{d}{dt}R^{D_t}\frac{d}{dt}\Big|_{t=0}D^t > dv$$
$$= \int_M < \delta^D R^D, \alpha > dv.$$

Let P be a principal fiber bundle over a manifold M with structure group G and canonical projection π, and $\mathcal{G}$ be the Lie algebra of G. A connection A can be considered locally as a $\mathcal{G}$-valued 1-form $A = A_\mu(x)dx^\mu$. The curvature of the connection A is defined to be the 2-form $F = F_{\mu\nu}dx^\mu dx^\nu$, with

$$F_{\mu\nu} = \partial_\mu A_\nu - \partial_\nu A_\mu + [A_\mu, A_\nu].$$

Then the Yang-Mills Lagrangian is

$$L(A) = \frac{1}{2}\int_M F_{\mu\nu}F^{\mu\nu}dv, \tag{3.3}$$

which corresponds to (3.1). The Euler-Lagrange equations have the form

$$D^\mu F_{\mu\nu} = 0, \tag{3.4}$$

which corresponds to (3.2). For $M = \mathbf{R}^{1+d}$ the Minkowski space, and letting $G = SO(d)$ be the group of orthogonal transformations on $\mathbf{R}^d$, we have that $A_\mu(x)$ is a $d \times d$ skew-symmetric matrix A_μ^{ij}. The appropriate equivariant ansatz is

$$A_\mu^{ij}(x) = (\delta_\mu^i x^j - \delta_\mu^j x^i)h(t, |x|),$$

where $h : M \to \mathbf{R}$ is a spatially radial function. Setting $u = r^2 h$, $r = |x|$ and $m = d - 2$, the Yang-Mills system reduces to the following scalar equation for u:

$$u_{tt} - u_{rr} - \frac{m-1}{r}u_r + \frac{2m}{r^2}(1-u)(1-\frac{1}{2}u) = 0, \tag{3.5}$$

which is a wave equation.

3.1.2 *The Bochner-Weitzenböck Formula*

We fix a compact Lie group G and a principal G-bundle P over a compact Riemannian m-manifold M. Let $E = P \times \mathbf{R}^N$ be a G-vector bundle, associated to P by a faithful orthogonal representation $\rho : G \to O(N)$. (Note that if P is given by transition functions $g_{\alpha\beta} : U_\alpha \cap U_\beta \to G$, where $\{U_\alpha\}$ is an open covering of M, then E is given by the transition functions $\rho \circ g_{\alpha\beta} : U_\alpha \cap U_\beta \to O(N)$.)

Let $\Delta^D = d^D\delta^D + \delta^D d^D$ be the *Hodge-de Rham Laplacian* for vector bundle valued exterior k-forms. Any form $\phi \in \Omega^k(\mathcal{G}_E)$ satisfying $\Delta^D\phi = 0$ is called *harmonic*. Hence, Yang-Mills connections are connections with harmonic curvature. Further, $D^*D\phi = -\sum_{j=1}^m (D^2_{e_j,e_j}\phi)$ is another second-order operator called, the *rough Laplacian*, defined on $\mathcal{G}_E$-valued differential forms, where $D^2_{X,Y} = D_X D_Y - D_{(\nabla_X Y)}$. Here D is the connection on E, and ∇ is the Levi-Civita connection on M. We try to establish the relationship between the Hodge-de Rham Laplacian and the rough Laplacian of $\mathcal{G}_E$-valued 1- and 2-forms. They involve both the curvature of the Riemannian base manifold and that of the bundle with connection.

We define an operator $\mathcal{R}^D : \Omega^1(\mathcal{G}_E) \to \Omega^1(\mathcal{G}_E)$ by

$$\mathcal{R}(\phi)_X = \sum_{j=1}^m [R^D_{e_j,X}, \phi_{e_j}], \tag{3.6}$$

where R^D is the curvature of the connection D on E and $\{e_1, \cdots, e_m\}$ is an orthonormal basis of the tangent space T_xM at a point x in M. Recall that the Ricci curvature $Ric : T_xM \to T_xM$ is defined by

$$Ric(X) = \sum_{j=1}^m R_{X,e_j} e_j,$$

where R is the Riemannian curvature. For $\phi \in \Omega^1(\mathcal{G}_E)$ we define

$$(\phi \circ Ric)_X = \phi_{Ric(X)}.$$

Remark that on the standard sphere $Ric(X) = (m-1)X$.

All the following theorems and results were obtained by Bourguignon and Lawson [37, 39].

Theorem 3.1.2. *For any $\phi \in \Omega^1(\mathcal{G}_E)$, we have*

$$\Delta^D\phi = D^*D\phi + \phi \circ Ric + \mathcal{R}^D(\phi). \tag{3.7}$$

Proof. For a fixed point x in M, choose X, $e_1, \cdots, e_m$ in T_xM such that $\{e_1, \cdots, e_m\}$ is an orthonormal basis. Extend X to a local vector field and $\{e_1, \cdots, e_m\}$ to a local orthonormal frame field such that $(\nabla X)(x) = (\nabla e_1)(x) = \cdots = (\nabla e_m)(x) = 0$, where ∇ is the Levi-Civita connection on M. Then at each x,

$$(d^D\delta^D\phi)_X = D_X(\delta^D\phi) = -D_X\left(\sum_{j=1}^m (D_{e_j}\phi)e_j\right) = -\sum_{j=1}^m (D^2_{X,e_j}\phi)e_j \tag{3.8}$$

and

$$(\delta^D d^D \phi)_X = -\sum_{j=1}^{m} (D_{e_j} d^D \phi)_{e_j,X} = -\sum_{j=1}^{m} D_{e_j} \left[(D_{e_j}\phi)_X - (D_X \phi)_{e_j} \right]$$

$$= -\sum_{j=1}^{m} \left[(D^2_{e_j,e_j}\phi)_X - (D^2_{e_j,X}\phi)_{e_j} \right]. \tag{3.9}$$

Summing (3.8) and (3.9), we have

$$(\triangle^D \phi)_X = (D^* D\phi)_X + \sum_{j=1}^{m} (R^D_{e_j,X}\phi)_{e_j} \tag{3.10}$$

(here R^D is the curvature of the connection on the bundle $T^*M \otimes \mathcal{G}_E$ induced by the Levi-Civita connection ∇ on the base M and the connection D on the bundle E).

□

The operator R^D acts as a derivation. Thus

$$(R^D_{X,Y}\phi)_Z = [R^D_{X,Y}, \phi_Z] - \phi_{R_{X,Y}Z}.$$

Applying this formula to the second term of (3.10) yields the result.

Corollary 3.1.3. *If ϕ is a $\mathcal{G}_E$-valued D-harmonic 1-form on the sphere S^m, then*

$$D^* D\phi = -(m-1)\phi - \mathcal{R}^D(\phi). \tag{3.11}$$

We can carry out the similar computation for arbitrary k-forms. In particular, when $k = 2$, similarly to (3.6), we define $\mathcal{R}^D : \Omega^2(\mathcal{G}_E) \to \Omega^2(\mathcal{G}_E)$ by

$$\mathcal{R}^D(\phi)_{X,Y} = \sum_{j=1}^{m} [(R^D_{e_j,X}, \phi_{e_j,Y}) - (R^D_{e_j,Y}, \phi_{e_j,X})]. \tag{3.12}$$

For a linear map ω on 2-vectors we define

$$(\phi \circ \omega)_{X,Y} = \frac{1}{2} \sum_{j=1}^{m} \phi_{e_j, \omega_{X,Y} e_j}$$

(this is the decomposition of ω and ϕ viewed as maps from $\Lambda^2 TM$ to $\mathcal{G}_E$). Note that $\phi_{X,Y} = \phi(X \wedge Y)$. Observe that the extension of the Ricci transformation to 2-forms as $Ric \wedge I$ is given by

$$(Ric \wedge I)_{X,Y} = Ric(X) \wedge Y + Y \wedge Ric(Y).$$

Applying the above formulas and arguing similarly to the proof as Theorem 3.1.2, we can obtain the following result.

Theorem 3.1.4. *For any $\phi \in \Omega^2(\mathcal{G}_E)$ we have*

$$\Delta^D \phi = D^* D\phi + \phi \circ (Ric \wedge I + 2R) + \mathcal{R}^D(\phi). \tag{3.13}$$

The sign in (3.13) in front of $\mathcal{R}^D$ may change to $-$ by convention. The above formula has the advantage of incorporating the influence of the curvature of the base manifold into one term. This is particularly interesting in dimension 4, where we can check by decompositing the Riemannian curvature into irreducible components under the action of $O(4)$ or $SO(4)$ that $Ric \wedge I + 2R$ does not involve the Ricci traceless part of the Riemannian curvature.

Corollary 3.1.5 (Bochner-Weitzenböck formula). *Let ϕ be a $\mathcal{G}_E$-valued harmonic 2 form on the sphere S^m. Then*

$$D^* D\phi = -2(m-2)\phi - \mathcal{R}^D(\phi). \tag{3.14}$$

3.1.3 Stability

Using the notions and notations in Sect. 3.1.1, we discuss the second variation of the Yang-Mills functional, which is precisely the first variation of bi-Yang-Mills functional. See the proof of Theorem 3.1.6 in Sect. 6.1, which is exactly the same as the proof of Theorem 6.1.2.

Theorem 3.1.6 ([39]). *Let $D \in \mathcal{C}(E,h)$ be a connection. For any smooth one-parameter family D^t ($|t| < \epsilon$) with $D^0 = D$. We have*

$$\frac{d^2}{dt^2}\Big|_{t=0} \mathcal{YM}(D^t) = \int_M < (\delta^D d^D + R^D)(\alpha), \alpha > dv, \tag{3.15}$$

where $\alpha = \frac{d}{dt}D^t\big|_{t=0}$. If $\delta^D\alpha = 0$, then

$$\frac{d^2}{dt^2}\Big|_{t=0} \mathcal{YM}(D^t) = \int_M < S^D(\alpha), \alpha > dv, \tag{3.16}$$

where $S^D = (d^D\delta^D + \delta^D d^D)(\alpha) + R^D(\alpha)$ and is a second-order self-adjoint elliptic differential operator acting on $\Omega^1(F)$.

If $D \in \mathcal{C}(E,h)$ is a Yang-Mills connection, then S^D is a self-adjoint elliptic differential operator. Let E_λ be the eigenspace of S^D on $\Omega^1(F)$ with eigenvalue λ. Clearly, S^D leaves $Ker(\delta^D)$ invariant. Thus the restriction of S^D to $Ker(\delta^D)$ has a discrete spectrum consisting of distinct eigenvalues $\lambda_1 < \lambda_2 < \cdots < \lambda_i < \cdots \to \infty$ corresponding to finite-dimensional eigenspaces E_{λ_i}. Then the *index* and *nullity* of D are defined by

$$index(D) = dim\left(\bigoplus_{\lambda<0} E_\lambda\right), \quad nullity(D) = dim(E_0).$$

Definition 3.1.7. A Yang-Mills field is *stable* if $i(D) = n(D) = 0$, i.e., $\frac{d^2}{dt^2}\Big|_{t=0} \mathcal{YM}(D^t) > 0$.

Observe that the weak stability $\frac{d^2}{dt^2}\big|_{t=0}\mathcal{YM}(D^t) \geq 0$ is equivalent to the condition $i(D) = 0$. In particular, stability (i.e. $i(D) = n(D) = 0$) implies weak stability (i.e., $i(D) = 0$).

From now on, we fix a compact Lie group G and a principal G-bundle P over a compact Riemannian manifold M. Let $E = P \times \mathbf{R}^N$ be a G-vector bundle, associated to P by a faithful orthogonal representation $\rho : G \to O(N)$.

Let $S^m = \{x \in \mathbf{R}^{m+1} | \, ||x|| = 1\}$ be the Euclidean m-sphere. On S^m there is a finite-dimensional family of vector fields which tend to decrease the energy of any Yang-Mills field. Using this family on can find some restrictions on any weakly stable field on S^4. Specially, when $G = SU_2, SU_3$ or U_2, one can show that any such field is either self dual or anti-self dual.

Fix a Yang-Mills connection on a Riemannian manifold M and consider a 2-form $\phi \in \Omega^2(\mathcal{G}_E)$. Then for each tangent vector field V on M, the contraction $i_V\phi \in \Omega^1(\mathcal{G}_E)$ defined by

$$(i_V\phi)_X = \phi_{V,X} \tag{3.17}$$

is an infinitesimal variation of the connection. Assume that V is of gradient type, i.e.,

$$< \nabla_X V, Y > = < \nabla_Y V, X > \text{ for all} X, Y, \tag{3.18}$$

where ∇ is the Levi-Civita connection on M. ((3.18) is equivalent to the dual 1-form $w(\cdot) = < V, \cdot >$ is closed.)

All the following theorems and results were obtained by Bourguignon and Lawson [39]. In order to prove Theorems 3.1.9 and 3.1.10, we require the following lemmas to prove Theorem 3.1.8 first.

(L1) *For $\phi \in \Omega^2(\mathcal{G}_E)$, let $B = i_V\phi$ such that $\delta^D\phi = 0$, where V is a vector field of gradient type. Then $\delta^D B = 0$.*

Consider the special finite-dimensional space of vector fields on S^m

$$\mathcal{V} = \{grad\, f : f = F|_{S^m} \text{ and } F : \mathbf{R}^{m+1} \to \mathbf{R} \text{ is linear}\}.$$

There is a natural isomorphism $\mathbf{R}^{m+1} \to \mathcal{V}$ which associates to each $V \in \mathbf{R}^{m+1}$ the vector field V given by

$$V(x) = v - < v, x > x \tag{3.19}$$

for $x \in S^m$. Remark that $V = grad\, f$, where $f(x) = < v, x >$.

(L2) *Each $V \in \mathcal{V}$ satisfies (a) $\nabla_X V = -fX$, (b) $\nabla^*\nabla V = V$, where ∇ denotes the Levi-Civita connection of the standard metric in S^m, and f is as above.*

Theorem 3.1.8. *Let (E, P, G, D) be any Yang-Mills set up over the Euclidean sphere S^m, and assume that $\phi \in \Omega^2(\mathcal{G}_E)$ is harmonic, i.e., $\delta^D\phi = d^D\phi = 0$. Associate to ϕ a quadratic form Q_ϕ on $\mathcal{V}$ by letting*

$$Q_\phi(V) = \frac{d^2}{dt^2}\Big|_{t=0} \mathcal{YM}(D^t),$$

where $D^t = D + t(i_V\phi)$. Then

$$trace(Q_\phi) = 2(4-m)\int_{S_m} ||\phi||^2.$$

Proof. By (L1), $\delta^D(i_V\phi) = 0$ for all $V \in \mathcal{V}$ and we can apply the second variation formula, namely

$$trace(Q_\phi) = \int_{S^m} trace(q_\phi) \tag{3.20}$$

where $q_\phi(V) = < \mathcal{J}^D(i_V\phi), i_V\phi >$, and $\mathcal{J}(\alpha) = \Delta^D\alpha + R^D(\alpha) = D^*D\alpha + \alpha \circ Ric + 2R^D(\alpha)$. (In particular, if M is an Einstein manifold with $Ric = k \cdot Id$, then

$$\frac{d^2}{dt^2}\Big|_{t=0} \mathcal{YM}(D^t) = \int_M < D^*D\alpha + k\alpha + 2\mathcal{R}^D(\alpha), \alpha > .) \tag{3.21}$$

To calculate the trace of q_ϕ at a point $x \in \mathcal{J}^m$ we choose an orthonormal basis $\{\epsilon_0, \epsilon_1, \cdots, \epsilon_m\}$ of $\mathcal{V}$ adapted to this point. Let $\epsilon_0, \epsilon_1, \cdots, \epsilon_m$ be the corresponding frame, under the isomorphism $\mathcal{V} \cong \mathbf{R}^{m+1}$ to the vectors $x, e_1, \cdots, e_m$ where $\{e_1, \cdots, e_m\}$ form an orthonormal basis of TS^m. We know from (3.19) that

$$\epsilon_0(x) = 0,\ \epsilon_1(x) = e_1, \cdots,\ \epsilon_m(x) = e_m. \tag{3.22}$$

In order to use the formula (3.21), we need to compute $D^*D(i_V\phi)$ for $V \in \mathcal{V}$. Choose local orthonormal tangent fields $\epsilon_1, \cdots, \epsilon_m$ on S^m such that $(\nabla\epsilon_j)(x) = 0$, and let $X = \sum a_j\epsilon_j$ be any linear combination of these fields. Then at the point x,

$$\begin{aligned}
[D^*D(i_V\phi)]X &= -\sum[D^2_{\epsilon_j,\epsilon_j}(i_V\phi)]X = -\sum[D_{\epsilon_j}D_{\epsilon_j}(i_V\phi)]X \\
&= -\sum\{D_{\epsilon_j}[(D_{\epsilon_j}i_V\phi)X] - (D_{\epsilon_j}i_V\phi)\nabla_{\epsilon_j}X\} \\
&= -\sum D_{\epsilon_j}[D_{\epsilon_j}(i_V\phi)X - (i_V\phi)\nabla_{\epsilon_j}X] \\
&= -\sum D_{\epsilon_j}[D_{\epsilon_j}[(\phi_{V,X}) - \phi_{V,\nabla_{\epsilon_j}X}]
\end{aligned}$$

$$= -\sum D_{\epsilon_j}[(D_{\epsilon_j}\phi)_{V,X} + \phi_{\nabla_{\epsilon_j},V,X}]$$
$$= (D^* D\phi)_{V,X} - 2\sum (D_{\epsilon_j}\phi)_{\nabla_{\epsilon_j},V,X} + \phi_{\nabla^*\nabla V,X}. \quad (3.23)$$

By applying (L2) and the fact that $\delta^D \phi = 0$, we arrive at

$$(D^* D(i_V\phi))_X = (D^* D\phi)_{V,X} + \phi_{V,X}. \quad (3.24)$$

Because $Ric = (m-1)Id$ on S^m, we derive from (3.21) and (3.24) that

$$\mathcal{J}^D(i_V\phi) = (D^* D\phi)_{V,\cdot} + m\phi_{V,\cdot} + 2\sum_{i=1}^{m}[R_{e_i,\cdot}, \phi_{V,e_i}].$$

Since ϕ is harmonic, we have the following Bochner-Weitzenböck formula (cf. Corollary 3.1.5):

$$(D^* D\phi)_{V,\cdot} = -2(m-2)\phi_{V,\cdot} - \sum_{i=1}^{m}\{[R^V_{e_i,V}, \phi_{e_i,\cdot}] - [R^D_{e_i,\cdot}, \phi_{e_i,V}]\}.$$

Combining the previous two equations we get

$$\mathcal{J}^D(i_V\phi) = (4-m)i_V\phi - \sum_{i=1}^{m}\{[R^V_{e_i,\cdot}, \phi_{e_i,\cdot}] + [R^D_{e_i,V}, \phi_{e_i,\cdot}]\}. \quad (3.25)$$

Thus we deduce at x that

$$trace(q_\phi) = \sum_{j=0}^{m} < \mathcal{J}^D(i_{\epsilon_j}, \phi), i_{\epsilon_j}\phi > = (4-m)\sum_{j=0}^{m}\sum_{k=1}^{m} < \phi_{\epsilon_j,e_k}, \phi_{\epsilon_j,e_k} > . \quad (3.26)$$

Remark that the contribution from the second term of (3.26) drops out since we take the inner product of a symmetric and a skew-symmetric form.

Applying (3.22), we derive at x that

$$trace(q_\phi) = 2(4-m)\sum_{j<k} ||\phi_{e_j,e_k}||^2 = 2(4-m)||\phi||^2,$$

and the result follows. □

Theorem 3.1.9 (J. Simons). *There are no weakly stable Yang-Mills fields over the Euclidean m-sphere S^m for $m \geq 5$.*

Proof. Notice that the quadratic form Q_ϕ in Theorem 3.1.8 defined through the second variation of $\mathcal{YM}$. In particular, in dimension 4 the density q_ϕ vanishes. Since D is a Yang-Mills connection, we can choose $\phi = R^D$ in Theorem 3.1.8. If $m \geq 5$, $\frac{d^2}{dt^2}\big|_{t=0}\mathcal{YM}(D^t) < 0$, and we obtain the result.

Notice that when $m = 4$, Theorem 3.1.8 asserts that $trace(Q_\phi) = 0$. Applying this fact, we can verify the following result.

Theorem 3.1.10. *Any weakly stable Yang-Mills field over S^4 with group SU_2, SU_3 or U_2 is either self-dual or anti-self-dual.*

(L3) *Let ϕ^+ and ϕ^- be a self-dual and anti-self-dual 2 form on $\mathbf{R}^4$, respectively. Then for vectors $X, Y \in \mathbf{R}^4$, then $\sum_{j=1}^4 \phi^+_{e_j,X} \otimes \phi^-_{e_j,Y}$ is symmetric in X and Y. For ordinary forms, the resulting map from $\Lambda^+\mathbf{R}^4 \otimes \Lambda^-\mathbf{R}^4 \cong Hom(\Lambda^+\mathbf{R}^4, \Lambda^-\mathbf{R}^4)$ to $S_0^2\mathbf{R}^4 \equiv \{h \in Hom(\mathbf{R}^4, \mathbf{R}^4) : h \text{ is traceless and symmetric}\}$ is an SO_4-isomorphism of SO_4-modules. Its inverse is given by $h \mapsto \frac{1}{2} h \wedge I$, where after restriction the homomorphism $h \wedge I$ gives a linear map from $\Lambda^+\mathbf{R}^4$ to $\Lambda^-\mathbf{R}^4$.*

Proof of Theorem 3.1.10. Consider a weakly stable Yang-Mills field with group G on S^4. Let $\mathcal{I}(\alpha, \alpha)$ be the quadratic form on $\Omega^1(\mathcal{G}_E)$ given by the second variation of the functional at the connection. Since $\mathcal{I}(\alpha, \alpha) \geq 0$, we know that $\mathcal{I}(\alpha, \alpha) = 0$ if and only if α is in the null space of the quadratic form. In particular, if $\mathcal{I}(\alpha, \alpha) = 0$ and $\delta^D \alpha = 0$, then $\mathcal{J}(\alpha) = 0$. It follows from Theorem 3.1.8 that if $\phi \in \Omega^2(\mathcal{G}_E)$ is harmonic and $V \in \mathcal{V}$, then

$$\mathcal{J}^D(i_V \phi) = 0. \tag{3.27}$$

Since R^D is harmonic, so are $R^+ = (R^D)^+$ and $R^- = (R^D)^-$. Therefore, by (3.27) we have that

$$\mathcal{J}^D(i_V R^+) = 0. \tag{3.28}$$

for all $V \in \mathcal{V}$. ((3.28) is also true for R^-.) Applying (3.23) and writing $R^D = R^+ + R^-$, we obtain from (3.28) that

$$(D^* D R^+)_{X,Y} + 4R^+_{X,Y} + 2\sum_{j=1}^4 [R^+_{e_j,X}, R^+_{e_j,Y}] = -2\sum_{j=1}^4 [R^+_{e_j,X}, R^-_{e_j,Y}] \tag{3.29}$$

for all tangent vector fields X, Y on S^4. The left-hand side of (3.29) is skew symmetric in X and Y. By (L3), the right-hand side is symmetric in X and Y. Hence, the $\mathcal{G}_E$-valued tensor

$$\tau_{X,Y} = \sum_{j=1}^4 [R^+_{e_j,X}, R^-_{e_j,Y}] \tag{3.30}$$

vanishes for all X, i.e., $\tau = 0$. We now need the following two lemmas.

(L4) *At each point $x \in S^4$, $[R^+_{X,Y}, R^-_{Z,W}] = 0$ for all $X, Y, Z, W \in T_x S^4$. Therefore, at each point $x \in S^4$, $[\alpha_x^+, \alpha_x^-] = 0$, where $\alpha^\pm \subset (\mathcal{G}_E)_x$ is the Lie sub-algebra generated by the curvature transformations $R^\pm_{X,Y}$ for $X, Y \in T_x S^4$.*

(L5) *Let α^+ and α^- be sub-algebras of a Lie algebra $\mathcal{G}$ such that $[\alpha^+, \alpha^-] = 0$. If $\mathcal{G} \cong su_2$, su_3, or u_2, then α^+ or α^- is abelian.*

Notice that if $\mathcal{G}$ is the Lie algebra of G_2 or if $rank(\mathcal{G}) \geq 3$ then (L5) is not true, since $\mathcal{G}$ contains an so_4 sub-algebra.

We now discuss the $\mathcal{G}_E$-valued 4 tensors $C^+ = [R^+, R^+]$ and $C^- = [R^-, R^-]$ on S^4, i.e., $C^{\pm}_{X,Y,Z,W} = [R^{\pm}_{X,Y}, R^{\pm}_{Z,W}]$ for $X, Y, Z, W \in T_x S^4$. (L4) and (L5) imply that either C^+ or C^- must vanish on some open subset $\mathcal{O} \subset S^4$. For simplicity, suppose C^+ vanishes on $\mathcal{O}$. Since C^+ is a algebraic function of a solution of the elliptic equation $\triangle R^+ = 0$, the Aronszajn theorem [10] on unique continuation of solutions to elliptic systems implies that $C^+ = 0$ on S^4.

We notice that, since $[R^+, R^-]$ and $[R^+, R^+]$ both vanish on S^4, (3.29) for R^+ becomes

$$D^* D R^+ + 4R^+ = 0. \tag{3.31}$$

Thus $D^* D \geq 0$ on S^4 implies $R^+ = 0$ (we can apply similar argument to see that $R^- = 0$ is also true). This finishes the proof of Theorem 3.1.10. □

Let M be a compact Riemannian manifold and P be a principal-G bundle over M, where G is a compact Lie group. Let E be a Riemannian vector bundle associated with P by a locally faithful, orthogonal representation of G. In particular, when M has dimension 4, this setting is of interest in physics. In this case the $*$-operator carries exterior two forms (with values in any bundle) into themselves and satisfies $(*)^2 = 1$. Thus, there is a decomposition, $\Lambda^2 \otimes \mathcal{G}_E = (\Lambda^2_+ \otimes \mathcal{G}_E) \oplus (\Lambda^2_- \otimes \mathcal{G}_E)$, into the $+1$ and the -1 eigenspaces of $*$. In particular, for any connection D the curvature tensor field R^D has a splitting $R^D = R^{D+} + R^{D-}$, in which $*(R^D_{\pm}) = \pm R^{D\pm}$. It follows from [39] that the field R^D is harmonic if and only if both R^{D+} and R^{D-} are harmonic.

Let p_1 be the first Pontryagin class of E (G being fixed). Then there is a universal constant $c > 0$ such that for any Riemannian connection

$$p_1(E)[M] = c \int_M (||R^{D+}||^2 - ||R^{D-}||^2) dv. \tag{3.32}$$

Since $\mathcal{YM}(D) = \int_M (||R^{D+}||^2 + ||R^{D-}||^2) dv$, it follows that when one of the conditions $R^{D-} = 0$ or $R^{D+} = 0$ holds, the functional attains its minimum. If $R^{D-} = 0$, the field is called *self-dual*. If $R^{D+} = 0$, the field is called *anti-self-dual*. Generally speaking, there are constraints on the integrand coming from topology. For instance, for SO_4 bundles, one has an Euler integrand. We have the following further result from [39].

Theorem 3.1.11. *Any weakly stable Yang-Mills field with group SU_2 on any compact orientable homogeneous Riemannian 4-manifold is either self-dual, or anti-self-dual, or reduces to an abelian field.*

3.1.4 Isolation Phenomena

We present a series of isolation results for Yang-Mills fields obtained by Bourguignon and Lawson [38, 39]. If $\phi = R^D$, Corollary 3.1.5 holds and the Bochner-Weitzenböck formula (3.14) implies

$$< D^* DR^D, R^D > = - < R^D \circ (Ric \wedge I + 2R), R^D > -\rho(R^D), \quad (3.33)$$

where

$$\rho = \sum_{i,j,k=1}^{m} < [R^D_{e_i,e_j}, R^D_{e_j,e_k}], R^D_{e_i,e_k} > . \quad (3.34)$$

If $m = 4$, a similar formula holds with R^D replaced by $R^{D+} \equiv R^+$ (or $R^{D-} \equiv R^-$). Since R^+ is harmonic, (3.14) holds. We then check the last term of $\mathcal{R}(R^+)$ of the formula. Writing $R^D = R^+ + R^-$ gives a splitting of this into two terms, the second of which is of the form

$$\mathcal{R}^-(R^+)_{X,Y} = \sum_{j=1}^{m}\{[R^-_{e_j,X}, R^+_{e_j,Y}] - \{[R^-_{e_j}, Y, R^+_{e_j,X}]\}.$$

By (L3), this expression vanishes. We conclude that $< \mathcal{R}(R^+), R^+ > = \rho(R^+)$ and so (3.33) holds for R^+. Similarly, it also holds for R^-.

When $m = 4$, $G = SO_4$, there is a further decomposition of the curvature into harmonic components, $R^D = R^+_+ + R^+_- + R^-_+ + R^-_-$. Since $[R^{\cdot}_+, R^{\cdot}_-] = 0$ and $R^+ = R^+_+ + R^+_-$, we see that $< \mathcal{R}(R^+_+), R^+_+ > = \rho(R^+_+)$. It follows that R^+_+ (also R^+_-, etc.) satisfies (3.33).

Let us examine the term ρ given by (3.34). For any Lie algebra $\mathcal{G}$ with a fixed invariant inner product $< \cdot, \cdot >$, we have the associated fundamental 3-form

$$\Phi_{\mathcal{G}}(U, V, W) = < [U, V], W >$$

for U, V, $W \in \mathcal{G}$. There is a canonical isometry $\Lambda^2 E \cong so_E$ (given by $(u \wedge v)(w) = < u, w > v - < v, w > u$ for u, v, $w \in E_x$). Let $E = TM$ and we consider $R : so_M \to \mathcal{G}_E$ as a linear map. In so_M we get

$$[e_i \wedge e_j, e_k \wedge e_l] = \delta_{il} e_k \wedge e_j + \delta_{jl} e_i \wedge e_k + \delta_{ik} e_j \wedge e_l + \delta_{jk} e_l \wedge e_i \quad (3.35)$$

for all i, j, l, k. Therefore, we may write (3.34) as

$$\begin{aligned}\rho(R^D) &= \sum_{i,j,k=1}^{m} \Phi_{\mathcal{G}_E}(R^D_{e_i,e_j}, R^D_{e_j,e_k}, R^D_{e_k,e_i}) \\ &= \sum_{i,j,k=1}^{m} (R^{D*}\Phi_{\mathcal{G}_E})(e_i \wedge e_j, e_j \wedge e_k, e_k \wedge e_i) = (R^{D*}\Phi_{\mathcal{G}_E}, \Phi_{so_M}),\end{aligned} \quad (3.36)$$

where we define the inner product on $\Lambda^3 so_M^*$ by

$$(\Phi, \Psi) = \sum_{\alpha,\beta,\gamma} \Phi(\alpha,\beta,\gamma)\Psi(\alpha,\beta,\gamma)$$

where α, β and γ run over an orthonormal basis of so_M. Combining (3.33) and (3.36) and integrating by parts, we obtain the following theorem.

Theorem 3.1.12. *Let R^D be a Yang-Mills field, and λ be the minimal eigenvalue of the operator $Ric \wedge I + 2R$ on 2-forms on a compact Riemannian manifold M. Then*

$$\int_M ||\triangle R^D||^2 \leq -\int_M [\lambda ||R^D||^2 + (R^{D*}\Phi_{\mathcal{G}_E}, \Phi_{so_M})]. \tag{3.37}$$

In particular, if $m = 4$, the above formula holds with R^D replaced by R^+ and with λ replaced by λ^+, the minimal eigenvalue of $Ric \wedge I + 2R$ on $\Lambda^+ TM$ (this also holds with $+$'s replaced by $-$'s). Moreover, if $G = SO_4$, the formula also holds with R replaced by R_+^+, R_-^+, etc.

We notice that the term $\rho(R^D) = (R^{D*}\Phi_{\mathcal{G}_E}, \phi_{so_M})$ is a homogeneous cubic function of R^D, whereas $||R^D||^2$ is homogeneous quadratic. For R^D small enough and $\lambda > 0$, this quadratic term will dominate the right-hand side of (3.37) so that it becomes negative, which is impossible. The point at which $||R^D||^2$ dominates the expression can be estimated concretely. To do this we want to estimate $(L^*\Phi_{\mathcal{G}}, \Phi_{so_m})$ in terms of $||L||^2$, where $L : so_m \to \mathcal{G}$ is a linear map and where $\mathcal{G}$ is any sub-algebra of so_N. Remark that the inner product on $\mathcal{G}$ is induced from the canonical one on so_N (defined by $<A, B> = \frac{1}{2}(A^t \circ B)$ for any two endormorphisms A and B of E_x). Thus $L^*_{\Phi_{\mathcal{G}}} = L^*\Phi_{so_N}$ and we may ignore $\mathcal{G}$ for the moment.

(L6) *Let $L : so_m \to so_N$ be any linear map. If $||L||^2 \leq \frac{1}{2}\binom{m}{2}$, then*

$$(L^*\Phi_{so_N}, \Phi_{so_m}) \leq 2(m-2)||L||^2.$$

When $m \geq 5$, this inequality is strict.

When $m = 4$ (respectively $m = 3$), equality holds if and only if there is an orthogonal splitting $\mathbf{R}^N = S_0 \oplus S_1$ ($dim S_1 = 4$) with respect to which $L = 0 \oplus \sigma$, where σ is one of the two irreducible spin representations of so_4 (respectively, where σ is the irreducible spin representation of so_3).

If $m = N = 3$, the inequality is true for $||L^2|| \leq \binom{m}{2} = 3$. In this case equality holds if and only if $L : so_3 \to so_3$ is a Lie algebra isomorphism.

Theorem 3.1.13. *Any Yang-Mills field R on S^m, $m \geq 5$, which satisfies the pointwise estimate $||R^2|| \leq \frac{1}{2}\binom{m}{2}$ must vanish identically.*

Proof. By (L6), the integrand on the right-hand side of (3.37) is ≥ 0 since $\lambda = 2(m-2)$ on S^m. Consequently, this integrand vanishes identically (so does DR). It follows from (L6) that $R = 0$. □

Let R be any Yang-Mills field over $S^m, m \geq 3$, which satisfies the pointwise condition

$$||R||^2 \leq \frac{1}{2}\binom{m}{2}.$$

Then either $R = 0$, or $||R||^2 = \frac{1}{2}\binom{m}{2}$ and R is parallel. If $m = 3$ or 4, then $R = 0$, or R is the curvature of the tangent frame bundle of S^m with its Levi-Civita connection. If $m \geq 5$, then $R = 0$.

Theorem 3.1.14. *Let R be any Yang-Mills field on S^4. If R^+ satisfies the pointwise condition $||R^+||^2 < 3$, then $R^+ = 0$. (This is also true for R^-).*

The proof is similar to the proof of Theorem 3.1.13, with R replaced by R^+, R^-, respectively.

Theorem 3.1.15. *Let (E, G, D, R) be a Yang-Mills set up on S^4 such that R satisfies the pointwise condition $||R||^2 \leq 3$. Then either E is flat, or $E = E_0 \oplus S$, where E_0 is flat and where S is one of the (two) 4-dimensional bundles of tangent spinors with the canonical Riemannian connection.*

Proof. We observe that, by the previous argument, either $R = 0$, or $||R||^2 = 3$ and $DR = 0$. In the latter case, it follows from (L6) that there is an orthogonal splitting $E = E_0 \oplus S$, where E_0 is flat, S is 4-dimensional, and $-R : so_M^+ \to so_S$ is one of the two fundamental spin representations σ^+ or σ^- at each point. Assume that it is σ^+ (the analysis is similar for σ^-) at each point. Then $R^- = 0$ and $-R : so_M^+ \to so_S$ is an isometric bundle injection, since $DR = [D, R] = 0$, which is connection preserving. Consequently,

$$so_S = so_M^+ \oplus \underline{so_3}, \tag{3.38}$$

where $\underline{so_3}$ is the flat bundle. (Remark that $\underline{so_3}$ corresponds to a parallel quaternion structure on S.) It follows from (3.38) that S, pulled back over the principal $Spin_4$-bundle, is canonically trivialized and transforms based on the representation σ^+. This finishes the proof. □

We can apply the similar argument to R^-. We conclude that if $||R^-||^2 \leq 3$, then either $R^- = 0$, or the following holds: there is an orthogonal splitting $E = E_0 \oplus V$, where V is 4-dimensional and

$$-R : so_M^- \to so_V^- \subset \mathcal{G}_E$$

is a connection preserving bundle isometry. (Note that since $dimV = 4$, there is a canonical splitting $so_V = so_V^+ \oplus so_V^-$.) That is, E_0 is self-dual and $so_V \cong so_V^+ \oplus so_M^-$, where so_V^+ is also self-dual. It follows that the principal $Spin_4$-bundle of V can be expressed as a Whitney sum

$$P_{Spin_4}(V) = P_{SU_2} \oplus P_{SU_2}(S^-), \tag{3.39}$$

where S^- is the canonical spin bundle above and the connection on P_{SU_2} is self-dual. This is one of the connections satisfying the minimum conditions in [39]. By the same paper, we have that $\frac{1}{2}p_1(V) - \chi(V) = 2$. Hence, we can conclude the following theorem.

Theorem 3.1.16. *Let (E, G, D, R) be a Yang-Mills set up on S^4 such that R satisfies the pointwise condition $||R^-||^2 \leq 3$. Then either E is self-dual, or $E = E_0 \oplus V$, where E_0 is flat and V is a 4-dimensional bundle satisfying (3.39) and the connection on the first factor is also self-dual.*

A corresponding statement for R^+ can be derived by reversing orientations. We next consider the 3-sphere S^3.

Theorem 3.1.17. *Let (E, G, D, R) be a Yang-Mills set up on S^3. If $||R||^2 < 3/2$, then E is flat. If $||R||^2 \leq 3/2$, then either E is flat, or $E = E_0 \oplus S$, where E_0 is flat and S is one of the 4-dimensional tangent spin bundles with the Riemannian connection. If $dim(E) = 3$ and $||R||^2 < 3$, then either E is flat, or $E = TS^3$ with the Riemannian connection.*

Proof. For the first part, the proof is similar to the proof of Theorem 3.1.15. For the second part, we use the last statement in (L6) to deduce that if $R \neq 0$, then $-R : so_M \to so_E$ is a connection preserving bundle isometry. Therefore, the composition

$$TM \xrightarrow{*} \Lambda^2 TM = so_M \xrightarrow{-R} so_E = \Lambda^2 E \xrightarrow{*} E$$

gives an equivalence $TM = E$. Hence, the theorem is proved. □

3.2 Weak and Strong Compactness

All the theorems and results presented in this section were obtained by Uhlenbeck [383, 384]. We first prove the weak compactness Theorems 3.2.1 and 3.2.2 for compact manifolds and non-compact manifolds, respectively. We then establish Theorem 3.2.5 for weak Yang-Mills connections, and the strong compactness Theorem 3.2.8.

3.2.1 Weak Compactness

Let G be a Lie group and $\mathcal{G} = T_e G$ be its Lie algebra. G acts on itself by the conjugation $c : G \to Aut(G)$ given by $c_g(h) = ghg^{-1}$, $\forall g, h \in G$. The adjoint representation on the Lie algebra, $Ad : G \to End(g)$, $g \mapsto Ad_g = d_e c_g$, is given by $Ad_g(\xi) = g\xi g^{-1}$, $\forall \xi \in \mathcal{G}$, $g \in G$. This utilizes the following notation: For $\xi \in \mathcal{G}$ and $g \in G$,

$$g\xi = d_e L_g(\xi) = \frac{d}{dt}\Big|_{t=0} g\, exp(t\xi) \in T_g G,$$

where L_g is the left multiplication by g and $exp : \mathcal{G} \to G$ is the usual exponential map. The notation ξg is defined similarly by right multiplication. This makes sense when $G \subset \mathbf{C}^{m\times m}$ is a matrix group, since $g\xi$ can be realized as matrix multiplication.

The adjoint representation of $\mathcal{G}$ is given by the Lie bracket of vector fields: We identify the Lie element $\xi \in \mathcal{G}$ with left-invariant vector field $g \mapsto g\xi$ on G, then for $\xi,\ \eta \in \mathcal{G}$

$$ad_\xi(\eta) = d_e Ad(\xi)\eta = \frac{d}{dt}\Big|_{t=0} exp(t\xi)\eta exp(t\xi)^{-1} = \mathcal{L}_\xi \eta(e) = [\xi, \eta]. \quad (3.40)$$

The Lie bracket is given by the commutator $[\xi, \eta] = \xi\eta - \eta\xi$ in the case of a matrix group.

The underlying object in gauge theory is a principal G-bundle $\pi : P \to M$. This is a manifold with a free right action $P \times G \to P$, $(p, g) \mapsto pg$, of a Lie group G, such that the orbits of this action are the fibres $\pi^{-1}(x) \cong G$ of a locally trivial fibre bundle $\pi : P \to M$. Here M is a smooth manifold and the G-action preserves the fibres, i.e., $\pi(pg) = \pi(p)$.

Let G be a compact Lie group and $P \to G$ be a principal G-bundle on a compact manifold M. Assume that $1 < p < \infty$ is such that $p > \frac{m}{2}$. Denote by $\mathcal{D}^{1,p}(P)$ the $W^{1,p}$-Sobolev space of connections and by $\mathcal{G}^{2,p}(P)$ the $W^{1,p}$ Sobolev space of gauge transformations on P. These and the action of $\mathcal{G}^{2,p}(P)$ on $\mathcal{D}^{1,p}(P)$ are well defined (cf. [401]).

Theorem 3.2.1 (Weak Uhlenbeck Compactness). *Let $(D^\nu)_{\nu\in\mathbf{N}} \subset \mathcal{D}^{1,p}(P)$ be a sequence of connections such that $||F_{D^\nu}||_p$ is uniformly bounded. Then there exist a subsequence (still denoted by $(D^\nu)_{\nu\in\mathbf{N}}$) and a sequence of gauge transformations $f^\nu \in \mathcal{G}^{2,p}$ such that $f^{\nu *} D^\nu$ converges weakly in $\mathcal{D}^{1,p}(P)$.*

In order to prove the main Theorem 3.2.1, we require the following lemmas (the proof is given in [383]).

(L1) (Uhlenbeck Gauge). *Let M be an m-dimensional Riemannian manifold and let G be a compact Lie group. Assume that $1 < q \leq p < \infty$ are such that $q \geq \frac{m}{2}$, $p > \frac{m}{2}$, and if $q < m$ assume that $p \leq \frac{mq}{m-q}$. Then there exist constants C_{U_h} and ϵ_{U_h} such that the following holds:*

For every point in M we can find a neighborhood $U \subset M$ such that for every connection $D \in \mathcal{D}^{1,p}(U)$ with $\mathcal{E}(D) \leq \epsilon_U$, there exists a gauge transformation $f \in \mathcal{G}^{2,p}(U)$ such that

(a) $d^*(f^*D) = 0$,
(b) $||f^*D||_{W^{1,q}} \leq C_{U_h}||F_D||_q$,
(c) $*(f^*D)|_{\partial U} = 0$,
(d) $||f^*D||_{W^{1,p}} \leq C_{U_h}||F_D||_p$.

Note that the domains U will be geodesic balls (if the given point lies on the boundary of M). The radius of these domains can be chosen arbitrarily small without affecting the constants C_{U_h} and ϵ_{U_h}. Here for every local trivialization $P|_U \to U \times G$ of a principal G-bundle P the space $\mathcal{D}^{1,p}(U) = W^{1,p}(U, T^*U \otimes \mathcal{G})$ means the connections on $P|_U$. The energy of a connection $D \in \mathcal{D}^{1,p}(U)$ is defined by

$$\mathcal{E}(D) = \int_U |F_D|^q dv = ||F_D||_q^q,$$

and $\mathcal{G}^{2,p}(U) = W^{2,p}(U, G)$ means the gauge transformations on $P|_U$.

Let $\Delta_{\exp} > 0$ be the radius of a *convex geodesic ball* $B_{\Delta_{\exp}}(e) \subset G$ around e with the following two properties: (i) The exponential map is a bijection between $B_{\Delta_{\exp}}(0) \subset \mathcal{G}$ and $B_{\Delta_{\exp}}(e)$. (ii) For all g, $h \in B_{\Delta_{\exp}}(e)$ there is a unique minimal geodesic from g to h and this lies within $B_{\Delta_{\exp}}(e)$. For the existence of such balls see [152]. Furthermore, since the left multiplications are isometries of G, there exist convex balls $B_{\Delta_{\exp}}(g)$ of the same radius around all $g \in G$.

(L2) *Let M be an m-dimensional Riemannian manifold and let $p \geq \frac{m}{2}$, and $M = \bigcup_{i \in \mathbf{N}} U_i$ be a locally finite open covering by precompact sets U_i. Then we can find open subsets $V_i \subset U_i$ with $\bigcup_{i \in \mathbf{N}} V_i = M$ with the following properties:*

(a) *Let $k \in \mathbf{N}$ and $g_{ij}, h_{ij} \in \mathcal{G}^{k+1,p}(U_i \cap U_j)$ be two sets of transition functions that satisfy*

$$g_{ii} = 1, \quad g_{ij} g_{jk} = g_{ik} \text{ on } U_i \cap U_j \cap U_k \tag{3.41}$$

and

$$d(g_{ij}, h_{ij}) \leq \Delta_{\exp}, \ \forall i, j \in N.$$

Then there exist local gauge transformations $h_i \in \mathcal{G}^{k+1,p}(V_i)$ for all $i \in \mathbf{N}$ such that on all intersections $V_i \cap V_j$,

$$h_i^{-1} h_{ij} h_j = g_{ij}. \tag{3.42}$$

(b) *Let the h_{ij} as in (a) run through a sequence h_{ij}^ν of sets of transition functions such that $g_{ij}, h_{ij}^\nu \in \mathcal{G}^{k+1,p}(U_i \cap U_j)$ for all $k < K$, where $K \geq 2$ is an integer or $K = \infty$. Suppose that for every $i, j \in \mathbf{N}$ and $k < K$ there is a uniform bound on $||(h_{ij}^\nu)^{-1} dh_i^\nu||_{W^{k,p}(V_i \cap V_j)}$.*

Then the gauge transformations h_i^ν in (a) are constructed in such a way that for each $i \in \mathbf{N}$ and $k < K$ they satisfy $h_i^\nu \in \mathcal{G}^{k+1,p}(V_i)$ and

$$\sup_{\nu \in \mathbf{N}} ||(h_i^\nu)^{-1} dh_i^\nu||_{W^{k,p}(V_i)} < \infty.$$

(L3) *Let $k \in \mathbf{N}$ and $1 \le p < \infty$ with $kp > m$. Then the gauge action is a continuous map*

$$\mathcal{G}^{k,p}(P) \times D^{k-1,p}(P) \to D^{k-1,p(P)}$$
$$(f, D) \mapsto f^*D.$$

Furthermore, for every trivialization over some $U \subset M$ there exists a constant C such that for all $f \in \mathcal{G}^{k,p}(U)$ and $D \in \mathcal{D}^{k-1,p}(U)$,

$$||f^*D||_{W^{k-1,p}} \le ||f^{-1}df||_{W^{k-1,p}} + C||D||_{W^{k-1,p}}(1 + ||f^{-1}df||_{W^{k-2,2p}})^{k-1}.$$

(L3A) *Let $k \in \mathbf{N}$ and $1 \le p \le \infty$ be such that $kp > m$ and $p > m/2$. Let $(D^\nu)_{\nu\in\mathbf{N}} \subset \mathcal{D}^{k-1,p}(P)$ and $(f^\nu)_{\nu\in\mathbf{N}} \subset \mathcal{G}^{k,p}(P)$ be two sequences such that both $||D^\nu||_{W^{k-1,p}}$ and $||f^{\nu*}D^\nu||_{W^{k-1,p}}$ are uniformly bounded. Then the following holds:*

(a) *In every trivialization over some domain $U_i \subset M$ there is a uniform bound on*

$$||(f_i^\nu)^{-1}df_i^\nu||_{W^{k-1,p}(U_i)}.$$

(b) *There exists a subsequence of the f^ν that converges in the C^0-topology to some $f^\infty \in \mathcal{G}^{k,p}(P)$.*

Proof of Theorem 3.2.1. Select $1 < q < p$ such that $q \ge \frac{m}{2}$ and $q \ge \frac{pm}{p+m}$. Since $p > \frac{m}{2}$ and $p > \frac{pm}{p+m}$, this is possible and then the local gauge lemma (L1) holds on M with the L^q-energy $\mathcal{E}$. Let C_{U_h} and ϵ_{U_h} be the constants from that lemma and consider the energy of the connections D^ν over some small trivialization chart $U \subset M$,

$$\mathcal{E}(D^\nu|_U) = \int_U |F_{D^\nu}|^q \le (Vol\, U)^{1-\frac{q}{p}}||F_{D^\nu}||_p^q.$$

This is less than ϵ_{U_h} if U has sufficiently small volume independently of $\nu \in \mathbf{N}$ due to the uniform bound on $||F_{D_\nu}||_p$. For every point in M we now fix a neighborhood of such small volume over which the bundle P is trivial. Then the local gauge lemma (L1) (and note) states that for every point in M there exists a trivialization over an even smaller neighborhood such that the lemma holds (i.e. all connections with sufficiently small energy can be put into Uhlenbeck gauge). Since M is compact, it is covered by finitely many of these neighborhoods, $M = \bigcup_{i=1}^l U_i$.

The trivializations over U_i form a bundle atlas of P. With respect to this atlas the connections D^ν are represented by 1-forms $D_i^\nu \in \mathcal{D}^{1,p}(U_i)$ with $\mathcal{E}(D_i^\nu) = \mathcal{E}(D^\nu|U_i) \le \epsilon_{U_h}$ thanks to the small volume of the U_i. Therefore, all D_i^ν can be put into Uhlenbeck gauge, i.e., for $i = 1, \cdots, l$ and $\nu \in \mathbf{N}$ there exists a local gauge transformation $f_i^\nu \in \mathcal{G}^{2,p}(U_i)$ such that $f_i^{\nu*}D_i^\nu$ satisfies the Uhlenbeck gauge conditions. Remark that $||f_i^{\nu *} D_i^\nu||_{W^{1,p}} \leq C_{U_h}||F_{D_i^\nu}||_p$ is uniformly bounded for all $\nu \in \mathbf{N}$ because of the uniform bound on $||F_{D^\nu}||_{L^p(M)} \geq ||F_{D_i^\nu}||_{L^p(U_i)}$. Thus on each U_i there exists a weakly convergent subsequence of the $f_i^{\nu *} D_i^\nu$. But, the f_i^ν do not define global gauge transformations (i.e. bundle isomorphisms). This would only be the case if on all intersections $U_i \cap U_j$, the functions $f_{ij}^\nu = (f_i^\nu)^{-1}\phi_{ij} f_j^\nu$ were equal to the transition functions ϕ_{ij} of the bundle atlas. If the action of f_i^ν is seen as a change of the trivialization over U_i then f_{ij}^ν are the new transition functions and hence satisfy the cocycle conditions.

General speaking, the f_{ij}^ν are not necessarily C^0-close to the ϕ_{ij}. The Uhlenbeck gauge conditions only fix the f_i^ν up to a constant gauge transformation; thus at a fixed point every value of f_{ij}^ν can be obtained by the choice of these constants. However, in order to use cutoff functions for the patching of the f_i^ν to a global gauge transformation we need to work in the Lie algebra (via the exponential map), in geodesic balls of the Lie group. In that way we can obtain C^0-small corrections of the f_{ij}^ν. The possible constant changes described above need to be compensated separately from the actual patching. The key issue is that on all intersections $U_i \cap U_j$ we have

$$f_{ij}^{\nu *}(f_i^\nu D_i^\nu) = f_j^{\nu *} D_j^\nu.$$

This follows from the transition identity $D_j^\nu = \phi_{ij}^* D_i^\nu$ for the representatives of the connections D^ν. Since $f_i^{\nu *} D_i^\nu$ and $f_j^{\nu *} D_j^\nu$ are uniformly bounded in the $W^{1,p}$-norm on $U_i \cap U_j$, we can apply (L3A) to the trivial bundle $P|_{U_i \cap U_j}$. It asserts that there are uniform bounds on $||(f_{ij}^\nu)^{-1} d f_{ij}^\nu||_{W^{1,p}}$ and that some subsequence of the f_{ij}^ν converge C^0-uniformly on all (finitely many) intersections $U_i \cap U_j$. Therefore, for every $\delta > 0$ there exists a subsequence (still indexed by $\nu \in \mathbf{N}$) such that all the transition functions lie within a geodesic δ-ball of one another: Let $g_i = f_i^1$, with the corresponding transition functions $g_{ij} = g_i^{-1}\phi_{ij} g_j = f_{ij}^1$. Then for all $i, j = 1, \cdots, l$ and $\nu \in \mathbf{N}$ we have $d(f_{ij}^\nu g_{ij}) \leq \delta$. Here d denotes the supremum over $U_i \cap U_j$ of the geodesic distance in G, for which purpose we have fixed an invariant metric on G (cf. Appendix A in [401]). In order to apply the patching lemma (L2), we select $\delta = \Delta_{\exp} > 0$ to be the radius of a convex geodesic ball in G. Then the f_i^ν can be modified to $f_i^\nu h_i^\nu$ defined on smaller set $V_i \subset U_i$ that still cover M, such that for all $\nu \in \mathbf{N}$

$$(f_i^\nu h_i^\nu)^{-1}\phi_{ij}(f_j^\nu h_j^\nu) = (h_i^\nu)^{-1} f_{ij}^\nu h_j^\nu = g_{ij}$$

on all intersections $V_i \cap V_j$. This defines no gauge transformation yet, but we only need to make another ν-independent change: Let $\tilde{f}_{ij}^\nu = f_i^\nu h_i^\nu g_i^{-1}$ on V_i; then this defines a gauge transformation $\tilde{f}^\nu \in \mathcal{G}^{2,p}(P)$ for all $\nu \in \mathbf{N}$. In fact, $\tilde{f}^\nu$ is defined on $\cup_{i=1}^l V_i = M$ and it is well-defined since on $V_i \cap V_j$

$$(\tilde{f}_i^\nu)^{-1}\phi_{ij}\tilde{f}_{ij}^\nu = g_i(f_i^\nu h_i^\nu)^{-1} f_{ij}^\nu (f_j^\nu h_j^\nu) g_j = g_i g_{ij} g_j^{-1} = \phi_{ij}.$$

For the regularity of $\tilde{f}^{\nu}$ we notice that (L2) ensures that $h_i^{\nu} \in \mathcal{G}^{2,p}(V_i)$. Moreover, f_i^{ν} and $g_i = f_i^1$ restricted to V_i lie in $\mathcal{G}^{2,p}(V_i)$ for all i, and now the regularity follows from the fact that $\mathcal{G}^{2,p}(V_i)$ is closed under group multiplication and inversion.

Lastly, we claim that $\tilde{f}_i^{\nu *} D_i^{\nu}$ is bounded in $\mathcal{D}^{1,p}(V_i)$ for all $i = 1, \cdots, l$. The h_i^{ν} are determined in (L2) from the functions f_{ij}^{ν}, which satisfy a uniform bound on $||(f_{ij}^{\nu})^{-1} df_{ij}^{\nu}||_{W^{1,p}}$. Hence, (L2) (b) with $K = 2$ states that there is a uniform bound on $||(h_i^{\nu})^{-1} dh_i^{\nu}||_{W^{1,p}}$. Furthermore, the $f_i^{\nu *} D_i^{\nu}$ are $W^{1,p}$-bounded by the Uhlenbeck gauge, as seen above. Thus, (L3) implies that $h^{\nu *} f_i^{\nu *} D_i^{\nu}$ is $W^{1,p}$-bounded as well. Finally, $(g_i^{-1})^*$ is a ν-independent continuous map on $\mathcal{D}^{1,p}(V_i)$ (by L3), and so for every i we have a $W^{1,p}$-bound on $(g_i^{-1})^* h_i^{\nu *} f_i^{\nu *} D_i^{\nu} = \tilde{f}_i^{\nu *} D_i^{\nu}$.

The Banach-Alaoglu Theorem (cf. Appendix B in [401]) ensures that for every i the sequence $\tilde{f}_i^{\nu *} D_i^{\nu}$ has a $W^{1,p}$-weakly convergent subsequence. This can be selected with the same subsequence $(\nu_i)_{i \in \mathbf{N}}$ for all $i = 1, \cdots, l$. Then the subsequence $\tilde{f}_i^{\nu *} D_i^{\nu}$ converges $W^{1,p}$-weakly on the entire M. □

Let G be a compact Lie group and $P \to G$ be a principal G-bundle over a non-compact manifold M (exhausted by compact sets). The following theorem is the generalization of the weak Uhlenbeck compactness theorem for a non-compact manifold.

Theorem 3.2.2. *Let $M = \bigcup_{k \in \mathbf{N}} M_k$ be exhausted by an increasing sequence of compact submanifolds M_k that are deformation retracts of M. Let $(D^{\nu})_{\nu \in \mathbf{N}} \subset \mathcal{D}_{loc}^{1,p}(P)$ be a given sequence and assume that for all $k \in \mathbf{N}$ there is a uniform bound on $||F_{D^{\nu}}||_{L^p(M_k)}$. Then there exist a subsequence (still denoted $(D^{\nu})_{\nu \in \mathbf{N}}$) and a sequence of gauge transformation $f^{\nu} \in \mathcal{G}_{loc}^{2,p}(P)$ such that $f^{\nu *} D^{\nu}|_{M_k}$ converges weakly in $\mathcal{D}^{1,p}(P|_{M_k})$ for all $k \in \mathbf{N}$.*

The above theorem follows from the weak Uhlenbeck compactness Theorem 3.2.1 for the compact submanifolds M_k combined with the following proposition, which is a general result for the sequences of connections and gauge transformations on manifolds that are exhausted by compact deformation retracts. It will be used again to generalize the strong Uhlenbeck compactness to non-compact manifolds. We fix here a reference connection $\hat{D} \in \mathcal{D}(P)$ with respect to which the Sobolev norms of connections on P are defined.

Proposition 3.2.3. *Let $M = \bigcup_{k \in \mathbf{N}} M_k$ be exhausted by an increasing sequence of compact submanifolds M_k that are deformation retracts of M, and let $I = \mathbf{N}$ or $I = \{1, \cdots, l_0\}$ for some $l_0 \in \mathbf{N}$. Let $(D^{\nu})_{\nu \in \mathbf{N}} \subset \mathcal{D}_{loc}^{1,p}(P)$ be a sequence of connections with the following property: For every k and every subsequence of $(D^{\nu})_{\nu \in \mathbf{N}}$ there exist a further subsequence $(\nu_{k,i})_{i \in \mathbf{N}}$ and gauge transformations $f^{k,i} \in \mathcal{G}^{2,p}(P|_{M_k})$ such that*

$$\sup_{i \in \mathbf{N}} ||f^{k,i *} D^{\nu_{k,i}}||_{W^{l,p}(M_k)} < \infty, \quad \forall l \in I.$$

Then there exist a subsequence $(\nu_i)_{i\in\mathbf{N}}$ *and a sequence of gauge transformations* $f^i \in \mathcal{G}^{2,p}_{loc}(P)$ *such that*

$$\sup_{i\in\mathbf{N}} \|f^{k*} D^{\nu_i}\|_{W^{l,p}(M_k)} < \infty, \ \forall k \in \mathbf{N}, \forall l \in I.$$

3.2.2 Weak Yang-Mills Connections

Next we present the regularity theorems for Yang-Mills connections in the weak sense on compact manifolds and non-compact manifolds that are exhausted by compact deformation retracts. All the following theorems and results were obtained by Uhlenbeck [383, 384].

Let G be compact Lie group and M be an m-dimensional Riemannian manifold. For a principal G-bundle $P \to M$, the Yang-Mills functional is defined by

$$\mathcal{YM}(D) = \int_M |F_D|^2 dv,$$

for smooth connections $D \in \mathcal{D}(P)$ with compact support, where the norm $|\cdot|$ on the fibres of $\Lambda^2 T^*(M) \otimes \mathcal{G}_P$ is determined by the metric of M and the inner product on $\mathcal{G}$. If M is compact, then the Yang-Mills functional can be generalized to $\mathcal{D}^{1,p}$ for $2 \le p < \infty$ and such that $p \ge \frac{4m}{4+m}$, (by the Sobolev embedding $W^{1,p} \hookrightarrow L^4$, which ensures that $[D \wedge D]$, and thus F_D are of class L^2). The Euler-Lagrange equations for the Yang-Mills functional are

$$d_D^* F_D = 0, \ *F_D|_{\partial M} = 0, \tag{3.43}$$

where d_D^* denotes the adjoint of the differential operator d_D. Solutions of this boundary value problem are called *Yang-Mills connections*. It is not clear whether every critical point of the Yang-Mills functional on $\mathcal{D}^{1,p}(P)$ solves the (strong) Yang-Mills equation (3.43) or not. General speaking, the critical points are the only weak Yang-Mills connections in the following sense.

Definition 3.2.4. Let $1 \le p < \infty$ be such that $p > \frac{m}{2}$ and if m = 2 assume $p \ge 4/3$. Then a connection $D \in \mathcal{D}^{1,p}_{loc}(P)$ is called a *weak Yang-Mills connection* if it satisfies

$$\int_M (F_D, d_D\beta) dv = 0, \ \forall \beta \in \Omega^1(M, \mathcal{G}_P). \tag{3.44}$$

If M is non-compact, then the test 1-forms β must have compact support. The Yang-Mills functional is not necessarily defined or finite for weak Yang-Mills connections (for $m \le 3$, we do not assume $p \ge 2$). But, in order for (3.44) make sense, the regularity assumption on D should at least ensure that

$(f_D, d_D\beta) \in L^1_{loc}(M)$. The curvature F_D is locally of class L^p since the Sobolev embedding $W^{1,p} \hookrightarrow L^{2p}$ holds for $p \geq m/2$. Thus $d_D\beta$ should be of class $L^{p^*}_{loc}$, i.e., we require the Sobolev embedding $W^{1,p} \hookrightarrow L^{p*}$ for $\frac{1}{p*} + \frac{1}{p} = 1$. The condition for the latter to hold is $p \geq \frac{2m}{m+1}$. For $m = 1$ this holds since $p \geq 1$ and for $m \geq 3$ this is met because $p \geq m/2$. For $m = 2$ this needs the further assumption $p \geq 4/3$. The inequality $p > m/2$ will be required in (L4) to show that the weak Yang-Mills equation (3.44) is preserved under the gauge action $\mathcal{G}^{2,p}_{loc}(P)$.

Theorem 3.2.5. *Let $1 < p < \infty$ such that $p > \frac{m}{2}$. When $m = 2$, assume $p \geq \frac{4}{3}$. Then the following statements hold:*

(1) *Suppose that M is compact. Then for every weak Yang-Mills connection $D \in \mathcal{D}^{1,p}(P)$ there exists a gauge transformation $f \in \mathcal{G}^{2,p}(P)$ such that f^*D is smooth.*
(2) *Suppose that $M = \bigcup_{k\in\mathbf{N}} M_k$ is exhausted by an increasing sequence of compact submanifolds M_k that are deformation retracts of M. Then for every weak Yang-Mills connection $D \in \mathcal{D}^{1,p}_{loc}(P)$ there exists a gauge transformation $f \in \mathcal{G}^{2,p}(P)$ such that f^*D is smooth.*

In order to prove the above main theorem, we require the following theorem and lemmas (proofs are given in [383]).

Theorem 3.2.6 (Local Slice Theorem). *Suppose that M is compact and let $1 < p \leq q < \infty$ be such that $p > \frac{m}{2}$ and $\frac{1}{m} > \frac{1}{q} > \frac{1}{p} - \frac{1}{m}$. Fix a reference connection $\hat{D} \in \mathcal{D}^{1,p}(P)$ and let a constant c_0 be given. Then there exist constants $\delta > 0$ and C_{CG} such that the following holds: For every $D \in \mathcal{D}^{1,p}$ with*

$$||D - \hat{D}||_q \leq \delta, \ ||D - \hat{D}||_{W^{1,p}} \leq c_0, \tag{3.45}$$

there exists a gauge transformation $f \in \mathcal{G}^{2,p}(P)$ satisfying

$$d^*_{\hat{D}}(f^*D - \hat{D}) = 0, \ *(f^*D - \hat{D})|_{\partial M} = 0$$

and

$$||f^*D - \hat{D}||_q \leq C_{CG}||D - \hat{D}||_q, \ ||f^*D - \hat{D}||_{W^{1,p}} \leq C_{CG}||D - \hat{D}||_{W^{1,p}}.$$

(L4) *Let $D \in \mathcal{D}^{1,p}_{loc}(P)$ be a weak Yang-Mills connection and fix a compact subset $K \subset M$. Then for every gauge transformation $f \in \mathcal{G}^{2,p}(P|_K)$ the connection $f^*D|_K \in \mathcal{D}^{1,p}(P|_K)$ also solves (3.44) for all $\beta \in \Omega^1(M, \mathcal{G}_P)$ supported in K. In particular, $f^*D \in \mathcal{D}^{1,p}_{loc}(P)$ also is a weak Yang-Mills connection for every $f \in \mathcal{G}^{2,p}_{loc}(P)$.*
(L5) *Let $k \in \mathbf{N}$ and $1 \leq p < \infty$ be such that $kp > m$. Then group multiplication and inversion are continuous maps on $\mathcal{G}^{k,p}(P)$.*

To obtain an easy expression for the weak Coulomb equation in the subsequent L^r-version of the local slice theorem, consider the dual operator of $d_{\hat{D}} : W^{1,r^*}(M,\mathcal{G}_p) \to L^{r^*}(M, T^*M \otimes \mathcal{G}_p)$, i.e.,

$$d'_{\hat{D}} : L^{r^*}(M, T^*M \otimes \mathcal{G}_p) \to (W^{1,r^*}(M,\mathcal{G}_p))^*.$$

For $\alpha \in L^{r^*}(M, T^*M \otimes \mathcal{G}_p)$ the linear form $d'_{\hat{D}}\alpha$ acts on $W^{1,r^*}(M,\mathcal{G}_p)$ by $\eta \mapsto \int_M (\alpha,\, d_{\hat{D}}\eta)$. In the general context of a vector bundle we deal with a covariant derivative $\nabla := d_{\hat{D}}$ on $\mathcal{G}_P$, its adjoint operator $\nabla^* = d^*_{\hat{D}}$, and the dual operator $\nabla D' = d'_{\hat{D}}$. With this identification, we have for all $\alpha \in \Gamma(T^*M \otimes \mathcal{G}_P)$ and $\eta \in \Gamma(\mathcal{G}_P)$

$$(d'_{\hat{D}}\alpha)\eta = \int_M (d^*_{\hat{D}}\alpha, \eta)dv + \int_{\partial M} (\alpha(\nu),\, \eta)dv,$$

where $*\alpha|_{\partial M} = \alpha(\nu) d\, vol_{\partial M}$.

Theorem 3.2.7 (**L^r-Local Slice Theorem).** *Suppose that M is compact, let $2 \le r \le \infty$ be such that $r > m$, and fix a reference connection $\hat{D} \in \mathcal{D}^{0,r}(P)$. Then there exist constants $\delta > 0$ and C_{CG} such that the following holds. For every $D \in \mathcal{D}^{0,r}(P)$ with $||D - \hat{D}||_r \le \delta$ there exists a gauge transformation $f \in \mathcal{G}^{1,r}(P)$ such that*

$$d'_{\hat{D}}(f^*D - \hat{D}) = 0, \;\; ||f^*D - \hat{D}||_r \le C_{CG}||D - \hat{D}||_r.$$

Remark that the weak Coulomb equation stated here is equivalent to the weak equation

$$\int_M (f^*D - \hat{D}, d_{\hat{D}}\eta)dv = 0, \;\; \forall \eta \in \Gamma(\mathcal{G}_p)$$

since $\Gamma(\mathcal{G}_p)$ is dense in $W^{1,r^*}(M,\mathcal{G}_p)$.

(L6) *We have the following statements:*

(a) *Let $D^1, D^2 \in D^{0,r}(P)$; then $d'_{D^1}(D^2 - D^1) = 0 \Longleftrightarrow d'_{D^2}(D^1 - D^2) = 0$.*
(b) *Let $D^1, D^2 \in D^{1,p}(P)$; then $d^*_{D^1}(D^2 - D^1) = 0 \Longleftrightarrow d^*_{D^2}(D^1 - D^2) = 0$.*
(c) *Let $\hat{D}, D \in D^{0,r}(P)$, $f \in \mathcal{G}^{1,r}(P)$, and $g = f^{-1}$; then $d_{D'}(g^*\hat{D} - D) = 0 \Longleftrightarrow d_{\hat{D}'}(f^*D - \hat{D}) = 0$.*
(d) *Let $\hat{D}, D \in D^{1,p}(P)$, $f \in \mathcal{G}^{2,p}(P)$, and $g = f^{-1}$; then*

$$d^*_D(g^*\hat{D} - D) = 0, \; *(g^*\hat{D} - D)|_{\partial M} = 0$$

if and only if

$$d^*_{\hat{D}}(f^*D - \hat{D}) = 0, \; *(f^*D - \hat{D})|_{\partial M} = 0.$$

(L7) *Let $k \in \mathbf{N}$ and $1 \leq p < \infty$ be such that either $kp > m$ or $k=1$ and $\frac{m}{2} < p < m$ (if $m=2$ assume $p \geq \frac{4}{3}$). In the first case let $q=p$, in the second case let $q = \frac{mp}{2m-p}$.*

(a) *Suppose M is compact and let $\hat{D} \in \mathcal{D}(P)$ be a smooth reference connection. Then there exists a constant C such that the following holds: Assume that the connection $D = \hat{D} + \alpha \in D^{k,p}(P)$ satisfies $d^*_{\hat{D}}\alpha = 0$, $*\alpha|_{\partial M} = 0$, and for all smooth $\beta \in \Omega^1(M, g_P)$, $\int_M (F_D, d_D\beta)dv = 0$. Then $D \in \mathcal{D}^{k+1,q}(P)$*

$$||\alpha||_{W^{k+1,q}} \leq C(1 + ||\alpha||_{W^{k,p}} + ||\alpha||^3_{W^{k,p}}).$$

(b) *Let $\hat{D} \in \mathcal{D}(P)$ be a smooth reference connection over the possibly non-compact manifold M. Let $M'' \subset M' \subset M$ be compact submanifolds such that M'' is contained in the interior of M'. Then there exists a constant C such that the following holds: Let $D = \hat{D} + \alpha \in \mathcal{D}^{k,p}(P|_{M'})$ be a connection over M'. Suppose that it satisfies $d^*_{\hat{D}}\alpha = 0$ on M', $*\alpha|_{\partial M'} = 0$ on $\partial M \cap \partial M'$, and that for all smooth $\beta \in \Omega^1(M, g_P)$ supported in M', $\int_M (F_D, d_D\beta)dv = 0$. Then $D|_{M''} \in \mathcal{D}^{k+1,q}(P|_{M''})$ and*

$$||\alpha||_{W^{k+1,q}(M'')} \leq C(1 + ||\alpha||_{W^{k,p}(M')} + ||\alpha||^3_{W^{k,p}(M')}).$$

(c) *The constants C in (a) and (b) can be selected so that they depend continuously on the metric with respect to the $W^{k+1,\infty}$-topology.*

(L8) *Suppose $\frac{m}{2} < p \leq m$. Define sequences (p_i) and (q_i) by $p_0 = p$ and for all $i \in \mathbf{N}$, $q_i = \frac{mp_i}{2m-p_i}$, if $p_i < m$; $q_i = p_i$, if $p_i \geq m$. In case $p_i \geq m$, terminate the sequence with this $q_i = p_i$; in case $p_i < m$, let $p_{i+1} = \frac{mq_i}{m-q_i}$. This defines a finite increasing sequence (p_i) that terminates with some $q_j \geq p$.*

(L9) *Suppose $M = \bigcup_{k\in\mathbf{N}} M_k$ is exhausted by an increasing sequence of compact submanifolds M_k that are deformation retracts of M. Let $D \in \mathcal{D}^{1,p}_{loc}(P)$ and assume that for each $k \in \mathbf{N}$ there is a gauge transformation $f_k \in \mathcal{G}^{2,p}_{loc}(P|_{M_k})$ such that $f^*D|_{M_k}$ is smooth. Then there exists a gauge transformation $f \in \mathcal{G}^{2,p}_{loc}(P)$ such that f^*D is smooth.*

Proof of Theorem 3.2.5: For $p > \frac{m}{2}$ we can fix a $p \leq q < \infty$ that satisfies the condition $\frac{1}{m} > \frac{1}{q} > \frac{1}{p} - \frac{1}{m}$ of Theorem 3.2.6. We prove (1) first. Let a weak Yang-Mills connection $D \in \mathcal{D}^{1,p}(P)$ be given, fix a constant c_0 and let $\delta > 0$ be the constant from Theorem 3.2.6 with the reference connection D. Then find a smooth connection $\hat{D} \in \mathcal{D}(P)$ such that $||D - \hat{D}||_q \leq \delta$, $||D - \hat{D}||_{W^{1,p}} \leq c_0$. This is possible since $||D-\hat{D}||_q \leq C||D-\hat{D}||_{W^{1,p}}$ for some finite Sobolev constant C and since $\mathcal{D}^{1,p}$ is the completion of the set of smooth connections. Theorem 3.2.6 gives a gauge transformation $\hat{f} \in \mathcal{G}^{2,p}(P)$ that sends $\hat{D}$ into relative Coulomb gauge with respect to D. Set $f := \hat{f}^{-1}$. Then $f \in \mathcal{G}^{2,p}(P)$ by (L5) and (L6) implies that $\alpha := f^*D - \hat{D}$ satisfies $d^*_{\hat{D}}\alpha = 0$, $*\alpha|_{\partial M} = 0$. This is the first differential equation for $f^*D = \hat{D} + \alpha$ in (L7) (a). The second (weak) equation is provided by the fact that D and thus f^*D satisfies the weak Yang-Mills equation (see (L4)):

$$\int_M (F_{f^*D}, d_{f^*d}\beta)dv = 0, \quad \forall \beta \in \Omega^1(M, \mathcal{G}_P).$$

We iterate (L7) (a) to prove that f^*D is smooth: We first have $f^*D \in \mathcal{D}^{1,p}(P)$. In case $p > m$ (L7) implies $f^*D \in \mathcal{D}^{2,p}(P)$. In case $p = m$ we can replace p by some $\frac{m}{2} < p < m$ due to the compactness of M and begin the iteration from that regularity. In case $p < m$ the iteration of the lemma and the Sobolev embeddings $W^{2,q_i} \subset W^{1,p_i}$ yield $f^*D \in \mathcal{D}^{2,q_i}(P)$ for the sequences q_i, p_i defined as in (L7). Since $q_j \geq p$ for some $j \in \mathbf{N}$, we also obtain $f^*D \in \mathcal{D}^{2,p}(P)$ after finitely many iterations. Thus in all cases we have shown that $f^*D \in \mathcal{D}^{2,p}(P)$, where $p > \frac{m}{2}$. Now iterate (L7)(a) again to deduce $f^*D \in \mathcal{D}^{k,p}(P)$ for $k \in \mathbf{N}$. This implies that f^*D is smooth and verifies (1).

To prove (2) we argue as in (1). For every compact submanifold M_{k+1} to find a smooth reference connection $\hat{D}_k \in \mathcal{D}(P|_{M_{k+1}})$ and a gauge transformation $f_k \in \mathcal{G}^{2,p}(P|_{M^{k+1}})$ such that $\alpha_k = f_k^*D|_{M_{k+1}} - \hat{D}_k$ meets the relative Coulomb gauge conditions $d^*_{\hat{D}_k}\alpha_k = 0$, $*\alpha_k|_{\partial M_{k+1}} = 0$. Furthermore, D is a weak Yang-Mills connection, and (L3) states that f_k^*D satisfies for all test 1-forms $\beta \in \Omega^1(M, \mathcal{G}_P)$ supported in M_{k+1} the relation $\int_M (F_{f_k^*D}, d_{f_k^*D}\beta)dv = 0$. Fix compact submanifolds $M_k \subset M_K^l \subset M_{k+1}$ such that $M_k^l = M_{k+1}$ and $M_k^{l+1} \subset int(M_k^l)$ for all $l \in \mathbf{N}$. This is possible because M_k is contained in the interior of M_{k+1}. Then we have $\partial M \cap \partial M_k^l \subset \partial M_{k+1}$ and ∂M_k^l agrees with ∂M_{k+1} near every point of this intersection. Thus, for all $l \in \mathbf{N}$, $*\alpha_k|_{\partial M_k^l} = 0$ on $\partial M \cap \partial M_k^l$. Hence we can iterate (L7) (b) to derive that $f_k^*D|_{M_k}$ is smooth. We will prove by induction that $\alpha_k|_{M_k^l} \in W^{l,p}(M_p^l, T^*M_k^l \otimes \mathcal{G}_P)$ for all $l \in \mathbf{N}$ (and this shows that $f_k^*D|_{M_k}$ is smooth because $M_k \subset M_k^l$ for all $l \in \mathbf{N}$).

Suppose that $\alpha_k|_{M_k^l} \in W^{l,p}(M_p^l, T^*M_k^l \otimes \mathcal{G}_P)$ for some $l \in \mathbf{N}$ (which is true for $l = 1$). Then (L7) (b) with $M' = M_k^l$ and $M'' = M_{k+1}^l$ implies $\alpha_k|_{M_k^{l+1}} \in W^{l+1,q}(M_k^{l+1}, T^*M_k^{l+1} \otimes \mathcal{G}_P)$. In case $l \geq 2$ or $p > m$ we get $q = p$, so this proves the iteration. In case $l = 1$ and $p < m$, a further iteration is needed (for $p = m$ we begin with a smaller $\frac{m}{2} < p < m$, and then we still get $W^{1,p}$-regularity on the compact manifold M_{k+1}).

Select a sequence of compact submanifolds $M_k^2 \subset N_i \subset M_{k+1}$ such that $N_{-1} = M_{k+1}$ and N_i is contained in the interior of N_{i-1} for all $i \in \mathbf{N}$. We then iterate (L7)(b) with $M' = N_i$ and $M'' = N_{i+1}$. (Remark that the boundary condition is satisfied as before.) This implies $\alpha_k|_{N_i} \in W^{2,q_i}(N_i, T^*N_i \otimes \mathcal{G}_P)$, with the sequence q_i as in (L8). Again this sequence terminates with some $q_j \geq p$ and thus $\alpha_k|_{M_k^2} \in W^{2,p}(M_k^2, T^*M_k^2 \otimes \mathcal{G}_P)$.

For every $k \in \mathbf{N}$ this proves that $\alpha_k|M_k$ and thus $f_k^* D|_{M_k} = (\hat{D}_k + \alpha_k)|_{M_k}$ is smooth. Apply (L9) to the gauge transformations $f_k \in \mathcal{G}^{2,p}(P|_{M_k})$ to obtain a gauge transformation $f \in \mathcal{G}^{2,p}_{loc}(P)$ such that $f^* D$ is smooth. □

3.2.3 *Strong Compactness*

We introduce Yang-Mills connections in the strong sense on a manifold with boundary and establish a regularity and estimates. All the following results were obtained by Uhlenbeck [383, 384].

Theorem 3.2.8 (Strong Uhlenbeck Compactness). *Let $(D^\nu)_{\nu\in\mathbf{N}} \subset \mathcal{D}^{1,p}(P)$ be a sequence of weak Yang-Mills connections on a compact manifold M and assume that $||F_{D^\nu}||_p$ is uniformly bounded. Then there exist a subsequence (still denoted by $(D^\nu)_{\nu\in\mathbf{N}}$) and a sequence of gauge transformations $f^\nu \in \mathcal{G}^{2,p}_{loc}$ such that $f^{\nu *} D^\nu$ converges uniformly with all its derivatives to a smooth connection $\hat{D} \in \mathcal{D}(P)$.*

In order to prove the above theorem, we need the following lemma.

(L10) *Suppose M is compact, let $\hat{D} \in \mathcal{D}(P)$ be a smooth reference connection, c be a fixed constant, and $p \le q < \infty$ be such that $\frac{1}{m} > \frac{1}{q} > \frac{1}{p} - \frac{1}{m}$. Then there exist constants $(C_l)_{l\in\mathbf{N}}$ and $\delta > 0$ such that for every weak Yang-Mills connection $D \in \mathcal{D}^{1,p}(P)$ that satisfies*

$$||D - \hat{D}||_q \le \delta,\ ||D - \hat{D}||_{W^{1,p}} \le c, \tag{3.46}$$

there exists a gauge transformation $f \in \mathcal{G}^{2,p}(P)$ such that

$$||f^* D - \hat{D}||_q \le C_0 ||D - \hat{D}||_q,\ ||f^* D - \hat{D}||_{W^{l,p}} \le C_l,\ \forall l \in \mathbf{N}. \tag{3.47}$$

Proof of Theorem 3.2.8. The weak compactness Theorem 3.2.1 gives a subsequence (still denoted $(D^\nu)_{\nu\in\mathbf{N}}$) and gauge transformations $f^\nu \in \mathcal{G}^{2,p}(P)$ such that $f^{\nu *} D^\nu$ converges in the weak $W^{1,p}$-topology to some $D \in \mathcal{D}^{1,p}(P)$ and $||f^{\nu *} D^\nu - D||_{W^{1,p}}$ is bounded (cf. the boundedness is due to the weak convergence by Yosida [427], V. 1, Theorem 3).

Let $q = sup\{2p, p^*\}$ where $\frac{1}{p^*} = 1 - \frac{1}{p}$. Then q satisfies the assumptions of (L10). Actually, $q \ge 2p > m$ and $\frac{1}{2p} > \frac{1}{p} - \frac{1}{m}$, because $p > \frac{m}{2}$. Moreover, $\frac{1}{p^*} > \frac{1}{p} - \frac{1}{m}$ is equivalent to $p > \frac{2m}{m+1}$; for $m = 1$ this holds since $p > 1$, for $m = 2$ this needs $p > \frac{4}{3}$, and for $m \ge 3$ this is met by $p > \frac{m}{2}$. The condition on q ensures that the embedding $W^{1,p} \subset L^q$ is compact, thus a subsequence of the sequence $f^{\nu *} D^\nu$, still denoted by $(D^\nu)_{\nu\in\mathbf{N}}$, also converges in the L^q-norm to D. This sequence in $\mathcal{D}^{1,p}(P)$ converges to $D \in \mathcal{D}^{1,p}(P)$ in the L^q-norm and in the weak $W^{1,p}$-topology.

We have only utilized the L^p-bound on the curvature so far. Furthermore, the D^ν are weak Yang-Mills connections (the weak Yang-Mills equation in invariant under gauge transformation, by (L4)). Thus the limit connection D also solves the weak Yang-Mills equation. For all $\beta \in \Omega^1(M, \mathcal{G}_P)$,

$$\int_M (F_D, d_D\beta) = \lim_{\nu\to\infty} \int_M (F_{D^\nu}, d_D^\nu\beta) = 0.$$

In fact, $d_{D^\nu}\beta$ converges in the L^{p*}-norm to $d_D\beta$ since $q \geq p^*$ and F_{D^ν} converges in the weak $W^{1,p}$-topology to F_D. The latter follows from the fact that $(F_{D^\nu})_\alpha = dD_\alpha^\nu + [D_\alpha^\nu \wedge +D_\alpha^\nu]$ is preserved under weak L^p-convergence in all bundle charts. The second term even converges strongly, indeed the D_α^ν converge in the L^{2p}-norm since $q \geq 2p$. For the weak convergence of dD_α^ν we first test only with smooth $\beta \in \Omega^2(U_\alpha, \mathcal{G})$ that vanish on ∂U_α. For these

$$\int_{U_\alpha} (dD_\alpha^\nu, \beta) = \int_{U_\alpha} (dD_\alpha^\nu, d^*\beta) \underset{\nu\to\infty}{\longrightarrow} \int_{U_\alpha} (D_\alpha, d^*\beta) = \int_{U_\alpha} (dD_\alpha, \beta).$$

Then

$$\lim_{\nu\to\infty} \int_{U_\alpha} (dD_\alpha^\nu, \beta) = \int_{U_\alpha} (dD_\alpha, \beta)$$

holds for all smooth $\beta \in \Omega^2(U_\alpha, \mathcal{G})$ because these can be L^{p*}-approximated by such forms that vanish on the boundary and because the dD^ν are L^p-bounded. Hence, D is a weak Yang-Mills connection (note that the compact Sobolev embedding $W^{1,p} \hookrightarrow L^{p^*}$ requires $p > \frac{4}{3}$ when $m = 2$).

Theorem 3.2.5 (1) gives a gauge transformation $\hat{f} \in \mathcal{G}^{2,p}(P)$ such that $\hat{D} = \hat{f}^*D$ is smooth. Thus $\hat{D}^\nu = \hat{f}^*D^\nu$ converge to $\hat{D}$ in the L^q-norm and satisfy a uniform bound $||\hat{D}^\nu - \hat{D}||_{W^{1,p}} \leq c$ for some constant c, which follows from the continuity of the gauge action in (L3). For the L^q-convergence use that lemma with p replaced by q and remark that we have the Sobolev embeddings $\mathcal{G}^{2,p}(P) \subset \mathcal{G}^{1,q}(P)$ and $\mathcal{D}^{1,q}(P) \subset \mathcal{D}^{0,q}(P)$. To prove the theorem we need to find gauge transformations $f^\nu \in \mathcal{G}^{2,p}(P)$ such that a subsequence of the $f^{\nu*}\hat{D}^\nu$ converges in the uniform C^∞-topology (note that $\mathcal{G}^{2,p}(P)$ is closed under composition by (L5)).

Let $\delta > 0$ be determined by $\hat{D}$, q and c as in (L10). Then there is $\nu_0 \in \mathbf{N}$ such that $||\hat{D}^\nu - \hat{D}||_q \leq \delta$ for all $\nu \geq \nu_0$ and thus (L10) applies to $\hat{D}^\nu$. Hence, we find gauge transformations $f^\nu \in \mathcal{G}^{2,p}(P)$ for all $\nu \geq \nu_0$ such that

$$||f^{\nu*}\hat{D}^\nu - \hat{D}||_q \leq C_0||\hat{D}^\nu - \hat{D}||_q.$$

Therefore, $f^{\nu*}\hat{D}^\nu$ converges to $\hat{D}$ in the L_q-norm. Furthermore, one has uniform bounds for all $l \in \mathbf{N}$:

$$||f^{\nu*}\hat{D}^\nu - \hat{D}||_{W^{1,p}} \leq C_l, \ \forall \nu \geq \nu_0.$$

For every $l \in \mathbf{N}$ there is a compact Sobolev embedding $W^{l+2,p} \hookrightarrow C^l$. Hence for all $l \in \mathbf{N}$ we find a further subsequence of the $f^{\nu*}\hat{D}^\nu$ that converges in the uniform C^l-topology. Fixing one further element of the sequence in each step we obtain a sequence that converges in the uniform C^∞-topology. The limit must be $\hat{D}$ since this already was the L^q-limit. □

The strong Uhlenbeck compactness Theorem 3.2.8 for compact manifolds can be generalized to non-compact manifolds (with possibly non-empty boundary) that are exhausted by compact deformation retracts as follows.

Theorem 3.2.9. *Suppose that $M = \bigcup_{k\in\mathbf{N}} M_k$ is exhausted by an increasing sequence of compact submanifolds M_k that are deformation retracts of M. Let $\{g_\nu)_{\nu\in\mathbf{N}}$ be a sequence of metrics on M that converges uniformly with all derivatives on every compact set. Let $D^\nu \in \mathcal{D}^{1,p}_{loc}(P)$ (for all $\nu \in \mathbf{N}$) be a weak Yang-Mills connection with respect to g_ν and suppose that for all $k \in \mathbf{N}$ $\sup_{\nu\in\mathbf{N}} ||F_{D^\nu}||_{L^p(M_k)} < \infty$. Then there exist a subsequence (still denoted by $(D^\nu)_{\nu\in\mathbf{N}}$) and a sequence of gauge transformations $f^\nu \in \mathcal{G}^{2,p}_{loc}(P)$ such that $f^{\nu *} D^\nu$ converges uniformly with all derivatives on every compact set to a smooth connection $\hat{D} \in \mathcal{D}(P)$.*

Similarly to the weak Uhlenbeck compactness, we can use Proposition 3.2.3 to generalize the strong Uhlenbeck compactness on compact manifolds to Theorem 3.2.9. However, we can not apply the strong Uhlenbeck compactness Theorem 3.2.8 to the given sequence of connections restricted to the compact manifolds M_k directly, since the boundary condition the given connections do not restrict to Yang-Mills connections on the subsets of M. In order to have the uniform bounds assumed in Proposition 3.2.3, we need the following lemma.

(L11) *Let $M'' \subset M' \subset M$ be compact submanifolds such that M'' is contained in the interior of M'. Fix a metric g on M and a smooth reference connection $\hat{D} \in \mathcal{D}(P)$, and let $p \le q < \infty$ be such that $1/m > 1/q > 1/p - 1/m$. Then for every constant c there exist a constant δ, ϵ, $(C_l)_{l\in\mathbf{N}}$, and $(\epsilon_l)_{l\in\mathbf{N}} > 0$ such that the following holds: Let g' be a metric on M with $||g' - g||_{W^{1,\infty}(M')} \le \epsilon$. Assume that $\hat{D} \in \mathcal{D}^{1,p}_{loc}(P)$ is a weak Yang-Mills connection with respect to the metric g' such that*

$$||D - \hat{D}||_{L^q(M')} \le \delta, \ ||D - \hat{D}||_{W^{1,p}(M')} \le c.$$

Then there exists a gauge transformation $f \in \mathcal{G}^{2,p}(P|_{M'})$ such that $f^* D|_{M''}$ is smooth. Furthermore, for all $l \in \mathbf{N}$ such that $||g' - g||_{W^{l,\infty}(M')} \le \epsilon_l$ we have

$$||f^* D - \hat{D}||_{W^{l,p}(M'')} \le C_l.$$

3.3 Monotonicity and Curvature Bounds

We first derive a monotonicity formula for Yang-Mills connections, which is important in establishing cone properties of blow-up loci. We then discuss Uhlenbeck's curvature bounds and singular Yang-Mills connections of a certain type. Sections 3.3–3.6 are based on Tian's work [374].

3.3.1 Monotonicity

Let M be an m-dimensional Riemannian manifold with a metric g and E be a vector bundle over M with compact structure group. From now on, we denote the connection of E by A, and save D for other use. For any connection A of E, its curvature form F_A takes values in the Lie algebra $T_e(G) = \mathcal{G}$. The norm of F_A at any point $p \in M$ is

$$|F_A|^2 = \sum_{i,j=1}^{m} < F_A(e_i, e_j), F_A(e_i, e_j) > \tag{3.48}$$

where $\{e_i\}$ is a local orthonormal basis of $T_p(M)$ and $< \cdot, \cdot >$ is the Killing form of $\mathcal{G}$. The Yang-Mills functional is defined by

$$YM(A) = \frac{1}{4\pi^2} \int_M |F_A|^2 dV_g,$$

where dV_g is the volume form determined by g (we drop the subscript g for convenience if no confusion, and denote dV instead of dv for the induced measure on a subspace V of the tangent space T_M of M purposely for later use), and it is modified slightly by a constant $\frac{1}{4\pi^2}$ than usual functional.

Let $\{\phi_t\}_{t\in\mathbf{R}}$ be a one-parameter family of diffeomorphisms of M, and A_0 be a fixed smooth connection of E and D be its associated covariant derivative. For any connection A we define a family of connections $\phi_t^*(A)$ as follows: Let τ_t^0 be the parallel transport of E associated to A_0 along the path $\phi_s(x)_{0\le s\le t}$, for $x \in M$. For any $u \in E_x$ over $x \in M$, let $\tau_x^0(u)$ be the section of E over the path $\phi_s(x)_{0\le s\le t}$ satisfying

$$D_{\frac{\partial}{\partial s}} \tau_s^0(u) = 0, \;\; \tau_0^0(u) = u. \tag{3.49}$$

We define $A^t = \phi_t^*(A)$ by its associated covariant derivative

$$D_X^t v = (\tau_t^0)^{-1}(D_{d\phi_t(X)} \tau_t^0(v)) \tag{3.50}$$

for $X \in TM, v \in \Gamma(M, E)$, where $\Gamma(M, E)$ is the space of sections of E over M.

To verify that A^t is a connection, it suffices to show that

$$\begin{aligned} D_X^t(fv)(x) &= (\tau_t^0)^{-1}(D_{d\phi_t(X)}((\phi_t^{-1})^* f \cdot \phi_t)\tau_t^0(v))(x) \\ &= (\tau_t^0)^{-1}\Big(f(x) D_{d\phi_t(X)}\tau_t^0(v)(\phi_t(x)) + d\phi_t(x)((\phi_t^{-1})^* f)\tau_t^0(v)(\phi_t(x))\Big) \\ &= f(x) D_X^t v + X(f)(x) v(x). \end{aligned} \tag{3.51}$$

Thus the curvature of A^t is given by

$$F_{A^t}(X, Y) = (\tau_t^0)^{-1} \cdot F_A(d\phi_t(X), d\phi_t(Y)) \cdot \tau_t^0. \tag{3.52}$$

It follows that

$$YM(A^t) = \frac{1}{4\pi^2}\int_M |F_{A^t}|^2 dV$$
$$= \frac{1}{4\pi^2}\int_M \sum_{i,j=1}^m |F_A(d\phi_t(e_i), d\phi_t(e_j))|^2 (d\phi_t(x))dV.$$

where $\{e_i\}$ is a local orthonormal basis of TM.

Changing variables, we obtain

$$YM(A^t) = \frac{1}{4\pi^2}\int_M \sum_{i,j=1}^m |F_A(d\phi_t(e_i\phi_t^{-1}(x))), d\phi_t(e_j(\phi_t^{-1}(x)))|^2 J(\phi_t^{-1})dV,$$

where J is the Jacobian. Let X be the vector field $\frac{\partial\phi_t}{\partial t}\big|_{t=0}$ on M. Then we arrive at

$$\frac{d}{dt}YM(A^t)\Big|_{t=0} = -\frac{1}{4\pi^2}\int_M \Big(|F_A|^2 divX + 4\sum_{i,j=1}^m < F_A([X,e_i],e_j),\ F_A(e_i,e_j) > \Big)dV, \tag{3.53}$$

where we used $\frac{d}{dt}(d\phi_t(e_i(\phi_t^{-1}(x))))\big|_{x=0} = -[X,\ e_i]$. Since $[X,e_i] = \nabla_X e_i - \nabla_{e_i} X$ (where ∇ is the Levi-Civita connection of g), we have

$$\sum_{i,j=1}^m < F_A([X,e_i],e_j),\ F_A(e_i,e_j) >$$
$$= -\sum_{i,j=1}^m \Big(< F_A(\nabla_{e_i}X,e_j),\ F_A(e_i,e_j) > - < F_A(\nabla_X e_i,e_j), F_A(e_i,e_j) > \Big)$$
$$= -\sum_{i,j=1}^m \Big(< F_A(\nabla_{e_i}X,e_j),\ F_A(e_i,e_j) > -g(\nabla_X e_i,e_k) < F_A(e_k,e_j),\ F_A(e_i,e_j) > \Big). \tag{3.54}$$

Since

$$g(\nabla_X e_i, e_k) = -g(e_i, \nabla_X e_k) = -g(\nabla_X e_k, e_i),$$

the second term of (3.54) vanishes.

If A is a Yang-Mills connection, then we obtain

$$0 = \int_M \left(|F_A|^2 divX - 4\sum_{i,j=1}^m < F_A(\nabla_{e_i}X_i,\ e_j), F_A(e_i,\ e_j) > \right) dV. \tag{3.55}$$

For any point $p \in M$, let r_p be a positive number satisfying: there are normal coordinates $x_1, \cdots, x_m$ in the geodesic ball $B_{r_p}(p)$ of (M, g) such that $p = (0, \cdots, 0)$ and for some constant $c(p)$,

$$|g_{ij} - \delta_{ij}| \leq c(p)(|x_1|^2 + \cdots + |x_m|^2), \tag{3.56}$$

$$|dg_{ij}| \leq c(p)\sqrt{|x_1|^2 + \cdots + |x_m|^2}, \tag{3.57}$$

where $g_{ij} = g(\frac{\partial}{\partial x_i}, \frac{\partial}{\partial x_j})$. Note that r_p and $c(p)$ can be selected depending only on the injectivity radius at p and the curvature of g. When $M = \mathbf{R}^m$ and g is flat, we can choose $r_p = \infty$ and $c(p) = 0$. Let $r(x) = \sqrt{|x_1|^2 + \cdots + |x_m|^2}$ be the distance from p, and ϕ be a positive function on the unit sphere S^{m-1}. Define

$$X(x) = \xi(\tau)\phi(\frac{x}{r})r\frac{\partial}{\partial r} = \xi(r)\phi(\frac{x}{r})\left(\sum_i x_i \frac{\partial}{\partial x_i}\right), \tag{3.58}$$

where ξ is a smooth function with compact support in $B_{r_P}(p)$.

Let$\{e_1, \cdots, e_m\}$ be an orthonormal basis at p such that $e_1 = \frac{\partial}{\partial r}$. We have $\nabla_{\frac{\partial}{\partial r}}\frac{\partial}{\partial r} = 0$, since $x_1, \cdots, x_m$ are normal coordinates. It implies that

$$\nabla_{\frac{\partial}{\partial r}} X = (\xi r)'\phi(\theta)\frac{\partial}{\partial r} = (\xi' r + \xi)\phi(\theta)\frac{\partial}{\partial r} \tag{3.59}$$

where $\theta = \frac{x}{r}$. Furthermore, if $i \geq 2$,

$$\nabla_{e_i} X = \xi r \nabla_{e_i}(\phi\frac{\partial}{\partial r}) = \xi r e_i(\phi)\frac{\partial}{\partial r} + \xi\phi\sum_{j=1}^m k_{ij}e_j, \tag{3.60}$$

where $|k_{ij} - \delta_{ij}| = O(1)c(p)r^2$, and $O(1)$ is a quantity bounded by a constant depending only on m.

Substituting (3.59) and (3.60) to (3.55), we arrive at

$$\begin{aligned} &\int_M |F_A|^2\Big(\xi' r + (m-4)\xi + O(1)c(p)r^2\xi\Big)\phi\, dV \\ &= 4\int_M \left(\xi' r\phi|\frac{\partial}{\partial r}\rfloor F_A|^2 + \xi r(\frac{\partial}{\partial r}\rfloor F_A, \nabla\phi\rfloor F_A)\right) dV, \end{aligned} \tag{3.61}$$

where $\frac{\partial}{\partial r}\rfloor F_A = F_A(\frac{\partial}{\partial r}, \cdot)$.

For sufficiently small τ we select $\xi(r) = \xi_\tau(r) = \eta(\frac{r}{\tau})$, where η is smooth and such that $\eta(r) = 1$ for $r \in [0, 1]$, $\eta(r) = 0$ for $r \in [1+\epsilon, \infty)$, $\epsilon > 0$, and $\eta'(r) \leq 0$. Thus we have

$$\tau\frac{\partial}{\partial \tau}(\xi_\tau(r)) = -r\xi'_\tau(r). \tag{3.62}$$

Applying this to (3.61), we deduce

$$\tau\frac{\partial}{\partial\tau}\Big(\int_M \xi_\tau\phi|F_A|^2dV + (4-m) + O(1)c(p)\tau^2\Big)\int_M \xi_\tau\phi|F_A|^2dV$$
$$= 4\tau\frac{\partial}{\partial\tau}\Big(\int_M \xi_\tau\phi|\frac{\partial}{\partial\tau}\rfloor F_A|^2\Big)dV - 4\int_M \xi_\tau r < \frac{\partial}{\partial r}\rfloor F_A, \nabla\phi\rfloor F_A > dV. \quad (3.63)$$

Selecting a non-negative number $a \geq O(1)c(p)$, we derive from the preceding formula that

$$\frac{\partial}{\partial\tau}\Big(\tau^{4-m}e^{\pm a\tau^2}\int_M \xi_\tau\phi|F_A|^2dV\Big) = 4\tau^{4-m}e^{\pm a\tau^2}\Big(\frac{\partial}{\partial\tau}(\int_M \xi_\tau\phi|\frac{\partial}{\partial\tau}\rfloor F_A|^2\, dV)$$
$$+ (-O(1)c(p) \pm 2a)\tau\int_M \xi_\tau\phi|F_A|^2\, dV - \tau^{-1}\int_M \xi_\tau < \frac{\partial}{\partial r}\rfloor F_A, \nabla\phi\rfloor F_A > dV\Big). \quad (3.64)$$

Integrating over τ and setting $\epsilon \to 0$, we obtain the following monotonicity formula.

Theorem 3.3.1. *Let r_p, c(p) and a as above. Then for* $0 < \sigma < \rho < r_p$,

$$\pm\, \rho^{4-m}e^{\pm a\rho^2}\int_{B_\rho(p)}\phi|F_A|^2dV \mp \sigma^{4-m}e^{\pm a\sigma^2}\int_{B_\sigma(p)}\phi|F_A|^2dV$$
$$\mp\, 4\int_{B_\rho(p)-B_\sigma(p)}\tau^{4-m}e^{\pm ar^2}\phi|\frac{\partial}{\partial r}\rfloor F_A|^2dV$$
$$\geq -4\int_\sigma^\rho \tau^{3-m}e^{\pm a\tau^2}d\tau\int_{B_\tau(p)}\Big|\frac{\partial}{\partial\tau}\rfloor F_A||\nabla\phi\rfloor F_A\Big|dV. \quad (3.65)$$

This inequality is required to establish the existence of tangent cones of blow-up loci. In particular, if $\phi = 1$, then we have the following result due to Price [304].

Theorem 3.3.2. *Let r_p, c(p) and a as above. Then for* $0 < \sigma < \rho < r_p$,

$$\rho^{4-m}e^{a\rho^2}\int_{B_\rho(p)}|F_A|^2dV - \sigma^{4-m}e^{a\sigma^2}\int_{B_\sigma(p)}|F_A|^2dV$$
$$\geq 4\int_{B_\rho(p)-B_\sigma(p)}\tau^{4-m}e^{ar^2}\Big|\frac{\partial}{\partial\tau}\rfloor F_A\Big|^2 dV \quad (3.66)$$

In particular, if $M = \mathbf{R}^m$ and g is flat, then the equality holds in (3.66) for $\rho \in (0,\infty)$ and $a = 0$.

3.3.2 Curvature Bounds

We provide a curvature bound for Yang-Mills connections that was obtained by Uhlenbeck [383], as follows.

Theorem 3.3.3. *Let A be any Yang-Mills connection of a G-bundle E over M. Then there are* $\epsilon = \epsilon(m) > 0$ *and* $C = C(m) > 0$*, which depend only on m and M, such that if*

$$\rho^{4-m} \int_{B_{\rho(p)}} |F_A|^2 dV \leq \epsilon,$$

for any $p \in M$ *and* $\rho < r_p$*, then*

$$|F_A|(p) \leq \frac{C}{\rho^2} \left(\rho^{4-m} \int_{B_{\rho(p)}} |F_A|^2 dV \right)^{1/2}. \tag{3.67}$$

Proof. By scaling, we may assume that $\rho = 1$. Consider the function

$$f(r) = (1-2r)^2 \sup_{x \in B_{r(p)}} |F_A|(x), \; r \in [0, 1/2]. \tag{3.68}$$

Then $f(r)$ is continuous in $[0, 1/2]$ with $f(1/2) = 0$, and so f attains its maximum at some $r_0 \in [0, 1/2]$. Our proof is based on [319].

We now show that $(\star)$ $f(r_0) \leq 64$ if ϵ is small enough. Suppose that $f(r_0) > 64$. Setting $q = \sup_{x \in B_{r_0(p)}} |F_A|(x) = |F_A|(x_0)$ and choosing $\sigma = \frac{1}{4}(1 - 2r_0)$, we have

$$\sup_{x \in B_{r_0}(p)} |F_A| \leq \sup_{x \in B_{r_0+\sigma}(p)} |F(A)|(x) \leq \frac{(1-2r_0)^2}{(1-2r_0-2\sigma)^2} \sup_{x \in B_{r_0}(p)} |F_A|(x) = 4q. \tag{3.69}$$

Clearly, $16\sigma^2 q \geq 64$, that is, $\sigma\sqrt{q} \geq 2$. Let $\tilde{g} = qg$ be a scaled metric. Then the norm $|F_A|_{\tilde{g}}$ of F_A is $q^{-1}|F_A|$ v with respect to $\tilde{g}$. Therefore,

$$\sup_{x \in B_2(x_0, \tilde{g})} |F_A|_{\tilde{g}} \leq 4, \tag{3.70}$$

where $B_2(x_0, \tilde{g})$ is the geodesic ball of $\tilde{g}$ with radius 2 and center at x_0.

Because A is a Yang-Mills connection, a straightforward calculation and the second Bianchi identity show that

$$\frac{1}{2}\Delta_{\tilde{g}} |F_A|^2 = |\tilde{\nabla} F_A|^2_{\tilde{g}} - 2F_A \# F_A \# R(\tilde{g}) - 2F_A * F_A * F_A, \tag{3.71}$$

where

$$F_A \# F_A \# R(\tilde{g}) = \sum_{i,j,k,l} \Big(< F_A(e_l, e_k), F_A(e_i, e_j) > \\ - \sum_m < F_A(e_l, e_m), F_A(e_i, e_m) > \delta_{jk} \Big) R(\tilde{g})(e_l, e_j, e_k, e_i), \quad (3.72)$$

$$F_A * F_A * F_A = \sum_{i,j,k} < [F_A(e_i, e_j), F_A(e_j, e_k)], F_A(e_k, e_i) >, \quad (3.73)$$

and $\{e_1, \cdots, e_m\}$ is an orthonormal basis of $\tilde{g}$. Then we can derive from (3.71) to (3.73) that

$$-\Delta_{\tilde{g}} |F_A|_{\tilde{g}} \leq c_1 |F_A|_{\tilde{g}} + c_2 |F_A|_{\tilde{g}}^2, \quad (3.74)$$

where the uniform constants c_1, c_2 depend only on m. Applying (3.70), we have in $B_2(x_0, \tilde{g})$

$$-\Delta |F_A|_{\tilde{g}} \leq (c_1 + 4c_2) |F_A|_{\tilde{g}}.$$

Now using the mean value theorem or a Moser iteration we get

$$1 = |F_A|_{\tilde{g}}(x_0) \leq c_3 \left(\int_{B_1(x_0, \tilde{g})} |F_A|_{\tilde{g}}^2 dV \right)^{1/2}, \quad (3.75)$$

where c_3 is a uniform constant.

Applying Theorem 3.3.1, we have

$$\int_{B_1(x_0, \tilde{g})} |F_A|_{\tilde{g}}^2 dV \leq (\sqrt{q})^{m-4} \int_{B_{1/\sqrt{q}}(x_0)} |F_A|^2 dV \\ \leq (1/2)^{4-m} e^{a/4} \int_{B_{1/2}(x_0)} |F_A|^2 dV \leq \epsilon 2^{m-4} e^{a/4}.$$

Combining this with (3.75), we have

$$1 \leq c_3 \epsilon 2^{m-4} e^{a/4},$$

which is impossible when $\epsilon = \epsilon(m)$ is small enough. So $(\star)$ is verified.

Consequently, we obtain

$$\sup_{x \in B_{1/4}(p)} |F_A|(x) \leq 4 f(r_0) \leq 256. \quad (3.76)$$

Thus by (3.76) and (3.71) with $\tilde{g}$ replaced by g, we have

$$-\Delta_g |F_A| \le c_4 |F_A|, \tag{3.77}$$

where c_4 is a uniform constant. Hence, we can conclude (3.67) by (3.77) and a Moser iteration. □

3.3.3 Admissible Yang-Mills Connections

An *admissible* Yang-Mills connection is a smooth connection A defined outside a closed subset $S(A)$ in M such that the following conditions hold: (i) $H^{m-4}(S(A)\cap K) < \infty$ for any compact subset $K \subset M$, where $H^{m-4}(\cdot)$ is the $(m-4)$-dimensional Hausdorff measure; (ii) A is a Yang-Mills connection on $M \setminus S(A)$; (iii) A obeys

$$\int_{M\setminus S(A)} |F_A|^2 dV < \infty. \tag{3.78}$$

It follows from (3.78) that for any smooth $\mathcal{G} = T_e(G)$-valued 1-form u on M with compact support,

$$\int_M (F_A, du) dV = 0. \tag{3.79}$$

Obviously, A is smooth on M if $S(A) = \emptyset$. $S(A)$ is called the *singular set* of A, and is not invariant under gauge transformations. Even if $S(A) \neq \emptyset$, there may be a gauge transformation σ on $M\setminus S(A)$ such that $\sigma(A)$ extends to a smooth connection on M. Two admissible connections A_1 and A_2 are *gauge equivariant* if there is a gauge transformation σ of E over $M \setminus (S(A_1) \cup S(A_2))$ such that $\sigma(A_1) = A_2$ outside $S(A_1) \cup S(A_2)$. This new gauge equivalence extends the previous one for smooth connections. Likewise, if A is Ω-anti-self-dual outside $S(A)$, we can define admissible Ω-anti-self-dual instantons.

Suppose that G is a unitary group. By Chern-Weil theory, for each smooth connection A, we have closed forms $\frac{i}{2\pi} tr(F_A)$ and $(\frac{i}{2\pi})^2 tr(F_A \wedge F_A)$ $(i = \sqrt{-1})$ of degree 2 and 4, respectively. If M is compact, they represent the first two Chern characters $Ch_1(E)$ and $Ch_2(E)$. We can extend these to admissible Yang-Mills connections.

Let A be an admissible Yang-Mills connection with the singular set $S = S(A)$. Then $tr(F_A)$ and $tr(F_A \wedge F_A)$ are closed forms on $M - S$. Thanks to condition (iii) above, we can extend them to forms on M in the sense of distributions. Indeed, these forms are invariant under gauge transformations.

Theorem 3.3.4. *The extended forms* $\frac{i}{2\pi} tr(F_A)$ *and* $(\frac{i}{2\pi})^2 tr(F_A \wedge F_A)$ *are closed on M; they are denoted by* $Ch_1(E)$ *and* $Ch_2(E)$, *respectively*

Proof. Since the first case is simple, we only show that $(\frac{i}{2\pi})^2 tr(F_A \wedge F_A)$. It suffices to claim that for any smooth form ϕ of degree $m-5$ with compact support in M,

$$\int_M d\phi \wedge tr(F_A \wedge F_A) = 0. \tag{3.80}$$

This is well-defined since F_A is L^2 integrable.

Without loss of generality, we may assume that M is a ball in R^m and E is a trivial bundle over M. Let K be a compact subset in M with $supp(\phi)$ in its interior. As in Theorem 3.3.3 and fixing $\epsilon \leq \epsilon(m)$, we define

$$E_r = \{x \in K \,|\, r^{4-m} e^{ar^2} \int_{B_r(x)} |F_A|^2 dV \geq \epsilon\}, \tag{3.81}$$

where a is as in Theorem 3.3.2. It follows from Theorem 3.3.2 that $E_{r'} \subset E_r$ if $r' \leq r$. Consider a finite covering $\{B_{2r}(x_k)\}_{1\leq k\leq L_r}$ of E_r such that (a) $x_k \in E_r$ and (b) $B_r(x_k) \cap B_r(x_l) = \emptyset$ for $k \neq l$. We then expand $\{B_{2r}(x_k)\}_{1\leq k\leq L_r}$ to a covering $\{B_{2r}(x_k)\}_{1\leq k\leq L'_r}$ $(L'_r \geq L_r)$ of $(S \cap K) \cup E_r$, such that $x_k \in (S \cap K) \cup E_r$ and $B_r(x_k) \cap B_r(x_l) = \emptyset$ for $k \neq l$. Remark that for any k, the number of x_l with $B_{8r}(x_k) \cap B_{8r}(x_l) \neq \emptyset$ is bounded by a constant depending on m and M.

For any $x \notin \cup_k^{L'_r} B_{2r}(x_k)$,

$$r^{4-m} \int_{B_r(x)} |F_A|^2 dV < \epsilon. \tag{3.82}$$

By Theorem 3.3.3, we have

$$|F_A|(x) \leq \frac{c}{r^2} \left(r^{4-m} \int_{B_r(x)} |F_A|^2 dV \right)^{1/2} \leq \frac{c\sqrt{\epsilon}}{r^2} \tag{3.83}$$

where c is a uniform constant. It follows from Theorem 1.2.7 [383] that we can construct a gauge transformation σ_x over $B_r(x)$ for any $x \in M \setminus N_{3r}((S \cap K) \cup E_r)$ satisfying

$$|\sigma_x(A)|(y) \leq \frac{c}{r} \left(r^{4-m} \int_{B_r(x)} |F_A|^2 dV \right)^{1/2}, \quad y \in B_r(x). \tag{3.84}$$

Remark that for any $\delta > 0$ and any subset $S' \subset M$, $N_\delta(S') = \{x \in M \,|\, d(x, S') \leq \delta\}$, where $d(\cdot,\cdot)$ is the distance corresponding to the metric g.

Patching these σ appropriately, we can construct a gauge transformation σ_x over each $B_{8r}(x_k) \setminus N_{3r}((S \cap K) \cup E_r)$ satisfying

$$|\sigma_k(A)|(x) \leq \frac{c}{r} \left(r^{4-m} \int_{B_r(x)} |F_A|^2 dV \right)^{1/2}, \tag{3.85}$$

for $x \in B_{8r}(x_k) \setminus N_{3r}((S \cap K) \cup E_r)$. By (3.85), we have on the intersection $B_{8r}(x_l) \cap B_{8r}(x_k) \setminus N_{3r}((S \cap K) \cup E_r)$,

$$|d\sigma_k \cdot \sigma_l^{-1}| \leq \frac{2c\sqrt{\epsilon}}{r}.$$

Thus by modifying σ_k slightly on the intersection, we may assume that $\sigma_k \cdot \sigma_l^{-1}$ is constant on each connected component of $B_{8r}(x_k) \cap B_{8r}(x_l) \setminus N_{3r}((S \cap K) \cup E_r)$ for any $k \neq l$.

Let $\nu : \mathbf{R} \to \mathbf{R}$ be a C^∞ cut-off function such that $\nu(t) = 0$ for $t \leq 1$, $\nu(t) = 1$ for $t \geq 2$ and $0 \leq \nu'(t) \leq 1$. Then

$$\int_M d\phi \wedge tr(F_A \wedge F_A) = \lim_{r \to 0} \int_M \nu(\frac{d(x, (S \cap K) \cup E_r)}{3r}) d\phi \wedge tr(F_A \wedge F_A). \tag{3.86}$$

For each $k \leq L'_r$, we have

$$\begin{aligned} tr(F_A \wedge F_A)(x) &= tr(F_{\sigma_k(A)} \wedge F_{\sigma_k(A)})(x) \\ &= d\, tr\left(F_{\sigma_k(A)} \wedge F_{\sigma_k(A)} + \frac{1}{3}\sigma_k(A) \wedge \sigma_k(A) \wedge \sigma_k(A)\right)(x), \end{aligned} \tag{3.87}$$

for all $x \in B_{8r}(x_k) \setminus N_{3r}((S \cap K) \cup E_r)$.

Because $\sigma_k \cdot \sigma_l^{-1}$ is piecewise constant, we obtain

$$\begin{aligned} & tr\left(\sigma_k(A) \wedge F_{\sigma_k(A)} + \frac{1}{3}\sigma_k(A) \wedge \sigma_k(A) \wedge \sigma_k(A)\right) \\ &= tr\left(\sigma_l(A) \wedge F_{\sigma_l(A)} + \frac{1}{3}\sigma_l(A) \wedge \sigma_l(A) \wedge \sigma_l(A)\right) \end{aligned} \tag{3.88}$$

on the intersection $B_{8r}(x_l) \cap B_{8r}(x_k) \setminus N_{3r}((S \cap K) \cup E_r)$. Thus there is a globally defined Chern-Simon transgression form Ψ outside $N_{3r}(S \cup E_r)$, satisfying $d\Psi = tr(F_A \wedge F_A)$ and

$$\Psi(x) = tr(\sigma_k(A) \wedge F_{\sigma_k(A)} + \frac{1}{3}\sigma_k(A) \wedge \sigma_k(A) \wedge \sigma_k(A)),$$

for $x \in B_{8r}(x_k)$. For each k and any $x \in B_{6r}(x_k) \setminus B_{3r}(x_k)$, we have

$$|\psi(x)| \leq cr^{-3}\left(r^{4-m}\int_{B_r(x)} |F_A|^2 dV\right)^{3/2} \leq cr^{1-m}\int_{B_{8r}(x_k)} |F_A|^2 dV.$$

It follows that

$$\left|\int_M d\phi \wedge tr(F_A \wedge F_A)\right| = \lim_{r\to 0}\left|\int_M \nu(\frac{d(x,(S\cap K)\cup E_r)}{3r})d\phi \wedge d\Psi\right|$$
$$\leq \lim_{r\to 0}\int_{3r\leq d(x,(S\cap K)\cup E_r)\leq 6r} \frac{1}{3r}|\Psi||d\phi|dV$$
$$\leq c\lim_{r\to 0}\left\{\sup_M|d\phi|\sum_{k=1}^{L'_r}\int_{B_{8r}(x_k)}|F_A|^2dV\right\}$$
$$\leq c\sup_M|d\phi|\lim_{r\to 0}\int_{N_{8r}(S\cup E_r)}|F_A|^2dV.$$

Since $\bigcap_{r>0} N_{8r}(S\bigcup E_r) \subset S$ and $N_{8r}(S\cup E_r)\subset N_{8r'}(S\cup E_{r'})$ for $r\leq r'$, the last integral converges to zero as r tends to 0. Hence, we obtain

$$\int_M d\phi \wedge tr(F_A\wedge F_A) = 0,$$

so that $tr(F_A \wedge F_A)$ is closed in the sense of distributions. □

Let CW_1 and CW_2 be the Chern-Weil polynomial defining the first two Chern classes. Then $CW_1(A) = Ch_1(A)$ is well-defined. On $M\setminus S(A)$, we have

$$CW_2(A) = \frac{1}{8\pi^2}(tr(F_A\wedge F - A) - tr(F_A)\wedge tr(F_A)). \tag{3.89}$$

Thus $CW_2(A)$ extends to a form on M, still called $CW_2(A)$, in the sense of distributions.

Corollary 3.3.5. *The extended form $CW_2(A)$ is closed.*

Proof. Since $tr(F)A$ is harmonic outside $S(A)$ and L_2-bounded, by the elliptic theory it extends to a smooth form on M. Then the last theorem implies the result. □

3.4 Rectifiability of Blow-Up Loci

We first study the blow-up set of a sequence of Yang-Mills connections that converges to an admissible Yang-Mills connection. We then investigate the geometry of blow-up loci. All the aruments are based on [374].

3.4.1 Convergence of Yang-Mills Connections

For a sequence of admissible Yang-Mills connections $\{A_i\}$, we say that $\{A_i\}$ converge weakly to an admissible Yang-Mills connection (modulo gauge transformations), if $\int_M |F_{A_i}|^2 dV \le c$ for some uniform constant c and there are a closed subset S and gauge transformations σ_i of the G-bundle E over $M \setminus S$, such that for any compact $K \subset M \setminus S$, $\sigma_i(A_i)$ extends smoothly across K for i large enough and converge in the C^∞-topology to A on K as $i \to \infty$. Clearly, S contains $S(A)$. It follows that for any smooth form ϕ with compact support in M,

$$\lim_{i\to\infty} \int_M (F_{\sigma_i(A_i)}, d\phi) dV = \int_M (F_A, d\phi) dV,$$

which is precisely the weak convergence. Clearly, we have the following result.

(L1) Weak limits of admissible connections $\{A_i\}$ are unique modulo gauge transformations.

Suppose that $\{A_i\}$ is a sequence of smooth Yang-Mills connections with $YM(A_i) \le \Lambda$. All the arguments in this section also work for admissible Yang-Mills connections with slight modification.

Proposition 3.4.1. *There exists a subsequence $\{A_{i_j}\}$ which converges weakly to some admissible Yang-Mills connection A on M.*

Proof. Let a be as in Theorem 3.3.2 and ϵ be as in Theorem 3.3.3. Consider the closed subset

$$E_{i,r} = \{x \in M \mid e^{ar^2} r^{4-m} \int_{B_r(x)} |F_{A_i}|^2 dV \ge \epsilon\}, \tag{3.90}$$

for each i and $r > 0$ small enough. Theorem 3.3.1 implies that $E_{i,r} \subset E_{i,r'}$ whenever $r \le r'$. By the classical diagonal process, we can select a subsequence $\{i_j\}$ of $\{i\}$ such that for each k, the sets $E_{i_j,2^{-k}}$ converge to a closed subset $E_{2^{-k}}$. Thus $E_{2^{-k}} \subset E_{2^{-l}}$ for $k \ge l$. Set $S = \bigcap_k E_{2^{-k}}$.

We first show that S is of Hausdorff co-dimension at least 4. Given $\delta > 0$ small enough and any compact subset K of M, let $\{B_{4\delta}(x_\alpha)\}$ be any finite covering of $S \cap K$ such that (a) $x_\alpha \in S \cap K$ and (b) $B_{2\delta}(x_\alpha) \cap B_{2\delta}(x_\beta) = \emptyset$ for $\alpha \ne \beta$. Choose k sufficiently large such that $2^{-k} < \delta$. Thus for j large enough, there are $y_\alpha \in E_{i_j,2^{-k}}$ such that $d(x_\alpha, y_\alpha) < \delta$. So $\{B_{5\delta}(y_\alpha)\}$ is a finite covering of $S \cap K$ and $B_\delta(y_\alpha) \cap B_\delta(y_\beta) = \emptyset$ for $\alpha \ne \beta$. By Theorem 3.3.2, we have

$$e^{a\delta^2} \delta^{4-m} \int_{B_\delta(y_\alpha)} |F_{A_{i_j}}|^2 dV \ge e^{a2^{-2k}} 2^{(m-4)k} \int_{B_{2^{-k}}(y_\alpha)} |F_{A_{i_j}}|^2 dV \ge \epsilon.$$

Therefore,

$$\sum_\alpha \delta^{m-4} \le \frac{e^a}{\epsilon} \sum_\alpha \int_{B_\delta(y_\alpha)} |F_{A_{i_j}}|^2 dV \le \frac{e^a}{\epsilon} \int_M |F_{A_{i_j}}|^2 dV \le \frac{ce^a}{\epsilon}.$$

This implies that $H^{m-4}(S \cap K)$, and thus $H^{m-4}(S)$ is no larger than $\frac{5^{m-4} e^a c}{\epsilon}$. Hence, S is of Hausdorff co-dimension at least 4.

We next claim that $\{A_{i_j}\}$ converges outside S to some A modulo gauge transformations. For simplicity, we denote i_j by i. Note that for any $r > 0$, there is an $i(r) > 0$, such that for any $i \geq i(r)$ and $x \in M$ with $d(x, E_{2^{-k}}) \geq r$, and $2^{-k-1} \leq r \leq 2^{-k}$,

$$e^{ar^2} r^{4-m} \int_{B_r(x)} |F_{A_i}|^2 dV < \epsilon, \tag{3.91}$$

i.e., $x \in M \setminus E_{i,r}$. By Theorem 3.3.3, (3.91) implies that for $x \in M \setminus B_r(E_r)$,

$$|F_{A_i}|(x) < \frac{c\sqrt{\epsilon}}{r^2}. \tag{3.92}$$

Theorem 3.6 in [383] implies that there exist a subsequence $\{i'\} \subset \{i\}$ and gauge transformations $\sigma(i')$, such that $\sigma(i')(A_{i'})$ converge in C^1 topology to a smooth connection A on any compact subset outside S. Because A_i are Yang-Mills connections, A is a Yang-Mills connection and $\{i'\} \subset \{i\}$ converge to A smoothly outside S by standard elliptic theory. □

In what follows, suppose that the sequence A_i converges to an admissible Yang-Mills connection A with $\int_M |F_{A_i}|^2 dV \leq \Lambda$. We need the following lemmas (proofs in [374]) for arguments below.

(L2) *Define*

$$S_b(\{A_i\}) = \bigcap_{r>0} \left\{ x \in M \,|\, \liminf_{i \to \infty} e^{ar^2} r^{4-m} \int_{B_r(x)} |F_{A_i}|^2 dV \geq \epsilon \right\}, \tag{3.93}$$

where ϵ is given in Theorem 3.3.3. Then (a) *$S_b(\{A_i\})$ is closed and contained in the preceding S;* (b) *Its Hausdorff measure $H^{m-4}(S_b(\{A_i\}) \leq c$ for some constant c depending only on M and Λ;* (c) *A extends to a smooth connection on $M \setminus S_b(\{A_i\})$.*

Because we can extend A smoothly to $M \setminus S_b(\{A_i\})$, we may assume that $S(A) \subset S_b(\{A_i\})$. If $S_b(\{A_i\}) = \emptyset$, then there is a subsequence of $\{A_i\}$ which converges to A smoothly on M.

Consider the Randon measures $\mu_i = |F_{A_i}|^2 dV_g$ $(i = 1, 2, \cdots)$. Choosing a subsequence if needed, we may assume that $\mu_i \to \mu$ weakly on M as Radon measures, that is, for any continuous function ϕ with compact support in M,

$$\lim_{i \to \infty} \int_M \phi |F_{A_i}|^2 dV_g = \int_M \phi \, d\mu. \tag{3.94}$$

Write $\mu = |F_A|^2 dV_g + \nu$ (by Fatou's lemma) for some nonnegative Radon measures ν on M.

(L3) *When $\nu(x) = \Theta(x) H^{m-4} \lfloor S_b(\{A_i\})$, $x \in M$, for H^{m-4}-a.e. $x \in S_b(\{A_i\})$, then*

$$\epsilon \leq \Theta \leq 4^{m-4} r_x^{4-m} e^{ar_x^2} \Lambda,$$

where r_x and a are provided in Theorem 3.3.2.

We have the following facts:

(i) *For any* $x \in M$, $e^{ar^2} r^{4-m}\mu(B_r(x))$ *is a non-decreasing function for* r *small enough, and the density*

$$\Theta(\mu, x) = \lim_{r\to 0+} r^{4-m}\mu(B_r(x)) \tag{3.95}$$

exists for every $x \in M$;

(ii) $x \in S_b(\{A_i\})$ *if and only if* $\Theta(\mu, x) \geq \epsilon$;

(iii) *For* H^{m-4}*-a.e.* $x \in S_b(\{A_i\})$,

$$\lim_{r\to 0+} \int_{B_r(x)} |F_A|^2 dV = 0.$$

Define

$$S_b = \overline{\{x \in S_b(\{A_i\}) | \, \Theta(\mu, x) > 0, \ \lim_{r\to 0+} r^{4-m} \int_{B_r(x)} |F_A|^2 dV = 0\}}. \tag{3.96}$$

Thus we have $S_b(\{A_i\}) = S_b \cup S(A)$. (S_b, Θ) is called the *blow-up locus* of the weakly convergent sequence $\{A_i\}$, where S_b is the support of the blow-up locus and Θ is its multiplicity. We simply say that S_b is the blow-up locus whenever there is no danger of confusion.

3.4.2 Tangent Cones

For convenience, we write $S = S_b$ for the blow-up locus. We study the properties of tangent cones of S. Recall that μ is the limit Randon measure of $\mu_i = |F_{A_i}|^2 dV_g$. For any $y \in M$ and sufficiently small λ, we define the scaled measure

$$\mu_{y,\lambda}(E) = \lambda^{4-m}\mu(exp_y(\lambda E)), \tag{3.97}$$

for any E in $T_y M$, where $exp_y : T_y M \to M$ is the exponential map of the metric g and

$$\lambda E = \{x \in T_y M \, | \, \lambda^{-1} x \in E\}. \tag{3.98}$$

(L4) *Let* $\{\lambda_k\}$ *be any sequence with* $\lim_{k\to\infty} \lambda_k = 0$. *Then there exist a subsequence* $\{\lambda'_k\}$ *and a Radon measure* η *on* $T_y M$ *such that* $\mu_y \lambda'_k$ *converges to* η *weakly. Furthermore,* $\eta_{0,\lambda} = \eta$ *for each* $\lambda > 0$, *i.e.,* η *is a cone measure.*

Define a connection on $T_y M$ for each y and λ by

$$A_{i,y,\lambda} = \tau_\lambda^* exp_y^* A_i, \tag{3.99}$$

where $\tau_\lambda : T_yM \to T_yM$ maps v to λv. Thus $A_{i,y,\lambda}$ is a Yang-Mills connection via the metric $\lambda^{-2}exp_y^* g$, which is denoted by $g_{y,\lambda}$. Clearly, $g_{y,\lambda}$ converges to the flat metric $g_{y,0} = g|_{T_yM}$ on T_yM as $\lambda \to 0+$.

We next study tangent cones η with support in T_yM for H^{m-4}-a.e. $y \in S$. In order to prove Theorem 3.4.2, we require the following lemmas.

(L5) *The density function* $\Theta(\mu, x)$ *is* H^{m-4}*-approximately continuous at* H^{m-4}*-a.e.* $x \in S$. *Here* $\Theta(\mu, \cdot)$ *is* H^{m-4}*-approximately continuous at* $x \in S$ *if for any* $\epsilon > 0$,

$$\lim_{r\to 0} \frac{H^{m-4}(\{y \in B_r(x) \cap S | \Theta(\mu, y) - \Theta(\mu, x)| > \epsilon\})}{r^{m-4}} = 0. \quad (3.100)$$

(L6) *Suppose* $x \in S$ *is such that* $\Theta(\mu, x) \geq \epsilon_0 > 0$ *and* $\Theta(\mu, \cdot)$ *is* H^{m-4}*-approximately continuous at* x. *Then there is a* $r_x > 0$, *such that for each* $r \in (0, r_x)$, *one can find* $m-4$ *points* $x_1, \cdots, x_{m-4}$ *in* $B_r(x) \cap S$ *such that*

(a) $\Theta(\mu, x_j) \geq \Theta(\mu, x) - \epsilon(r)$ *for* $j = 1, 2, \cdots, m-4$, *where* $\epsilon(r) \to 0$ *as* $r \to 0$.

(b) *Let* exp_x *be the exponential map of* (M, g) *at* x. *Then for some* $s \in (0, 1/2)$ *depending only on* m, $d(x_1, x) \geq sr$ *and* $d(x_k, exp_x(V_{k-1})) \geq sr$ *for* $k \geq 2$, *where* V_{k-1} *is the subspace in* T_xM *spanned by* $(exp_x|_{B_r(0)})^{-1}(x_1), \cdots, (exp_x|_{B_r(0)})^{-1}(x_{k-1})$.

The proof is based on the arguments in [253]. By the assumption, there exists a positive function $\epsilon(r)$ for $0 < r < r_x$ such that $lim_{r\to 0}\epsilon(r) = 0$ and

$$\frac{H^{m-4}(\{y \in B_r(x) \cap S| \ |\Theta(\mu, y) - \Theta(\mu, x)| \geq \epsilon(r)\})}{r^{m-4}} \leq \frac{s(m)}{2} \leq \frac{1}{2}, \quad (3.101)$$

(where $s(m)$ will be determined later). This lemma can be proved by using a contradiction argument (cf. [374]).

The following theorem is useful in proving the rectifiability of blow-up loci.

Theorem 3.4.2. *Let* μ *be the Radon measure introduced above. Then for* H^{m-4}*-a.e.* $x \in S \subset M$, *any tangent cone measure* η *on* T_xM *of* μ *is of the form* $\Theta(\mu, x)H^{m-4}\lfloor F$ *for some* $(m-4)$*-dimensional subspace* F *in* T_xM.

Proof. First note that the existence of η is ensured by (L4). By (L3), $\mu = |F_A|^2 dV_g + \Theta(\mu, \cdot)H^{m-4}\lfloor S$, where A is the weak limit of a sequence $\{A_i\}$. By (L3) and the fact (iii) stated above, for H^{m-4}-a.e. $x \in S$,

$$\Theta(\mu, x) \geq \epsilon_0 > 0, \quad \lim_{r\to 0} r^{4-m} \int_{B_r(x)} |F_A|^2 dV = 0. \quad (3.102)$$

Moreover, (L5) implies that $\Theta(\mu, \cdot)$ is H^{4-m}-approximately continuous at H^{m-4}-a.e. $x \in S$.

Fix a point $x \in S$ such that (3.102) holds and $\Theta(\mu, \cdot)$ is H^{m-4}-approximately continuous at x. Suppose that η is the weak limit of μ_{x,r_k}, where $lim_{k\to\infty} r_k = 0$. For k large enough, by (L6), we can find $m-4$ points $x_1^k, \cdots, x_{m-4}^k$ in $B_{r_k}(x) \cap S$, such that for $j = 1, 2, \cdots, m-4$,

$$\Theta(\mu, x_j^k) \geq \Theta(\mu, x) - \epsilon(r_k) \tag{3.103}$$

$$d(x_j^k, exp_x(V_{j-1}^k)) \geq s r_k, \tag{3.104}$$

where V_{j-1}^k is the 0-dimensional space $\{0\}$ if $j = 1$, and the subspace in $T_x M$ spanned by the vectors $\xi_1^k = exp_x^{-1}(x_1^k), \cdots, \xi_{j-1}^k = exp_x^{-1}(x_{j-1}^k)$ for $j \geq 2$.

As before, let g_{x,r_k} be the scaled metric $r_k^{-2} exp_x^* g$ on $T_x M$, which converges to the flat cone metric $g_{x,0}$ as $k \to \infty$. Then $r_k^{-1}\xi_j^k \in B_1(0, g_{x,r_k})$ for each j. Choosing a subsequence of r_k if needed, we may assume that as $k \to \infty$, $r_k^{-1}\xi_j^k \in B_1(0, g_{x,0})$ converges to ξ_j via the fixed metric $g_{x,0}$. It follows from (3.104) that $\xi_1, \cdots, \xi_{m-4}$ span an $(m-4)$-dimensional subspace F in $T_x M$, which is the limit of V_{m-4}^k. Furthermore, $d_{g_{x,0}}(\xi_i, 0) \geq s$ and $d_{g_{x,0}}(\xi_i, \xi_j) \geq s$ for $i \neq j$.

By (3.103), we can derive that for any $r > 0$

$$\begin{aligned} r^{4-m} \mu_{x,r_k}(B_r(\xi_j^k, g_{x,r_k})) &= (r r_k)^{4-m} \mu(B_{r r_k}(x_j^k)) \\ &\geq \Theta(\mu, x_j^k) \geq \Theta(\mu, x) - \epsilon(r_k). \end{aligned}$$

Therefore, for all $r < 0$

$$r^{4-m}\eta(B_r(\xi_j, g_{x,0})) \geq \Theta(\mu, x) = \Theta(\eta, 0). \tag{3.105}$$

Then we have

$$\Theta(\eta, \xi_j) \geq \Theta(\eta, 0) \tag{3.106}$$

Next, for any $r, \tilde{r} > 0$, the monotonicity implies

$$\begin{aligned} r^{4-m}\eta(B_r(\xi_j, g_{x,0})) &= \lim_{k\to\infty} r^{4-m}\mu_{x,r_k}(B_r(\xi_j^k, g_{x,r_k})) = \lim_{k\to\infty} r r_k^{4-m}\mu(B_{r r_k}(x_j^k)) \\ &\leq \lim_{k\to\infty} (e^{a\tilde{r}^2}\tilde{r}^{4-m}\mu(B_{\tilde{r}}(x_j^k))) = e^{a\tilde{r}^2}\tilde{r}^{4-m}\mu(B_{\tilde{r}}(x)). \end{aligned}$$

Since we can take $\tilde{r}$ sufficiently small, we have

$$r^{4-m}\eta(B_r(\xi_j, g_{x,0})) = \Theta(\eta, 0)$$

for any $r > 0$. By applying Theorem 3.3.1, we can show that η is a cone measure with center at ξ_j for each $j = 1, \cdots, m-4$, i.e.,

$$d\eta(r_j, \theta) = r_j^{m-5} dr_j d\xi(\theta)$$

for some Radon measure $d\xi_j(\theta)$ on the unit sphere $\{\xi \in T_xM \mid r_j(\xi) = 1\}$, with $r_j(\xi) = |\xi - \xi_j|$. It follows that

$$\eta(y_1, \cdots, y_{m-4}, y_{m-3}, \cdots, y_m) = \eta(y_{m-3}, \cdots, y_m),$$

where $y_1, \cdots, y_m$ are Euclidean coordinates of T_xM such that $y_1, \cdots, y_{m-1}$ are in F. By the second relation in (3.102), we have $supp(\eta) \subset F$. Hence, $\eta = \Theta(\mu, x)H^{m-4}\lfloor F$. □

3.4.3 Rectifiability

We have verified that tangent cones exist at H^{m-4}-a.e. $x \in S$. Furthermore, if (3.102) is met and $\Theta(\mu, \cdot)$ is H^{m-4}-approximately continuous at $x \in S$, then by Theorem 3.4.2 any tangent cones at x are $(m-4)$-dimensional subspaces in T_xM. Here, we will show that S is rectifiable, i.e., the tangent cones are unique at H^{m-4}-a.e. $x \in S$. Indeed, this follows from [302], since $\Theta(\nu, \cdot)$ exists almost everywhere and ν is Borel regular. We give below a direct proof using the structure theorem of Federer [141, 253]. We write $S = S_r \cup S_u$, where S_r is a rectifiable set and S_u is a unrectifiable set. Let $G(T_xM, m-4)$ be the Grassmannian of all $(m-4)$-dimensional subspaces in T_xM.

(L7) *For any $x \in M$ and V in $G(T_xM, m-4)$,*

$$H^{m-4}(P_V(exp_x^{-1}(B_r(x) \cap S_u))) = 0$$

where $r > 0$ is sufficiently small and P_V is the orthogonal projection of T_xM onto V with respect to $g_{x,0}$. (The proof is similar to 3.3.5 in [141], with some modifications).

We show that $H^{m-4}(S_u) = 0$ using the contradiction method. Assume that this is not true, i.e., $H^{m-4}(S_u) > 0$. Then for H^{m-4}-a.e. $x \in S_u$, $r > 0$ small and any $V \in G(T_xM, m-4)$,

$$H^{m-4}(P_V(exp_x^{-1}(S_u \cap B_r(x)))) = 0 \tag{3.107}$$

and

$$\overline{\lim_{\lambda \to 0+}} \frac{H^{m-4}(S_r \cap B_\lambda(x))}{\lambda^{m-4}} = 0. \tag{3.108}$$

Since $H^{m-4}(S_u) > 0$, we can select x in S_u such that (3.102), (3.107) and (3.108) hold, and $\Theta(\mu, \cdot)$ is H^{m-4}-approximately continuous at x. As before, we define

$$\mu_{x,\lambda}(E) = \lambda^{m-4}\mu(exp_x(\lambda E)), \tag{3.109}$$

where $E \subset T_xM$. Let $\{\lambda_k\}$ be a sequence of positive numbers such that $\lim_{k\to\infty}\lambda_k = 0$ and μ_{x,λ_k} converges weakly to a tangent measure η on T_xM. By the proof of Theorem 3.4.2 and the choice of x, we get $\eta = \Theta(\mu, x)H^{m-4}\lfloor V$ for some $(m-4)$-subspace V in T_xM. We want to show that

$$\overline{\lim_{k\to\infty}} \frac{H^{m-4}(P_V(exp_x^{-1}(S \cap B_{\lambda_k}(x))))}{\lambda_k^{m-4}} > 0 \tag{3.110}$$

If this is true, then

$$\overline{\lim_{k\to\infty}} \frac{H^{m-4}(P_V(exp_x^{-1}(S_u \cap B_{\lambda_k}(x))))}{\lambda_k^{m-4}} > 0, \tag{3.111}$$

due to (3.108). But, this contradicts (3.107), thus proving the result.

So let us show (3.110). As before, we may find a sequence of Yang-Mills connections A_{i,x,λ_k} (see (3.99)) such that the $|F_{A_{i,x,\lambda_k}}|^2 dV_{x,\lambda}$ converge to μ_{x,λ_k} weakly as $i \to \infty$. Remark that for k sufficiently large, the A_{i,x,λ_k} are well defined in $B_4(0, g_{x,\lambda_k}) \subset T_xM$. We identify T_xM with $V \times V^\perp$, so that any point $z \in T_xM$ is represented as $z = (z', z'')$ with $z' \in V$ and $z'' \in V^\perp$, where $V^\perp$ is the orthogonal complement of V in T_xM. Select orthonormal coordinates $x_1, \cdots, x_m$ of T_xM with respect to $g_{x,0}$, so that $z_1, \cdots, z_{m-4}$ are coordinates in V and $z_{m-3}, \cdots, z_m$ are coordinates in $V^\perp$. We denote z' by $(z_1, \cdots, z_{m-4})$ and z'' by $(z_{m-3}, \cdots, z_m)$. We set

$$B_2^2(0) = \{z'' \in V^\perp \,|\, |z''| < 2\}.$$

If k is large enough (g_{x,λ_k} is sufficiently closed to the flat metric $g_{x,0}$), we get $(z', 0) + \{0\} \times B_2^2(0) \subset B_4(0, g_{x,\lambda_k})$ for any $(z', 0) \in V \times \{0\} \cap B_2(0, g_{x,\lambda_k})$.

Let

$$m_{i,k}(z') = \int_{B_2^2(0)} |F_{A_{i,x,\lambda_k}}|^2(z', z'')\phi^2(z'')dV_k(z''), \tag{3.112}$$

where $dV_k(z'')$ is the volume form on $B_2^2(0)$ induced by the metric g_{x,λ_k}, and $\phi \in C_0^\infty(B_2^2(0))$ with $\int_{B_2^2(0)} \phi^2 dV_{g_{x,0}} = 1$. Thus $m_{i,k}$ is a smooth function of z' in $V \cap B_2(0, g_{x,\lambda_k})$.

(L8) *Let $\{A_{i,x,\lambda,k}\}$ and x be defined as before. Then for $\alpha \le m-4$,*

$$\lim_{k\to\infty}\lim_{i\to\infty}\int_{B_4(0,g_{x,\lambda_k})} \left|\frac{\partial}{\partial z_\alpha} \rfloor F_{A_{i,x,\lambda_k}}\right|^2 dV_{x,\lambda_k} = 0. \tag{3.113}$$

Remark that the integral

$$\int_{B_4(0,g_{x,\lambda_k})} |F_{A_{i,x,\lambda_k}}|^2 dV_{x,\lambda_k}$$

is uniformly bounded. Then we have

$$grad\, m_{i,k} = f_{i,k} + div\,(u_{i,k}), \tag{3.114}$$

where $f_{i,k} : V \cap B_2(0, g_{x,\lambda_k}) \to V$ and $u_{i,k} : V \cap B_2(0, g_{x,\lambda_k}) \to V \times V$ are functions such that

$$\lim_{k\to\infty}\lim_{i\to\infty}\int_{V\cap B_2(0,g_{x,\lambda_k})}(|f_{i,k}| + |u_{i,k}|)dV_{x,0} = 0. \tag{3.115}$$

It follows (see [5]) that there are constants $C_{i,k}$, such that

$$\lim_{k\to\infty}\lim_{i\to\infty}||m_{i,k} - C_{i,k}||_{L^1(V\cap B_{4/3}(0,g_{x,\lambda_k}))} = 0. \tag{3.116}$$

Since $\lim_{k\to\infty}\lim_{i\to\infty}|F_{A_{i,x,\lambda_k}}|^2 dV_{x,\lambda_k} = \eta$ and $\eta + \Theta(\mu, x)H^{m-4}\lfloor V$, we have

$$\lim_{k\to\infty}\lim_{i\to\infty} C_{i,k} = \Theta(\mu, x) > 0.$$

The ball $B_{3/2}(0, g_{x,0})$ is contained in every ball $B_{4/3}(0, g_{x,\lambda_k})$ for k large enough. Therefore, for any $\xi \in C_0^\infty(V \cap B_{3/2}(0, g_{x,0}))$, we have

$$\begin{aligned}\Theta(\mu,x)\int_{V\cap B_{3/2}(0,g_{x,0})}\xi(z')dz' &= \lim_{k\to\infty}\lim_{i\to\infty}\int_{V\cap B_{3/2}(0,g_{x,0})}\xi(z')m_{i,k}(z')dz'\\ &= \lim_{k\to\infty}\lim_{i\to\infty}\int_{B_2(0,g_{x,\lambda_k})}|F_{A_{i,x,\lambda_k}}|^2(z', z'')\xi(z')\phi^2(z'')dV_{x,\lambda_k}\\ &= \lim_{k\to\infty}\int_{B_2(0,g_{x,\lambda_k})}\xi(z')\phi^2(z'')d\mu_{x,\lambda_k}(z', z'').\end{aligned} \tag{3.117}$$

As a weak limit of Radon measures $|F_{A_i}|^2 dV_g$, the measure μ is of the form $|F_A|^2 dV_g + \nu$. We obtain, after scaling,

$$\mu_{x,\lambda_k} = |F_{A_{x,\lambda_k}}|^2 dV_{x,\lambda_k} + \nu_{x,\lambda_k}, \tag{3.118}$$

where A_{x,λ_k} is a connection on $T_x M \setminus \lambda_k^{-1} exp_x^{-1}(S)$ as defined in (3.99), and ν_{x,λ_k} is a Radon measure on $T_x M$ of the form

$$\Theta(\mu_{x,\lambda_k},\cdot)H^{m-4}\lfloor\lambda_k^{-1} exp_x^{-1}(S). \tag{3.119}$$

By the second relation in (3.102),

$$\lim_{k\to\infty}\int_{B_2(0,g_{x,\lambda_k})}\xi(z')\phi^2(z'')|F_{A_{i,x,\lambda_k}}|^2 dV_{x,\lambda_k} = 0, \tag{3.120}$$

whence

$$\Theta(\mu, x) \int_{V \cap B_{3/2}(0, g_{x,0})} \xi(z') dz'$$

$$= \lim_{k\to\infty} \int_{B_{3/2}(0,g_{x,0}) \cap \lambda_k^{-1} exp_x^{-1}(S)} \xi(z') \Theta(\mu_{x,\lambda_k}, (z', z'')) dH^{m-4}(z', z'')$$

$$= \Theta(\mu, x) \lim_{k\to\infty} \int_{B_{3/2}(0,g_{x,0}) \cap \lambda_k^{-1} exp_x^{-1}(S)} \xi(z') dH^{m-4}(z', z''). \quad (3.121)$$

Since $\Theta(\mu, x) > 0$, it follows from (3.121) that

$$\overline{\lim_{k\to\infty}} \frac{H^{m-4}(P_V(exp_x^{-1}(S \cap B_{\lambda_k}(x))))}{\lambda_k^{m-4}}$$

$$= \overline{\lim_{k\to\infty}} H^{m-4}(P_V(\lambda_k^{-1} exp_x^{-1}(S \cap B_1(0, g_{x,\lambda_k})))) \geq Vol(V \cap B_{1/2}(0, g_{x,0})) > 0$$

Therefore, (3.110) is verified and reached a contraction with (3.107). Hence, $H^{m-4}(S_u) = 0$. We can conclude the following theorem.

Theorem 3.4.3. *If (S_b, Θ) is the blow-up locus of a weakly convergent sequence $\{A_i\}$, then its support S_b is H^{m-4}-rectifiable. In particular, for H^{m-4}-a.e. $x \in S_h$, there is a unique tangent subspace $T_x S_b \subset T_x M$.*

3.5 Structure of Blow-Up Loci

In this section, we discuss the bubbling Yang-Mills connections, the blow-up loci of anti-dual instantons, application of calibrated geometry to blow-up loci, and general blow-up loci, based on [374].

3.5.1 *Bubbling Yang-Mills Connections*

Suppose that $\{A_i\}$ converges to an admissible Yang-Mills connection A with the blow-up locus (S, Θ) (see (L3)). We have verified that S is H^{m-4}-rectifiable in the preceding section. When $m = 4$, S consists of finitely many points. Moreover, K. Uhlenbeck showed that if i is sufficiently large, A_i approaches a connected sum of A with certain Yang-Mills connections on the unit sphere S^4, which are called the bubbling connections.

We now analyze the structure of A_i near S when i is sufficiently large, and then construct bubbling connections on $\mathbf{R}^m$ as A_i approaches A. Notice that μ is the weak limit of the Radon measures $|F_{A_i}|^2 dV_g$ and is of the form $|F_A|^2 dV_g + \Theta(\mu, \cdot) H^{m-4} \lfloor S$.

Proposition 3.5.1. *Suppose that $x \in S$ fulfills:*

(a) *The tangent plane $V = T_x S \subset T_x M$ exists and is unique;*
(b) *(3.102) holds for μ and A.*

Then there are linear transformations $\sigma_i : T_x M \to T_x M$ such that a subsequence of $\sigma_i^* exp_x^* A_i$ converges to a Yang-Mills connection D on $T_x M$ such that $F_D \neq 0$ and $v \rfloor F_D = 0$ for any $v \in V$. Such a connection D is called a *bubbling connection* at $x \in S$.

Let $A_{i,x,\lambda}$ be the scaled connections on $T_x M$ defined in (3.99), that is,

$$A_{i,x,\lambda} = \tau_\lambda^* exp_x^* A_i, \tag{3.122}$$

where $\tau_\lambda(v) = \lambda v$ for any v in $T_x M$. Each $A_{i,x,\lambda}$ is a Yang-Mills connection with respect to the scaled metric $g_{x,\lambda}$. As $i \to \infty$, $|F_{A_{i,x,\lambda}}|^2 dV_{x,\lambda}$ converges to $\mu_{x,\lambda}$ weakly. On the other hand, as $\lambda \to 0$, $\mu_{x,\lambda}$ converges to $\Theta(\mu, x) H^{m-4} \lfloor V$ weakly. Thus there is a sequence λ_i such that the Radon measures $|F_{A_{i,x,\lambda}}|^2 dV_{x,\lambda}$ converge to $\Theta(\mu, x) H^{m-4} \lfloor V$ weakly. Furthermore, $A_{i,x,\lambda}$ converges to 0 uniformly on any compact subset in $T_x M \setminus V$, modulo gauge transformations. It follows that for i sufficiently large,

$$|F_{A_{i,x,\lambda}}|(v) \leq \frac{\epsilon(r)}{r^2}. \tag{3.123}$$

We also get (see (L8))

$$\lim_{i \to \infty} \sum_{\alpha=1}^{m-4} \int_{B_2(0, g_{x,0})} \left| \frac{\partial}{\partial z_\alpha} \rfloor F_{A_{i,x,\lambda_i}} \right|^2 dV_{x,\lambda_i} = 0, \tag{3.124}$$

where $\{z_1, \cdots, z_{m-4}\}$ is an orthonormal coordinate system in V.

As before, we denote a point in $T_x M$ by $z = (z', z'')$ with $z' \in V$, $z'' \in V^\perp$. We identify V and $V^\perp$ with $V \times \{0\}$ and $\{0\} \times V^\perp$ in $T_x M$.

(L9) *There are points z'_i in $V \cap B_{1/2}(0, g_{x,0})$ with* $\lim_{i \to \infty} z'_i = 0$, *such that*

$$\lim_{i \to \infty} \left(\sup_{0 < r \leq 1/2} r^{4-m} \int_{V \cap Br(z'_1, g_{x,0})} dx' \int_{V^\perp \cap B_{1/2}(0, g_{x,0})} \sum_{\alpha=1}^{m-4} \left| \frac{\partial}{\partial z_\alpha} \rfloor F_{A_{i,x,\lambda_i}} \right|^2 dV_{x,\lambda_i} \right) = 0. \tag{3.125}$$

Remark that for any $\delta > 0$,

$$\max_{z'' \in V^\perp \cap B_{1/2}(0, g_{x,0})} \delta^{4-m} \int_{B_\delta(z'_i + z'', g_{x,0})} |F_{A_{i,x,\lambda_i}}|^2 dV_{x,\lambda_i} \geq \epsilon, \tag{3.126}$$

where ϵ is as in Theorem 3.3.3. Otherwise, A_{i,x,λ_i} converge to a smooth Yang-Mills connection on $(V \cap B_\delta(z'_i, g_{x,0})) \times (V^\perp \cap B_{1/2}(0, g_{x,0}))$, which contradicts our hypothesis on $A_{i,x,\lambda}$.

Thanks to (3.126), we can find $\delta_i \in (0, 1/2)$ and $z_i'' \in (V^\perp \cap B_{1/4}(0, g_{x,0}))$, such that

$$\delta_i^{4-m} \int_{B_\delta(z_i'+z_i'', g_{x,0})} |F_{A_{i,x,\lambda_i}}|^2 dV_{x,\lambda_i}$$
$$= \max_{z'' \in V^\perp \cap B_{1/2}(0, g_{x,0})} \delta_i^{4-m} \int_{B_\delta(z_i'+z_i'', g_{x,0})} |F_{A_{i,x,\lambda_i}}|^2 dV_{x,\lambda_i} = \frac{\epsilon}{4}. \quad (3.127)$$

We may take z_i'' with $\lim_{i\to\infty} z_i'' = 0$. We define new connections

$$D_i(y) = A_{i,x,\lambda_i}(z_i' + z_i'' + \delta_i y). \quad (3.128)$$

Each D_i is a Yang-Mills connection via the scaled metric $g_i' = \delta_i^{-2} g_{x,\lambda_i}$ on $B_{4R_i}(0, g_{x,0})$ where $R_i = (4\delta_i)^{-1}$. Observe that the based manifolds $(T_x M, g_i', z_i' + z_i'')$ converge to $(T_x M, g_{x,0}, 0)$ as $i \to \infty$.

Applying (3.125) and (3.127), we arrive at

$$\lim_{i\to\infty} \left(\sum_{\alpha=1}^{m-4} \int_{B_{R_i}(0, g_{x,0})} \left| \frac{\partial}{\partial z_\alpha} \rfloor F_{D_i} \right|^2 dV_{g_i'} \right) = 0, \quad (3.129)$$

$$\int_{B_1(0, g_{x,0})} |F_{D_i}|^2 dV_{g_i'} = \max_{y \in V^\perp \cap B_{R_i - 1}(0, g_{x,0})} \int_{B_1((0,y), g_{x,0})} |F_{D_i}|^2 dV_{g_i'} = \frac{\epsilon}{4}. \quad (3.130)$$

The monotonicity implies that

$$\sup_i \left\{ \int_{B_R(0, g_{x,0})} |F_{D_i}|^2 dV_{g_i'} \right\} \le C(\Lambda) R^{m-4}, \quad (3.131)$$

for $0 < R < R_i$, where $C(\Lambda)$ is a constant depending only on Λ.

Applying (3.131) and Proposition 3.4.1, we may assume that D_i converges to an admissible Yang-Mills connection D, by taking a subsequence if needed. We know from (3.130) that D is a smooth Yang-Mills connection on $(V \cap B_1(0, g_{x,0})) \times V^\perp \subset T_x M$ with respect to $g_{x,0}$. Furthermore, it follows from (3.129) that for any $v \in V$,

$$v \rfloor F_D = 0, \quad (3.132)$$

if D is well defined.

On $((V \cap B_1(0, g_{x,0}))) \times V^\perp$, we denote $D = \sum_{\alpha=1}^m D^\alpha dy_\alpha$, where $D^\alpha \in T_e(G) = \mathcal{G}$ and $y_1, \cdots, y_m$ are Euclidean coordinates such that $y_1, \cdots, y_{n-4}$ are tangent to V along V. We discard D^α for $\alpha \le m-4$ inductively. Firstly, by a gauge transformation, we may assume that $D^1 = 0$; then (3.132) implies that all D^α are independent of y_1. Again applying a gauge transformation, we can get rid of D^2, and so on. Lastly, by finitely many gauge transformations, we obtain a connection, still denoted by D, which is a pull-back of some connection on $V^\perp$. It follows that D extends to a smooth connection on $T_x M$. This completes the proof.

3.5.2 Blow-Up Loci of Anti-self-dual Instantons

Suppose that $\{A_i\}$ is a sequence of Ω-anti-self-dual instantons which converge to an admissible Ω-anti-self-dual instanton A, where Ω is a form on M of degree $m-4$. The closedness of Ω is not required here. Let $S \subset M$ be the blow-up locus of $\{A_i\}$. We claim that as Ω restricts to the induced form on S, if Ω is a calibrating form as in [174], then S is calibrated by Ω and is minimal. There is a bubbling connection similar to that constructed in Proposition 3.5.1, now in terms of anti-dual instantons, as follows.

Proposition 3.5.2. *Let M, g, Ω, $\{A_i\}$, A and S be as above. Suppose that $x \in S$ fulfills:*

(a) *The tangent cone $T_xS \subset T_xM$ exists and is unique;*
(b) *(3.102) holds for μ and A, where μ is the weak limit of the Radon measures $|F_{A_i}|^2 dV_g$.*

Then there is an Ω_x-anti-self-dual intanton D on T_xM, where $\Omega_x = \Omega|_{T_xM}$, such that $F_D \neq 0$, $tr(F_D) = 0$ and $v \rfloor F_D = 0$ for any $v \in T_xS$.

Proof. The proof is similar to the proof of Proposition 3.5.1. We first notice that $tr(F_{D_i})$ converges to zero uniformly as $i \to \infty$, where D_i are the scaled connections defined in (3.128), since $Tr(F_{A_i})$ are harmonic 2-forms with uniformly bounded L^2-norm. We then notice that D_i are Ω'-anti-self-dual with respect to the metric g_i' and the closed form Ω_i' of degree $m-4$ on $B_{4R_i}(0, g_{x,0})$ defined by

$$\Omega_i' = \tau^{\delta_i *}_{(z_i', z''_i)} exp_x^* \Omega,$$

where $\tau^{\delta_i *}_{(z_i', z''_i)} : T_xM \to T_xM,\ y \mapsto (z_i', z''_i) + \delta_i y.$

Because (z', z'') tends to zero as $i \to \infty$, Ω_i' converges to Ω_x. Thus, the limit connection D in Ω_x-anti-self-dual with respect to $g_{x,0}$, and $tr(F_D) = 0$. The rest of the proof is the same as the proof of Proposition 3.5.1. □

Corollary 3.5.3. *If $x \in S$ is as in the previous proposition, then Ω_x restricts to a volume form on $T_xS \subset T_xM$ which is induced by the flat metric $g_{x,0}$.*

Proof. We may identify T_xM with $\mathbf{R}^m$, where m is the dimension of M, so that $g_{x,0}$ is the standard Euclidean metric g_0. Let $*$ be the Hodge operator of g_0. Then the connection D satisfies

$$F_D = - * (\Omega_x \wedge F_D). \tag{3.133}$$

Let $x_1, \cdots, x_m$ be any Euclidean coordinates of T_xM such that $x_1, \cdots, x_{m-4}$ are tangent to T_xS. Then we define a constant form $\Phi_{S,x}$ of degree $m-4$ on T_xM as follows:

$$\Phi_{S,x} = dx_1 \wedge \cdots \wedge dx_{m-4}.$$

Now decompose Ω_x as $\Omega_x = \alpha\Phi_{S,x} + \Omega_0$, where α is a constant and $\Omega_0|_{T_xS} = 0$.

Since $v\rfloor F_D = 0$ for any $v \in T_xS$, by taking a gauge transformation if needed, we may assume that $D = \pi_L^* D_L$ for some non-trivial connection D_L, where L is the orthogonal complement of T_xS and π_L is the orthogonal projection from T_xM onto L. Thus (3.133) reduces to

$$F_{D_L} = -\alpha *_L F_{D_L}, \tag{3.134}$$

$$0 = *(\Omega_0 \wedge F_D), \tag{3.135}$$

where $*_L$ is the Hodge operator of L. Since $F_{D_L} \neq 0$, we derive from (3.134) that $\alpha = \pm 1$. Then we can conclude the corollary. □

Theorem 3.5.4. *Let (M,g) be a compact Riemannian manifold, Ω be a closed form of degree $m-4$ and $\{A_i\}$ be a sequence of Ω-anti-self-dual instantons. Then by taking a subsequence if needed, A_i converges to an admissible Ω-anti-self-dual instanton A with the blow-up locus (S,Θ), such that (a) S is rectifiable and $\Omega|_S$ is one of its volume forms induced by g. In particular, S carries a natural orientation; (b) $\frac{1}{8\pi^2}$ is integer-valued; (c) $CW_2(S,\Theta)$ is closed in M, where $CW_2(S,\Theta)$ is the integral current defined by*

$$CW_2(S,\Theta)(\phi) - \frac{1}{8\pi^2}\int_S (\phi, \Omega|_S)\Theta d(H^{m-4}\lfloor S), \tag{3.136}$$

where ϕ is any smooth form with compact support in M. Furthermore, in terms of currents, we have

$$\lim_{i\to\infty} CW_2(A_i) = CW_2(A) + CW_2(S,\Theta), \tag{3.137}$$

where $CW_2(A)$ is as given in Corollary 3.3.5.

Notice that by applying (3.136) to the smooth form $4\pi^2\Omega$, we have the conservation of the action:

$$\lim_{i\to\infty}\int_M |F_i|^2 dV_g = \int_M |F_A|^2 dV_g + \int_S \Theta(H^{m-4}\lfloor S).$$

Proof of Theorem 3.5.4. (a) follows from Theorem 3.4.3, Proposition 3.5.2 and Corollary 3.5.3 and results of the last section. We only need to show (b) and (c).

We first claim that the density $\frac{1}{8\pi}\Theta(\mu,\cdot)$ is integer-valued. Let x be any point in S such that (3.101) holds and there is a unique tangent space T_xS. Therefore, (3.124) is true. Then we have

$$\Theta(\mu,x) = \lim_{i\to\infty}\int_{B_1(0,g_{x,0})} |F_{A_{i,x,\lambda_i}}|^2 dV_{x,\lambda_i}. \tag{3.138}$$

Because A_{i,x,λ_i} converges to zero uniformly on any compact subset away from $V = T_x S$, for any $z' \in V \cap B_1(x, g_{x,0})$, $A_{i,x,\lambda_i}|_{\{z'\}\times V^{\perp}\cap B_{\sqrt{1-|z'|^2}}(0,g_{x,0})}$ converges to zero uniformly away from $(z', 0)$. Then by a standard argument

$$\lim_{i\to\infty} \frac{1}{8\pi^2} \int_{z'\times V^{\perp}\cap B_{\sqrt{1-|z'|^2}}(0,g_{x,0})} tr(F_{A_{i,x,\lambda_i}} \wedge F_{A_{i,x,\lambda_i}}) \in \mathbf{Z}. \tag{3.139}$$

Indeed, the limit on the right side of (3.139) is a topological number which does not depend on z'.

For convenience, let $F^V_{A_{i,x,\lambda_i}}$ be the curvature of the restricted connection $F_{A_{i,x,\lambda_i}}|_{z'\times V^{\perp}}$. Since A_{i,x,λ_i} is $\tau_\lambda^* exp^*\Omega$-anti-self-dual with respect to g_{x,λ_i} and $\lim_{i\to\infty} g_{x,\lambda_i} = g_{x,0}$, we arrive at

$$\begin{aligned}\frac{1}{8\pi^2}|F_{A_{i,x,\lambda_i}}|^2 dV_{x,\lambda_i} &= -\frac{1}{8\pi^2} tr(F_{A_{i,x,\lambda_i}} \wedge F_{A_{i,x,\lambda_i}}) \wedge \tau_\lambda^* exp^*\Omega \\ &= \frac{1}{8\pi^2}\Big[- tr(F^V_{A_{i,x,\lambda_i}} \wedge F^V_{A_{i,x,\lambda_i}}) \\ &\quad + (O(1) \sum_{\alpha=1}^{m-4} |\frac{\partial}{\partial z_\alpha} \lfloor F_{A_{i,x,\lambda_i}}| + o(1)|F_{A_{i,x,\lambda_i}}|)|F_{A_{i,x,\lambda_i}}|\Big] dV_{x,\lambda_i},\end{aligned} \tag{3.140}$$

where o(1) is a quantity which converges to zero as $i \to \infty$. Combining (3.139) and (3.124), it follows that

$$\begin{aligned}\frac{1}{8\pi^2}\Theta(\mu, x) &= \lim_{i\to\infty} \int_{B_1(0,g_{x,0})} |F_{A_{i,x,\lambda_i}}|^2 dV_{x,\lambda_i} \\ &= \lim_{i\to\infty} \int_{V\cap B_1(0,g_{x,0})} d(H^{m-4}\lfloor V) \cdot \left[\frac{1}{8\pi^2} \int_{\{z'\}\times V^{\perp}\cap B_{\sqrt{1-|z'|^2}}(0,g_{x,0})} tr(F_{A_{i,x,\lambda_i}} \wedge F_{A_{i,x,\lambda_i}})\right]\end{aligned}$$

Thus by (3.138), $\frac{1}{8\pi^2}\Theta(\mu, \cdot)$ is integer-valued.

We next claim that $CW_2(S, \Theta)$ is closed, that is, for any smooth form ψ of degree $m-5$ and with compact support in M,

$$\partial CW_2(S, \Theta)(\psi) = CW_2(S, \Theta)(d\psi) = 0. \tag{3.141}$$

This follows from (3.136) and Corollary 3.3.5, since

$$\int_M d\psi \wedge tr(F_{A_i} \wedge F_{A_i}) = 0 \tag{3.142}$$

for any i. We also get

$$\lim_{i\to\infty} \int_M tr(F_{A_i}) \wedge tr(F_{A_i}) = \int_M tr(F_A) \wedge tr(F_A). \tag{3.143}$$

Thus it suffices to show that, by taking a subsequence if needed, for any smooth ϕ of degree $m-4$,

$$\frac{1}{8\pi^2}\lim_{i\to\infty}\int_M \phi\wedge tr(F_{A_i}\wedge F_{A_i}) = \frac{1}{8\pi^2}\int_M \phi\wedge tr(F_A\wedge F_A) + CW_2(S,\Theta)(\phi). \tag{3.144}$$

We define the currents

$$Q_i(\phi) = \frac{1}{8\pi^2}\lim_{i\to\infty}\int_M \Big(\phi\wedge tr(F_{A_i}\wedge F_{A_i}) - tr(F_A\wedge F_A)\Big).$$

By Theorem 3.3.4, $\partial Q_i = 0$. Furthermore, the total mass of Q_i is uniformly bounded, that means, for any ϕ with $||\phi||_{C^0}\le 1$,

$$|Q_i(\phi)| \le \frac{1}{8\pi^2}\int_M \Big(|F_{A_i}|^2 - |F_A|^2\Big)dV_g \le \Lambda. \tag{3.145}$$

Taking a subsequence if needed, it follows that Q_i converges weakly to a closed current Q. Obviously, the mass of Q is also bounded by Λ and $\partial Q = 0$. Therefore, by Theorem 3.2.1 in [339], Q is rectifiable. This means that there are a rectifiable set S' with orientation vector $\eta : S' \to \Lambda^{m-4}T^*S'$ and a density function $\Theta'(x)$, such that

$$Q(\phi) = \frac{1}{4\pi^2}\int_{S'}(\phi,\eta)\Theta' d(H^{m-4}\lfloor S').$$

Take ϕ to be $f\Omega$, where f is a smooth function with compact support. Then

$$Q(f\Omega) = \int_{S'} f(\Omega,\eta)\Theta' d(H^{m-4}\lfloor S'). \tag{3.146}$$

On the other hand, since $tr(F_{A_i})$ converges to $tr(F_A)$ uniformly on M, we get

$$\begin{aligned} Q(f\Omega) &= \lim_{i\to\infty} Q_i(f\Omega) = \frac{1}{8\pi^2}\lim_{i\to\infty}\int_M f\Omega\wedge\Big(tr(F_{A_i}\wedge F_{A_i}) - tr(F_A\wedge F_A)\Big) \\ &= \frac{1}{8\pi^2}\lim_{i\to\infty}\int_M f(|F_{A_i}|^2 - |F_A|^2)dV_g = \frac{1}{8\pi^2}\int_S f(x)\theta(\mu,x)d(H^{m-4}\lfloor S). \end{aligned} \tag{3.147}$$

Comparing this with (3.146), we deduce that $S' = S$ and $\Theta(\mu,\cdot) = (\Omega,\eta)\Theta'$. Since Ω_S is one of the volume forms of S, we have that $(\Omega,\eta) = 1$. Hence, $Q = CW_2(S,\Theta)$. □

3.5.3 Application of Calibrated Geometry to Blow-Up Loci

Let (M, g) be an m-dimensional Riemannian manifold and Ω be a closed form of degree $m-4$. Suppose that for any $x \in M$ and any subspace F of $T_x M$ of codimension 4, $\Omega|_F \leq dV_F$, where dV_F is the volume form on F induced by g. Following [174], we say that (F, dV_F) is calibrated by Ω if $\Omega|F = dV_F$. Furthermore, if $\Phi = (S, \xi, \Theta)$ is an integral current with orientation ξ and density Θ, where S is the support of Φ and rectifiable, then we say that Φ is Ω-calibrated if $(T_x S, \xi(x))$ is calibrated by Ω for H^{m-4}-a.e. $x \in S$.

(L10) *Any integral current calibrated by Ω is minimizing in its homology class. In particular, its generalized mean curvature vanishes.*

Proof. Let $\Phi = (S, \xi, \Theta)$ be an integral current calibrated by Ω, and $\Psi = (S', \xi', \Theta')$ be another integral current homologous to Φ, that means, there is a current C of degree $m-5$ such that for any smooth form ϕ on M,

$$\int_S (\phi, \xi)\, \Theta\, dH^{m-4} - \int_{S'} (\phi, \xi')\Theta'\, dH^{m-4} = C(d\phi).$$

By our hypothesis, $(\Omega, \xi') \leq 1$ and $(\Omega, \xi) = 1$. Therefore,

$$\int_S \Theta dH^{m-4} \leq \int_{S'} \Theta'\, dH^{m-4} + C(d\Omega) = \int_{S'} \Theta' dH^{m-4},$$

which implies that Φ is minimal. □

Such a Φ is obviously determined by S with multiplicity Θ. (S, Θ) is called an Ω-calibrated cycle. We know from geometric measure theory that for such a cycle, S is regular in an open and dense subset. It follows from [6] that S can be decomposed as $\bigcup_\alpha S_\alpha$, such that each S_α is closed and smooth outside a closed subset of Hausdorff codimension at least two and Θ restricts to a positive integer on each S_α. Theorem 3.5.4 and the above arguments imply the following theorem.

Theorem 3.5.5. *Let (M, g) be a compact Riemannian manifold, Ω be as above, and $\{A_i\}$ be a sequence of Ω-anti-self-dual instantons. Suppose that either M is compact or the $YM(A_i)$ are uniformly bounded. Then by taking a subsequence if needed, A_i converges to an admissible Ω-anti-self-dual instanton A with the blow-up locus (S, Θ), such that (S, Θ) is an Ω-calibrated cycle, and*

$$\lim_{i\to\infty} CW_2(A_i) = CW_2(A) + CW_2(S, \Theta), \tag{3.148}$$

The following theorem involves the Hermitian Yang-Mills connections on a unitary bundle of a Kähler manifold.

Theorem 3.5.6. *Let (M, g) be a complex m-dimensional compact Kähler manifold with Kähler form ω, and $\{A_i\}$ be a sequence of Hermitian Yang-Mills connections* *on a given unitary bundle E. Then by taking a subsequence if needed, A_i converges weakly to an admissible Hermitian Yang-Mills connection A with the blow-up locus (S, Θ), such that $S = \bigcup_\alpha S_\alpha$ and $\Theta|_{S_\alpha} = 8\pi^2 m_\alpha$, where each S_α is a holomorphic sub-variety in M and m_α is a positive integer. Furthermore, for any smooth ϕ,*

$$\lim_{i\to\infty} \int_M \phi \wedge CW_2(A_i) = \int_M \phi \wedge CW_2(A) + \sum_\alpha m_\alpha \int_{S_\alpha} \phi. \tag{3.149}$$

Proof. By Theorem 3.5.5, we may assume that A_i converges to an admissible Hermitian Yang-Mills connection A with an $\frac{\omega^{m-2}}{(m-2)!}$-calibrated cycle (S, Θ) as its blow-up locus. It is sufficient to verify that (S, θ) is a holomorphic cycle.

By a straightforward calculation, we know that for any $x \in M$ and any subspace $F \subset T_x M$ of co-dimension 4, $\frac{\omega^{m-2}}{(m-2)!}|_F \le dV_F$, and the equality holds if and only if F is a complex subspace in $T_x M$. Thus $T_x S$ is a complex subspace in $T_x M$ for H^{2m-4}-a.e. $x \in S$. Because $CW_2(S, \Theta)$ is a closed integral current, a result of King [221] or Harvey and Shiffman [175] implies that there are holomorphic subvarieties S_α and positive integers m_α such that

$$CW_2(S, \Theta)(\phi) = \sum_\alpha m_\alpha \int_{S_\alpha} \phi$$

for any ϕ. This concludes the proof of the theorem. □

Note. In the previous theorem, let A be the Hermitian Yang-Mills connection. A result of [27] implies that there is a gauge transformation σ on $M \setminus S$ such that $\sigma(A)$ extends to a smooth Hermitian Yang-Mills connection outside a holomorphic subvariety in M of codimension at least three. Actually, the (0,1)-part of A induces a holomorphic structures on the underlying complex vector bundle. Then the induced holomorphic bundle on $M \setminus S$ extends to a coherent sheaf which is locally free outside a subvariety of codimension at least three. Similarly, we can discuss Cayley cycles and complex anti-self-dual instantons (cf. [374]).

3.5.4 General Blow-Up Loci

Let $\{A_i\}$ be a sequence of smooth Yang-Mills connections which converges to an admissible Yang-Mills connection A with blow-up locus (S, Θ).

Theorem 3.5.7. *For any vector field X with compact support in M,*

$$-\int_S div_S X \Theta \, dH^{m-4} = \int_M (|F_A|^2 div X - 4(F_A(\nabla X, \cdot), F_A)) dV = 0, \tag{3.150}$$

where $(F_A(\nabla X, \cdot), F_A)$ is defined in a local orthonormal basis $\{e_i\}$ of M as

$$\sum_{i,j=1}^m (F_A(\nabla_{e_i} X, e_j), F_A(e_i, e_j))$$

and $div_S X$ *is the divergence of* X *along* S *(i.e., if* T_pS *exists and* $\{v_i\}$ *is an orthonormal basis of* T_pS*, then* $div_S X(p) = \sum_{i=1}^{m-4}(\nabla_{v_i} X, v_i)(p)$*).*

Proof. As before, c is a uniform constant. Because S is rectifiable, we can find a countable set of submanifolds $\{M_\alpha\}$ such that $S = S_0 \cup \bigcup_\alpha S_\alpha$, where $S_\alpha = M_\alpha \cup S$ and $H^{m-4}(S_0) = 0$ (see [339]). Furthermore, we may assume that $T_xS = T_xM_\alpha$ for H^{m-4}-a.e. $x \in S_\alpha$.

For any sufficiently small $\delta > 0$, we can arrange M_α such that for some $\alpha_\delta > 0$,

$$S_\alpha \cap S_{\alpha'} = \emptyset, \text{ for } \alpha, \alpha' \leq \alpha_\delta \text{ and } H^{m-4}(\cup_{\alpha > \alpha_\delta} S_\alpha) \leq \delta. \tag{3.151}$$

It follows (taking a subsequence if needed) that

$$\lim_{\epsilon \to 0} \lim_{i \to \infty} \int_{B_\epsilon(\cup_{\alpha > \alpha_\delta} S_\alpha)} |F_{A_i}|^2 dV \leq 2\delta. \tag{3.152}$$

Since δ is sufficiently small, it is sufficient to show that for each $\alpha \leq \alpha_\delta$,

$$\lim_{\epsilon \to 0} \lim_{i \to \infty} \int_{B_\epsilon(S_\alpha)} \left(|F_{A_i}|^2 div\, X - 4 \sum_{k,l} (F_{A_i}(\nabla_{e_k} X, e_l), F_{A_i}(e_k, e_l)) \right) dV$$
$$= \int_{S_\alpha} div_S X \theta dH^{m-4}. \tag{3.153}$$

We may assume, without loss of generality, that $e_1, \cdots, e_{m-4}$ are tangent to M_α, while $e_{m-3}, \cdots, e_m$ are normal to M_α. Thus (L8) implies that (3.153) is the same as

$$\lim_{\epsilon \to 0} \lim_{i \to \infty} \int_{B_\epsilon(S_\alpha)} \left(|F_{A_i}|^2 div^\perp X - 4 \sum_{k,l=m-3}^{m} (F_{A_i}(\nabla_{e_k} X, e_l), F_{A_i}(e_k, e_l)) \right) dV = 0. \tag{3.154}$$

where $div^\perp X = \sum_{k=m-3}^m g(\nabla_{e_k} X, e_k)$ is the divergence of X in the normal direction to M_α.

Denote $\nabla_{e_k} X = X_{i,k} e_i$; then $div^\perp X = \sum_{l=m-3}^m X_{l,l}$ and (3.154) becomes

$$\lim_{\epsilon \to 0} \lim_{i \to \infty} \int_{B_\epsilon(S_\alpha)} \sum_{k,l=m-3}^{m} X_{k,l} \cdot \left(|F_{A_i}|^2 \delta_{kl} - 4 \sum_{j=m-3}^{m} (F_{A_i}(e_k, e_j), F_{A_i}(e_l, e_j)) \right) dV = 0. \tag{3.155}$$

Taking a subsequence if needed, we may assume that there are measures μ_{kl}, $k, l = m-3, \cdots, m$ defined by

$$\mu_{kl}(h) = \lim_{i \to \infty} \int_{B_\epsilon(S_\alpha)} h \left(|F_{A_i}|^2 \delta_{kl} - 4 \sum_{j=m-3}^{m} (F_{A_i}(e_k, e_j), F_{A_i}(e_i, e_j)) \right) dV = 0, \tag{3.156}$$

where h is any function with compact support in $B_\epsilon(M_\alpha)$. It follows from Theorem 3.3.2 that for any $x \in S$ and r small enough,

$$\mu_{kl}(B_r(x)) \leq ce^{ar^2} r^{m-4},$$

where c is a uniform constant. Instead of proving (3.155), we only need to show that the upper-density

$$\overline{\Theta}(\mu_{kl}, x) = \limsup_{r \to 0} r^{4-m} |\mu_{kl}(B_r(x))| \tag{3.157}$$

vanishes for H^{m-4}-a.e. $x \in S_\alpha$.

We want to prove (3.157) by contradiction. If (3.157) is not true, then there is an $S'_\alpha \subset S_\alpha$ such that $H^{m-4}(S'_\alpha) > 0$ and for some k, l, $\overline{\Theta}(\mu_{kl}, x) > 0$ for any $x \in S'_\alpha$. We may assume that $k = l = m$ by applying orthogonal transformations. We also have that for $x \in S'_\alpha$, the tangent space $T_x S = T_x S'_\alpha$ exists and

$$\lim_{r \to 0} r^{4-m} \int_{B_r(x)} |F_A|^2 dV = 0. \tag{3.158}$$

Then using the arguments in the proof of (L9) (cf. [374]) and taking a subsequence if needed, we can find ϵ_i, $r_i > 0$ with $\lim \epsilon_i = 0$ and $\lim \frac{r_i}{\epsilon_i} = 0$, $x_i \in S'_\alpha$, such that

$$r_i^{4-m} \left| \int_{B_{r_i}(x_i)} \left(|F_{A_i}|^2 - 4 \sum_{j=m-3}^{m} (F_{A_i}(e_m, e_j), F_{A_i}(e_m, e_j)) \right) dV \right| \geq \eta_0, \tag{3.159}$$

$$\lim_{i \to \infty} \epsilon_i^{4-m} \int_{B_{\epsilon_i}(x_i)} \sum_{j=1}^{m-4} |e_j \lrcorner F_{A_i}|^2 dV = 0. \tag{3.160}$$

We may assume that $M \subset \mathbf{R}^m$ and g is flat, for simplicity. We can treat the general case with slight modifications. Set $D_i(y) = r_i A_i(x_i + r_i y)$. Then D_i converges to zero outside a subspace $\mathbf{R}^{m-4} \times \{0\} = \lim_{i \to \infty} T_{x_i} M_\alpha$.

Let X be a vector field with compact support in $B_2(0) \subset \mathbf{R}^m$. Since D_i is Yang-Mills, we have, for any $j \leq m-4$,

$$\begin{aligned}
\int_{B_2(0)} |F_{D_i}|^2 X_{j,j} dV &= -2 \int_{B_2(0)} \sum_{k,l=1}^{m} (F_{D_i}(e_k, e_l), \nabla_{e_j} F_{D_i}(e_k, e_l)) X_j dV \\
&= -4 \int_{B_2(0)} \sum_{k,l=1}^{m} (F_{D_i}(e_k, e_l), \nabla_{e_l} F_{D_i}(e_k, e_j)) X_j dV \text{ (by Bianchi identity)} \\
&= 4 \int_{B_2(0)} \sum_{k,l=1}^{m} (F_{D_i}(e_k, e_l), F_{D_i}(e_k, e_j)) X_{j,l} dV_g \to 0, \quad \text{as } i \to \infty.
\end{aligned}$$

Consequently,

$$0 = \int_{B_2(0)} \left(|F_{D_i}|^2 divX - 4 \sum_{k,l=1}^{m} (F_{D_i}(\nabla_{e_k} X, e_l), F_{D_i}(e_k, e_l)) \right) dV$$

$$= \int_{B_2(0)} \sum_{k,l=m-3}^{m} X_{k,l} \left(|F_{D_i}|^2 \delta_{kl} - 4 \sum_j (F_{D_i}(e_k, e_j), F_{D_i}(e_l, e_j)) \right) dV = 0. \tag{3.161}$$

Let η be a non-negative function on $\mathbf{R}^1$ such that $\eta(t) = 1$ for $t \leq 1$ and $\eta(t) = 0$ for $t > 4/3$. Set

$$X = \eta(|y'|)\eta(|y''|) y_m e_m,$$

where $y' = (y_1, \cdots, y_{m-4})$, $y'' = (y_{m-3}, \cdots, y_m)$. It follows from the above that

$$\lim_{i \to \infty} \int_{B_2(0)} \left(|F_{D_i}|^2 - 4 \sum_j (F_{D_i}(e_m, e_j), F_{D_i}(e_m, e_j)) \right) dV = 0. \tag{3.162}$$

This contradicts (3.159) and the theorem is verified. □

A is *stationary* if for any vector field X with compact support in M one has that

$$\int_M \left(|F_A|^2 divX - 4 \sum_{i,j=1}^{m} (F_A(\nabla_{e_i} X, e_j), F_A(e_i, e_j)) \right) dV = 0, \tag{3.163}$$

where $\{e_i\}$ is an orthonormal basis of M. If A is a smooth Yang-Mills connection, this follows from the first variation of the Yang-Mills functional. If A is stationary, then the right-hand side of (3.150) vanishes for any X. Hence, we obtain the following:

Corollary 3.5.8. *If A is stationary, then S is stationary, i.e., S has no boundary in M and its generalized mean curvature vanishes.*

3.6 Removable Singularities

Let us study the extension problem of admissible Yang-Mills connections. Since this problem is local, we may assume that M is an open subset in $\mathbf{R}^m$ with a metric g, which is allowed to be non-flat.

3.6.1 Stationary Properties of Yang-Mills Connections

Let A be an admissible Yang-Mills connection as in Sect. 3.3.3 and $r_p, c(p)$ and a be as in Theorem 3.3.2. By the discussion of Sect. 3.3.1, we obtain the following:

Proposition 3.6.1. *Suppose that A is an arbitrary admissible Yang-Mills connection satisfying (3.151), that is,*

$$\int_M \left(|F_A|^2 divX - 4 \sum_{i,j=1}^{m} (F_A(\nabla_{e_i} X, e_j), F_A(e_i, e_j)) \right) dV = 0 \tag{3.164}$$

where $\{e_i\}$ is an orthonormal basis of M. Then, for any $0 < \sigma < \rho < r_p$,

$$\rho^{4-m} e^{a\rho^2} \int_{B_\rho(p)} |F_A|^2 dV - \sigma^{4-m} e^{a\sigma^2} \int_{B_\sigma(p)} |F_A|^2 dV \geq 4 \int_{B_\rho(p)\backslash B_\sigma(p)} r^{4-m} |\frac{\partial}{\partial r} \rfloor F_A|^2 dV. \tag{3.165}$$

Furthermore, if $M = \mathbf{R}^m$ and g is flat, the equality holds in (3.165) for $\rho \in (0, \infty)$ and $a = 0$.

We next show that all admissible Ω-anti-dual-instantons are stationary, that is, they obey (3.164). Let A be an admissible Ω-anti-dual instanton with singular set $S = S(A)$. For any vector field X with compact support in M, let $\phi_t : M \to M$ be its flow. As in Sect. 3.3.1, we define A^t to be the connection $\phi_t^*(A)$. Then by arguments similar to those in Sect. 3.3.3, we can show that $Ch_2(A^t)$ defines a closed 4-forms on M in the sense of distributions.

We first show that $Ch_2(A^t)$ is independent of t, i.e., for any closed $(m-4)$-form ϕ,

$$\int_M \phi \wedge (Ch_2(A^t) - Ch_2(A)) = 0. \tag{3.166}$$

Since ϕ_t is the identity near the boundary ∂M of M,

$$Ch_2(A^t) - Ch_2(A) = 0 \text{ near } \partial M.$$

We can assume that the bundle E is trivial over M (without loss of generality). As in Sect. 3.3.3, we can construct a Chern-Simon 3-form Ψ, such that

$$d\Psi = Ch_2(A) \text{ on } M \setminus S, \tag{3.167}$$

and

$$|\Psi(x)| \leq \frac{c}{d(x,S)^3}, \quad x \in M \setminus S, \tag{3.168}$$

where c is a uniform constant. Note that $Ch_2(A^t) = \phi_t^* Ch_2(A)$, and so for $\Psi_t = \phi_t^* \Psi$, we obtain

$$d(\Psi_t - \Psi) = Ch_2(A^t) - Ch_2(A) \text{ in } M \setminus (S \cup \phi_t(S)) \qquad (3.169)$$

and

$$|\Psi_t - \Psi|(x) \leq \frac{2c}{d(x, S \cup \phi_t(S))^3}, \quad x \in M \setminus (S \cup \phi_t(S)). \qquad (3.170)$$

Moreover, $\Psi_t - \Psi = 0$ near ∂M and for H^{m-4}-a.e. $x \in S \cup \phi_t(S)$,

$$\lim_{x \to x_0} d(x, S \cup \phi_t(S))^3 (\Psi_t - \Psi)(x) = 0. \qquad (3.171)$$

Hence, (3.166) follows from (3.169) to (3.171) and the same discussions as in the proof of Proposition 3.3.4. □

Proposition 3.6.2. *If Ω is a closed form of degree $m - 4$, then any admissible Ω-anti-self-dual instanton A on M is stationary.*

Proof. We may assume that $tr(F_A) = 0$ for simplicity (we can treat the general case using the same arguments, since $tr(F_A)$ is smooth on M).

We first have

$$\int_M tr(F_{A^t} \wedge F_{A^t}) \wedge \Omega = \int_M tr(F_A \wedge F_A) \wedge \Omega. \qquad (3.172)$$

This is the same as

$$\int_M (F_{A^t}, T(F_{A^t}))dV = \int_M (F_A, T(F_A))dV, \qquad (3.173)$$

where T is the operator $- * \cdot \Omega \wedge$ acting on 2-forms. Then the Ω-anti-self-duality of A gives $T(F_A) = F_A$. It follows that

$$\begin{aligned} YM(A^t) &= \frac{1}{4\pi^4} \int_M |F_{A^t}|^2 dV \\ &= \frac{1}{4\pi^2} \int_M (F_{A^t}, (Id - T)(F_{A^t}))dV + \frac{1}{4\pi^2} \int_M (F_{A^t}, T(F_{A^t}))dV \\ &= \frac{1}{4\pi^2} \int_M (F_{A^t}, (Id - T)(F_{A^t}))dV - Ch_2(A). \end{aligned}$$

Since $(Id - T)(F_A) = 0$ and T is symmetric, the third integral above is of order t^2. Hence, A is stationary, which follows from $\frac{d}{dt} YM(A^t)|_{t=0} = 0$. □

Any admissible Yang-Mills connection (subject perhaps to certain mild conditions) may be stationary. If this is true, then we may conclude form Corollary 3.5.8 that the blow-up locus of any Yang-Mills connection is stationary, that means, it is a generalized minimal variety.

3.6.2 A Removable Singularity Theorem

Suppose that A is an admissible Yang-Mills connection on M that is stationary. Pick a point $p \in S = S(A)$, where $S(A)$ is the singular set of A. Let $r_p, c(p)$ and a be as in Theorem 3.3.2. We want to prove a removable singularity theorem under suitable conditions. Suppose that $S \cap B_{\frac{r_p}{2}}$ satisfies the following uniform covering (UC) property: for any $y \in S \cap B_{\frac{r_p}{2}}$ and $\delta \le r < \frac{r_p}{2}$, there are balls $B_\delta(x_i)$, $i = 1, \cdots, l$, such that $x_i \in S$, $S \bigcap B_r(y) \subset \bigcup_i B_\delta(x_i)$ and $l\delta^{n-4} \le cr^{m-4}$ for some uniform constant $c > 0$. We can easily check that the (UC) holds, whenever there is a measure μ with support S such that the total measure $\mu(S \cap B_{r_p}(p)) < \infty$, and for each $x \in S \cap B_{r_p}(p)$, $r^{4-m}\mu(S \cap B_r(x))$ is decreasing in r, and the density $\Theta(x) = \lim_{r\to\infty} r^{4-m}\mu(S \cap B_r(x)) > 0$. In particular, if A is the limit of a sequence of smooth Yang-Mills connections A_i outside S, then S has the (UC) property, since $\mu = \lim_{i\to\infty} |F_{A_i}|^2 dV$ obeys the above conditions.

Theorem 3.6.3. *Let A and S be as above. Then there is an $\epsilon > 0$ depending only on $m = dimM$ such that for any $p \in S$ and $0 < r < r_p$, if*

$$r^{4-m} \int_{B_r(p)} |F_A|^2 dV < \epsilon, \tag{3.174}$$

then there is a unique transformation σ near p such that $\sigma(A)$ extends to a smooth connection near p.

In order to prove Theorem 3.6.3, we need the following lemmas (proofs given in [374]). By scaling, we may suppose that $r = 5$, $M = B_5(p)$ and E is trivial over M. We may further suppose that the metric g is flat, for simplicity. The general case can be proved by an analogous argument. Let c be a uniform constant and $S = S(A)$ be the singular set of A as above.

(L1) *There is a gauge transformation σ on $M \backslash S$ such that for any $x \in B_3(p) \backslash S$,*

$$\rho(x)^{m-2} |A^\sigma|^2(x) \le c \int_{B_{\frac{1}{2}\rho(x)}(x)} |F_A|^2 dV, \tag{3.175}$$

$$\int_{B_{\frac{2}{5}\rho(x)}(x)} \left(\frac{|A^\sigma|^2}{\rho(x)^2} + |\nabla A^\sigma|^2 \right) dV \le c \int_{B_{\frac{1}{2}\rho(x)}(x)} |F_A|^2 dV, \tag{3.176}$$

where $\rho(x) = d(x, S)$ and $D_{\sigma(A)} = d + A^\sigma$, with $A^\sigma \in \Omega(M \backslash S, \mathcal{G})$ ($\mathcal{G} = T_e G$).

(L2) *Let A as above. Then*

$$\int_{B_1(x)} \left(\frac{|A|^2}{\rho(y)^2} + |\nabla A|^2\right) dV \le c \int_{B_3(x)} |F_A|^2 dV, \qquad (3.177)$$

where $\rho(y) = d(y, S)$.

(L3) *Let A as above. Then there exist a function* α *and a 2-form* β *such that*

$$A = d\alpha + d^*\beta,\ d\beta = 0 \ on\ B_1(x), \qquad (3.178)$$

$$||\alpha||_{H^{1,2}(B_1(x))} + ||\beta||_{H^{1,2}(B_1(x))} \le c||A||_{L^2(B_2(x))}. \qquad (3.179)$$

(L4) *Let* $\tilde{A} = A - d\alpha$. *Then we have*

$$||\tilde{A}_1||_{H^{1,2}(B_1(x))} \le c\sqrt{\epsilon}||F_A||_{L^2(B_3(x))}, \qquad (3.180)$$

where ϵ *is as given in (3.174).*

(L5) *For any function* f *vanishing on* $\partial B_1(x)$,

$$\int_{B_1(x)} \frac{|f|^2}{\rho(y)^2} dV \le c \int_{B_1(x)} |\nabla f|^2 dV. \qquad (3.181)$$

Proof of Theorem 3.6.3. Let $\theta \in (0,1)$ be fixed. Because $\tilde{A}_0$ is harmonic, the standard elliptic estimates imply that

$$\frac{1}{\theta^{m-4}} \int_{B_\theta(x)} |d\tilde{A}_0|^2 dV \le \theta^4 \int_{B_1(x)} |d\tilde{A}_0|^2 dV \le \int_{B_1(x)} |d\tilde{A}|^2 dV. \qquad (3.182)$$

Therefore,

$$\begin{aligned}
\theta^{4-m} \int_{B_\theta(x)} |F_A|^2 dV &= \theta^{4-m} \int_{B_\theta(x)} \left(|dA|^2 + 2(F_A, A \wedge A) - |A \wedge A|^2\right) dV \\
&\le \theta^{4-m} \int_{B_\theta(x)} (|d\tilde{A}|^2 + 2(F_A, A \wedge A)) dV \\
&\le \theta^{4-m} \int_{B_\theta(x)} \left(|d\tilde{A}|^2 + \frac{c\sqrt{\epsilon}|A||F_A|}{\rho(y)}\right) dV \quad \text{(by (3.175), (3.174))} \\
&\le \theta^{4-m} \int_{B_\theta(x)} |d\tilde{A}|^2 dV + c\sqrt{\epsilon}\theta^{4-m} \int_{B_3(x)} |F_A|^2 dV \quad \text{(by (3.177))}. \qquad (3.183)
\end{aligned}$$

Likewise, we have

$$\int_{B_1(x)} |d\tilde{A}_0|^2 dV \le \int_{B_1(x)} |d\tilde{A}|^2 dV + c\sqrt{\epsilon} \int_{B_3(x)} |F_A|^2 dV. \qquad (3.184)$$

On the other hand, applying (L2) and (L4), we derive

$$\int_{B_\theta(x)} |d\tilde{A}|^2 dV = \int_{B_\theta(x)} \Big(|d\tilde{A}_0|^2 + |d\tilde{A}_1|^2 + 2(d\tilde{A}_0, d\tilde{A}_1)\Big) dV$$

$$\le \int_{B_\theta(x)} |d\tilde{A}_0|^2 dV + 2||\tilde{A}_1||_{H^{1,2}(B_1(x))} \cdot \left(\int_{B_1(x)} |d\tilde{A}_0|^2 dV\right)^{1/2} + ||\tilde{A}_1||^2_{H^{1,2}(B_1(x))}$$

$$\le \int_{B_\theta(x)} |d\tilde{A}_0|^2 dV + c\sqrt{\epsilon} \int_{B_3(x)} |F_A|^2 dV. \tag{3.185}$$

The previous four inequalities imply that

$$\theta^{4-m} \int_{B_\theta(x)} |F_A|^2 dV \le \theta^4 \int_{B_1(x)} |F_A|^2 dV + c\sqrt{\epsilon}\theta^{4-m} \int_{B_3(x)} |F_A|^2 dV.$$

Hence, for $r \le 1$ and $y \in B_1(p)$, we have, by scaling,

$$(\theta r)^{4-m} \int_{B_{\theta_r}(x)} |F_A|^2 dV \le \theta^4 r^{4-m} \int_{B_r(x)} |F_A|^2 dV + c\sqrt{\epsilon}\theta^{4-m} r^{4-m} \int_{B_{3r}(x)} |F_A|^2 dV. \tag{3.186}$$

Thus, by the monotonicity of A, we get

$$(\lambda r)^{4-m} \int_{B_{\lambda_r}(x)} |F_A|^2 dV \le \left(3^4 + c\sqrt{\epsilon(r)}\lambda^{-m}\right) \lambda^4 r^{4-m} \int_{B_r(x)} |F_A|^2 dV, \tag{3.187}$$

where $\lambda = \frac{\theta}{3} < 1/3$ and

$$\epsilon(r) = r^{4-m} \int_{B_r(x)} |F_A|^2 dV \le 8\epsilon.$$

An iteration gives

$$(\lambda^k r)^{4-m} \int_{B_{\lambda^k r}(y)} |F_A|^2 dV \le \prod_{i=0}^{k-1} \left(1 + c\sqrt{\epsilon(\lambda^i r)}\lambda^{-m}\right) (3\lambda)^{4k} r^{4-m} \int_{B_r(x)} |F_A|^2 dV, \tag{3.188}$$

where $k \ge 1$.

Select λ and ϵ such that $6^4\lambda < 1$ and $8c\sqrt{\epsilon}\lambda^{-m} < 1$. It follows that for any $i \le k-1$, $(1 + c\sqrt{\epsilon(\lambda^i r)}\lambda^{-m})3^4\lambda < 1$. For any $r \le 1$, we define $k, r_0 \in (1/3, 1]$ by $\lambda^k r_0 = r$. Then

$$r^{4-m} \int_{B_r(y)} |F_A|^2 dV \le \lambda^{3k} r_0^{4-m} \int_{B_{r_0}(y)} |F_A|^2 dV$$

$$\le r^3 r_0^{-3} \int_{B_1(y)} |F_A|^2 dV \le c r^3. \tag{3.189}$$

Substituting $\epsilon(r)$ in (3.186) by cr^3, we get

$$(\theta r)^{4-m}\int_{B_{\theta r}(x)}|F_A|^2 dV \le \theta^4 r^{4-m}\int_{B_r(x)}|F_A|^2 dV + c\theta^{4-m}r^{9/2}. \quad (3.190)$$

Select $\theta = 1/2$ and c_1 so that $c(1/2)^{4-m} + c_1(1/2)^{9/2} \le c_1(1/2)^4$. Therefore,

$$(r/2)^{4-m}\int_{B_{r/2}(x)}|F_A|^2 dV + c_1(r/2)^{9/2} \le (1/2)^4\left(r^{4-m}\int_{B_r(x)}|F_A|^2 dV + c_1 r^{9/2}\right).$$

It implies that, after an iteration,

$$r^{4-m}\int_{B_r(x)}|F_A|^2 dV_g \le c_2 r^4,$$

where c_2 is a uniform constant.

Thus the curvature F_A is bounded in $B_1(p)$. Applying [384], we can construct a gauge transformation σ such that $d^*A_\sigma = 0$ and $||A_\sigma||_{C^1(B_1(p))}$ is bounded. Since $D^*_{\sigma(A)}F_{\sigma(A)} = 0$, A_σ is smooth, and thus $\sigma(A)$ extends to a smooth connection near p. This completes the proof. □

We can conclude the following theorem from Theorem 3.6.3.

Theorem 3.6.4. *Let A and S be as in Theorem 3.6.3. Then there is a gauge transformation σ such that $\sigma(A)$ is a smooth outside a closed subset S' of H^{m-4}-measure zero.*

Proof. Let ϵ be given as in Theorem 3.6.3. Thus for any $x \in M$, the limit

$$\lim_{r\to 0} r^{4-m}e^{ar^2}\int_{B_r(x)}|F_A|^2 dV$$

exists. Define

$$S' = \{x \in M \mid \lim_{r\to 0} r^{4-m}e^{ar^2}\int_{B_r(x)}|F_A|^2 dV \ge \epsilon\}. \quad (3.191)$$

It follows from Proposition 3.6.1 that S' is closed. Furthermore, using the same arguments as the proof of (L3)(c) (cf. [374]), we can verify that $H^{n-4}(S') = 0$.

By Theorem 3.6.3, there is a countable covering $\{U_\alpha\}$ of $M\backslash S'$, with the property that for each α there is a gauge transformation σ_α on $(M\backslash S')\cap U_\alpha$ such that E is trivial over U_α and $D_{\sigma_\alpha(A)} = d + A_\alpha$ for some smooth A_α. It follows that for any α, β we have the transition function $g_{\alpha\beta} = \sigma_\alpha\cdot\sigma_\beta^{-1} : U_\alpha \cup U_\beta\backslash S' \to G$, where G is the structure group of E, such that

$$A_\alpha = g_{\alpha\beta}^{-1}dg_{\alpha\beta} + g_{\alpha\beta}^{-1}A_\beta g_{\alpha\beta}. \quad (3.192)$$

Thus $g_{\alpha\beta}$ extends to a smooth map on $U_\alpha \cap U_\beta$, because $g_{\alpha\beta}$ takes values in a compact group G. Moreover, $\{g_{\alpha\beta}\}$ satisfies the cocycle condition

$$g_{\alpha\beta} \cdot g_{\beta\gamma} = g_{\alpha\gamma} \text{ on } U_\alpha \cap U_\beta \cap U_\gamma.$$

Hence, $\{g_{\alpha\beta}\}$ defines a G-bundle E' over $M \backslash S$ extending $E|_{M\backslash S(A)}$, and $\{A_\alpha\}$ defines a Yang-Mills connection for E', and the theorem is verified. □

3.7 Brief Overview of Taubes' Work

We make a brief overview of the main results of Taubes' two well-known papers [368, 369] on Yang-Mills fields, and one paper [370] about the Seiberg-Witten equations for pseudo-holomorphic curves. Due to the length of these papers and technical methods, it is impossible to present all the details. The readers interested in the proofs are referred to his original papers. Recently, Taubes has studied the Seiberg-Witten equations further in his papers [371–373].

3.7.1 Self-dual Connections on Non-self-dual 4-Manifolds

Let M be a compact connected oriented 4-manifold, G be a compact connected semi-simple Lie group, and P be a principal G-bundle over M. Recall that the Yang-Mills functional on the space of smooth connections $\mathcal{C}(P)$ on P is given by

$$YM(A) = \frac{1}{2}\int_M |F_A|^2 dv = \frac{1}{2}||F_A||^2_{L^2}, \tag{3.193}$$

where F_A is the curvature of A, dv is the volume form of M, and $|| \cdot ||$ is a norm defined in terms of the Riemannian metric on M and the Cartan metric on the Lie algebra $\mathcal{G}$ of G. The critical points of (3.193) on $\mathcal{C}(P)$ are called Yang-Mills connections. The condition for $A \in \mathcal{C}(P)$ to be a critical point is expressed by the harmonic curvature, in the sense that

$$D_A^* F_A = 0, \tag{3.194}$$

which is Yang-Mills equation; here D_A^* is the adjoint of the covariant exterior derivative D_A. By the Bianchi identity, we have

$$D_A F_A = 0. \tag{3.195}$$

Let $\tilde{\mathcal{G}} = P \times_{Ad_G} \mathcal{G}$ be the vector bundle which is associated to P by the adjoint representation. The Hodge duality operator $*$ acts on sections of $\tilde{\mathcal{G}} \otimes \Lambda^p$, and defines an automorphism of $\tilde{\mathcal{G}} \otimes \Lambda^2$ with eigenvalues ± 1. Then $\tilde{\mathcal{G}} \otimes \Lambda^2 = (\tilde{\mathcal{G}} \otimes P_+\Lambda^2) \oplus (\tilde{\mathcal{G}} \otimes P_-\Lambda^2)$, where

$$P_\pm = \frac{1}{2}(1 \pm *). \tag{3.196}$$

The curvature $F_A \in \Gamma(\tilde{\mathcal{G}} \otimes \Lambda^2)$. The connection A is *self-dual*, if $P_- F_A = 0$, and *anti-self-dual*, if $P_+ F_A = 0$. If $A \in \mathcal{C}(P)$ is self-dual, then (3.195) implies (3.194), and thus every self-dual connection is a Yang-Mills connection. Actually, self-dual connections minimize $YM(\cdot)$ over all $A \in \mathcal{C}(P)$.

A descent technique for finding the global minima of $YM(\cdot)$ works in dimensions 2 and 3, but not in dimension 4. The problem of finding critical points of $YM(\cdot)$ in 4 dimensions is similar to the problem of harmonic maps in 2 dimensions. Few cases of self-dual connections are known to exist and for these a high degree of symmetry in the base manifold is utilized. This symmetry is encoded by the vanishing of the traceless, anti-self-dual Weyl tensor $\mathcal{W}_-$, which is part of the Riemann curvature. Namely, the Riemann curvature defines a self-adjoint transformation

$$R : \Lambda^2 \to \Lambda^2, \tag{3.197}$$

and $\mathcal{W}_-$ is the restriction of R to the traceless endormorphisms of $P_-\Lambda^2$. Atiyah, Hitchin and Singer [16] investigated the properties of self-dual connections over base manifolds M which have positive scalar curvature and satisfy $\mathcal{W}_- = 0$ (self-dual spaces). In this situation, the bundle of projective anti-self-dual spinors PV_- has a complex structure and the following *Ward correspondence* holds:

Let E be a hermitian vector bundle with self-dual connection over a self-dual space M, and $F = p^*E$ be the pull-back bundle. Then

(a) *F is holomorphic on PV_- with holomorphically trivial fibre.*
(b) *There is a holomorphic isomorphism $\sigma : \tau^*\bar{F} \to F^*$, where $\tau : PV_- \to PV_-$ is the real structure, and σ induces a positive definite Hermitian structure on the space of holomorphic sections of F on each fibre.*
(c) *Every such bundle on PV_- is the pull-back of a bundle $E \to M$ with self-dual connection.*

In particular, if $M = S^4$, the Ward correspondence leads to the construction of all self-dual connections on G-bundles over S^4 [14, 16, 105]. In this case, PV_- is identified with $P\mathbf{C}^3$, and algebraic techniques are utilized to construct certain important complex structures [105].

Taubes [368] studied self-dual connections using analytic techniques, which does not require the self-duality of the Riemannian curvature of the base manifold M, but requires that there be no anti-self-dual harmonic two-forms on M, i.e.,

$$P_- H^2_{de\,Rham}(M) = 0, \tag{3.198}$$

where $H^2_{de\,Rham}(M)$ is the second cohomology group of the de Rham complex:

$$0 \to \Gamma(\Lambda^0) \xrightarrow{d} \Gamma(\Lambda^1) \xrightarrow{d} \Gamma(\Lambda^2) \xrightarrow{d} \Gamma(\Lambda^3) \xrightarrow{d} \Gamma(\Lambda^4) \to 0,$$

and d is the exterior derivative. Taubes [368] obtained all the following theorems concerning the existence and classification.

Theorem 3.7.1. *Let M be a compact oriented Riemannian manifold of dimension 4 such that $P_- H^2_{de\,Rham}(M) = 0$, and G be a compact semi-simple Lie group. Then there exist principal G-bundles $P \to M$ which admit smooth irreducible self-dual connections.*

We discuss the classification as follows. For G semi-simple and compact, principal G-bundles over M are classified up to isomorphism by the set of homotopy classes of maps from M into the classifying space for G, BG. This set is denoted by $[M; BG]$ and there is a surjection

$$\phi : [M, BG] \to \mathbf{Z}^l \to 0, \tag{3.199}$$

where l is the number of non-trivial simple ideals composing the Lie algebra of G. Let $P \to M$ be a principal G-bundle. Thus the Pontryjagin classes $\{P_l^k(\tilde{\mathcal{G}})\}_{k=1}^l$ of the associated vector bundle $\tilde{\mathcal{G}} = P \times Ad_G \mathcal{G}$ specify the map ϕ. If G is simply connected, then ϕ is a bijection. If G is not simply connected, then there is a map $\eta : [M; BG] \to H^2(M; \pi_1(G))$, and the map ϕ is a bijection on the kernel of η.

Theorem 3.7.2. *Suppose that M is a compact oriented Riemannian manifold of dimension 4 such that $P_- H^2_{de\,Rham}(M) = 0$. Let G be a compact semi-simple Lie group. Let $P \to M$ be a principal G-bundle, all of whose Pontryjagin classes $\{P_l^k(\tilde{\mathcal{G}})\}_{k=1}^l$ are nonnegative. Moreover, suppose that the image of the isomorphism class of P under η in $H^2(M; \pi_1(G))$ is trivial. Then we have*

(a) *The space $\mathcal{C}(P)$ contains a smooth self-dual connection*
(b) *If the principal G-bundle over S^4 with the same Pontryjagin classes admits an irreducible self-dual connection, then so does $\mathcal{C}(P)$.*
(c) *If M is a real analytic manifold, then there is a real analytic principal G-bundle P' which is isomorphic to P, and on which (a) and (b) above are satisfied by real analytic connections.*

The conditions which make (b) of Theorem 3.7.2 applicable have been worked out by Atiyah, Hitchin and Singer [16]. When the image of the isomorphism class of P under η in $H^2(M; \pi_1(G))$ is non-trivial, nothing can be said. The combination of Taubes' techniques and Uhlenbeck's method [381] may produce some results.

To count self-dual connections on P, we must consider the gauge group $Aut\ P = \Gamma(P \times_{Ad_G} \mathcal{G})$ has a natural action on $\mathcal{C}(P)$. We denote this action by $(g, A) \mapsto g(A)$ for $(g, A) \in Aut P \times \mathcal{C}(P)$. The action leaves invariant for (3.194), (3.195) and the condition of self-duality. Therefore, it is natural to consider the space of orbits in $\mathcal{C}(P)$ under the action of $Aut\ P$. The set of irreducible self-dual connections in

$\mathcal{C}(P)$ modulo this action is called the space of moduli of self-dual connections in $\mathcal{C}(P)$. Atiyah, Hitchin and Singer verified that when M is a self-dual manifold, these moduli spaces are finite-dimensional manifolds. The extension to those M which satisfy (3.198) is given in the following theorem.

Theorem 3.7.3. *Under the assumptions of Theorem 3.7.2, suppose that $P \to M$ is a principal G bundle with G compact and semi-simple. Let A be a connection given by item* (b) *of Theorem 3.7.2. Then in a neighborhood of A in $\mathcal{C}(P)/Aut P$, the space of moduli of irreducible self-dual connections is a manifold of dimension*

$$p_l(\tilde{\mathcal{G}}) - \frac{1}{2}(dim\, G)(\chi - \tau), \tag{3.200}$$

where $p_l(\tilde{\mathcal{G}}) = \sum_{j=1}^{l} p_l^j(\tilde{\mathcal{G}})$ is the sum of the l Pontryjagin classes of $\tilde{\mathcal{G}}$, χ is the Euler characteristic of M and τ is the signature of M.

Theorem 3.7.3 is a local result on the space of moduli. That means, there may be irreducible self-dual connection in $\mathcal{C}(P)$ for which the conclusions of the theorem are not met. We need further assumptions to obtain a stronger result. Recall that two metrics g, g' are said to be pointwise conformal if $g' = v^2(x)g$ with $v(x)$ a smooth, strictly positive function on M.

Theorem 3.7.4. *Besides the assumptions on M and P in Theorem 3.7.3, suppose that the Riemannian metric g on M is pointwise conformal to a metric g' on M whose curvature satisfies*

$$s' - 3w' > 0 \tag{3.201}$$

where $s'(x)$ is the scalar curvature of g' and $w'_-(x) = \sup_{\xi \in S^2 \subset \mathbf{R}^3} \mathcal{W}_-^{ij} \xi^i \xi^j$ is the largest eigenvalue of the traceless anti-self-dual Weyl tensor of g'. Then the space of moduli of irreducible self-dual connections is globally a Hausdorff manifold of dimension given by (3.200).

Notice that (3.198) implies that $\chi - \tau \leq 2$, so a corollary of Theorems 3.7.3 and 3.7.4 is as follows: Let $p_1(\tilde{\mathcal{G}})$ be fixed. The (3.200) is a function on the set of 4-manifolds that satisfy the conditions of Theorem 3.7.4. This function is minimized by S^4, since $\chi(S^4) = 2$ and $\tau(S^4) = 0$. Applying the Ward correspondence and Theorems 3.7.2 and 3.7.3, we obtain the following theorem on complex structures.

Theorem 3.7.5. *Let M be a 4-dimensional compact orientable Riemannian manifold with positive scalar curvature and $\mathcal{W}_- = 0$. Let $p : PV_- \to M$ be the bundle of projective anti-self-dual spinors. Let G be a compact semi-simple Lie group which has a unitary representation on a vector space L. Then there are holomorphic vector bundles F with fibre L over PV_- satisfying the following properties:*

(a) *F is holomorphically trivial on each fibre.*
(b) *$\sigma : \tau^* \bar{F} \to F^*$ is a holomorphic isomorphism.*

It follows from Theorems 3.7.1–3.7.4 that the self-dual connections on S^4 are stable with respect to all deformations of the standard Riemannian structure. Moreover, (3.198) holds on $S^3 \times S^1$, where the product metric satisfies (3.201), and on $P\mathbf{C}^2$, where the Fubini-Study metric fulfills (3.201). Thus these spaces admit bundles with irreducible self-dual connections as given by the previous theorem.

We don't know whether irreducible self-dual connections exist when (3.198) is not satisfied. We don't know for $S^2 \times S^2$ and the K^3 manifolds either. However, we know that there are self-dual and anti-self-dual $SU(2)$ connections on $\mathbf{R}^2 \times S^2$. Taubes proved the following approximation theorem.

Theorem 3.7.6. *Let M be a compact oriented Riemannian 4-manifold, with no assumption on its Riemannian curvature. Let G be a compact semi-simple Lie group. Let $P \to M$ be a principal G-bundle all of whose first Pontryjagin classes are non-negative. Furthermore, assume that the isomorphism class of P has trivial image under η in $H^2(M;\pi_1(G))$. Then given $\delta > 0$, there exists $A \in \mathcal{C}(P)$ with $||P_- F_A||_{L_2} < \delta$.*

Concerning anti-dual connections, note that reserving the orientation of the base manifold interchanges self-dual and anti-dual forms. Hence, Theorems 3.7.1–3.7.6 and the previous arguments are valid, if self-dual P_-, $p_l(\tilde{\mathcal{G}})$, τ and $\mathcal{W}_-$ are replaced by anti-self-dual P_+, $-p_l(\tilde{\mathcal{G}})$, $-\tau$ and $\mathcal{W}_+$, respectively. For the detailed proofs of Theorems 3.7.1–3.7.6, see [368].

3.7.2 *Morse Theory for the Yang-Mills Functionals on 4-Manifolds*

Let M be a compact, connected, oriented, 4-dimensional Riemannian manifold, and let G be a compact, simple Lie group. The space of isomorphism class of pairs $[P, A]$, where P is a principal G-bundle over M and A is a connection on P, has a countable number of connected components. A component $\mathcal{B}$ is labeled by data (k, η), where k is the first Pontryjagin class of the associated vector bundle, $Ad\ P = P \times_{AdG} \mathcal{G}$ ($\mathcal{G} = T_e(G)$ is the Lie algebra of G). The characteristic class η is in $H^2(M, \pi_1(G))$.

For a given Riemannian metric on *TM*, there is a natural and non-negative functional on $\mathcal{B}$, namely, the Yang-Mills functional. This energy functional measures the extent to which the horizontal sub-bundle in TP of a given connection is not involutive. The Yang-Mills functional assigns to an orbit $[A] \in \mathcal{B}$ of a connection A the number

$$a(A) = \int_M |F_A|^2 dv, \tag{3.202}$$

where F_A is the curvature of the connection, a section over M of the vector bundle $\Omega^2 AdP = AdP \otimes \Lambda^2 T^* M$. The above norm is induced by the metric's inner product on *TM*, and a normalized Killing form on $\mathcal{G}$.

We select the normalization of the Killing form on $\mathcal{G}$ so that the Yang-Mills functional takes values in $[|k|, \infty)$ on $\mathcal{B}$. The functional can attain its minimal value, $|k|$; these minimum points form exactly the set $\mathcal{M} = \mathcal{M}(k, \eta)$ of points in $\mathcal{B}$ that are orbits of connections whose curvature is self- or anti-self-dual with respect to the Hodge star operator on $\Lambda^2 T^* M$. The Hodge star $* : \Lambda^p T^* \to \Lambda^{4-p} T^*$ is uniquely defined for p-forms ω by $\omega \wedge *\omega = (\omega, \omega) dv$, where $(\cdot, \cdot)$ is the given metric on $\Lambda^p T^* M$ and dv is the volume 4-form. The orbit $[A] \in \mathcal{B}$ of a connection A lies in $\mathcal{M}$ if and only if the curvature of A obeys

$$F_A = \pm * F_A, \tag{3.203}$$

where the $\pm$ depends on whether $\pm k \geq 0$. The set $\mathcal{M}$ is called the moduli space of (anti-) self-dual connections. As the Yang-Mills functional, one might hope that the moduli space of (anti-) self-dual connections, $\mathcal{M}$, the manifold of minimal points of the Yang-Mills functional on $\mathcal{B}$, is a special set of connection orbits, and is closely tied to the topology of M and of the Lie group G. Such a nice relationship was studied by Donaldson [99–101].

Because the properties of the orbit space $\mathcal{B}$ are related to properties of M and G, we can conjecture a relationship between the topology of M and G in one way, and with the non-minimum critical points of the Yang-Mills functional in another way. Morse theory provides such a relationship, which links these two.

In Yang-Mills theory, we can follow Atiyah and Bott [13] and consider the equivariant Morse theory of the Yang-Mills functional on a space $\mathcal{B}'$ which maps to $\mathcal{B}$. The space $\mathcal{B}'$ is constructed by selecting a base point $x_0 \in M$. Then $\mathcal{B}'$ is the space of isomorphism classes of triples $[P, p, A]$, where P is a principal G-bundle over M, $p \in P|_{x_0}$ and A is a connection on P. Let EG be the universal G-bundle and $BG = EG/G$ be the corresponding classifying space. We know from [13, 100] that $\mathcal{B}'(k, \eta)$ has the homotopy type of the space $Maps_{P0}(M, BG)$ of smooth based maps from M into the classifying space BG which pull back a fixed bundle P with characteristic classes (k, η).

The action of G on P induces an action on $\mathcal{B}'$, and $\mathcal{B}'/G \equiv \mathcal{B}$. This action factors through $G/Center(G)$. The advantage of using $\mathcal{B}'$ is that $\mathcal{B}'$ can be given the structure of a smooth Hilbert manifold, but $\mathcal{B}$ doesn't have such structure in many cases. When a group G acts on a space X (e.g. $X = \mathcal{B}'$), we can define the G-invariant homology (H_{G*}) and homotopy (π_{G*}) of X. These are the ordinary homology and homotopy groups of the space $EG \times_G X$. When G acts with the same stabilizer (up to conjugacy) at each point, then X/G has a natural manifold structure, and $EG \times_G X$ is homotopy equivalent to X/G. Then the equivariant homology and homotopy of X are the ordinary homology and homotopy of X/G. A G-invariant functional f on X defines a functional f on $EG \times_G X$. The Morse theory of f on $EG \times_G X$ gives the G-invariant Morse theory of f on X (cf. [298]). We know that no Morse theory will work for the Yang-Mills functional on $\mathcal{B}'$, since $\mathcal{B}'$ is infinite-dimensional and non-compact.

When $M = S^4$ with the standard metric, $G = SU(2)$ and $P = S^4 \times SU(2)$, the Yang-Mills functional fails to be proper: In this situation, $\pi_1(EG \times_G \mathcal{B}')$ is non-trivial, and were the usual Morse theory to hold, the min-max arguments in [368] would imply the existence of a critical point of the functional a in (3.202) which has Morse index less than two. On the other hand, every non-minimal critical point of a (for S^4 and $G = SU(2)$) should have Morse index at least two [369]. However, when a functional fails to be proper, Morse theory can still be recovered by examining the restriction of the functional to a countable set of finite-dimensional, non-compact varieties.

There are three unusual facts about this recovery process: 1. While working below a fixed energy E, a finite number (independent of the base 4-manifold) of these varieties need to be considered. 2. These varieties are naturally parameterized with topological data from the 4-manifold (configurations of points in fiber bundles over the manifold) and with the critical manifolds of the Yang-Mills functional on S^4. 3. The connections on these varieties satisfy those a priori estimates which are obeyed by genuine solutions to the Yang-Mills equations. Taubes [369] obtained all the following theorems and results.

Theorem 3.7.7. *Let M be a compact, oriented, Riemannian 4-manifold, and G be a compact, simple Lie group. Suppose that $P \to M$ is a principal G-bundle with first Pontryjgin class k and characteristic class $\eta \in H^2(M; \pi_1(G))$. Fix a real number $E \geq 0$ (an energy for the Yang-Mills functional) and $\delta \in (0, 1]$. Let $\mathcal{B}'_E = \{[p, A] \in \mathcal{B}'(k, \eta) : a(A) < E\}$; this is a space of orbits of connections on P with Yang-Mills energy less than E. There are an integer $d = d(E)$ and a number $z(E) \geq 1$, and there exist G-invariant, real analytic varieties $\{\Sigma_k\}_{k \in \{0,1,\cdots,dim\, G\}} \subset \mathcal{B}'(k, \eta)$ satisfying the following:*

(a) *$\Sigma_k \cap \mathcal{B}'_{E+\delta}$ is a union of smooth, G-invariant submanifolds, each of dimensions d or less. The intersection of any pair of submanifolds of Σ_k is an embedding of an open subset of one of the pair into the other. (Σ_k is the subset of $\mathcal{B}'_{E+\delta}$ along which the Yang-Mills functional can fail to be proper, and along which the stabilizers (in $\mathcal{G}(P)$) of connections have limiting dimension k.)*
(b) *For $k > 0$, there exist an open subset $\Sigma'_k \subset \Sigma_k$ and a G-invariant map $\phi_{k-1} : [0, 1] \times \Sigma'_k \to \mathcal{B}'(k, \eta)$ with $\phi_{k-1}(0, .) = Id$, and $\phi_{k-1}(1, \cdot) : \Sigma'_k \to \Sigma_{k-1}$. Moreover, ϕ_{k-1} does not increase a on the interval [0,1].*
(c) *Let $\Sigma \equiv (\bigcup_{k \in \{0,1,\cdots,dim\, G\}} \Sigma_k) \bigcup_{k'>0} \{\phi_{k'-1} : (t, b) \in [0, 1] \times \Sigma'_k\}$. Each orbit $[A] \in \Sigma$ obeys an elliptic, integro-differential equation which gives a priori estimates.*

   Let $\Sigma_E \equiv \Sigma \cap \mathcal{B}'_E$. The pair $(\Sigma_{E+\delta}, \Sigma_E)$ calculates the change of homology between $\mathcal{B}'_{E+\delta}$ and $\mathcal{B}'_E$ in the following sense:
(d) *If $\Sigma_{E+\delta} \setminus \Sigma_E$ is empty, then the inclusion $\mathcal{B}'_E \subset \mathcal{B}'_{E+\delta}$ is a G-equivariant homotopy equivalence.*
(e) *If $\Sigma_{E+\delta} \setminus \Sigma_E$ is not empty, let j be the inclusion map of pairs, $j : (\Sigma_{E+\delta}, \Sigma_E) \to (\mathcal{B}'_{E+\delta}, \mathcal{B}'_E)$. Then j induces epimorphisms $j_* : H_{G^*}(\Sigma_{E+\delta}, \Sigma_E) \to H_{G^*}(\mathcal{B}'_{E+\delta}, \mathcal{B}'_E)$, and $j_* : \pi_{G^*}(\Sigma_{E+\delta}, \Sigma_E) \to \pi_{G^*}(\mathcal{B}'_{E+\delta}, \mathcal{B}'_E)$.*

   To express the kernel of j_*, introduce the inclusion $i : \mathcal{B}'_{E+\delta} \to \mathcal{B}'_{E+z\delta}$.
(f) *$j_* i_*(H_{G^*}(\Sigma_{E+\delta}, \Sigma_E)) \approx i_*(H_{G^*}(\mathcal{B}'_{E+\delta}, \mathcal{B}'_E))$.*

Assume that $a|_{\mathcal{B}}$ has a non-degenerate critical point at the orbit b of an irreducible connection (thus $\mathcal{B}$ has a manifold structure at b). Assume that $a|_b = E$, and that b has Morse index $j \geq 0$. Also, assume that $\delta > 0$ exists such that the norm of the gradient of a is bounded away from zero on the complement in $\mathcal{B}_{E+\delta} \setminus \mathcal{B}_{E-\delta}$ of a neighborhood of b. Let $\pi : \mathcal{B}' \to \mathcal{B}$ be the projection. Then $\pi(\Sigma_{E+\delta} \setminus \Sigma_{E-\delta})$ is the descending j-dimensional disk from the critical point b. The following theorem is an application of Theorem 3.7.7.

Theorem 3.7.8. *Let M be a compact and oriented 4-dimensional Riemannian manifold with a compact and simple Lie group G. Let $(k, \eta) \in \mathbf{Z} \times H^2(H; \pi_1(G))$ be allowable characteristic classes for a principal G-bundle over M. Then the space $\mathcal{B}'(k, \eta)$ does not retract onto any subspace where the Yang-Mills functional is bounded.*

3.7.3 Seiberg-Witten Equations and Pseudo-holomorphic Curves

Here we explain how pseudo-holomorphic curves in a symplectic 4-manifolds can be constructed from solutions to the Seiberg-Witten equations. The main Theorem 3.7.11, obtained by Taubes [370], is an existence theorem for pseudo-holomorphic curves.

Let X be a compact, oriented, 4-dimensional manifold with the second Betti number

$$b_2^+ = \frac{1}{2}(rank(H^2(X; \mathbf{R})) + signature) \tag{3.204}$$

at least 2. Then these invariants define a diffeomorphism-invariant map, Seiberg-Witten (SW), from the set of equivalence classes, *Spin*, of $Spin^{\mathbf{C}}$ structures on X to $\mathbf{Z}$. Observe that the set *Spin* has the structure of a principal $H^2(X; \mathbf{Z})$ bundle over a point.

A symplectic 4-manifold is a pair of (X, ω), where X is a 4-manifold and ω is a closed 2-form with $\omega \wedge \omega$ nowhere zero. So a symplectic 4-manifold has a canonical orientation. A symplectic 4-manifold also has a complex line bundle, K (called the *canonical* bundle), which is canonical up to isomorphism. By Taubes [368, 369], a symplectic 4-manifold has a canonical equivalence class of $Spin^{\mathbf{C}}$ structure. The latter endows *Spin* with a base point and thus provides the identification

$$Spin \cong H^2(X; \mathbf{Z}). \tag{3.205}$$

(The identification of *Spin* and the choice of orientation do not change under a continuous deformation of the symplectic form.) The canonical orientation for a symplectic manifold and the identification in (3.205) will be assumed implicitly.

Therefore, by (3.205), SW defines a map

$$SW : H^2(X; \mathbf{Z}) \to \mathbf{Z}. \tag{3.206}$$

Remark that *SW*, viewed as a map from *Spin* to $\mathbf{Z}$ is diffeomorphism-invariant, but the identification in (3.205) is not. The effect of a diffeomorphism in (3.206) depends on the behavior of $c_1(K)$.

The Seiberg-Witten invariants of a compact, oriented, 4-dimensional manifold X constitute a map from the set of equivalence classes of $Spin^{\mathbf{C}}$ structures on X (covering the frame bundle) to the integers. They are defined if the characteristic number b_2^+ in (3.204) is greater than 1. (If $b_2^+ = 1$, one has a complicated structure.) A submanifold Σ of a symplectic manifold is called a *symplectic submanifold* if the restriction of the symplectic form to $T\Sigma$ is non-degenerate. Let us review some concepts of 4-dimensional geometry as follows.

1. **Spin Geometry.** First note that the Lie groups $SO(4)$ and $Spin^{\mathbf{C}}(4)$ can be expressed as

$$SO(4) = (SU(2) \times SU(2))/\{\pm 1\}, \tag{3.207}$$

and

$$Spin^{\mathbf{C}}(4) = (U(1) \times SU(2)) \times SU(2)/\{\pm 1\}, \tag{3.208}$$

where ± 1 acts on all factors in the obvious way.

(a) For a given Riemannian metric on X, one defines the principal $SO(4)$ bundle of orthonormal frames on X. A $Spin^{\mathbf{C}}$ structure (denoted by L) is a lift of $SO(4)$ principal bundle to a $Spin^{\mathbf{C}}(4)$ principal bundle. The set of equivalence classes of such lifts has naturally the structure of a principal $H^2(X; \mathbf{Z})$ bundle over a point. This principal $H^2(X; \mathbf{Z})$ bundle, *Spin*, is canonically defined and independent of the original choice of metric on X. We can consider the Seiberg-Witten invariant as a map from *Spin* to $\mathbf{Z}$.

(b) $SO(4)$ has two representations in $SO(3) = SU(2)/\{\pm 1\}$, which are denoted by λ_+ and λ_-. They are distinguished by the fact that the associated $\mathbf{R}^3$ bundles to the frame bundle of X are isomorphic to the bundles Λ_+ of self-dual 2-forms and Λ_- of anti-self-dual 2-forms, respectively. Similarly, $Spin^{\mathbf{C}}(4)$ has two representations s_+ and s_- in $U(2) = (U(1) \times SU(2))/\{\pm 1\}$. The convention is that the composition of s_+ with the quotient homomorphism $U(2) \to U(2)/Center = SO(3)$ factors through $SO(4)$ via λ_+. Given a $Spin^{\mathbf{C}}$ structure L on X, introduce the C^2-vector bundles

$$S_+,\ S_- \to X, \tag{3.209}$$

which are associated to L via the representations s_+ and s_-, respectively. These bundles inherit natural fibre metrics.

Let L and $L \cdot e$ be elements in *Spin*, where $e \in H^2(X;\mathbf{Z})$. So the bundles S_+ for these two $Spin^{\mathbf{C}}$ structures are related by $S_+(L,e) = S_+(L) \otimes E$, where E is the complex line bundle with the first Chern class $c_1(E) = e$.

(c) Clifford multiplication c maps T^*X into the skew-adjoint endormorphism of $S_+ \oplus S_-$, it is formed by the equality $c(v)^2$ multiplied by $-|v|^2$. In particular, c induces maps

$$\sigma : S_+ \otimes T^*X \to S_- \tag{3.210}$$

(by duality) and $c_+ : \Lambda_+ \to End(S_+)$. The adjoint of the latter is denoted by

$$\tau : End(S_+) \to \lambda_+ \otimes C; \tag{3.211}$$

it maps each self-adjoint endomorphism into an imaginary valued form. Let $\{e^i\}_{i=1}^4$ be an oriented orthonormal frame at a point of X. Then $\tau(\eta \otimes \eta^*) = -2^{-1} \cdot < \eta, c(e^i)c(e^j)\eta > (e^i \wedge e^j)$, where $< \cdot, \cdot >$ is the Hermitian inner product on S^+.

(d) Let A be a connection on $L = det(S_+)$. Then A in conjunction with the Levi-Civita connection on T^*X induces a covariant derivative ∇_A on S_+. This maps sections of S_+ into sections of $S_+ \otimes T^*X$. The composition of this last map with σ in (3.210) defines the Dirac operator D_A, a first-order elliptic operator mapping sections of S_+ to sections of S_-. Explicitly, if ψ is a section of S_+, then the action of D_A on ψ is given by

$$D_A\psi = \sigma(\nabla_A\psi). \tag{3.212}$$

2. **Seilberg-Witten equations.** With the above discussion in mind, we note that the Seiberg-Witten equations [326, 327, 407] are equations for a pair (A, ψ), where A is a connection on $L = det(S_+)$, and ψ is a section of S_+. These equations read

$$D_A\psi = 0 \text{ and } P_+F_A = \frac{1}{4}\tau(\psi \otimes \psi^*), \tag{3.213}$$

where $P_+ : \Lambda^2 T^*X \to \Lambda_+$ is the orthogonal projection. It is useful to consider perturbations of (3.213), i.e.,

$$D_A\psi = 0 \text{ and } P_+F_A = \frac{1}{4}\tau(\psi \otimes \psi^*) + \mu, \tag{3.214}$$

where μ is a fixed, imaginary valued, anti-dual 2-form on X.

The Seiberg-Witten invariant for a given $Spin^{\mathbf{C}}$ structure $L \in Spin$ is obtained by a appropriate count of solutions of (3.212) and (3.213). Note that the group $C^\infty(X, S^1)$ (S^1 is the unit circle in $\mathbf{C}$) acts on the space of solutions of (3.213): a map ϕ sends (A, ψ) to $(A - 2\phi^{-1}d\phi, \phi\psi)$. (This group acts freely at solutions where ψ is not identically zero.) The quotient of the space of solutions to (3.213) by $C^\infty(X, S^1)$ is denoted by M. (The dependence on the $Spin^{\mathbf{C}}$ structure and on the choice of μ in (3.213) is normally pressed down.) Then we have the following facts:

(a) If $b_2^+ \geq 1$, the space of solutions to (3.213) and (3.214) contain no points where $\psi = 0$ for a generic metric or choice of μ as $c_1(L)$ is rationally non-zero (generic means off a set of codimension b_1^+, by a theorem of Uhlenbeck [383]).

(b) The space M has the structure of a real analytic variety. If $b_2^+ \geq 1$, M is a smooth manifold for a generic choice of μ in (3.214) (generic means a Baire subset of $C^\infty(\Lambda_+)$). The dimension of this manifold is calculated by Atiyah-Singer index theorem as

$$d = -\frac{1}{4}(2\chi(X) + 3\,sign(X)) + \frac{1}{4}c_1(L)\cdot c_1(L). \tag{3.215}$$

where χ is the Euler characteristic of X and $sign(X)$ is the signature, and for $a, b \in H^2(X, \mathbf{Z})$, $a \cdot b$ means the evaluation of their cup product on the fundamental class of X.

(c) M is oriented by a choice of orientation for the line

$$det(H^0(X;\mathbf{R})) \otimes det(H^1(X;\mathbf{R})) \otimes det(H^{2+}(X;\mathbf{R})) \tag{3.216}$$

(the orientation of a point is a choice of ± 1 assigned to the point).

(d) Fix a base point in X and let $C_0^\infty(S^1; X)$ be the subset of maps which map the base point to 1. Let M^0 be the quotient of the space of solutions to (3.214) by the latter group. Here M is a smooth manifold and the projection $M^0 \to M$ defines a principal S^1 bundle.

(e) The space M is compact.

Definition 3.7.9. Let X be a compact, oriented 4-dimensional manifold with $b_2^+ \geq 1$ and let $L \in Spin$ be a $Spin^C$ structure on X. Choose an orientation d for (3.215). The *Seiberg-Witten invariant* $SW(L)$ for L is defined as follows:

(i) If $d < 0$ in (3.215), one puts $SW(L) = 0$.

(ii) If $d = 0$ in (3.215), choose μ in (3.214) to make M a smooth manifold. Then M is a finite union of signed points and the $SW(L)$ is the sum over these points of the corresponding ± 1's.

(iii) If $d > 0$ in (3.215), choose μ in (3.214) to make M a smooth manifold. This M is compact and oriented thus has a fundamental class. Then $SW(L)$ is obtained by pairing this fundamental class with the maximum cup product of the first Chern class of the line bundle $M^0 \times_{S^1} \mathbf{C}$.

Based on the above arguments, we derive the following proposition.

Proposition 3.7.10 ([369]). *Let X be a compact, oriented, connected, 4-dimensional manifold with $b_2^+ > 1$. Then SW defines a map from Spin to* $\mathbf{Z}$ *which depends only on the underlying smooth structure of X. That means, the value of SW(L) is independent of the choice of metric and perturbation form μ in (3.214). It* *depends only on L up to isomorphism. Moreover, the assignment of SW to a* $Spin^{\mathbf{C}}$ *structure is invariant under self-diffeomorphisms of X in the following sense: if* ϕ *is a diffeomorphism of X, then the value of SW on* $\phi^* L$ *is, up to sign, the same as the value of SW on L.*

3. **Symplectic manifolds.** Recall that a 2-form ω on an oriented 4-manifold X is symplectic if

$$d\omega = 0 \text{ and } \omega \wedge \omega \neq 0 \tag{3.217}$$

everywhere. If ω is symplectic, then the 4-form $\omega \wedge \omega$ provides an orientation for X. A symplectic 4-manifold is a pair (X, ω).

Every symplectic manifold has a canonical complex line bundle K, called the *canonical bundle*. For a given Riemannian metric on X, K can be identified as the orthogonal 2-plane bundle to the projection of ω into Λ_+. ($\omega \wedge \omega \neq 0$ implies that the above projection is nowhere zero.) Thus K can be defined by selecting an almost complex structure TX which is compatible (in the sense of Gromov [162]) for ω. In this situation, K is $det(T^{1,0}X)$. The specification of such an almost complex structure TX is equivalent to the specification of a metric on X for which ω is self-dual. Remark that when $t \mapsto \omega_t$ is a continuous, one-parameter family of symplectic forms on X, then the canonical bundles for (X, ω_0) and (X, ω_1) are isomorphic.

A symplectic manifold also has a canonical $Spin^{\mathbf{C}}$ structure (cf. [369]). In fact, use a metric for which ω is self-dual and of length $\sqrt{2}$. For this metric, the canonical $Spin^{\mathbf{C}}$ structure is characterized by the fact that its associated bundle S_+ is naturally isomorphic to $I \oplus K^{-1}$, where I is the trivial complex line bundle. Here, ω acts by Clifford multiplication on the I summand with eigenvalue $-2i$, and it acts on the K^{-1} summand with eigenvalue $+2i$. (If $t \mapsto \omega_t$ is a continuous, one-parameter family of symplectic forms on X, then the canonical $Spin^{\mathbf{C}}$ structures for ω_0 and ω_1 can be identified.)

Notice that the line bundle K^{-1} has a canonical (up to gauge equivalence) connection A_0 which is characterized as follows: If A is a covariant derivative on K^{-1}, then the spin covariant derivative ∇_A induces a covariant derivative on the I summand of S_+, namely,

$$\nabla_A \equiv \frac{1}{2}(1 + \frac{i}{2}c_+(\omega))\, \nabla_A : C^\infty(I) \to C^\infty(I \otimes T^*X). \tag{3.218}$$

With (3.218) in mind, notice that A_0 is characterized by the requirement that ∇_{A_0} annihilates a non-trivial section u_0. This u_0 is taken to have norm 1. Remark that

$$D_{A_0} u_0 = 0, \tag{3.219}$$

since $d\omega = 0$ (cf. [369]).

The definition of the canonical $Spin^{\mathbf{C}}$ structure permits us to identify *Spin*, the set of equivalence classes of $Spin^{\mathbf{C}}$ structure on X, with the set of equivalence classes of complex line bundles over X. (The latter is the same as $H^2(X; \mathbf{Z})$.) The $Spin^{\mathbf{C}}$

structure which corresponds to a given complex line bundle E is characterized by the fact that

$$S_+ = E \oplus (K^{-1} \otimes E). \tag{3.220}$$

Remark that Clifford multiplication by ω on S_+ in (3.220) preserves the splitting with the summand E with eigenvalue $-2i$.

Observe that the line bundle $L = det(S_+)$ for (3.220) is $K^{-1} \otimes E^2$. Therefore, a connection A on L is determined by the canonical connection A_0 on K^{-1} and by the choice of connection a on E. The relationship between A and a is characterized as follows: Let α be a section of E, and $\alpha \cdot u_0$ be the corresponding section of the E summand in S_+ as given in (3.220). Then the spin covariant derivative (∇_A) of $\alpha \cdot u_0$ is related to the ∇_a covariant derivative of α by

$$\nabla_A(\alpha \cdot u_0) = (\nabla_a \alpha) \cdot u_0 + \alpha \cdot \nabla_{A_0} u_0, \tag{3.221}$$

where $\nabla_{A_0} u_0$ is a section of the $K^{-1} \otimes T^*X$ summand of $S_+ \otimes T^*X$. If β is a section of the $K^{-1} \otimes T^*E$ summand in (3.220), we introduce the notation

$$\nabla'_A \beta = \frac{1}{2}(1 - \frac{i}{2} c_+(\omega)) \nabla_A \beta, \tag{3.222}$$

which is a section of the $(K^{-1} \otimes E) \otimes T^*X$ summand of $S_+ \otimes T^*X$.

4. **Perturbation.** There is a natural one-parameter family of choices for the 2-form μ in (3.214) on a symplectic manifold. The family is parameterized by a real number $r \geq 0$ and is given by

$$\mu = -\frac{i \cdot r}{4} \omega + P_+ F_{A_0}, \tag{3.223}$$

where $P_+ : \Lambda^2 T^*X \to \Lambda_+$ is the metric's orthogonal projection onto the self-dual forms. With this choice of perturbation, the Seiberg-Witten equations are

$$D_A \psi = 0 \text{ and } P_+ F_A = P_+ F_{A_0} + \frac{1}{4}(\tau(\psi \otimes \psi^*) - i \cdot r \cdot \omega). \tag{3.224}$$

When analyzing (3.224), it is useful to write S_+ as in (3.220) and to write the section ψ as

$$\psi = r^{1/2} \cdot (\alpha u_0 + \beta), \tag{3.225}$$

where α is a section of E and β is a section of the $K^{-1} \otimes E$ summand in (3.220). Thus with ψ given as in (3.225), the Seiberg-Witten equations (3.224) are equivalent to

$$\sigma(u_0 \otimes \nabla_a \alpha) + D_A \beta = 0, \tag{3.226}$$

$$P_+ F_a = -\frac{i}{8} r \cdot (1 - |\alpha|^2 + |\beta|^2) \cdot \omega + \frac{i \cdot r}{4} (\alpha\beta^* + \alpha^* \beta), \tag{3.227}$$

where $\alpha\beta^*$ and $\alpha^*\beta$ are sections of K and K^{-1}, identified as summands of $\Lambda_+ \otimes \mathbf{C}$. With the previous equations in mind, we can state the following main results obtained by Taubes [370].

Theorem 3.7.11. *Let X be a compact 4-manifold with symplectic form ω. Fix a Riemannian metric which makes the symplectic form anti-self-dual and of length $\sqrt{2}$. Fix a complex line $E \to X$. Let $\{\Omega_i \subset X\}$ be a finite collection of closed sets. Suppose that there exists an unbounded sequence of values for the parameter r in (3.226) such that the equation has a solution for the $\mathrm{Spin}^{\mathbf{C}}$ structure in (3.220). Assume that for each of these r values, there is such a solution with $\Omega_i \cap \alpha^{-1}(0) \neq \emptyset$ for all i. Then there exist a smooth, compact, complex (not necessary connected) curve Σ and a pseudo-holomorphic map $\phi : \Sigma \to X$ with $\phi_*[\Sigma]$ equal to the Poincaré dual to $c_1(E)$ and with $\Omega_i \cap image(\phi) \neq \emptyset$ for all i. (The almost complex structure on TX is defined by the metric and the symplectic form.)*

Corollary 3.7.12. *Let X be a compact, oriented, 4-dimensional manifold with $b_2^+ > 1$. Let $e \neq 0 \in H^2(X; \mathbf{Z})$ be a class with $SW(e) \neq 0$. Then the Poincaré dual to e is represented by the fundamental class of an embedded, symplectic curve with genus $g = 1 + e \cdot e$.*

Corollary 3.7.13. *Let X be a compact, oriented, 4-dimensional manifold with $b_2^+ > 1$ and with a symplectic form ω. Then*

(a) *The Poincaré dual of $c_1(K)$ is represented by the fundamental class of an embedded, symplectic curve.*
(b) *Let $e \in H_2(X; \mathbf{Z})$ be a homology class which is represented by an embedded sphere with self-intersection number -1. Then e is represented by a symplectically embedded 2-sphere and $< c_1(K), e > = \pm 1$.*
(c) *If $c_1(K)$ has negative square, then X can be blown down along a symplectic sphere of self-intersection number -1.*
(d) *Assume that X cannot be blown down along a symplectic sphere of self-intersection number -1. Then the signature of the intersection form of X is no smaller than $-\frac{4}{3}(1 - b_1) - \frac{2}{3} b_2$. (The b_i's are the Betti numbers of X.)*
(e) *If $c_1(K)$ has square zero and X has no symplectically embedded 2-spheres with self-intersection number -1, then $c_1(K)$ is Poincaré dual to a disjoint union of embedded, symplectic tori with zero self-intersection number. Actually, any class in $H^2(X; \mathbf{Z})$ with non-zero Seiberg-Witten invariant is represented by disjoint, symplectically embedded tori with square zero.*
(f) *Symplectic manifolds have "simple type" in that only the dimension zero Seiberg-Witten invariants are non-zero. That means, $SW(e) = 0$ if $c_1(K) \cdot e - e \cdot e \neq 0$.*

Another main theorem obtained by Taubes [370] asserts an equivalence between the Seiberg-Witten invariants for a symplectic manifold and a certain Gromov invariant which counts (with signs) the number of pseudo-holomorphic curves in a given homotopy class. Since the proof is lengthy and technical, interested readers should refer to the original paper for a complete and detailed description.

Chapter 4
Biharmonic Maps

Biharmonic maps between Riemannian manifolds are the critical points of bienergy functionals, and they satisfy fourth-order elliptic PDE systems. Biharmonic maps, which generalize harmonic maps, were first studied by Jiang [196–198] in 1986. In the last two decades, there has been progress in biharmonic maps made by Balmuç, Caddeo, Montaldo, Loubeau, Oniciuc, and Piu [24–26, 54–57], Ou, Lu, Tang, and Wang [254, 256, 284, 285, 289–291, 400], Chiang, Wolak, and Sun [80–83], Chang, Wang, and Yang [63], Ichiyama, Inoguchi, and Urakawa [191, 192], and Wang [398, 399], etc. In this chapter, we discuss these new developments in the theory of biharmonic maps.

4.1 Definition and Examples

4.1.1 Definition and a Theorem

A biharmonic map $f : (M^m, g_{ij}) \to (N^n, h_{\alpha\beta})$ from an m-dimensional Riemannian manifold M into an n-dimensional Riemannian manifold N is a critical point of the bienergy functional

$$E_2(f) = \frac{1}{2}\int_M ||(d+d^*)^2 f||^2 dv = \frac{1}{2}\int_M ||(d^*d)f||^2 dv = \frac{1}{2}\int_M ||\tau(f)||^2 dv, \tag{4.1}$$

where dv is the volume form on M, d^* is the adjoint of d, and $\tau(f) = trace(Ddf) = (Ddf)(e_i, e_i) = (D_{e_i} df)(e_i)$ is the tension field. Here, D is the Riemannian connection on $T^*M \otimes f^{-1}TN$ induced by the Levi-Civita connections on M and N, and $\{e_i\}$ is the local frame at a point of M. Recall that the tension field has components

Y.-J. Chiang, *Developments of Harmonic Maps, Wave Maps and Yang-Mills Fields into Biharmonic Maps, Biwave Maps and Bi-Yang-Mills Fields*, Frontiers in Mathematics, DOI 10.1007/978-3-0348-0534-6_4, © Springer Basel 2013

$$\tau(f)^\alpha = g^{ij} f^\alpha_{i|j} = g^{ij}(f^\alpha_{ij} - \Gamma^k_{ij} f^\alpha_k + \Gamma'^\alpha_{\beta\gamma} f^\beta_i f^\gamma_j),$$

where Γ^k_{ij} and $\Gamma'^\gamma_{\alpha\beta}$ are the Christoffel symbols on M and N, respectively.

In order to compute the Euler-Lagrange equation for the bienergy functional, we consider a one-parameter family of maps $\{f_t\} \in C^\infty(M \times [0, 1], N)$ from a compact manifold M (without boundary) into a Riemannian manifold N such that $f_t(x)$ is the endpoint of a segment starting at $f(x)(=f_0(x))$ determined in length and direction by the vector field $\dot{f}(x)$ along $f(x)$. For a non-closed manifold M, we assume that the compact support of $\dot{f}(x)$ is contained in the interior of M. Then we have

$$\frac{d}{dt}E_2(f_t)\Big|_{t=0} = \dot{E}_2(f) = \int_M (D_t\tau f, \tau f)\Big|_{t=0} dv. \tag{4.2}$$

Similar to the second variation of the energy functional in the Introduction, let $\xi = \frac{\partial f_t}{\partial t}$. The components of $D_t\tau f$ are $f^\alpha_{i|j|t} = \frac{\partial f^\alpha_{i|j}}{\partial t} + \Gamma'^\alpha_{\mu\gamma} f^\mu_{i|j} \xi^\gamma$. We can use the curvature formula on $M \times [0, 1] \to N$ and get $f^\alpha_{i|j|t} = f^\alpha_{i|t|j} + R'^\alpha_{\beta\gamma\mu} f^\beta_i f^\gamma_j \xi^\mu$, where R' is the Riemannian curvature of N. But $f^\alpha_{i|t} = f^\alpha_{t|i} = \xi^\alpha_{|i}$, therefore, $D_t\tau f$ has components $\xi^\alpha_{|i|j} + R'^\alpha_{\beta\gamma\mu} f^\beta_i f^\gamma_j \xi^\mu$. We can rewrite (4.2) as

$$\frac{d}{dt}E_2(f_t)\Big|_{t=0} = \int_M (J_f(\tau f), \tau f) dv, \tag{4.3}$$

where

$$J^\alpha_f(\xi) = g^{ij}\xi^\alpha_{|i|j} + g^{ij} R'^\alpha_{\beta\gamma\mu} f^\beta_i f^\gamma_j \xi^\mu = \Delta\xi^\alpha + R'^\alpha(df, df)\xi \tag{4.4}$$

is a linear expression in $\xi(=\tau(f))$ and $\Delta(\xi) = D^*D(\xi)$ is an operator from $f^{-1}TN$ to $f^{-1}TN$. Solutions of the equation $J_f(\xi) = 0$ are called Jacobi fields. Hence, we obtain the following definition from (4.2) to (4.4).

Definition 4.1.1. $f : M \to N$ is a *biharmonic map* iff the bitension field

$$\begin{aligned}\tau_2(f)^\alpha &= J_f(\tau f)^\alpha = \Delta\tau(f)^\alpha + R'^\alpha(df, df)\tau(f) \\ &= g^{ij}(f^\alpha_{ij} - \Gamma^k_{ij} f^\alpha_k + \Gamma'^\alpha_{\beta\gamma} f^\beta_i f^\gamma_j + R'^\alpha_{\beta\gamma\mu} f^\beta_i f^\gamma_j \tau(f)^\mu) = 0,\end{aligned} \tag{4.5}$$

i.e., the tension field $\tau(f)$ is a Jacobi field.

If $\tau(f) = 0$, then $\tau_2(f) = 0$. Thus, harmonic maps are obviously biharmonic. The equations of biharmonic maps form a fourth-order elliptic system of PDEs, which generalize those for harmonic maps. Our computation for the first variation of the bienergy functional presented here using tensor technique is different, but much easier than Jiang's [196] original computation.

Theorem 4.1.2 ([196]). *If $f : (M, g) \to (N, h)$ is a biharmonic map from a compact manifold M into a Riemannian manifold N with the Riemannian curvature $R^N \leq 0$, then f is harmonic.*

Proof. Suppose that f is biharmonic, i.e., $\tau_2(f) = 0$. Then it follows from (4.5) that

$$\frac{1}{2}\Delta|\tau(f)|^2 = (D_{e_k}\tau(f), D_{e_k}\tau(f)) + (D^* D\tau(f), \tau(f))$$
$$= (D_{e_k}\tau(f), D_{e_k}\tau(f)) - (R^N(df(e_i), \tau(f))df(e_i), \tau(f)) \geq 0.$$

Applying Bochner's technique, we deduce that $\tau(f)$ is constant, since M is compact. Therefore, $d\tau(f) = 0$. Using the identity

$$div(df, \tau(f)) = |\tau(f)|^2 + (df, d\tau(f)),$$

and integrating both sides by applying the divergence theorem, we conclude that $\tau(f) = 0$. □

4.1.2 Curves on Surfaces

Let $\gamma : I \to (N, h)$ be a curve parameterized by arc length from $I = [0, 1]$ to an n-dimensional Riemannian manifold (N^n, h). Then the tension field and bitension field of γ are, respectively,

$$\tau(\gamma) = \nabla_T T, \ \tau_2(\gamma) = \nabla_T^3 T - R(T, \nabla_T T)T, \tag{4.6}$$

where $T = \gamma'$ is the unit tangent vector field along the curve. It follows from Laugwitz [240] that the Frenet frame $\{e_i\}_{i=1,\cdots,n}$ associated to the curve γ parameterized by arc length is the orthonormalization of $(n+1)$-tuple $\{\nabla^{(k)}_{\frac{\partial}{\partial t}d\gamma}(\frac{\partial}{\partial t})\}_{k=0,\cdots,n}$ expressed as follows:

$$\begin{aligned} e_1 &= d\gamma(\frac{\partial}{\partial t}), \\ \nabla^{\gamma}_{\frac{\partial}{\partial t}} e_1 &= k_1 e_2, \\ \nabla^{\gamma}_{\frac{\partial}{\partial t}} e_i &= -k_{i-1}e_{i-1} + k_i e_{i+1}, \ i = 2, \cdots, n-1, \\ \nabla^{\gamma}_{\frac{\partial}{\partial t}} e_n &= -k_{n-1}e_{n-1}, \end{aligned} \tag{4.7}$$

where $k_1 = k > 0$, $k_2 = -\tau$, $k_3, \cdots, k_{n-1}$ are called the curvatures of γ and ∇^γ is the connection on the pull-back bundle $\gamma^{-1}TN$. By (4.7), the curve γ is biharmonic non-harmonic if and only if

$$
\begin{aligned}
k_1 &= constant \neq 0, \\
k_1^2 + k_2^2 &= R(e_1, e_2, e_1, e_2), \\
k_2' &= -R(e_1, e_2, e_1, e_3), \\
k_2 k_3 &= -R(e_1, e_2, e_1, e_4), \\
R(e_1, e_2, e_1, e_i) &= 0, \ \ i = 5, \cdots, n.
\end{aligned} \tag{4.8}
$$

In particular, if γ is a differentiable curve parameterized by arc length and (N^2, h) is an oriented surface, then (4.8) reduces to

$$k_g = constant \neq 0, \ k_g^2 = G,$$

where k_g is the curvature (with sign) of γ, $G = R(T, N, T, N)$ is the Gaussian curvature of the surface, and T and N are the unit tangent vector field and the unit normal vector field of the surface.

Let $\beta(u) = (f(u), 0, g(u))$ be a curve parameterized by arc length in the xz-plane and let $Z(u, v)$ be the surface of revolution obtained by rotating the curve about z-axis, with the standard parameterization given by

$$Z(u, v) = (f(u)\cos(v), \ f(u)\sin(v), \ g(u))$$

where v is the rotation angle. Oniciuc [269] showed the following:

Proposition 4.1.3. *A parallel* $u = u_0 = const$ *is biharmonic if and only if* u_0 *satisfies*

$$f'^2(u_0) + f''(u_0) f(u_0) = 0. \tag{4.9}$$

Example 1. On a torus of revolution with its standard parameterization

$$Z(u, v) = \left((a + r\ \cos(\frac{u}{r}))\cos\ v, \ (a + r\ \cos(\frac{u}{r}))\sin\ v, \ r\ \sin(\frac{u}{r})\right), \ \ a > r,$$

the biharmonic parallels are

$$u_1 = r\ \arccos\left(\frac{-a + \sqrt{a^2 + 8r^2}}{4r}\right), \ \ u_2 = 2r\pi - r\ \arccos\left(\frac{-a + \sqrt{a^2 + 8r^2}}{4r}\right).$$

There is a geometric way to understand the behavior of biharmonic curves on a sphere. Indeed, the torsion τ and curvature κ (without sign) of a curve γ in the ambient space $\mathbf{R}^3$ obey $k_g(k'_g + \tau\kappa^2 r) = 0$. It follows that γ is a non-trivial biharmonic curve if and only if $\tau = 0$ and $\kappa = \sqrt{2}/r$, i.e., γ is the circle of radius $r/\sqrt{2}$.

Caddeo, Montaldo and Piu [57] showed that a biharmonic curve on a surface of non-positive Gaussian curvature is a geodesic (i.e., harmonic, by Theorem 4.1.2) and gave examples of biharmonic non-harmonic curves on spheres, ellipses, unduloids and nodoids.

4.2 Riemannian Immersions and Submersions

4.2.1 *Curves of the Heisenberg Group H_3*

The Heisenberg group H_3 can be described as the Euclidean space $\mathbf{R}^3$ endowed with the multiplication

$$(\tilde{x}, \tilde{y}, \tilde{z})(x, y, z) = (\tilde{x} + x, \tilde{y} + y, \tilde{z} + z + \frac{1}{2}\tilde{x}y - \frac{1}{2}\tilde{y}x). \tag{4.10}$$

It is equipped with the left-invariant Riemannian metric

$$g = dx^2 + dy^2 + (dz + \frac{y}{2}dx - \frac{x}{2}dy)^2. \tag{4.11}$$

Let $\gamma : I \to H_3$ be a differentiable curve parameterized by arc length. It follows from (4.8) that γ is a non-trivial biharmonic curve if and only if

$$k = constant \neq 0,\ k^2 + \tau^2 = \frac{1}{4} - B_3^2,\ \tau' = N_3 B_3 \tag{4.12}$$

where $T = T_1e_1 + T_2e_2 + T_3e_3$, $N = N_1e_1 + N_2e_2 + N_3e_3$, $B = T \times N = B_1e_1 + B_2e_2 + B_3e_3$ and $\{e_1, e_2, e_3\}$ is the left-invariant orthonormal basis with respect to the metric (4.11).

Similarly to curves in $\mathbf{R}^3$, a *helix* is a curve in Riemannian manifold with constant geodesic curvature and constant torsion. Applying (4.12), Caddeo, Montaldo and Piu [57] showed that a non-trivial biharmonic curve in H_3 is a helix as follows.

Theorem 4.2.1. *The parametric equations of all non-trivial biharmonic curves of H_3 are*

$$x(t) = \frac{1}{K}\sin\alpha_0 \sin(Kt + a) + b,$$

$$y(t) = -\frac{1}{K}\sin\alpha_0 \cos(Kt + a) + c,$$

$$z(t) = \Big(\cos\alpha_0 + \frac{(\sin\alpha_0)^2}{2K}\Big)t - \frac{b}{2K}\sin\alpha_0 \cos(Kt + a) - \frac{c}{2K}\sin\alpha_0 \sin(Kt + a) + d,$$

where $2K = \cos\alpha_0 \pm \sqrt{(5\cos\alpha_0)^2 - 4}$, $\alpha_0 \in (0, \arccos(\frac{2\sqrt{5}}{5})] \cup [\arccos(\frac{-2\sqrt{5}}{5}), \pi)$ *and* $a, b, c, d \in \mathbf{R}$.

In fact, non-trivial biharmonic curves in H_3 can be obtained by intersecting a minimal helicoid and a round cylinder. Furthermore, they are geodesics of this round cylinder.

We can generalize the above technique to construct biharmonic curves in Cartan-Vranceanu 3-manifolds $(N^3, ds^2_{m,l})$, where $N = \mathbf{R}^3$ if $m \geq 0$, $N = \{(x,y,z) \in \mathbf{R}^3 : x^2 + y^2 < -\frac{1}{m}\}$ if $m < 0$ and the Riemannian metric $ds^2_{m,l}$ is given by

$$ds^2_{m,l} = \frac{dx^2 + dy^2}{[1 + m(x^2 + y^2)]^2} + \left(dz + \frac{l}{2}\frac{ydx - xdy}{[1 + m(x^2 + y^2)]}\right)^2, \quad l, m \in \mathbf{R} \tag{4.13}$$

This two-parameter family of metric reduces to the Heisenberg metric if $m = 0$ and $l = 1$. The system for non-trivial biharmonic curves corresponding to the metric $ds^2_{m,l}$ can be derived by using the same method and reads

$$\begin{aligned} k &= constant \neq 0, \\ k^2 + r^2 &= \frac{l^2}{4} - (l^2 - 4m)B_3^2, \\ \tau' &= (l^2 - 4m)N_3 B_3. \end{aligned} \tag{4.14}$$

It follows from (4.14) that non-trivial biharmonic curves of $(N, ds^2_{m,l})$ are helices. The explicit forms of biharmonic curves of $(N, ds^2_{m,l})$ were given by Cho, Inoguchi, and Lee [89] for $l = 1$, and by Caddeo, Montaldo, and Piu [57] in the general case.

Fetcu [143] studied biharmonic curves in the $(2n + 1)$-dimensional Heisenberg group H_{2n+1} and obtained two families of non-trivial biharmonic curves. Balmuç [26] studied biharmonic curves on Berger spheres S^3_ϵ and provided explicit forms.

4.2.2 Biharmonic Submanifolds

Theorem 4.2.2 ([196]). *Let $f : M^m \to S^{m+1}(1)$ be an isometric embedding from an m-dimensional compact Riemannian manifold M into an (m+1)-dimensional unit sphere $S^{m+1}(1)$ with non-zero parallel mean curvature (i.e., the norm of the mean curvature H is constant). The map f is biharmonic if and only if $||B(f)||^2 = m$, where $B(f)$ is the second fundamental form of f.*

Example 1. In $S^{m+1}(1)$, compact hypersurfaces whose Gauss maps are isometric embeddings are the Clifford surfaces [426]:

$$M_k^m(1) = S^k(1/\sqrt{2}) \times S^{m-k}(1/\sqrt{2}), \quad 0 \leq k \leq m. \tag{4.15}$$

Let $f : M_k^m(1) \to S^{m+1}(1)$ be a standard embedding such that $k \neq m/2$. Because $||B(f)||^2 = k + m - k = m$ and $\tau(f) = k - (m - k) = 2k - m \neq 0$, f is a biharmonic non-harmonic map, by Theorem 4.2.2.

Example 2. Let m_1, m_2 be two positive integers such that $m = m_1 + m_2$ and let r_1, r_2 be two positive real numbers such that $r_1^2 + r_2^2 = 1$. Then the generalized Clifford torus $S^{m_1}(r_1) \times S^{m_2}(r_2)$ is a hypersurface of S^{m+1}. A simple calculation shows that

$$|H| = \frac{1}{m r_1 r_2}|m_2 r_1^2 - m_1 r_2^2|, \quad |B|^2 = m_1\left(\frac{r_2}{r_1}\right)^2 + m_2\left(\frac{r_1}{r_2}\right)^2.$$

Then we have

(a) If $m_1 \neq m_2$, then $S^{m_1}(r_1) \times S^{m_2}(r_2)$ is a non-trivial biharmonic submanifold of S^{m+1} if and only if $r_1 = r_2 = \frac{1}{\sqrt{2}}$.
(b) If $m_1 = m_2 = q$, then the following three statements are equivalent: (i) $S^q(r_1) \times S^q(r_2)$ is a biharmonic submanifold of S^{2q+1}; (ii) $S^q(r_1) \times S^q(r_2)$ is a minimal submanifold of S^{2q+1}; (iii) $r_1 = r_2 = \frac{1}{\sqrt{2}}$.

B. Y. Chen [66, 68] proposed a conjecture: *Any biharmonic submanifold in a Euclidean space is minimal.* This conjecture has been proved true in the following cases: (1) biharmonic surfaces in $\mathbf{R}^3$ by Jiang [196], and by Chen and Ishikawa [69], independently; (2) biharmonic hypersurfaces in $\mathbf{R}^4$ by Hasanis and Vlachos [176]; (3) any biharmonic curve, any finite biharmonic submanifold of finite type, any pseudo-umbilical biharmonic submanifold $M^m \subset \mathbf{R}^n$ with $m \neq 4$, and any biharmonic hypersurface in $\mathbf{R}^n$ with at most two distinct principal curvatures, by Dimitric [97].

Caddeo, Montaldo, and Oniciuc [56] showed that any biharmonic submanifold in hyperbolic 3-space $H^3(-1)$ is minimal, and any pseudo-umbilical biharmonic submanifold $M^m \subset H^n$ with $m \neq 4$ is minimal. Balmuç, Montaldo and Oniciuc [25] proved that any biharmonic hypersurface of H^n with at most two distinct principal curvatures is minimal. Therefore, Caddeo, Montaldo and Oniciuc proposed the following generalized Chen's conjecture: *Any biharmonic submanifold of a Riemannian manifold (N, h) with* $Riem^N \leq 0$ *is minimal.* For the study of this conjecture, please refer to [24, 26, 54–56]. However, Ou and Tang [290] recently showed that the generalized Chen's conjecture is false.

The Sasakian space form is a generalization of the concept of a Riemannian manifold with constant sectional curvature. (N, η, ξ, ϕ, g) is a *contact Riemannian manifold* if N is a $(2n+1)$-dimensional manifold, η is an one-form such that $(d\eta)^n \wedge \eta \neq 0$, ξ is the vector field defined by $\eta(\xi) = 1$ and $d\eta(\xi, \cdot) = 0$, ϕ is an endomorphism field, g is a Riemann metric on N such that for $X, Y \in C(TN)$ (a) $\phi^2 = -I + \eta \oplus \xi$, (b) $g(\phi X, \phi Y) = g(X, Y) - \eta(X)\eta(Y), g(\xi, \cdot) = \eta$, (c) $d\eta(X, Y) = 2g(X, \phi Y)$.

A contact Riemannian manifold (N, η, ξ, ϕ, g) is a *Sasaki manifold* if

$$(\nabla_X \phi)(Y) = g(X, Y)\xi - \eta(Y)(X).$$

If the sectional curvature is constant on all ϕ-invariant tangent 2-planes of N, then we say that N is of *constant holomorphic sectional curvature*. Furthermore, if a Sasakian manifold N is connected, complete and with constant holomorphic sectional curvature, then it is called a *Sasakian space form.*

Theorem 4.2.3 ([30]). *Any simply connected 3-dimensional Sasakian space form is isomorphic to one of the following:*

(i) *The special unitary group* $SU(2)$;
(ii) *The Heisenberg group* H_3;
(iii) *The universal covering group of* $SL_2(\mathbf{R})$.

In particular, any simply connected three-dimensional Sasakian space form of constant holomorphic sectional curvature 1 is isomorphic to S^3.

Recall that a curve $\gamma : I \to N$ parameterized by arc length is *Legendre* if $\eta(\gamma') = 0$; also, a *Hopf cylinder* is a set of the form $S_{\bar{\gamma}} = p^{-1}(\bar{\gamma})$, where $p : N \to \bar{N} = N/G$ is the projection of N onto the orbit space $\bar{N}$ of the action of the one-parameter group of isometries generated by ξ when the action is transitive. Inoguchi [193] classified non-trivial biharmonic Legendre curves and Hopf cylinder in three-dimensional Sasakian space forms, as follows.

Theorem 4.2.4. *Let* $N^3(\epsilon)$ *be a Sasakian space form of constant holomorphic sectional curvature* ϵ *and* $\gamma : I \to N$ *a biharmonic Legendre curve parameterized by arc-length.*

(i) *If* $\epsilon \leq 1$, *then* γ *is a Legendre geodesic.*
(ii) *If* $\epsilon > 1$, *then* γ *is a Legendre geodesic or a Legendre helix of curvature* $\sqrt{\epsilon - 1}$.

Theorem 4.2.5. *Let* $S_{\bar{\gamma}} \subset N^3(\epsilon)$ *be a biharmonic Hopf cylinder in a Sasakian space form.*

(i) *If* $\epsilon \leq 1$, *then* $\bar{\gamma}$ *is a Legendre geodesic.*
(ii) *If* $\epsilon > 1$, *then* $\bar{\gamma}$ *is a geodesic or a Riemannian circle of curvature* $\bar{k} = \sqrt{\epsilon - 1}$.

In particular, there exists non-trivial biharmonic Hopf cylinders in Sasakian forms of holomorphic sectional curvature greater than 1.

Sasahara [317, 318] classified non-trivial biharmonic Legendre surfaces in Sasakian space forms when the ambient space is the unit five-dimensional sphere S^5, and obtained the following result.

Theorem 4.2.6 ([317]). *If* $f : M^2 \to S^5$ *is a non-trivial biharmonic Legendre immersion, then the position vector field* $x_0 = x_0(u, v)$ *of* M *in* $\mathbf{R}^6$ *is given by*

$$x_0(u, v) = \frac{1}{\sqrt{2}}(\cos u,\ \sin u\ \sin(\sqrt{2}v), -\sin u\ \cos(\sqrt{2}v),\ \sin u,\ \cos u\ \sin(\sqrt{2}v), -\cos u\ \cos(\sqrt{2}v)).$$

For more results on biharmonic Legrendre curves and biharmonic anti-invariant surfaces in Sasakian space forms and (k, μ)-manifolds, see [11, 12].

Let $\mathbf{C}P^m$ be a complex projective space of holomorphic sectional curvature 4 of complex dimension m, and M be an m-dimensional totally real submanifold. If the isometric immersion of M is biharmonic, then M is called an m-dimensional totally real biharmonic submanifold immersed in $\mathbf{C}P^m$. Based on [197], the notion of totally real biharmonic submanifold immersed in $\mathbf{C}P^m$ is an extension of the notion of totally real minimal submanifold immersed in $\mathbf{C}P^m$ (cf. [336]). Let σ be the second fundamental form of M, S be the square norm of σ, and H be the norm of the mean curvature vector $\eta = (1/m) trace\, \sigma$. If $H \neq 0$, set $\sigma_H = < \sigma(X, Y), \eta/H >$ for X, $Y \in T_x M$; σ_H is called the second fundamental form of M with respect to η. Denote the square norm of σ_H by S_H. The following theorems were obtained by Chiang and Sun [80], and generalized the results of A. M. Li, J. M. Li, and Y. B. Shen [251, 336].

Theorem 4.2.7 ([80]). *Let M be an m-dimensional totally real biharmonic submanifold immersed in CP^m. If $S_H \geq m + 3$, then either M is minimal, or M has parallel mean curvature vector.*

Theorem 4.2.8 ([80]). *Let M be an m-dimensional totally real biharmonic submanifold with parallel mean curvature immersed in $\mathbf{C}P^m$. Then either M is minimal, or $S_H = m + 3$.*

Theorem 4.2.9 ([80]). *Let M be an m-dimensional totally real biharmonic submanifold with parallel mean curvature immersed in $\mathbf{C}P^m$. If the Ricci curvature of M is not less than $(m - 2 - (3/m) + mH^2)$, then M is minimal.*

4.2.3 *Riemannian Submersions*

Let $f : (M, g) \to (N, h)$ be a Riemannian submersions with basic tension field τ (i.e., $\tau(f) = W \circ f$ for $W \in \Gamma(TN)$). The bitension field was computed by Onicius [280] as follows:

$$\tau_2(f) = trace^N (\nabla)^2 \tau(f) + \nabla^N_{\tau(f)} \tau(f) + Ricc^N \tau(f). \tag{4.16}$$

He obtained the following three propositions.

Proposition 4.2.10. *A biharmonic Riemannian submersion $f : M \to N$ with basic tension field is harmonic in the following cases:*

(i) *If M is compact, orientable and $Ricc^N \leq 0$;*
(ii) *If $Ricc^N < 0$ and $|\tau(f)|$ is constant;*
(iii) *If N is compact and $Ricc^N < 0$.*

Proposition 4.2.11. *Let $f : M \to N$ be a biharmonic Riemannian submersion with non-zero basic tension field. Then f is a non-trivial biharmonic map if*

(i) $\nabla^N \tau(f) = 0$;
(ii) *$\tau(f)$ is a unit Killing vector field on N.*

Let (M, g) be an m-dimensional Riemannian manifold and let $\pi : TM \to M$ be the tangent bundle projection. Let $V(TM)$ be the vertical distribution on TM, given by $V_v(TM) = ker\, d\pi_v$, where $v \in TM$. We consider the nonlinear connection on TM defined by the distribution $H(TM)$ on TM, complementary to $V(TM)$, i.e., $H_v(TM) \oplus V_v(TM) = T_v(TM)$, $v \in TM$. For any induced local chart $(\pi^{-1}(U); x^i, y^j)$ on TM we have a local adapted frame in $H(TM)$ defined by the local vector fields

$$\frac{\delta}{\delta x^i} = \frac{\partial}{\partial x^i} - N_i^j(x, y)\frac{\partial}{\partial y^j}, \quad i = 1, 2, \cdots, m,$$

where the local functions $N_j^i(x, y)$ are the connection coefficients of the nonlinear connection given by $H(TM)$. If we endow TM with the Riemannian metric S defined by

$$S(X^V, Y^V) = S(X^H, Y^H) = g(X, Y), \quad S(X^V, Y^H) = 0,$$

then the canonical projection $\pi : (TM, S) \to (M, g)$ is a Riemannian submersion (cf. [282]).

Proposition 4.2.12 ([280, 282]).

(i) *Let ξ be a unit Killing vector field and let $N_j^i = (\Gamma^i_{jk} + \delta^i_j\xi_k + \delta^i_k\xi_j)y^k$ be a projective change of the Levi-Civita connection D on (M,g). Then π is a non-trivial biharmonic map.*

(ii) *Let $\rho \in C^\infty(M)$, $\rho \neq$ constant, be an affine function and let $N_j^i = (\Gamma^i_{jk} + \delta^i_j\alpha_k + \delta^i_k\alpha_j - g_{jk}\alpha^i)y^k$, $\alpha_k = \frac{\partial\rho}{\partial x^k}$, be a conformal change of the connection ∇. Then π is a non-trivial biharmonic map.*

Jiang [196] and Chen-Ishikawa [69] independently showed that *an isometric immersion $(M^2, g) \hookrightarrow \mathbf{R}^3$ into an Euclidean space is biharmonic if and only if it is harmonic*. Wang and Ou obtained the following dual theorem about a Riemannian submersion recently.

Theorem 4.2.13 ([400]). *Let $p : (M^3(c), g) \to (N^2, h)$ be a Riemannian submersion from a space form of constant sectional curvature c. Then p is biharmonic if and only if it is harmonic (similarly, p is a biharmonic morphism if and only if it is a harmonic morphism).*

Ou [288] followed a few methods of Eells and Lemaire [119] to construct many examples of non-trivial biharmonic maps including biharmonic tori of any dimension in Euclidean spheres, biharmonic maps between spheres and into spheres via orthogonal multiplication and eigenmaps, etc.

4.3 Conformally Biharmonic Immersions, Morphisms and Second Variation

4.3.1 Conformal Changes and Conformally Biharmonic Immersions

Let $f : (M^m, g) \to (N^n, h)$ be a harmonic map and let $\tilde{g} = e^{2\rho} g$ for some smooth function ρ. If $m = 2$, we know that the energy of f is conformally invariant, i.e., $f : (M, \tilde{g}) \to (N, h)$ is still harmonic. However, if $m \neq 2$, then f is not necessarily harmonic. The following proposition on biharmonic maps was obtained by Baird and Kamissoko.

Proposition 4.3.1 ([19]). *Let $f : (M^m, g) \to (N^n, h)$, $m \neq 2$, be a harmonic map, and $\tilde{g} = e^{2\rho} g$ be a metric conformally equivalent to g. Then $f : (M, \tilde{g}) \to (N, h)$ is biharmonic if and only if*

$$-\Delta df(grad\,\rho) + (m-6)\nabla_{grad\,\rho} df(grad\,\rho) + 2(\Delta\rho - (m-4)|d\rho|^2) df(grad\,\rho)$$
$$+ trace R^N(df(grad\,\rho), df)\, df = 0.$$

When $f : (M, g) \to (M, g)$ is the identity map *id*, we say that a conformally equivalent metric $\tilde{g} = e^{2\rho} g$ for which *id* turns out to be biharmonic, is a *biharmonic metric* with respect to g. The following theorem was obtained using the maximum principle by Baird and Kamissoko.

Theorem 4.3.2 ([19]). *Let (M^m, g) $(m \neq 2)$ be a compact manifold with negative Ricci curvature. Then there is no biharmonic metric conformally related to g other than a constant multiple of g.*

There is a relationship between biharmonic metrics and isoparametric functions. A smooth function $\phi : M \to \mathbf{R}$ is *isoparametric* if for each $x \in M$ with *grad* $\phi_x \neq 0$, there are real functions λ and σ such that

$$|d\phi|^2 = \lambda \circ \phi, \quad \Delta\phi = \sigma \circ \phi,$$

in some neighborhood of x.

Theorem 4.3.3 ([19]). *Let (M^m, g) $(m \neq 2)$ be an Einstein manifold and $\tilde{g} = e^{2\rho} g$ be a biharmonic metric conformally equivalent to g. Then the function $\rho : M \to \mathbf{R}$ is isoparametric. Conversely, if $\phi : M \to \mathbf{R}$ is an isoparametric function, then there is a reparameterization $\rho = \rho \circ \phi$ away from critical points of ϕ such that $\tilde{g} = e^{2\rho} g$ is a biharmonic metric.*

Likewise, [24] studied conformal changes on the codomain and proved a result similar to Theorem 4.3.3. It is well known that minimal surfaces in $\mathbf{R}^3$ are equivalent to confomally harmonic immersions of $M^2 \to \mathbf{R}^3$. Ou [286,287] studied confomally biharmonic immersions and obtained the following results.

Proposition 4.3.4. *Let $f : (M^m, g) \to (N^n, h)$ be a conformal immersion with $f^*h = \lambda^2 g$, and $f : (M^m, \bar{g}) \to (N^n, h)$ be the associated isometric immersion with mean curvature vector η and $\bar{g} = f^*h = \lambda^2 g$. Then the conformal immersion $f : (M^m, g) \to (N^n, h)$ is biharmonic if and only if*

$$\lambda^4 \tau_2(f, \bar{g}) = -(m-2) J_g^f (df(grad\, ln\lambda)) + 2m\lambda^2(-\Delta ln\lambda - 2|grad\, ln\lambda|^2)\eta$$
$$+ m(m-6)\lambda^2 \nabla^f_{grad\, ln\lambda} \eta,$$

where J_g^f is the Jacobi field of f with respect to g.

Theorem 4.3.5. *A conformal immersion $f : (M^2, g) \to (\mathbf{R}^3, h_0)$ (h_0 is a standard metric and $f^*h_0 = \lambda^2 g$) is biharmonic if and only if*

$$A_\xi(grad\, H) + \frac{1}{2} grad(H^2) + 2H\, A_\xi(grad\, ln\lambda) = 0,$$
$$\Delta H - H\,|B|^2 + 2H(\Delta ln\lambda + 2|grad\, ln\lambda|^2) + 4g(grad\, ln\lambda,\, grad\, H) = 0,$$

where ξ is the unit normal vector field of the surface $f(M) \subset \mathbf{R}^3$, and A_ξ and H are the shape operator and the mean curvature of the surface $f(M) \subset \mathbf{R}^3$, respectively.

Theorem 4.3.6. *A conformal immersion $f : (M^2, g) \to (N^3, h)$ into a 3-dimensional Riemannian manifold (with $f^*h = \lambda^2 g$) is biharmonic if and only if*

$$\Delta H - H\left[|A|^2 - Ric^N(\xi, \xi) - \lambda^{-2}\Delta(\lambda^2)\right] + 4g(grad\, ln\lambda,\, grad\, H) = 0,$$
$$A(grad H) + H\left[grad H - (Ric^N(\xi))^T + 2A(grad\, ln(\lambda)\right] = 0,$$

where ξ, A and H are the unit normal vector field, the shape operator, and the mean curvature of the surface $f(M) \subset (N^3, h)$, respectively, and the operators Δ, grad, $|\ |$ are given with respect to the induced metric $\bar{g} = f^*h = \lambda^2 g$ on the surface.

4.3.2 Biharmonic Morphisms

A map $f : (M, g) \to (N, h)$ is a *biharmonic morphism* if for any biharmonic function $\phi : U \subset N \to \mathbf{R}$, its pull-back by f, $\phi \circ f : f^{-1}(U) \subset M \to \mathbf{R}$ is a biharmonic function. Loubeau and Ou [256] gave the following characterization of the biharmonic morphisms: a map is a biharmonic morphism if and only if it is a horizontally weakly conformal biharmonic map and its dilation factor satisfies some condition. They obtained the following results.

Theorem 4.3.7. *A map $f : (M, g) \to (N, h)$ is a biharmonic morphism if and only if there exists a function $\lambda : M \to \mathbf{R}$ such that*

$$\Delta^2(\phi \circ f) = \lambda^4 \Delta^2(\phi) \circ f$$

for all functions $\phi : U \subset N \to \mathbf{R}$.

If M is compact, biharmonic morphisms have a simple structure.

Theorem 4.3.8 ([256]). *Let $f : (M, g) \to (N, h)$ be a non-constant map. If M is compact, then f is a biharmonic morphism if and only if it is a harmonic morphism of constant dilation, thus a homothetic submersion with minimal fibres.*

Loubeau and Ou [254,256] using the theory of p-harmonic morphisms, obtained the following results.

Theorem 4.3.9. *The radial projection of $f : \mathbf{R}^m - \{0\} \to S^{m-1}$, $f(x) = \frac{x}{|x|}$ is a biharmonic morphism if and only if $m = 4$.*

Theorem 4.3.10. *The projection of $f : M \times_\eta N \to (N, h)$, $f(x, y) = y$ of a warped product onto its second factor is a biharmonic morphism if and only if $\frac{1}{\eta}$ is a harmonic function on M.*

Theorem 4.3.11. *Let $f : \mathbf{R}^m \to \mathbf{R}^n$ be a polynomial biharmonic morphism, i.e., a biharmonic morphism whose component functions are polynomials for $m > n \geq 2$. Then f is an orthogonal projection followed by a homothety.*

Theorem 4.3.12 ([254]). *Let $f : (M^m, g) \to (N^n, h)$ be a smooth map between Riemannian manifolds. Then f is a biharmonic morphism iff it is a horizontally weakly conformal biharmonic 4-harmonic map of dilation λ such that*

$$|\tau(f)|^4 - 2\Delta\lambda^2|\tau(f)|^2 + 4\Delta\lambda^2 div(df, \tau(f)) + n(\Delta\lambda^2)^2$$
$$+ 2(df, \tau(f))\Delta|\tau(f)|^2 + |S|^2 = 0,$$

where $S \in \odot^2 f^{-1}TN$ is the symmetrization of the g-trace of $df \otimes \nabla^f \tau(f)$ and $(df, \tau(f))X = (df(X), \tau(f))$.

4.3.3 Second Variation

Let $f : M \to N$ be a biharmonic maps from a compact Riemannian manifold M into a Riemannian manifold N, and $\{f_t\} \in C^\infty(M \times (-\epsilon, \epsilon), N)$ be a one-parameter family of maps such that $f_0 = f$. We assume that the compact supports of $\frac{\partial f_t}{\partial t}$ and $\nabla_{e_i} \frac{\partial f_t}{\partial t}$ are contained in the interior of M, where $\{e_i\}$ is a local orthonormal frame in M. Let $V \in \Gamma(f^{-1}TN)$ be a vector field such that $\frac{\partial f_t}{\partial t}\big|_{t=0} = V$ on M. Jiang [196] computed the second variation of bienergy from (4.3) and obtained the following result.

Theorem 4.3.13. *If $f : M \to N$ is a biharmonic map, then*

$$\frac{1}{2}\frac{d^2}{dt^2}E_2(f_t)\Big|_{t=0} = \int_M ||\bar{D}^*\bar{D}V + R'(df(e_i),V)df(e_i)||^2 dv$$
$$+ \int_M < V, (D'_{df(e_i)}R')(df(e_i),\tau(f))V + (D'_{\tau(f)}R')(df(e_i),V)df(e_i)$$
$$+ R'(\tau(f),V)\tau(f) + 2R'(df(e_i),V)\nabla_{e_i}\tau(f)$$
$$+ 2R'(df(e_i),\tau(f))D_{e_i}V > dv, \tag{4.17}$$

where $\bar{D}^*\bar{D} = \bar{D}_{e_i}\bar{D}_{e_i} - \bar{D}_{D_{e_i}e_i}$, D, $\bar{D}$ are the connections of $T(M \times (-\epsilon,\epsilon))$ and $f_t^{-1}TN$, D' is the Levi-Civita connection of TN, and R' is the Riemannian curvature of N.

Applying the above theorem, we can find the Hessian as follows and discuss the index and nullity of f.

Lemma 4.3.14. *Let $f : M \to S^n(1)$ be a biharmonic map. The Hessian of the bienergy functional E_2 of f is*

$$H(E_2)_f(X,Y) = \int_M (I_f(X,Y))dv,$$

where

$$I_f(X) = \Delta^f(\Delta^f X) + \Delta^f(trace(X, df\cdot)df\cdot - |df|^2 X) + 2(d\tau(f), df)X$$
$$+ |\tau(f)|^2 X - 2\,trace(X, d\tau(f)\cdot)df - 2trace(\tau(f), dX\cdot)df\cdot$$
$$- (\tau(f),X)\tau(f) + trace(df\cdot, \Delta^f X)df\cdot + trace(df,\, trace(X, df\cdot)df\cdot)df\cdot$$
$$- 2|df|^2 trace(df\cdot, X)df\cdot + 2(dX, df)\tau(f) - |df|^2\Delta^f X + |df|^4 X,$$

for $X, Y \in \Gamma(f^{-1}TS^n(1))$.

If f is the identity map $id : S^m \to S^m$, then

$$I^{id}(X) = \Delta(\Delta X) - 2(m-1)\Delta X + (m-1)^2 X.$$

The following results were obtained by Loubeau and Oniciuc [255, 281] and see [78].

Proposition 4.3.15. *The identity $id : S^m \to S^m$ is biharmonic stable and*

(i) *If $m = 2$, then nullity $(id) = 6$;*
(ii) *If $m > 2$, then nullity $(id) = \frac{m(m-1)}{2}$.*

Proposition 4.3.16. *The biharmonic index of the canonical inclusion $i : S^{m-1}(\frac{1}{\sqrt{2}}) \to S^m$ is 1 and its nullity is $\frac{m(m-1)}{2} + m$.*

Proposition 4.3.17. *Let $\psi : S^m(r) \to S^{n-1}(\frac{1}{\sqrt{2}})$ be a minimal immersion for $r \geq \frac{1}{\sqrt{2}}$, and $\phi = i \circ \psi : S^m(r) \to S^{n-1}(\frac{1}{\sqrt{2}}) \to S^n$. Then*

(i) *$Index(\phi) \geq m + 2$ if either $r^2 > \frac{1+\sqrt{m^2+1}}{2m}$ or $m \geq 5$ and $r^2 \geq \frac{(m-2)^2}{2m(m-4)}$;*
(ii) *$Index(\phi) \geq 2m + 3$ if $m \geq 5$ and $r^2 > \frac{(m-2)(1+\sqrt{m^2-4m+1})}{2m(m-4)}$.*

Proposition 4.3.18. *The biharmonic map derived from the generalized Veronese map $\psi : S^m(\sqrt{\frac{m+1}{m}}) \to S^{m+p}(\frac{1}{\sqrt{2}})$, $p = \frac{(m-1)(m+2)}{2}$ has index at least $m + 2$, if $m \leq 4$, and at least $2m + 3$, if $m > 4$.*

4.4 Biharmonic Homogeneous Real Hypersurfaces

In this section, we classify all the biharmonic homogeneous real hypersurfaces in the complex projective space $\mathbf{C}P^n(4)$ with positive constant holomorphic sectional curvature 4, and all the biharmonic homogeneous real hypersurfaces in the quarternionic projective space $\mathbf{H}P^n(4)$.

4.4.1 Hypersurfaces in a Complex Projective Space

We first can apply the arguments similar to those used in Theorem 4.2.2 by Jiang [196] and obtain the following result.

Theorem 4.4.1. *Let (M, g) be a real $(2n-1)$-dimensional compact Riemannian manifold and $f : (M, g) \to \mathbf{C}P^n(c)$ be an isometric immersion with non-zero constant mean curvature. Then f is biharmonic if and only if $||B(f)||^2 = \frac{n+1}{2}c$, where $B(f)$ is the second fundamental form of f.*

Next we present the classification of all the homogeneous real hypersurfaces in $\mathbf{C}P^n(c)$ by Takagi [357] based on the work of Hsiang and Lawson [190]. Let U/K be a symmetric space of rank two of compact type, and $u = \mathcal{K} + \eta$, the Cartan decomposition of the Lie algebra u of U, and the Lie subalgebra $\mathcal{K}$ corresponding to K. Let $< X, Y > = -B(X, Y)$ $(X, Y \in \eta)$ be the inner product on η, $||X||^2 = < X, X >$ and $S = \{X \in \eta : ||X|| = 1\}$, the unit sphere in the Euclidean space $(\eta, < \cdot, \cdot >)$, where B is the Killing form of u. Taking the adjoint action of K on η, the orbit $\tilde{M} = Ad(K)/A$ through any regular element $A \in \eta$ with $||A|| = 1$, yields a homogeneous hypersurface in the unit sphere S. Conversely, any homogeneous hypersurface in S can be given in this way [190].

Let U/K be a Hermitian symmetric space of compact type of rank two of complex dimension $n + 1$, and identify η with $\mathbf{C}^{n+1}$. Thus the adjoint orbit $\tilde{M} = Ad(K)A$ of K through any regular element A in η is again a homogeneous hypersurface in the unit sphere S. Let $\pi : \mathbf{C}^{n+1} \to \{0\}$ $(= \eta - \{0\}) \to \mathbf{C}P^n$ be the natural projection. Then the projection induces the Hopf fibration of S onto $\mathbf{C}P^n$, still denoted by π, and $f : M = \pi(\tilde{M}) \hookrightarrow \mathbf{C}P^n$ provides a homogeneous real hypersurface in the complex projective space $\mathbf{C}P^n(4)$ with constant holomorphic sectional curvature 4. Conversely, any homogeneous [357] real hypersurface M in $\mathbf{C}P^n(4)$ is obtained in this way [357]. Moreover, we can classify all the biharmonic homogeneous real spaces in the complex projective spaces as follows.

1. Type I: $u = su(p+2) \oplus su(q+2)$, $\mathcal{K} = s(u(p+1)+u(1)) \oplus s(u(q+1)+u(1))$, where $0 \leq p \leq q$, $0 < q$, $p+q = n-1$ and $dim(M) = 2n-1$.
2. Type II: $u = o(m+2)$, $\mathcal{K} = o(m) \oplus \mathbf{R}$, where $3 \leq m$ and $dim(M) = 2m-3$.
3. Type III: $u = su(m+2)$, $\mathcal{K} = s(o(m)+o(2))$, where $3 \leq m$ and $dim(M) = 4m-3$.
4. Type IV: $u = o(10)$, $\mathcal{K} = u(5)$ and $dim(M) = 17$.
5. Type V: $u = e_6$, $\mathcal{K} = o(10) \oplus \mathbf{R}$ and $dim(M) = 29$.

Takagi [358, 359] provided a list of the principal curvatures and their multiplicities of these spaces as follows:

1. Type I: Let

$$U/K = \frac{SU(p+2) \times SU(q+2)}{S(U(p+1) \times U(1)) \times S(U(q+1) \times U(1))}.$$

Then the adjoint orbit of K, $Ad(K)A$ is given by the Riemannian product of two odd-dimensional spheres,

$$\tilde{M} = \tilde{M}_{p,q} = S^{2p+1}(\cos u) \times S^{2n+1}(\sin u) \subset S^{2n+1}, \tag{4.18}$$

where $0 < u < \frac{\pi}{2}$. The projection $M_{p,q} = \pi(\tilde{M}_{p,q}(u))$ is a homogeneous real hypersurface of $\mathbf{C}P^n(4)$. The principal curvatures of $M_{p,q}$ with $0 \leq p \leq q$, $0 < q$ are

$$\begin{aligned}
\lambda_1 &= -\tan u \quad \text{(with multiplicity } m_1 = 2p,\ m_1 = 0 \text{ if } p = 0),\\
\lambda_2 &= \cot u \quad \text{(with multiplicity } m_2 = 2q),\\
\lambda_3 &= 2\cot u \quad \text{(with multiplicity } m_3 = 1).
\end{aligned} \tag{4.19}$$

Therefore, the mean curvature H of $M_{p,q}(u)$ is

$$\begin{aligned}
H &= \frac{1}{2n-1}[2q \cot u - 2p \tan u + 2\cot(2u)]\\
&= \frac{1}{2n-1}[(2q-1)\cot u - (2p+1)\tan u].
\end{aligned} \tag{4.20}$$

The constant $||B(f)||^2$ is the sum of the squares of all principal curvatures with their multiplicities, i.e.,

$$
\begin{aligned}
||B(f)||^2 &= 2q \cot^2 u + 2p \tan^2 u + 4 \cot^2(2u) \\
&= (2q+1) \cot^2 u + (2p+1) \tan^2 u - 2.
\end{aligned}
\tag{4.21}
$$

2. Type II: Let

$$
U/K = \frac{SO(m+2)}{SO(m) \times SO(2)} \quad (m = n+1).
$$

Then the adjoint orbit of K, $Ad(K)A$, is

$$
\tilde{M} = \{SO(n+1) \times SO(2)\}/\{SO(n-1) \times \mathbf{Z}_2\} \subset S^{2n+1}.
$$

The real hypersurface $f : M \hookrightarrow \mathbf{C}P^n$ is a tube over a complex quadric with radius $\frac{\pi}{4} - u$ $(0 < u < \frac{\pi}{4})$ or a tube over a totally geodesic real projective space $\mathbf{R}P^n$ with radius $(0 < u < \frac{\pi}{4})$. The principal curvatures of M are

$$
\begin{aligned}
\lambda_1 &= -\cot u \quad (\text{with multiplicity } m_1 = n-1), \\
\lambda_2 &= \tan u \quad (\text{with multiplicity } m_2 = n-1), \\
\lambda_3 &= 2\tan(2u) \quad (\text{with multiplicity } m_3 = 1).
\end{aligned}
\tag{4.22}
$$

Therefore, the mean curvature of M is

$$
\begin{aligned}
H &= \frac{1}{2n-1}[-(n-1)\cot u + (n-1)\tan u + 2\cot(2u)] \\
&= -\frac{1}{2n-1} \cdot \frac{(n-1)t^4 - 2(n+1)t^2 + n - 1}{t(t^2-1)},
\end{aligned}
\tag{4.23}
$$

where $t = \cot u$. Then we have the constant

$$
\begin{aligned}
||B(f)||^2 &= (n-1)\cot^2 u + n - \tan^2 u + 4\tan^2(2u) \\
&= (n-1)t^2 + \frac{n-1}{t^2} + \frac{16t^2}{(t^2-1)^2} \\
&= \frac{(n-1)(a-1)^2(a^2+1) + 16a^2}{a(a-1)^2},
\end{aligned}
\tag{4.24}
$$

where $a = t^2$.

3. Type III: Let

$$
U/K = \frac{SU(m+2)}{SU(m) \times U(2)} \quad (n = 2m+1).
$$

Then the adjoint orbit of $K, Ad(K)A$, is

$$\tilde{M} = S(U(m)(n+1) \times U(2))\}/(T^2 \times SU(m-2)) \subset S^{2n+1}.$$

The real hypersurface $f : M \hookrightarrow \mathbf{C}P^n$ is a tube over the Segre embedding of $\mathbf{C}^1 \times \mathbf{C}P^m$ with radius $(0 < u < \frac{\pi}{4})$. Hence, the principal curvatures of M are

$$\begin{aligned}
\lambda_1 &= -\cot u \quad (\text{with multiplicity } m_1 = n-3),\\
\lambda_2 &= \cot\,(\frac{\pi}{4} - u) \quad (\text{with multiplicity } m_2 = 2),\\
\lambda_3 &= \cot\,(\frac{\pi}{2} - u) \quad (\text{with multiplicity } m_3 = n-3),\\
\lambda_4 &= \cot\,(\frac{3\pi}{4} - u) \quad (\text{with multiplicity } m_4 = 2),\\
\lambda_5 &= -2\tan\,(2u) \quad (\text{with mltiplicity } m_5 = 1).
\end{aligned} \tag{4.25}$$

Thus,

$$\lambda_1 = -t,\ \lambda_2 = \frac{t+1}{t-1},\ \lambda_3 = \frac{1}{t},\ \lambda_4 = -\frac{t-1}{t+1},\ \lambda_5 = -t + \frac{1}{t},$$

where $t = \cot u$. Then the mean curvature of M is

$$\begin{aligned}
H &= \frac{1}{2n-1}\left[(n-3)(-t) + 2\frac{t+1}{t-1} + (n-3)\frac{1}{t} - 2\frac{t-1}{t+1} - t + \frac{1}{t}\right]\\
&= -\frac{(n-2)t^4 - 2(n+2)t^2 + n - 2}{t(t^2-1)}.
\end{aligned} \tag{4.26}$$

Therefore, we have the constant

$$\begin{aligned}
||B(f)||^2 &= (n-3)^2 + 2\left(\frac{t+1}{t-1}\right)^2 + (n-3)\frac{1}{t^2} + 2\left(\frac{t-1}{t+1}\right)^2 + \left(-t + \frac{1}{t}\right)^2\\
&= \frac{C(a)}{a(a-1)^2},
\end{aligned} \tag{4.27}$$

where $C(a) = (n-2)a^2(a-1)^2 + (n-2)(a-1)^2 + 4a(a^2+6a+1) - 2a(a-1)^2$ and $a = t^2$.

4. Type IV: Let

$$U/K = O(10)/U(5).$$

Then the adjoint orbit of $K, Ad(K)A$, is

$$\tilde{M} = U(5)/(SU(2) \times SU(2) \times U(1)) \subset S^{19}.$$

The real hypersurface $f : M \hookrightarrow \mathbf{C}P^n$ is a tube over the Plücker embedding of $Gr_2(\mathbf{C}^5)$ with radius u $(0 < u < \frac{\pi}{4})$. The principal curvatures of M are

$$\begin{aligned}
\lambda_1 &= -\cot u \quad (\text{with multiplicity } m_1 = 4),\\
\lambda_2 &= \cot(\frac{\pi}{4} - u) \quad (\text{with multiplicity } m_2 = 4),\\
\lambda_3 &= \cot(\frac{\pi}{2} - u) \quad (\text{with multiplicity } m_3 = 4),\\
\lambda_4 &= \cot(\frac{3\pi}{4} - u) \quad (\text{with multiplicity } m_4 = 4),\\
\lambda_5 &= -2\tan(2u) \quad (\text{with mltiplicity } m_5 = 1).
\end{aligned} \tag{4.28}$$

Thus,

$$\lambda = -t,\ \lambda_2 = \frac{t+1}{t-1},\ \lambda_3 = \frac{1}{t}, \lambda_4 = -\frac{t-1}{t+1},\ \lambda_5 = -t + \frac{1}{t},$$

where $t = \cot u$. The mean curvature of M is

$$\begin{aligned}
H &= \frac{1}{17}\left[4(-t) + 4\frac{t+1}{t-1} + \frac{4}{t} - 4\frac{t-1}{t+1} - t + \frac{1}{t}\right]\\
&= -\frac{5t^4 - 26t^2 + 5}{17t(t^2-1)} = -\frac{(5t^2-1)(t^2-5)}{17t(t^2-1)}.
\end{aligned} \tag{4.29}$$

Therefore, we have the constant

$$\begin{aligned}
\|B(f)\|^2 &= 4t^2 + 4\left(\frac{t+1}{t-1}\right)^2 + \frac{4}{t^2} + 4\left(\frac{t-1}{t+1}\right)^2 + \left(-t + \frac{1}{t}\right)^2\\
&= \frac{D(a)}{a(a-1)^2},
\end{aligned} \tag{4.30}$$

where $D(a) = 11a^3 + 63a^2 + a + 5$ and $a = t^2$.

5. Type V: Let

$$U/K = E_6/Spin(10) \times U(1).$$

Then the adjoint orbit of K, $Ad(K)A$, is

$$\tilde{M} = (Spin(10) \times (U(1))/(SU(4) \times U(1)) \subset S^{31}.$$

The real hypersurface $f : M \hookrightarrow \mathbf{C}P^{15}$ is a tube over the canonical embedding of $SO(10)/U(5) \subset \mathbf{C}P^{15}$ with radius $(0 < u < \frac{\pi}{4})$. The principal curvatures of M are given by

$$\lambda_1 = -\cot u \quad \text{(with multiplicity } m_1 = 8\text{)},$$
$$\lambda_2 = \cot\left(\frac{\pi}{4} - u\right) \quad \text{(with multiplicity } m_2 = 6\text{)},$$
$$\lambda_3 = \cot\left(\frac{\pi}{2} - u\right) \quad \text{(with multiplicity } m_3 = 8\text{)},$$
$$\lambda_4 = \cot\left(\frac{3\pi}{4} - u\right) \quad \text{(with multiplicity } m_4 = 6\text{)},$$
$$\lambda_5 = -2\tan(2u) \quad \text{(with mltiplicity } m_5 = 1\text{)}. \tag{4.31}$$

Thus,

$$\lambda_1 = -t,\ \lambda_2 = \frac{t+1}{t-1},\ \lambda_3 = \frac{1}{t},\ \lambda_4 = \frac{t-1}{t+1},\ \lambda_5 = -t + \frac{1}{t},$$

where $t = \cot u$. The mean curvature of M is

$$\begin{aligned} H &= \frac{1}{29}\left[8(-t) + 6\frac{t+1}{t-1} + \frac{8}{t} - 6\frac{t-1}{t+1} - t + \frac{1}{t}\right] \\ &= -\frac{9t^4 - 42t^2 + 9}{29t(t^2-1)}. \end{aligned} \tag{4.32}$$

Then we have the constant

$$\begin{aligned} \|B(f)\|^2 &= 8t^2 + 6\left(\frac{t+1}{t-1}\right)^2 + \frac{8}{t^2} + 6\left(\frac{t-1}{t+1}\right)^2 + \left(-t + \frac{1}{t}\right)^2 \\ &= \frac{E(a)}{a(a-1)^2} - 2, \end{aligned} \tag{4.33}$$

where $E(a) = 21a^3 + 99a^2 - 9a + 9$ and $a = t^2$.

By combining Theorem 4.4.1 and the above classifications, Ichiyama, Inoguchi and Urakawa [191, 192] obtained the following result.

Theorem 4.4.2. *Let M be any homogeneous real hypersurface in* $\mathbf{C}P^n(4)$ *so that M is a tube of any of the types I–V.*

A. *For each type, there is a unique u with* $0 < u < \frac{\pi}{4}$ *such that M is a tube of radius u and is minimal.*
B. *Suppose that M is biharmonic, but not minimal. Then M is of type I, IV or V as follows:*

   1. *Type I: M is a tube* $M_{p,q}(u)$ *of* $\mathbf{C}P^p \subset \mathbf{C}P^n (p \geq 0$ *and* $q = (n-1) - p)$ *of radius u with* $0 < u < \frac{\pi}{2}$ *such that* $t = \cot u$ *is a solution of the equation*

$$\cot u = \left\{\frac{p+q+3 \pm \sqrt{(p-q)^2 + 4(p+q+2)}}{1+2q}\right\}^{1/2}. \tag{4.34}$$

2. *Type IV: M is a tube of over the Plücker embbeding* $Gr_2(\mathbf{C}^5) \subset \mathbf{C}P^9$ *with radius* u $(0 < u < \frac{\pi}{4})$ *such that* $t = \cot u$ *is the unique solution of the equation* $41t^6 + 43t^4 + 41t^2 - 15 = 0$, *i.e.,* $u = 1.0917\cdots$.
3. *Type V: M is a tube of the embeddig* $SO(10)/U(5) \subset \mathbf{C}P^{15}$ *of radius* u *with* $0 < u < \frac{\pi}{4}$ *such that* $t = \cot u$ *is the unique solution of the equation* $13t^6 - 107t^4 + 43t^2 - 9 = 0$, *i.e.,* $u = 0.343448\cdots$.

4.4.2 Hypersurfaces in a Quarternionic Projective Space

We can apply arguments similar to those in Theorem 4.2.2, and obtain the following result.

Theorem 4.4.3. *Let* (M, g) *be a real* $(4n - 1)$*-dimensional compact Riemannian manifold and* $f : (M, g) \to \mathbf{H}P^n(c)$ *be an isometric immersion into a quarternion space with non-zero constant mean curvature* $(n \geq 2)$. *Then f is biharmonic if and only if*

$$||B(f)||^2 = (n + 2)c, \tag{4.35}$$

where B(f) is the second fundamental form of f.

Let us present Berndt's classification [29] of all the real hypersurfaces (M, g) in the quarternionic space projective space $\mathbf{H}P^n(4)$ which are curvature adapted, i.e., $J_\alpha(\xi)$ is a direction of principal curvature for all $\alpha = 1, 2, 3$, where ξ is the unit normal vector field along M.

Theorem 4.4.4 ([29]).

A. *Any curvature adapted real hypersurface in* $\mathbf{H}P^n(4)$ *is one of the following:*
   1. *a geodesic sphere* $M(u)$ *of radius* $(0 < u < \frac{\pi}{2})$;
   2. *a tube* $M(u)$ *of radius* u $(0 < u < \frac{\pi}{4})$ *of the complex projective space* $\mathbf{C}P^n \subset \mathbf{H}P^n(4)$;
   3. *a tube* $M_k(u)$ *of radius* u $(0 < u < \frac{\pi}{4})$ *of the quaternionic projective subspace* $\mathbf{H}P^k \subset \mathbf{H}P^n(4)$ *with* $1 \leq k \leq n - 1$.

B. *Moreover, the principal curvatures are given as follows:*
   1. *The geodesic sphere* $M(u)$:

$$\begin{aligned} \lambda_1 &= \cot u \quad (\textit{with multiplicity } m_1 = 4(n-1)), \\ \lambda_2 &= 2\cot(2u) \quad (\textit{with multiplicity } m_2 = 3). \end{aligned} \tag{4.36}$$

   2. *The cube* $M(u)$ *of the complex projective space:*

$$\begin{aligned}
\lambda_1 &= \cot u \quad (\textit{with multiplicity}\, m_1 = 2(n-1)),\\
\lambda_2 &= -\tan u \quad (\textit{with multiplicity}\, m_2 = 2(n-1),\\
\lambda_3 &= 2\cot(2u) \quad (\textit{with mutiplicity}\ m_3 = 1),\\
\lambda_4 &= -2\tan(2u)\ (\textit{with multiplicity}\ m_4 = 2).
\end{aligned} \tag{4.37}$$

3. *The tube $M_k(u)$ of the quarternionic projective space:*

$$\begin{aligned}
\lambda_1 &= \cot u \quad (\textit{with multiplicity}\ m_1 = 4(n-k-1)),\\
\lambda_2 &= -\tan u \quad (\textit{with multiplicity}\ m_2 = 4k),\\
\lambda_3 &= 2\cot(2u) \quad (\textit{with multiplicity}\,, m_3 = 3).
\end{aligned} \tag{4.38}$$

Likewise, combining Theorem 4.4.3 and the above theorem of Berndt [29], Ichiyama, Inoguchi and Urakawa [191, 192] have generalized Theorem 4.4.2 to biharmonic homogeneous real hypersurfaces in the quarternionic projective space in the following theorem.

Theorem 4.4.5. *All three classes of Theorem 4.4.4 are harmonic (i.e., minimal), and the biharmonic non-harmonic real hypersurfaces $M(u)$ or $M_k(u)$ in $HP^n(4)$ with radii u are given as follows:*

1. *The geodesic spheres $M(u)$ of radius ($0 < u < \frac{\pi}{2}$): $M(u)$ is harmonic iff $t = \cot u\ (0 < u < \pi/2)$ satisfies*

$$t = \sqrt{\frac{3}{4n-1}}. \tag{4.39}$$

$M(u)$ is biharmonic non-harmonic iff $t = \cot u\ (0 < u < \pi/2)$ satisfies

$$(4n-1)t^4 - 2(2n+7)t^2 + 3 = 0. \tag{4.40}$$

Both (4.39) and (4.40) have solutions.

2. *The tube $M(u)$ of radius u ($0 < u < \pi/4$) of the complex projective space: $M(u)$ is harmonic iff*

$$(2n-1)t^4 - (4n+5)t^2 + 2(n-1) = 0. \tag{4.41}$$

$M(u)$ is biharmonic non-harmonic iff

$$(2n-1)t^8 - 8(n+1)t^6 - (6n+11)t^4 - 2(2n-1)t^2 - 12 = 0. \tag{4.42}$$

Both (4.41) and (4.42) have solutions.

3. *The tubes $M_k(u)$ of radius u ($0 < u < \frac{\pi}{4}$) of the quarternionic projective subspaces: $M_k(u)$ is harmonic iff*

$$t = \sqrt{\frac{4k+3}{4n-4k-1}}. \tag{4.43}$$

$M_k(u)$ is biharmonic non-harmonic iff

$$(4n-4k-1)t^4 - 2(2n+4)t^2 + 4k + 3 = 0. \tag{4.44}$$

Both (4.43) and (4.44) always have solutions. (See the proof in [191]).

4.5 Regularity of Biharmonic Maps

We discuss the regularity of biharmonic maps into spheres obtained by Chang, Wang, and Yang [63], and the regularity of biharmonic maps into Riemannian manifolds obtained by C. Wang [398, 399], which generalized the regularity of harmonic maps in Chang et al. [62] and Schoen and Uhlenbeck [320]. We also present the results on the removable singularities and bubbling of biharmonic maps obtained by Nakauchi and Urakawa [277], which generalized results on removable singularities and bubbling of harmonic maps obtained in [310, 320].

4.5.1 Maps into Spheres

Let $f : (M^m, g) \to (S^n, h)$ be a map from an m-dimensional Riemannian manifold into an n dimensional unit sphere, where h is the standard canonical metric on S^n. Suppose that $f = (f^1, \cdots, f^{n+1})$ is a critical point of the bienergy functional $E_2(f) = \int_M \sum_{\alpha=1}^{n+1} (\Delta_g f^\alpha)^2 dv_g$. The Euler-Lagrange equation for f is described in the following proposition. All the following theorems and results were obtained by Chang, Wang and Yang [63].

Proposition 4.5.1. *If $f \in W^{2,2}$ is a critical point of E_2, then f satisfies*

$$\Delta^2 f^\alpha = -f^\alpha \lambda, \ \alpha = 1, 2, \cdots, n+1, \tag{4.45}$$

where $\lambda = \sum_{\beta=1}^{n+1} [(\Delta f^\beta)^2 + \Delta(|\nabla f^\beta|^2) + 2\nabla f^\beta \cdot \nabla \Delta f^\beta]$ and $\nabla \Delta f^\beta$ exists in the L^p sense for all $p < \frac{3}{4}$.

Theorem 4.5.2. *Any biharmonic map in $W^{2,2}$ defined on a disk of dimension 4 with values in the unit sphere S^n is Hölder-continuous.*

This theorem is similar to the following result for harmonic maps: *Any harmonic map from a two-dimensional disk to the unit sphere S^n* is *Hölder-continuous.*

For any ball B_r of radius r in $\mathbf{R}^m$ and any $p > 1$ and q with $\frac{1}{q} = \frac{1}{2} - \frac{1}{m}$, set

$$E(f)(B_r) = \left(r^4 \int_{B_r} |\nabla^2 f|^2\right)^{1/2} + \left(r^q \int_{B_r} |\nabla f|^q\right)^{1/q},$$

$$M_p(f)(B_r) = \left(\int_{B_r} |f - \bar{f}|^p\right)^{1/p}, \tag{4.46}$$

where $\bar{f} = \int_{B_r} f$ and $D_p(f)(B_r) = (r^p \int_{B_r} |\nabla f|^p)^{1/p}$.

Lemma 4.5.3. *Let f be as in Theorem 4.5.2 and $m = 4$. Then for given $0 < \beta < 1$, there exists some $\tau < \frac{1}{4}$ and $\epsilon > 0$ such that if $E(f)(B_1) < \epsilon$, we have*

$$(M_{p_0}(f) + D_{p_1}(f)(B_\tau) < \tau^\beta (M_{p_0}(f) + D_{p_1}(f))(B_1), \tag{4.47}$$

where $2 < p_1 < 4$ is any fixed number and $\frac{1}{p_0} = \frac{1}{p_1} - \frac{1}{4}$.

Proof (Proof of Theorem 4.5.2). We can apply Lemma 4.5.3 iteratively to the function f. If $E(f)(B_1) < \epsilon$, then

$$(M_{p_0}(f) + D_{p_1}(f))(B_{\tau^i}) < \tau^{i\beta} (M_{p_0}(f) + D_{p_1}(f))(B_1) \tag{4.48}$$

for each i. Equation (4.48) and Morrey's estimate imply that f is Höder-continuous.

To do the iteration, we only need to show that $E(f)(B_r) < \epsilon$ if $E(f)(B_1) < \epsilon$, where $r = \tau^i$ for all $i = 1, 2, 3, \cdots$. When $m = 4$,

$$E(f)(B_r) = \left(\int_{B_r} |\nabla^2 f|^2\right)^{1/2} + \left(\int_{B_r} |\nabla f|^4\right)^{1/4},$$

it is understood that $E(f)(B_r) < \epsilon$ whenever $E(f)(B_1) < \epsilon$. This establishes (4.48), and thus the theorem. □

Let f be a biharmonic map from a manifold M (possibly with boundary) to another compact manifold N. We say that f is *stationary* if

$$\frac{d}{dt} E(f(\phi(t)) = 0 \text{ at } t = 0,$$

where $\phi(t) : M \to M$ is a smooth, one-parameter family of diffeomorphisms such that $\phi(0) =$ *identity*.

Theorem 4.5.4. *A stationary biharmonic map from an m-dimensional Euclidean disk ($m \geq 5$) to the sphere S^n is Hölder-continuous except on a set of (m-4)-dimensional Hausdorff measure zero.*

The above theorem is similar to the stationary harmonic map case: *A stationary harmonic map from an m-dimensional Euclidean disk to the sphere S^n is Hölder-continuous except on a set of (m-2)-dimensional Hausdorff measure zero* (cf. [62]).

Fixing $0 < r \le 1$, if $1/2^{k+1} \le r < 1/2^k$ for some k, set $r^* = 1/2^k$. We say that ∂B_r is a *good slice* if the following conditions hold

$$r \int_{\partial B_r} |\nabla^2 f|^2 d\sigma \le 8 \int_{B_{r^*}} |\nabla^2 f|^2 dx,$$

$$r \int_{\partial B_r} |\nabla f| d\sigma \le 8 \int_{B_{r^*}} |\nabla f| dx.$$

Note that such a good slice always exists for all $k \ge 0$.

Lemma 4.5.5. *There exists a constant c such that for all good slices $\partial B_\rho, \partial B_r$, $\rho < r < \frac{1}{2}$ for all $\eta > 0$, η sufficiently small, we have*

$$E(f)(B_\rho) \le cE(f)(B_{r^*}) + \eta E(f)(B_{\rho^*}) + C_\eta M(f)(B_{\rho^*}) \tag{4.49}$$

where $C_\eta = c\eta^{-(3+m)}$, $M(f) = M_1(f)$, and $E(f)$ and $M_1(f)$ are given in (4.46).

Lemma 4.5.6. *There exists a $\tau < \frac{1}{4}$ and a dimensional constant c such that, for all $r < 1$,*

$$\begin{aligned}(M_{p_0}(f) + D_p(f))(B_{\tau_r}) &\le c\tau^{1-\frac{m}{p}} E^2(f)(B_r) M_s(f)(B_r) \\ &\quad + c\tau^{1-m/p} D_q(f)(B_r) M_t(f)(B_r) + \tau(M_s(f) + D_p(f))(B_r), \end{aligned} \tag{4.50}$$

where $\frac{1}{q} = \frac{1}{2} - \frac{1}{m}$, $\frac{1}{s} = \frac{1}{p} + \frac{3}{m} - 1$, $\frac{1}{t} = \frac{1}{p} + \frac{1}{m} - \frac{1}{2}$, $\frac{1}{p_0} = \frac{1}{p} - \frac{1}{m}$ and p is a constant chosen larger than 1.

Proof (Proof of Theorem 4.5.4). The proof is similar to the arguments in the proof of Theorem 2.5 for harmonic maps in [62]. For $E(f)(B_1) < \epsilon$ ($\epsilon > 0$ small enough), Lemma 4.5.5 yields

$$E(f)(B) \le CE(f)(B_1) \le C\epsilon, \quad B \subset B_{\rho_0}. \tag{4.51}$$

By (4.50), there exists a $\rho < 1$ such that

$$\sup_{B \subset B_\rho} (M_{p_0}(f) + D_p(f))(B) \le (C\rho^{1-\frac{m}{p}}\epsilon + C\rho)(M_s(f) + D_p(f)(B_1)).$$

It follows from the John-Nirenberg inequality [199] that there exists a universal constant M such that

$$||f||_{BMO_s(B_\rho)} + D_{p_1}(f)(B_\rho) \le M(C\rho^{1-\frac{m}{p}}\epsilon + C\rho)(||f||_{BMO_s(B_1)} + D_{p_1}(f))(B_1),$$

where

$$||f||_{BMO_s(B)} = \sup_{B_1 \subset B} \inf_{const\, C} \left(\int_{B^1} |f - C|^s \right)^{1/s} dx.$$

For any $\beta > 1$ there is a $\rho = \rho_0$ small enough so that $MC\rho_0 < \frac{\rho^\beta}{2}$ and an ϵ small enough so that $MC\rho_0^{1-m/p}\epsilon \leq \frac{\rho_0^\beta}{2}$. Therefore, we have

$$||f||_{BMO_s(B_{\rho_0})} + D_{p_1}(f)(B_{\rho_0}) \leq \rho_0^\beta(||f||_{BMO_s(B_1)} + D_{p_1}(f)(B_1)). \tag{4.52}$$

An iteration of (4.52) gives

$$||f||_{BMO_s(B_{\rho_0^k})} + D_{p_1}(f)(B_{\rho_0^k}) \leq \rho_0^{k\beta}(||f||_{BMO_s(B_1)} + D_{p_1}(f)(B_1)), \tag{4.53}$$

for $k = 1, 2, \cdots$. The above inequality shows that

$$||f||_{BMO_s(B_r)} + D_{p_1}(f)(B_r) \leq Cr^\beta(||f||_{BMO_s(B_1)} + D_{p_1}(f)(B_1)), \tag{4.54}$$

for all $0 \leq r \leq 1$. Equation (4.54) and the standard covering discussion as in Evans [139] imply that the singularity set of the stationary map f is a set of $(m-4)$-Hausdorff dimension 0, as claimed. □

In fact, if f is a weak solution of the biharmonic map equation and f is continuous on B_1, then f is smooth. Based on the classical regularity theory, it suffices to show that the solution is $C^{2,\alpha}$ for some $\alpha > 0$.

Let f be a Hölder-continuous with exponent $\alpha > 1$, i.e.,

$$|f(x) - f(y)| \leq C|x - y|^\alpha$$

and

$$D_{p_1}(f)(B_{\rho_0}) \leq C\rho_0^\alpha,$$

where the center of the ball B_{ρ_0} is in the regular set of f for some $p_1 < 4$ (note that the second condition implies the first).

Theorem 4.5.7. *If f is a Hölder-continuous biharmonic map satisfying the above two conditions in B_1, then f is locally smooth.* (See the proof in [63]).

Theorem 4.5.7 is similar to the harmonic map case: *If $f : M^m \to N^n$ is a weakly harmonic map for $m \geq 2$, and f is continuous in an open set in M^m, then f is locally smooth there* (cf. [62]).

4.5.2 *Maps into Manifolds*

Let $\Omega \subset \mathbf{R}^m$ be a bounded domain and $(N^n, h) \subset \mathbf{R}^k$ be a compact Riemannian manifold without boundary. The Sobolev space $W^{2,2}(\Omega, N)$ is given by

$$W^{2,2}(\Omega, N) = \{f \in W^{2,2}(\Omega, \mathbf{R}^k) | f(x) \in N, \text{ a.e., } x \in \Omega\}.$$

There are two second-order energy functionals on $W^{2,2}(\Omega, N)$ defined as follows:

$$H(f) = \int_\Omega |\Delta f|^2 dx, \quad T(f) = \int_\Omega |(\Delta f)^T|^2 dx,$$

where Δ is the Laplace operator on $\mathbf{R}^m$ and $(\Delta f)^T$ is the component of Δf tangent to N at f, which is called the tension field of f. Remark that the Hessian energy $H(\cdot)$ measures the degree of the bending of f and $T(\cdot)$ is the L^2-norm of the tension field of f, which vanishes if $f \in W^{2,2}(\Omega, N)$ is a harmonic map. A map $f \in W^{2,2}(\Omega, N)$ is called an *extrinsic* (resp. *intrinsic*) biharmonic map [239] if it is a critical point of $H(\cdot)$ (resp. $T(\cdot)$) over $W^{2,2}(\Omega, N)$.

Chang, Wang, and Yang [62] and Ku [237] rewrote the borderline nonlinearities of the biharmonic map equations into divergence forms, which relied heavily on the special structures of spheres and it was not clear how to generalize their work to usual target manifolds. C. Wang [398, 399] extended their theorems to the case that $N \subset \mathbf{R}^k$ is a compact Riemannian manifold without boundary by adopting Uhlenbeck's [383, 384] constructions of Coulomb gauge frames and obtained the following results.

Theorem 4.5.8. *For $m = 4$, if $f \in W^{2,2}(\Omega, N)$ is an extrinsic (or intrinsic) biharmonic map, then $f \in C^\infty(\Omega, N)$.*

Lemma 4.5.9. *There exist ϵ_0, $\theta_0 \in (0, \frac{1}{2})$ such that if $f \in W^{2,2}(B_2, N)$ is an intrinsic (or extrinsic) biharmonic map such that $\int_{B_2}(|\nabla f|^4 + |\nabla^2 f|^2)dx \le \epsilon_0^2$, then*

$$||\nabla f||_{L^{4,\infty}(B_{\theta_0})} + ||\nabla^2 f||_{L^{2,\infty}(B_{\theta_0})} \le \frac{1}{2}(||\nabla f||_{L^{4,\infty}(B_1)} + ||\nabla^2 f||_{L^{2,\infty}(B_1)}). \tag{4.55}$$

Proof (Proof of Theorem 4.5.8). For simplicity, we only verify the theorem for intrinsic biharmonic maps. Because $\int(|\nabla f|^4+|\nabla^2 f|^2)dx$ is invariant under scaling in dimension 4 and $\int(|\nabla f|^4+|\nabla^2 f|^2)dx$ is absolutely continuous, for given $\epsilon_0 > 0$ there exists a r_0 such that

$$\sup_{x_0\in\Omega} \int_{B_{r_0}(x_0)} (|\nabla f|^4 + |\nabla^2 f|^2)dx \le \epsilon_0^2. \tag{4.56}$$

Remark that $f_{x_0,r_0}(x) = f(x_0 + r_0 x) : B_1 \to N$ is an intrinsic biharmonic map satisfying the assumption of Lemma 4.5.9, and so there exists a $\theta_0 \in (0, \frac{1}{2})$ such that

$$||\nabla f||_{L^{4,\infty}(B_{\theta_0 r_0}(x_0))} + ||\nabla^2 f||_{L^{2,\infty}(B_{\theta_0 r_0}(x_0))} \le \frac{1}{2}(||\nabla f||_{L^{4,\infty}(B_{r_0}(x_0))} + ||\nabla^2 f||_{L^{2,\infty}(B_{r_0}(x_0))}). \tag{4.57}$$

By iterating (4.57), we deduce, for $x_0 \in \Omega$ and any $l \ge 1$,

$$\|\nabla f\|_{L^{4,\infty}(B_{\theta_0^l r_0}(x_0))} + \|\nabla^2 f\|_{L^{2,\infty}(B_{\theta_0^l r_0}(x_0))} \leq 2^{-l}\Big(\|\nabla f\|_{L^{4,\infty}(B_{r_0}(x_0))} + \|\nabla^2 f\|_{L^{2,\infty}(B_{r_0}(x_0))}\Big). \tag{4.58}$$

Thus, there exists an $\alpha_0 \in (0, 1)$ such that for any $1 < p < 4$, $x \in \Omega$, and $0 < r \leq \frac{r_0}{2}$,

$$\left(r^{p-4} \int_{B_r(x)} |\nabla f|^p\right)^{1/p} dx \leq \|\nabla f\|_{L^{4,\infty}(B_r(x))} \leq C r^{\alpha_0}. \tag{4.59}$$

It follows from the Morrey's Lemma that $f \in C^{\alpha_0}(\Omega, N)$.

To derive higher order regularity from this Hölder continuity estimate, we notice that although Theorem 5.1 of [62] is for $N = S^n \subset \mathbf{R}^k$, its proof works for any equation of the form

$$\Delta^2 f = \Delta(u(f)(Df, Df) + 2\nabla \cdot ((\Delta f, \nabla(v(f))) - (\Delta(w(f), \Delta f), \tag{4.60}$$

if u, v, w are smooth functions. Because the equation of biharmonic maps into any compact Riemannian manifold $N \subset \mathbf{R}^k$ is a special case of (4.60), Theorem 5.1 of [62] also holds for biharmonic maps into any compact Riemannian manifold $N \subset \mathbf{R}^k$. Hence, the proof is complete. □

We have the following fourth order PDE with borderline nonlinearity which as a byproduct of the Lorentz space estimate,

$$\Delta^2 f = Q(x, f, \nabla f), \quad x \in \Omega \subset \mathbf{R}^4, \tag{4.61}$$

has better regularity, provided $Q : \mathbf{R}^4 \times \mathbf{R}^k \times \mathbf{R}^{4k} \to \mathbf{R}^k$ satisfies

$$Q(x, y, p) \leq C|p|^4, \quad \forall (x, y, p) \in \mathbf{R}^4 \times \mathbf{R}^k \times \mathbf{R}^{4k}.$$

Theorem 4.5.10. *There is an $\alpha \in (0, 1)$ such that any weak solution $f \in W^{2,2}(\Omega, \mathbf{R}^k)$ to the equation (4.61) is $C^\alpha(\Omega, \mathbf{R}^k)$. Furthermore, if $Q \in C^\infty(\Omega \times \mathbf{R}^k \times \mathbf{R}^{4k}, \mathbf{R}^k)$, then $f \in C^\infty(\Omega, \mathbf{R}^k)$.*

In contrast to Theorem 4.5.10, in dimension 2 there exist singular solutions to the following second order PDE with quadratic nonlinearity:

$$\Delta f = G(x, f, \nabla f), \ x \in \Omega \subset \mathbf{R}^2, \tag{4.62}$$

where $G : \mathbf{R}^2 \times \mathbf{R}^k \times \mathbf{R}^{2k} \to \mathbf{R}$ satisfies

$$|G(x, y, p) \leq C|p|^2, \forall (x, y, \rho) \in \mathbf{R}^2 \times \mathbf{R}^k \times \mathbf{R}^{2k}.$$

(cf. [146, 319]).

The proof of Theorem 4.5.10 is based on the following lemma.

Lemma 4.5.11. *There exist ϵ_0, $\theta_0 \in (0, \frac{1}{2})$ such that if $f \in W^{2,2}(B_1, N)$ is a weak solution of the equation (4.61) satisfying $\int_{B_1}(|\nabla f|^4 + |\nabla^2 f|^2)dx \leq \epsilon_0^2$, then*

$$||\nabla f||_{L^{4,\infty}(B_{\theta_0})} + ||\nabla^2 f||_{L^{2,\infty}(B_{\theta_0})} \leq \frac{1}{2}(||\nabla f||_{L^{4,\infty}(B_1)} + ||\nabla||^2 f||_{L^{2,\infty}(B_1)}). \tag{4.63}$$

4.5.3 Removable Singularities

Recall from Sects. 1.2 and 1.3 of Chap. 1 that Sacks and Uhlenbeck [310] proved a removable singularity theorem for harmonic maps in 1981 as follows: *If $f : D^2 - \{0\} \to N$ is a harmonic map from a 2-disc omitting the origin 0 into an arbitrary manifold N with finite energy, then f extends to a smooth harmonic map $f : D \to N$.* Likewise, we discuss the removable singularities of biharmonic maps based on the work of Nakauchi and Urakawa [277]. In order to prove Theorem 4.5.15, we require the following lemma and proposition.

Lemma 4.5.12. *If $f : M - S \to N$ is a biharmonic map from a Riemannian manifold M omitting S into a Riemannian manifold N with non-positive sectional curvature, then*

$$0 \leq |\tau(f)|\triangle|\tau(f)|, \tag{4.64}$$

where S is a closed set of M, and $\triangle$ is the Laplace-Beltrami operator.

Proposition 4.5.13. *Suppose that (M, g) is a compact Riemannian manifold and (N, h) is a Riemannian manifold with non-positive sectional curvature. Then for a biharmonic map $f : (M - S, g) \to (N, h)$ with $S = \{x_1, \cdots, x_k\}$, there exists a positive constant $C > 0$ depending on dim(M) such that for each positive number $r > 0$ and each point $x_i \in S$,*

$$\sup_{B_r(x_i)} |\tau(f)| \leq \frac{C}{r^{m/2}} \int_{B_{2r}(x_i)} |\tau(f)|^2 dv, \tag{4.65}$$

where $B_r(x_i) = \{x \in M : r(x, x_i) < r\}$ is the metric ball around x_i with radius r in M such that $B_r(x_i) \cap B_r(x_j) = \emptyset\ (i \neq j)$ for every sufficient small $r > 0$.

Proof. (1) For a point $x_i \in S$ and $0 < \rho_1 < \rho_2 < \infty$, we consider the following cut-off smooth function μ on M:

$$\begin{cases} 0 \leq \mu(x) \leq 1, & x \in M, \\ 1, & x \in B_{\rho_1}(x_i), \\ 0, & x \notin B_{\rho_2}(x_i), \\ |\nabla\mu| \leq \frac{2}{\rho_2 - \rho_1}, & x \in M. \end{cases} \tag{4.66}$$

For $2 \leq p < \infty$, multiply (4.64) by $|\tau(f)|^{p-2}\mu^2$ on both sides and integrate over M. We get

$$\begin{aligned} 0 \leq \int_M |\tau(f)|^{p-1}\mu^2\Delta(|\tau(f)|)dv &= -\int_M (\nabla(|\tau(f)|^{p-1}\mu^2), \nabla\tau(f))dv \\ &= -(p-1)\int_M |\tau(f)|^{p-2}\mu^2|\nabla(|\tau(f)|)|^2dv - 2\int_M |\tau(f)|^{p-1}\mu(\nabla|\tau(f)|, \nabla\mu)dv \\ &= -\frac{4(p-1)}{p^2}\int_M |\nabla(|\tau(f)|^{p/2})|^2\mu^2dv - \frac{4}{p}\int_M (\mu\nabla|\tau(f)|^{p/2}, |\tau(f)|^{p/2}\nabla\mu)dv. \end{aligned} \tag{4.67}$$

Applying Young's inequality, we obtain, for each $\epsilon > 0$,

$$\begin{aligned} \int_M |\nabla(|\tau(f)|^{p/2})|^2\mu^2dv &\leq \frac{p}{p-1}\int_M (\mu\nabla\tau(f)^{p/2}, |\tau(f)|^{p/2}\nabla\mu)dv \\ &\leq \frac{p}{2(p-1)}\left\{\epsilon\int_M \mu^2|\nabla\tau(f)^{p/2}|^2dv + \frac{1}{\epsilon}\int_M |\tau(f)|^p|\nabla\mu|^2dv\right\}. \end{aligned} \tag{4.68}$$

It follows that

$$\left(1 - \frac{p}{2(p-1)}\epsilon\right)\int_M \mu^2|\nabla\tau(f)^{p/2}|^2dv \leq \frac{p}{2(p-1)}\frac{1}{\epsilon}\int_M |\tau(f)|^p|\nabla\mu|^2dv. \tag{4.69}$$

Choosing $\epsilon = \frac{p-1}{p}$ in the last inequality, we get

$$\int_M \mu^2|\nabla\tau(f)^{p/2}|^2dv \leq \frac{p^2}{(p-1)^2}\int_M |\tau(f)|^p|\nabla\mu|^2dv. \tag{4.70}$$

Using $\nabla(|\tau(f)|^{p/2}\mu) = \mu\nabla(|\tau(f)|^{p/2}) + |\tau(f)|^{p/2}\nabla\mu$, $|C+D|^2 \leq 2|C|^2+2|D|^2$ and (4.70), and by (4.66), we derive

$$\begin{aligned} \int_M |\nabla(|\tau(f)|^{p/2}\mu)|^2dv &\leq 2\int_M \mu^2|\nabla(|\tau(f)|^{p/2})|^2dv + 2\int_M |\tau(f)|^p|\nabla\mu|^2dv \\ &\leq 4\frac{p^2}{(p-1)^2}\int_M |\tau(f)|^p|\nabla\mu|^2dv \\ &\leq \frac{p^2}{(p-1)^2}\frac{16}{(\rho_2-\rho_1)^2}\int_{B_{\rho_2}(x_i)} |\tau(f)|^pdv. \end{aligned} \tag{4.71}$$

For the left-hand side of (4.71), note that, by the Sobolev embedding theorem,

$$H_1^2(M) \subset L^\alpha, \tag{4.72}$$

where $\alpha = \frac{m}{m-2}$, that is, there exists a positive constant $C > 0$ such that

$$\left(\int_M |\phi|^\alpha dv\right)^{1/\alpha} \leq C\left(\int_M |\nabla\phi|^2 dv\right)^{1/2}, \quad \phi \in H_1^2(M). \tag{4.73}$$

When $dim\, M = m = 2$, (4.72) and (4.73) still hold, but the left-hand side of (4.73) must be replaced by the supremum norm $\sup_M |f|$. Hence, by (4.66) we have

$$\begin{aligned}\int_M |\nabla(|\tau(f)|^{p/2}\mu)|^2 dv &\geq \frac{1}{C}\left(\int_M (|\tau(f)|^{p/2}\mu)^\alpha dv\right)^{2/\alpha}\\ &\geq \frac{1}{C}\left(\int_{B_{\rho_1}(x_i)} (|\tau(f)|^{p/2})^\alpha dv\right)^{2/\alpha}.\end{aligned} \tag{4.74}$$

Combining (4.71) and (4.74), we obtain the following lemma.

Lemma 4.5.14. *Let $f : (M - S, g) \to (N, h)$ be a biharmonic map from a compact Riemannian manifold M omitting S into a Riemannian manifold N with non-positive sectional curvature such that $S = \{x_1, \cdots, x_k\} \subset M$. Then for $1 < \rho_1 < \rho_2 < \infty$ and $2 \leq p < \infty$, we have*

$$\left(\int_{B_{\rho_1}(x_i)} (|\tau(f)|^{p/2})^\alpha dv\right)^{1/\alpha} \leq \frac{p}{p-1}\frac{C_1}{\rho_2 - \rho_1}\left(\int_{B_{\rho_2}(x_i)} (|\tau(f)|^{p/2})^2 dv\right)^{1/2}, \tag{4.75}$$

for $i = 1, \cdots, k$, where $C_1 = 4\sqrt{C}$ and $C > 0$ is the Sobolev constant in (4.73) and $\alpha = \frac{2m}{m-2}$, $m = dimM = 2$. When $m = 2$, the left-hand side of (4.75) is replaced by $\sup_{B_{\rho_1}(x_i)} |\tau(f)|^{p/2}$.

(2) We define

$$\begin{cases} \bar{\alpha} = \frac{m}{m-2} = \frac{1}{2}\alpha, \\ p_j = 2\bar{\alpha}^{j-1} \to \infty, \quad as\, j \to \infty, \\ r_j = (1 + \frac{1}{2^{j-1}})r \to r, \quad as\, j \to \infty, \end{cases} \tag{4.76}$$

and in (4.75) we set $p = p_j$, $\rho_1 = r_{j+1}$, $\rho_2 = r_j$. Thus we have

$$\begin{cases} \frac{p\alpha}{2} = p_j\bar{\alpha} = 2\bar{\alpha}^j = p_{j+1}, \\ \rho_2 - \rho_1 = r_j - r_{j+1} = (\frac{1}{2^{j-1}} - \frac{1}{2^j})r = \frac{1}{2^j}r \end{cases} \tag{4.77}$$

and so (4.75) can be recast as

$$\left(\int_{B_{r_{j+1}}(x_i)} |\tau(f)|^{p_{j+1}}\,dv\right)^{1/\alpha} \leq \frac{2\bar{\alpha}^{j-1}}{2\bar{\alpha}^{j-1}-1}\frac{2^j}{r}\left(\int_{B_{rj}(x_i)} |\tau(f)|^{p_j}\,dv\right)^{1/2}. \tag{4.78}$$

Applying the power $\frac{1}{\bar{\alpha}^{j-1}}$ to (4.78), we arrive at

$$||\tau(f)||_{L^{p_{j+1}}(B_{r_{j+1}}(x_i))} \leq \left(\frac{2\bar{\alpha}^{j-1}}{2\bar{\alpha}^{j-1}-1}\right)^{2/p_j}\frac{2^{(j/\bar{\alpha}^{j-1})}}{r^{(1/\bar{\alpha}^{j-1})}}||\tau(f)||_{L^{p_j}(B_{r_j}(x_i))}, \tag{4.79}$$

because the power of the left-hand side of (4.78) can be computed as

$$\frac{1}{\alpha}\frac{1}{\bar{\alpha}^{j-1}} = \frac{1}{2\bar{\alpha}\bar{\alpha}^{j-1}} = \frac{1}{2\bar{\alpha}^j} = \frac{1}{p_{j+1}}.$$

(3) We iterate (4.79) and derive

$$||\tau(f)||_{L^{p_{j+1}}(B_{r_{j+1}}(x_i))} \leq \prod_{j=1}^{\infty}\left(\frac{2\bar{\alpha}^{j-1}}{2\bar{\alpha}^{j-1}-1}\right)^{2/p_j}\frac{2^{(j/\bar{\alpha}^{j-1})}}{r^{(1/\bar{\alpha}^{j-1})}}||\tau(f)||_{L^2(B_{2r}(x_i))}, \tag{4.80}$$

where $p_1 = 2$ and $r_1 = 2r$. Here, firstly note that

$$\prod_{j=1}^{\infty}\frac{1}{r^{(1/\bar{\alpha}^{j-1})}} = \frac{1}{r^{(\sum_{j=1}^{\infty}\frac{1}{\bar{\alpha}^{j-1}})}} = \frac{1}{r^{m/2}}, \tag{4.81}$$

because

$$\sum_{j=1}^{\infty}\frac{1}{\bar{\alpha}^{j-1}} = \frac{1}{1-\frac{1}{\bar{\alpha}}} = \frac{1}{1-\frac{m-2}{m}} = \frac{m}{2}.$$

Secondly, note that

$$\prod_{j=1}^{\infty}\frac{1}{(2\bar{\alpha}^{j-1}-1)^{2/p_j}} \leq 1, \tag{4.82}$$

since $2\bar{\alpha}^{j-1} - 1 > 2 - 1 = 1$ as $\bar{\alpha} = \frac{m}{m-2} > 1$ $(m \geq 3)$, and the left-hand side of (4.82) is 1 as $\bar{\alpha} = \infty$ $(m = 2)$. Thirdly, note that

$$\prod_{j=1}^{\infty} 2^{(j/\bar{\alpha}^{j-1})} = 2^{\sum_{j=1}^{\infty}\frac{j}{\bar{\alpha}^{j-1}}} < \infty, \tag{4.83}$$

$$\prod_{j=1}^{\infty}(2\bar{\alpha})^{2(j-1)/p_j} = \alpha^{2\sum_{j=1}^{\infty}\frac{j-1}{p_j}} = \alpha^{\sum_{j=1}^{\infty}\frac{j-1}{\bar{\alpha}^{j-1}}} < \infty. \tag{4.84}$$

Hence, (4.80) becomes

$$||\tau(f)||_{L^{p_{j+1}}(B_{r_{j+1}}(x_i))} \leq C_2 \frac{1}{r^{m/2}} ||\tau(f)||_{L^2(B_{2r}(x_i))}, \tag{4.85}$$

where the positive constant C_2 depends only on $dim\, M = m$.
(4) As $j \to \infty$, by (4.74) the norm $||\tau(f)||_{L^{p_{j+1}}(B_{r_{j+1}}(x_i))}$ goes to

$$||\tau(f)||_{L^\infty(B_r(x_i))} = \sup_{B_r(x_i)} |\tau(f)|. \tag{4.86}$$

Consequently,

$$\sup_{B_r(x_i)} |\tau(f)| \leq \frac{C_2}{r^{m/2}} ||\tau(f)||_{L^2(B_{2r}(x_i))}, \tag{4.87}$$

which is exactly (4.65), and we conclude Proposition 4.5.13. □

Theorem 4.5.15. *Let (M, g) be a compact Riemannian manifold and (N, h) be a Riemannian manifold with non-positive sectional curvature. If $f : (M - S, g) \to (N, h)$ is a biharmonic map where S is a finite set of points in M, and if f has finite bienergy, i.e.,*

$$E_2(f) = \frac{1}{2}\int_M |\tau(f)|^2 dv < \infty, \tag{4.88}$$

then $|\tau(f)|$ is bounded on M, and $|\tau(f)|$ has a unique continuous extension to (M, g).

Proof. It follows from Proposition 4.5.13 immediately. □

4.5.4 Bubbling

Recall from Sect. 1.3 that Sacks and Uhlenbeck [310, 311] obtained what is now a well-known result about the bubbling of harmonic maps from Riemann surfaces into Riemannian manifolds. In 2011, Nakauchi and Urakawa [277] showed the bubbling of biharmonic maps between compact Riemannian manifolds, which is the main theme of this section. In order to prove Theorem 4.5.17, we require the following lemma.

Lemma 4.5.16. *Suppose that the sectional curvature of (N, h) is bounded above by a positive constant $C > 0$. Then there exists a positive number $\epsilon_0 > 0$ depending only on the Sobolev constant of M and C such that, for each $f \in C^\infty(M, N)$, if*

$$\int_{B_r(x_0)} |df|^m dv \leq \epsilon_0, \tag{4.89}$$

then

$$\sup_{B_{r/2}(x_0)} |\tau(f)|^2 \le \frac{C_1}{r^{m/2}} \int_{B_r(x_0)} |\tau(f)|^2 dv, \tag{4.90}$$

for some positive constant $C_1 > 0$ depending only on C and $m = \dim M$.

Theorem 4.5.17. *For each positive constant $C > 0$, consider the following family of biharmonic maps between two compact Riemannian manifolds (M,g) and (N,h):*

$$\mathcal{BH} = \{f : M \to N \text{ is biharmonic} : \int_M |df|^m dv \le C \text{ and } \int_M |\tau(f)|^2 dv \le C\}, \tag{4.91}$$

where $m = \dim M$. Then for any sequence $\{f_i\} \in \mathcal{BH}$ there exist a finite set $S = \{x_1, \cdots, x_k\}$ and a smooth biharmonic map $f_\infty : M - S \to N$ such that

(1) *A subsequence $\{f_{i_j}\}$ converges to f_∞ in the C^∞ on $M - S$ as $j \to \infty$;*
(2) *The Radon measure $|df_{i_j}|^m dv$ converges (as $j \to \infty$) to a measure*

$$|df_\infty|^m dv + \sum_{l=1}^k a_l \delta_{x_l}, \tag{4.92}$$

where a_l is a constant and δ_{x_l} is the Dirac measure whose support is $\{x_l\}$ ($l = 1, \cdots, k$).

Proof. (1) Let $\{f_i\}$ be an arbitrary sequence in $\mathcal{BH}$. For ϵ_0 as in Lemma 4.5.16, and let

$$S = \{x \in M : \epsilon_0 \le \liminf_{i\to\infty} \int_{B_r(x)} |df_i|^m dv, \ \forall r > 0\}. \tag{4.93}$$

We claim that the set S is finite. Indeed, for each finite subset $\{x_l\}$ in S, we choose a sufficiently small positive number $r_0 > 0$ such that $B_{r_0(x_l)} \cap B_{r_0(x_h)} = \emptyset$ $(l \ne h)$. Thus for a sufficiently large l,

$$\begin{aligned} k\epsilon_0 &\le \sum_{l=1}^k \int_{B_{r_0}(x_l)} |df_i|^m dv = \int_{\bigcup_{l=1}^k B_{r_0}(x_l)} |df_i|^m dv \\ &\le \int_M |df_i|^m dv \le C < \infty, \end{aligned} \tag{4.94}$$

by the definition of $\mathcal{BH}$. Therefore, we have $\#S \le \frac{C}{\epsilon_0} < \infty$, because $k \le \frac{C}{\epsilon_0}$.

We may choose a subsequence of $\{f_i\}$ if needed, and we assume that

$$S = \{x \in M : \epsilon_0 \le \limsup_{i\to\infty} \int_{B_r(x)} |df_i|^m dv\}. \tag{4.95}$$

Otherwise, we denote the right-hand side of (4.95) by S'. Then by definition S is a proper subset of S'. Choose a point $x' \in S' - S$ and pick a subsequence of $\{f_i\}$ using the same notation, such that

$$\epsilon_0 \leq \liminf_{i\to\infty} \int_{B_r(x')} |df_i|^m dv.$$

For this $\{f_i\}$, $x' \in S$. This process ends in a finite number of steps since S is a finite set, and eventually we have $S' = S$.

If $x \in M - S$, then

$$\limsup_{i\to\infty} \int_{B_r(x)} |df_i|^m dv < \epsilon_0. \tag{4.96}$$

It follows from Lemma 4.5.16 and the definition of $\mathcal{HB}$ that

$$\sup_{B_{r/2}(x)} |\tau(f_i)|^2 \leq \frac{C}{r^{m/2}} \int_{B_r(x)} |\tau(f_i)|^2 dv \leq \frac{C^2}{r^{m/2}}. \tag{4.97}$$

Therefore, we obtain (C^0): *the C^0-estimate on $B_r(x)$ of $\tau(f_i)$ uniformly on i.*

On the other hand, each f_i is biharmonic since $f_i \in \mathcal{HB}$, i.e. f_i obeys

$$\tau_2(f_i) = \bar{\Delta}\tau(f_i) - R(\tau(f_i)) = 0 \tag{4.98}$$

if and only if

$$\text{(a) } \bar{\Delta}\sigma_i = R(\sigma_i), \text{ (b) } \tau(f_i) = \sigma_i, \tag{4.99}$$

where $\bar{\Delta} = \bar{D}^*\bar{D} = (-)(\bar{D}_{e_i}\bar{D}_{e_i} - \bar{D}_{D_{e_i}e_i})$ [196]. Note that (a) and (b) are non-linear elliptic PDEs. By (a) the C^0 estimate for σ_i implies the C^∞-estimate of σ_i, and by (b) the C^∞-estimate of σ_i implies the C^∞-estimate of f_i. Therefore, the above (C^0) estimate implies the C^∞ estimate on $B_r(x)$ of f_i uniformly on i. Hence, there exist a subsequence $\{f_{i_j}\}$ of $\{f_i\}$ and a smooth map $f_\infty : M - S \to N$ such that $\{f_{i_j}\}$ converges to f_∞ in the C^∞ topology as $j \to \infty$. Consequently, $f_\infty : M - S \to N$ is also biharmonic.

(2) We consider the Radon measures $df_{i_j}^m dv$, which have a weak limit which is also a Randon measure, say ν. By definition, ν is a *Radon measure* if (i) ν is locally finite, i.e. $\nu(K) < \infty$ for every compact subset K of M; (ii) ν is Borel regular, i.e. for each Borel subset F of M,

$$\nu(F) = \sup\{\nu(K) : K \subset F, \; K \text{ compact}\};$$

$$\nu(F) = \inf\{\nu(O) : F \subset O, \; O \text{ open}\}.$$

Since f_{i_j} converges to f_∞ in $M - S$ in the C^∞-topology as $j \to \infty$, we have

$$\nu = |df_\infty|^m dv \quad \text{on} \quad M - S, \tag{4.100}$$

where $S = \{x_1, \cdots, x_k\}$ is a finite subset of M. Thus the Radon measure $\nu - |df_\infty|^m$ has its support in S. Hence,

$$\nu - |df_\infty|^m dv = \sum_{l=1}^{k} a_l \delta_{x_l}, \tag{4.101}$$

for some non-negative real numbers $a_l, l = 1, \cdots, k$, where δ_{x_l} is the Dirac measure, i.e.,

$$\delta_{x_l}(F) = \begin{cases} 1, & x_l \in F, \\ 0, & x_l \notin F, \end{cases}$$

for every Borel subset F of M. Note that $a_l < \infty$ for $l = 1, \cdots, k$. So ν is locally finite since ν is a Radon measure. Hence, the Radon measure $|df_{i_j}|^m dv$ converges weakly to ν, and

$$\nu = |df_\infty| dv + \sum_{l=1}^{k} a_l \delta_{x_l}, \tag{4.102}$$

by (4.101), which is the second part (2) of the theorem. □

Corollary 4.5.18. *Let (M, g) and (N, h) be compact Riemannian manifolds and let $C > 0$ be a positive constant. Denote*

$$\mathcal{H} = \{f : M \to N \text{ is harmonic} : \int_M |df|^m dv \le C\}, \tag{4.103}$$

where $m = \dim M$. Then for any sequence in $\{f_i\} \in \mathcal{H}$ there exist a finite set $S = \{x_1, \cdots, x_k\}$ and a smooth harmonic map $f_\infty : M - S \to N$ such that

(1) *A subsequence $\{f_{i_j}\}$ converges to f_∞ in the C^∞ topology on $M - S$ as $j \to \infty$;*
(2) *The Radon measure $|df_{i_j}|^m dv$ converges to a measure*

$$|df_\infty|^m dv + \sum_{l=1}^{k} a_l \delta_{x_l}, \tag{4.104}$$

as $j \to \infty$, where a_l is a constant and δ_{x_l} is the Dirac measure whose support is $\{x_l\}$ $(l = 1, \cdots, k)$.

4.6 Transversally Biharmonic Maps

We generalize the notion of transversally harmonic map by Konderak and Wolak [230] to that of transversally biharmonic map, which is based on the work of Chiang and Wolak [83] in 2008. Since V-manifolds are special cases of Riemannian foliations, transversally biharmonic maps [83] also generalize biharmonic maps of V-manifolds [82].

4.6.1 General Results

We follow the notations and notions in Sect. 1.9. Let $(M_1, \mathcal{F}_1, g_1)$ and $(M_2, \mathcal{F}_2, g_2)$ be two foliated Riemannian manifolds. Suppose that $f : (M_1, \mathcal{F}_1) \to (M_2, \mathcal{F}_2)$ is a smooth foliated leaf preserving map. Let $U_i \subset M_i$ be open subsets and let $\phi_i : (U_i, g_i) \to (\bar{U}_i, \bar{g}_i)$ be Riemannian submersions on U_i which define locally the Riemannian foliations $\mathcal{F}_i$ for $i = 1, 2$. Suppose that $f(U_1) \subset U_2$. Let $X_1, \cdots, X_{q1}$ and $Y_1, \cdots, Y_{q_2}$ be two local bases of foliated sections of $T\mathcal{F}_1{}^{\perp}$ and $T\mathcal{F}_2{}^{\perp}$ over U_1 and U_2, respectively. Then $X_1, \cdots, X_{q_1}$ are projectable via ϕ_1 on the frame sections $\bar{X}_1, \cdots, \bar{X}_{q1}$, and $Y_1, \cdots, Y_{q_2}$ are projectable via the map ϕ_2 on the frame sections $\bar{Y}_1, \cdots, \bar{Y}_{q2}$. Then there exists a unique map $\bar{f} : \bar{U}_1 \to \bar{U}_2$ such that the diagram

$$\begin{array}{ccc} U_1 & \xrightarrow{\ f\ } & U_2 \\ {\scriptstyle \phi_1}\downarrow & & {\scriptstyle \phi_2}\downarrow \\ \bar{U}_1 & \xrightarrow{\ \bar{f}\ } & \bar{U}_2 \end{array}$$

Diagram 4.6.1.

commutes.

Let X, Y, ξ be foliated sections of $T\mathcal{F}_2^{\perp}$, and $D' = D^2$ be the basic partial connection on $T\mathcal{F}_2^{\perp}$. Then the Riemann curvature $R'(X, Y)\xi = D'_X D'_Y \xi - D'_Y D'_X \xi - D'_{[X,Y]}\xi$ is a section of the bundle $T\mathcal{F}_2^{\perp} \to M_2$. Following the concept of transversal tension field in Sect. 1.9, we define the transversal bitension field as

$$(\tau_2)_b(f) = \Delta\tau_b(f) + R'(df, df)\tau_b(f), \tag{4.105}$$

where $\Delta\xi = D^*D(\xi)$, so Δ is an operator from sections of $f^{-1}T\mathcal{F}_2^{\perp}$ to sections of $f^{-1}T\mathcal{F}_2^{\perp}$, and D is the connection on $T\mathcal{F}_1^{\perp *} \otimes f^{-1}T\mathcal{F}_2^{\perp}$. Therefore, $(\tau_2)_b(f)$ is a section of the bundle $f^{-1}T\mathcal{F}_2^{\perp} \to M_1$.

We consider a one-parameter family of maps $\{f_t\} \in C^{\infty}((M_1, \mathcal{F}_1), (M_2, \mathcal{F}_2))$, $t \in I_\epsilon = (-\epsilon, \epsilon)$ from a compact foliated Riemannian manifold $(M_1, \mathcal{F}_1)$ into a foliated Riemannian manifold $(M_2, \mathcal{F}_2)$ such that $f_t(x)$ is the endpoint of the segment starting at $f(x)$ determined in length and direction by the vector field $\dot{f}$ along f. These induce a one-parameter family of maps $\{\bar{f}_t\} \in C^{\infty}(N_1, N_2)$ such that $\bar{f}_t(x)$ is the end point of the segment starting at $\bar{f}(x)$ determined in length and direction by the vector field $\dot{\bar{f}}$ along $\bar{f}$. The transversal bienergy of f is

$$E_2(\bar{f}) = \frac{1}{2}\int_{N_1} ||(d+d^*)^2 \bar{f}||^2 dv = \frac{1}{2}\int_{N_1} ||d^* d\bar{f}||^2 dv = \frac{1}{2}\int_{N_1} ||\tau \bar{f}||^2 dv, \tag{4.106}$$

where $N_1 = \coprod(\bar{U}_1)_i$ is the transverse manifold of M_1. Assume that the compact supports of $\frac{\partial \bar{f}_t}{\partial t}$ and $\nabla_{\bar{X}_i}\frac{\partial \bar{f}_t}{\partial t}$ are contained in the interior of each $\bar{U}_1$. Then by applying computation similar to those in Sect. 4.1 and the concepts of foliations, we have

$$\frac{d}{dt}E_2(\bar{f}_t)|_{t=0} = \int_{\coprod(\bar{U}_1)_i} (J(\tau\bar{f}), \tau(\bar{f}))dv, \tag{4.107}$$

where

$$\tau_2(\bar{f}) = J(\tau\bar{f}) = \Delta\tau(\bar{f}) + \bar{R}'(d\bar{f}, d\bar{f})\tau(\bar{f}) \tag{4.108}$$

in each $\bar{U}_1$, $\Delta = \nabla^*\nabla$ is an operator between local sections of $\bar{f}^{-1}T\bar{U}_2 \to \bar{U}_1$, ∇ is the connection on $T^*\bar{U}_1 \otimes f^{-1}T\bar{U}_2$, and the Riemannian curvature $\bar{R}'$ is a local section of $\bar{f}^{-1}T\bar{U}_2 \to \bar{U}_1$, which is the transverse Riemann curvature of $(M_2, \mathcal{F}_2)$.

There is a close relationship between the transversal bitension field of f and the bitension fields of the induced maps $\bar{f}$, obtained by using the local submersions defining the foliations $\mathcal{F}_1$ and $\mathcal{F}_2$. Then, by Diagram 4.6.1,

$$d\phi_2(\tau_2)_b(f)_x = \tau_2(\bar{f})_{\phi_1(x)} \tag{4.109}$$

holds for each of the foliation defining local submersions $\phi_i : U_i \to \bar{U}_i, i = 1, 2$, such that $f(U_1) \subset U_2$. The definition of a transversally biharmonic map does not depend on the choices of local Riemannian submersions defining the Riemannian foliations.

Theorem 4.6.1. *Let $f : (M_1, \mathcal{F}_1) \to (M_2, \mathcal{F}_2)$ be a smooth foliated map between two foliated Riemannian manifolds. Then f is transversally biharmonic if and only if the induced map $\bar{f}$ is biharmonic in each $\bar{U}_1$.*

Proof. It follows from the Diagram 4.6.1 and (4.109). □

Theorem 4.6.2. *If $f : (M_1, \mathcal{F}_1) \to (M_2, \mathcal{F}_2)$ is a transversally biharmonic map from a compact foliated Riemannian $(M_1, \mathcal{F}_1)$ manifold into a foliated Riemannian manifold $(M_2, \mathcal{F}_2)$ with non-positive transverse Riemann curvature, then f is transversally harmonic.*

Proof. Since f is transversally biharmonic, for any U_1, by Theorem 4.6.1, the induced map $\bar{f}$ is biharmonic in $\bar{U}_1$. Then from (4.108) we have

$$\tau_2(\bar{f}) = \Delta\tau(\bar{f}) + \bar{R}'(d\bar{f}, d\bar{f})\tau(\bar{f})$$

in each $\bar{U}_1$. Let $\bar{X}_1, \cdots, \bar{X}_{q_1}$ be the local frame sections over $\bar{U}_1$. We obtain

$$\frac{1}{2}\Delta||\tau(\bar{f})||^2 = (\nabla_{\bar{X}_i}\tau(\bar{f}), \nabla_{\bar{X}_i}\tau(\bar{f})) + (\nabla^*\nabla\tau(\bar{f}), \tau(\bar{f})$$
$$= (\nabla_{\bar{X}_i}\tau(\bar{f}), \nabla_{\bar{X}_i}\tau(\bar{f})) - (\bar{R}'(d\bar{f}, d\bar{f})\tau(\bar{f}), \tau(\bar{f})) \geq 0, \quad (4.110)$$

by (4.108) and the fact that $(M_2, \mathcal{F}_2)$ is a foliated Riemannian manifold of non-positive transverse Riemann curvature. Applying Bochner's techniques and the assumption that the compact supports of $\frac{\partial \bar{f}_t}{\partial t}$ and $\nabla_{\bar{X}_i}\frac{\partial \bar{f}_t}{\partial t}$ are contained in the interior of $\bar{U}_1$, it follows that $||\tau(\bar{f})||^2$ is constant in each $\bar{U}_1$. Using this in (4.110), we have

$$\nabla_{\bar{X}_i}\tau(\bar{f}) = 0, \quad \forall i = 1, 2, \cdots, q_1.$$

According to Eells and Lemaire [119], we can conclude that $\tau(\bar{f}) = 0$ in each $\bar{U}_1$, i.e., f is transversally harmonic on $(M, \mathcal{F}_1)$. □

4.6.2 Examples

Since any transversally biharmonic map into a foliated Riemannian manifold of non-positive transverse Riemann curvature is transversally harmonic, we construct the following examples of transversally biharmonic non-harmonic maps into foliated Riemannian manifolds with positive transverse Riemann curvature.

Example 1. In S^{q_1+1}, there is a compact hypersurface whose Gauss map is an isometric immersion, namely, the Clifford surface $M_k^{q_1}(1) = S^k(\sqrt{1/2}) \times S^{q_1-k}(\sqrt{1/2})$, where $0 \leq k \leq q_1$ [426]. Let $(F_1, h_1), (F_2, h_2)$ be two Riemannian manifolds. Consider the foliations on $M_k^{q_1}(1) \times F_1$ and $S^{q_1+1} \times F_2$ given by the projections on the first component $\pi_1 : M_k^{q_1}(1) \times F_1 \to M_k^{q_1}(1)$ and $\pi_2 : S^{q_1+1} \times F_2 \to S^{q_1+1}$, respectively. The projections π_1 and π_2 define Riemannian foliations. Let $f : M_k^{q_1}(1) \times F_1 \to S^{q_1+1} \times F_2$ be a smooth leaf preserving map. Then f has to be of the form $f(x, y) = (f_1(x), f_2(x, y)), x \in M_k^{q_1}(1), y \in F_1$, where $f_1 : M_k^{q_1}(1) \to S^{q_1+1}$, $f_2 : M_k^{q_1}(1) \times F_1 \to F_2$ are smooth, and the diagram

$$\begin{array}{ccc} M_k^{q_1}(1) \times F_1 & \xrightarrow{f} & S^{q_1+1} \times F_2 \\ \pi_1 \downarrow & & \pi_2 \downarrow \\ M_k^{q_1}(1) & \xrightarrow{f_1} & S^{q_1+1} \end{array}$$

commutes. Let $f_1 : M_k^{q_1}(1) \to S^{q_1+1}$ be the standard isometric immersion such that $k \neq \frac{q_1}{2}$. Then we have

$$||B(f_1)||^2 = k + q_1 - k = q_1, \ ||\tau(f_1)|| = |k - (q_1 - k)| = |2k - q_1| \neq 0. \tag{4.111}$$

It follows from Theorem 4.2.2 that f_1 is a biharmonic non-harmonic map. By the construction of the transversal bitension field of a map between foliated manifolds, f is a transversally biharmonic non-harmonic map.

Let $f : (M_1, \mathcal{F}_1) \to (M_2, \mathcal{F}_2)$ be a smooth foliated map between foliated Riemannian manifolds with $N_1 = M_k^{q_1+1}(1) = S^k(\sqrt{1/2}) \times S^{q_1-k}(\sqrt{1/2}) = \coprod \bar{U}_1, \ k \neq \frac{q_1}{2}, N_2 = S^{q_1+1} = \coprod \bar{U}_2$, which induces $\bar{f} : \bar{U}_1 \to \bar{U}_2$ such that $\bar{f} \circ \phi_1 = \phi_2 \circ f$ as in Diagram 4.6.1. Note that $f_1 : M_k^{q_1}(1) \to S^{q_1+1}$ is the standard isometric immersion such that $k \neq \frac{q_1}{2}$. Consider each induced $\bar{f}|_{\bar{U}_1} : \bar{U}_1 \to \bar{U}_2$ as the restriction of $f_1 : M_k^{q_1}(1) \to S^{q_1+1}$. Similarly to (4.111), we have $||B(\bar{f}|_{\bar{U}_1})||^2 = q_1, \ ||\tau(\bar{f}|_{\bar{U}_1})|| = ||2k - q_1|| \neq 0$. By Theorem 4.6.1, each induced map $\bar{f}|_{\bar{U}_1} : \bar{U}_1 \to \bar{U}_2$ is biharmonic non-harmonic if and only if $f : (M_1, \mathcal{F}_1) \to (M_2, \mathcal{F}_2)$ is transversally biharmonic non-harmonic.

Example 2. Let F_1, F_2, N be Riemannian manifolds, and $I \subset \mathbf{R}$ be an open interval. Consider the foliations on $I \times F_1$ and $N \times F_2$ given by the projections $\pi_1 : I \times F_1 \to I$ and $\pi_2 : N \times F_2 \to N$ on the first component. Let $f : I \times F_1 \to N \times F_2$ be a smooth foliated map preserving the leaves of the foliations. Then $f(x, y) = (f_1(x), f_2(x, y)), \ x \in I, \ y \in F_1$, where $f_1 : I \to N$ and $f_2 : I \times F_1 \to F_2$ are smooth, and we have $\pi_2 \circ f = f_1 \circ \pi_1$. Let $f_1 = \gamma : I \to (N, h)$ be a curve parameterized by arc length from I to a n-dimensional Riemannian manifold (N^n, h). In particular, if (N^2, h) is an oriented surface, then by (4.8) in Sect. 4.1,

$$k_g = constant, \ k_g^2 = G,$$

where k_g is the curvature (with sign) of γ, $G = R(T, N, T, N)$ is the Gaussian curvature of the surface, and T and N are the unit tangent vector field and the unit normal vector field of the surface. Therefore, $f_1 = \gamma : I \to N$ is a biharmonic non-harmonic map iff $f : I \times F_1 \to N \times F_2$ is a transversally biharmonic non-harmonic map.

Example 3. We can use the suspension construction in [230] to produce a transversally biharmonic map. Let $S_1 = S_2 = S^1$ be the unit circle and $F_1 = F_2 = S^1$. Consider a smooth map $f : S_1 \to S_2$ such that $\pi_1(f)(m) = mn$ for $n \in \mathbf{Z}^*$. Let $\phi : F_1 \to F_2$ be the map given by $\phi(z) = cz\bar{z}$, where $c \in S^1 \subset \mathbf{C}, c \neq 0$. We know that ϕ is biharmonic non-harmonic. Then we define two homomorphisms $h_i : \mathbf{Z} \to Isom(F_i), \ i = 1, 2$ by $h_1(m)z = q^{nm}z$ and $h_2(m)z = (q\bar{q})^m z\bar{z}$, where $q \in S^1$. It follows that ϕ is (h_1, h_2) equivariant, and we take

$$M_1 = \mathbf{R} \times S^1 / h_1, \ M_2 = \mathbf{R} \times S^1 / h_2.$$

Then we get the map $\tilde{\psi} = \tilde{f} \times \phi$, which in turn induces a map $\psi : M_1 \to M_2$. Applying Lemma 1.9.4 to biharmonic maps, we obtain the following proposition:

$\phi : F_1 \to F_2$ *is biharmonic iff* ψ *is transversally biharmonic.*

Therefore, $\psi([x, y]) = [\tilde{f}(x), cz\bar{z}]$ is transversally biharmonic non-harmonic. In fact, f is homotopic to the map $f_0 : S^1 \to S^1$ with $f_0(z) = z^n$, where $z \in S^1 \subset \mathbf{C}$. If we consider a particular case of $f = f_0$, then the suspension map is given by $\psi([x, y]) = [2xn\pi, cz\bar{z}]$.

Example 4 shows that the biharmonicity of a map does not imply the transversal biharmonicity of the map. Example 5 below shows that the transversal biharmonicity does not imply biharmonicity either.

Example 4. Let $f : B_1 \times F_1 \to B_2 \times F_2$ be a smooth map preserving the leaves of the foliations such that $f(x, y) = (f_1(x), f_2(x, y))$, where $B_1 = B_2 = F_1 = F_2 = \mathbf{R}$. By (1.58) and (1.59), choose $\alpha_1(x) = 0$ and $\alpha_2(x) = 3x$ as two warping functions in $\mathbf{R}$ and let $f_1(x) = x^4$, $f_2(x, y) = x^2$. We have

$$\tau(f) = \tau(f_1) + \tau(f_2|B_1) + \tau(f_2|F_1) - ||df_2||^2(grad_{g_2}\alpha_2) \circ f_1 = 12x^2 + 2 - 12x^2 = 2 \neq 0,$$

where the third term vanishes. Then we get $\tau_2(f) = 0$, and therefore, f is biharmonic non-harmonic. However, $\tau_2(f_1) = 24 \neq 0$, which implies that f is not transversally biharmonic. It follows that the biharmonicity of the map f does not imply the transversal biharmonicity of the map.

Example 5. Let (B_1, g_1), (B_2, g_2), (F_1, h_1) and (F_2, h_2) be Riemannian manifolds. Consider the foliations on $B_1 \times F_1$ and $B_2 \times F_2$ given by the projections on the first component $\pi_1 : B_1 \times F_1 \to B_1$ and $\pi_2 : B_2 \times F_2 \to B_2$, respectively. The projections π_1 and π_2 are Riemannian submersions, and the foliations are also Riemannian. Let $f : B_1 \times F_1 \to B_2 \times F_2$ be a smooth map which preserves the leaves of the foliations. Then f must be of the form $f(x, y) = (f_1(x), f_2(x, y)), x \in B_1, y \in F_1$, where $f_1 : B_1 \to B_2$, $f_2 : B_1 \times F_1 \to F_2$ are smooth. For the product Riemannian metrics on $B_1 \times F_1$ and $B_2 \times F_2$, the bitension field of f can be expressed as

$$\tau_2(f) = (\tau_2(f_1), \tau_2(f_2|_{B_1}) + \tau_2(f_2|_{F_1})), \tag{4.112}$$

where $\tau_2(f_1)$ is the bitension field at x of $f_1 : B_1 \to B_2$, $\tau_2(f_2|_{B_1})$ is the bitension field at x of the map $x \mapsto f_2(x, y)$ with y fixed, and $\tau_2(f_2|_{F_1})$ is the bitension field at y of the map $y \mapsto f_2(x, y)$ with x fixed. On one hand, by (4.112), the biharmonicity of $f = (f_1, f_2)$ is equivalent to f_1 is biharmonic and $\tau_2(f_2|_{B_1}) + \tau_2(f_2|_{F_1}) = 0$, i.e., the vertical and horizontal contributions to the bitension field annihilate each other. On the other hand, if f_1 is biharmonic and $f_2|_{B_1}$, $f_2|_{F_1}$ are biharmonic for $x \in B_1, y \in F_1$, then f is biharmonic. Hence, it follows that there are maps f which are transversally biharmonic, but not biharmonic.

4.6.3 Transversally Biharmonic Maps and Holonomy Pseudogroups

Let $f : (M_1, \mathcal{F}_1) \to (M_2, \mathcal{F}_2)$ be a smooth foliated map. Suppose that $\mathcal{U} = \{U_i, \phi_i, g_{ij}\}_I$ is a cocycle defining the foliation $\mathcal{F}_1$, and $\mathcal{V} = \{V_\alpha, \psi_\alpha, h_{\alpha\beta}\}_A$ is a cocycle defining the foliation $\mathcal{F}_2$, such that for any $i \in I$ there exists $\alpha(i) \in A$ for which $f(U_i) \subset V_{\alpha(i)}$. Let $\bar{U}_i = \phi_i(U_i)$ and $\bar{V}_\alpha = \psi_\alpha(V_\alpha)$. Then $N_1 = \coprod \bar{U}_i$ is a transverse manifold of the foliation $\mathcal{F}_1$, and $N_2 = \coprod \bar{V}_\alpha$ is a transverse manifold of the foliation $\mathcal{F}_2$ [169, 230]. The transformations g_{ij} generate a pseudogroup $\mathcal{H}_1$, which is called the holonomy pseudogroup of $\mathcal{F}_1$ associated to the cocycle $\mathcal{U}$, and similarly the transformations $h_{\alpha\beta}$ generate a pseudogroup $\mathcal{H}_2$, the holonomy pseudogroup of $\mathcal{F}_2$ associated with the cocycle $\mathcal{V}$.

On the level of transverse manifolds, the map f induces a smooth map $\bar{f}$ for any $i \in I$ and the following diagram commutes:

$$\begin{array}{ccc} U_i & \xrightarrow{f|U_i} & V_{\alpha(i)} \\ {\scriptstyle\phi_i}\downarrow & & \downarrow{\scriptstyle\psi_{\alpha(i)}} \\ \bar{U}_i & \xrightarrow{\bar{f}_{\alpha(i)i}} & \bar{V}_{\alpha(i)} \end{array}$$

The map $\bar{f} : N_1 \to N_2$ is defined as follows:

$$\bar{f}|\bar{U}_i = \bar{f}_{\alpha(i)i}.$$

The map $\bar{f}$ has the following property. Take two open sets U_i and U_j such that $U_i \cap U_j \neq \emptyset$; then $f(U_i \cap U_j) \subset V_{\alpha(i)} \cap V_{\alpha(j)}$. The intersection $U_i \cap U_j$ covers the open subset $\bar{U}_{ji}$ in $\bar{U}_i$ and the open subset $\bar{U}_{ij}$ in $\bar{U}_j$. Likewise, $V_{\alpha(i)} \cap V_{\alpha(j)}$ covers $\bar{V}_{\alpha(j)\alpha(i)}$ in $\bar{V}_{\alpha(i)}$ and $\bar{V}_{\alpha(i)\alpha(j)}$ in $\bar{V}_{\alpha(j)}$. Moreover, the map $g_{ji} : \bar{U}_{ji} \to \bar{U}_{ij}$ is a diffeomorphism and $h_{\alpha(j)\alpha(i)} : \bar{V}_{\alpha(j)\alpha(i)} \to \bar{V}_{\alpha(i)\alpha(j)}$ is also a diffeomorphism. Then

$$h_{\alpha(j)\alpha(i)} \bar{f}_{\alpha(i)i} | \bar{U}_{ji} = \bar{f}_{\alpha(j)j} g_{ji} | \bar{U}_{ji}.$$

In order to describe the properties of the induced map $\bar{f}$ better, we need to recall the notion of a morphism between pseudogroups in [169]. A family $\mathcal{K}$ of smooth local maps from N_1 to N_2 is called a *morphism* of $(N_1, \mathcal{H}_1)$ into $(N_2, \mathcal{H}_2)$ if the following conditions hold: (1) Each $k \in \mathcal{K}$ is a smooth map $k : W \to N_2$, $W \subset N_1$. (2) The domains of $k \in \mathcal{K}$ form an open covering of N_1. (3) For any $k \in \mathcal{K}, k : W \to N_2$, and any open subset $W' \subset W$, the restriction $k|W'$ lies in $\mathcal{K}$. (4) For any family of maps $k_i \in \mathcal{K}$ such that the map $k = \bigcup k_i$ is well defined, $k \in \mathcal{K}$. (5) For any $h_1 \in \mathcal{H}_1$ and $h_2 \in \mathcal{H}_2$, the map $h_2^{-1} \circ k \circ h_1 \in \mathcal{K}$. (6) For any $h \in \mathcal{H}_1$, and any $k_1, k_2 \in \mathcal{K}$, the map $k_2 h k_1^{-1} \in \mathcal{H}_2$.

A morphism Φ is an equivalence of the pseudogroups $\mathcal{H}_1$ and $\mathcal{H}_2$ if Φ^{-1} is also a morphism of $\mathcal{H}_2$ into $\mathcal{H}_1$. Let $\tilde{N}_1, \tilde{N}_2$ two smooth manifolds with two pseudogroups $\tilde{\mathcal{H}}_1, \tilde{\mathcal{H}}_2$ which are equivalent to $\mathcal{H}_1, \mathcal{H}_2$, respectively. Then for any $\phi, \phi' \in \Phi$ and $g \in \mathcal{H}_1, g' \in \tilde{H}_1, \phi' \circ g \circ \phi^{-1} \in \tilde{\mathcal{H}}_1$ and $\phi'^{-1} \circ g' \circ \phi \in \mathcal{H}_1$. Similarly, for any $\psi, \psi' \in \Psi$ and $g \in \mathcal{H}_2, g' \in \tilde{H}_2, \psi' \circ g \circ \psi^{-1} \in \tilde{\mathcal{H}}_2$ and $\psi'^{-1} \circ g' \circ \psi \in \mathcal{H}_2$. Let $\mathcal{K}$ be a morphism of $(N_1, \mathcal{H}_1)$ into $(N_2, \mathcal{H}_2)$. Then the maps $\psi \circ k \circ \phi^{-1}$ for $\phi \in \Phi, k \in \mathcal{K}, \psi \in \Psi$, define a morphism of $(\tilde{N}_1, \tilde{H}_1)$ into $(\tilde{N}_2, \tilde{H}_2)$, denoted by $\Psi \circ \mathcal{K} \circ \Phi^{-1}$. Finally, we have the following:

Proposition 4.6.3. *Let $f : (M_1, \mathcal{F}_1) \to (M_2, \mathcal{F}_2)$ be a foliated map. Let $\mathcal{U}, \tilde{\mathcal{U}}$ be two cocycles defining the foliation $\mathcal{F}_1$ and $\mathcal{V}, \tilde{\mathcal{V}}$ be two cocycles defining the foliation $\mathcal{F}_2$. Let $(N_1, \mathcal{H}_1), (\tilde{N}_1, \tilde{H}_1)$, $(N_2, \mathcal{H}_2), (\tilde{N}_2, \tilde{H}_2)$ be the corresponding transverse manifolds and holonomy pseudogroups. Let $\Phi : (N_1, \mathcal{H}_1) \to (\tilde{N}_1, \tilde{H}_1)$ and $\Psi : (N_2, \mathcal{H}_2) \to (\tilde{N}_2, \tilde{H}_2)$ be equivalences of pseudogroups. Let $\mathcal{K}(f) : (N_1, \mathcal{H}_1) \to (N_2, \mathcal{H}_2)$ and $\tilde{\mathcal{K}}(f) : (\tilde{N}_1, \tilde{\mathcal{H}}_1) \to (\tilde{N}_2, \tilde{\mathcal{H}}_2)$ be the morphisms induced by f. Then*

$$\tilde{\mathcal{K}}(f) = \Psi \circ \mathcal{K}(f) \circ \Phi^{-1}.$$

Now let $\mathcal{H}_1$ be a pseudogroup of local isometries of a Riemannian manifold (N_1, g_1). Let $\tilde{\mathcal{H}}_1$ be a pseudogroup of local transformations of the manifold $\tilde{N}_1$ that is equivalent to the pseudogroup $\mathcal{H}_1$. Then there is a Riemannian metric $\tilde{g}_1$ on $\tilde{N}_1$ for which $\tilde{H}_1$ is a pseudogroup of local isometries and the equivalence between $\mathcal{H}_1$ and $\tilde{\mathcal{H}}_1$ consists of local isometries of (N_1, g_1) into $(\tilde{N}_1, \tilde{g}_1)$ [42]. By applying Proposition 4.6.3, we can obtain the following:

Theorem 4.6.4. *Let $(M_1, \mathcal{F}_1)$, $(M_2, \mathcal{F}_2)$, $(\tilde{M}_1, \tilde{\mathcal{F}}_1)$, $(\tilde{M}_2, \mathcal{F}_2)$ be four foliated Riemannian manifolds. Let $f : (M_1, \mathcal{F}_1) \to (M_2, \mathcal{F}_2)$ and $\tilde{f} : (\tilde{M}_1, \tilde{\mathcal{F}}_1) \to (\tilde{M}_2, \tilde{\mathcal{F}}_2)$ be two foliated maps. Suppose that the holonomy pseudogroups of $(M_1, \mathcal{F}_1)$ and $(\tilde{M}_1, \tilde{\mathcal{F}}_1)$ are equivalent and so are those of $(M_2, \mathcal{F}_2)$ and $(\tilde{M}_2, \mathcal{F}_2)$. If $\tilde{\mathcal{K}}(f) = \mathcal{K}(\tilde{f})$, then the map f is transversally biharmonic if and only if the map $\tilde{f}$ is.*

Proof. The map f is transversally biharmonic if and only if the induced morphism $\mathcal{K}(f)$ consists of biharmonic maps between the transverse manifolds for some (and then for any) choice of the cocycles defining the foliations. Since the corresponding pseudogroups are equivalent, the second induced map consists of biharmonic maps for the transported Riemannian metric. Therefore, the map $\tilde{f}$ is transversally biharmonic for any bundle-like metric inducing the given Riemannian metrics on the transverse manifolds. □

Corollary ([230]). *Let $(M_1, \mathcal{F}_1)$, $(M_2, \mathcal{F}_2)$, $(\tilde{M}_1, \tilde{\mathcal{F}}_1)$, $(\tilde{M}_2, \mathcal{F}_2)$ be four foliated Riemannian manifolds. Let $f : (M_1, \mathcal{F}_1) \to (M_2, \mathcal{F}_2)$ and $\tilde{f} : (\tilde{M}_1, \tilde{\mathcal{F}}_1) \to (\tilde{M}_2, \tilde{\mathcal{F}}_2)$ be two foliated maps. Suppose that the holonomy pseudogroups of $(M_1, \mathcal{F}_1)$ and $(\tilde{M}_1, \tilde{\mathcal{F}}_1)$ are equivalent and so are those of $(M_2, \mathcal{F}_2)$ and $(\tilde{M}_2, \mathcal{F}_2)$. If $\tilde{\mathcal{K}}(f) = \mathcal{K}(\tilde{f})$, then the map f is transversally harmonic if and only if the map $\tilde{f}$ is.*

4.7 Conservation Law

We discuss the conservation law of the stress bienergy tensor of a biharmonic map. All the theorems and results were obtained by Jiang [198]. Following Jiang's work, Loubean, Montaldo and Oniciuc [259] have also studied the stress bienergy tensor of a biharmonic map and have obtained similar results as [198].

4.7.1 Stress Bienergy Tensor

Let $f : (M, g) \to (N, h)$ be a smooth map between two Riemannian manifolds, and $e(f) = \frac{1}{2}|df|^2$ be the energy density of f. Eells and Baird [18] proved that if $S_f = e(f)g - f^*h$ is the stress energy tensor of f, then $div\, S_f(X) = -(\tau(f), df(X))$, $\forall X \in \Gamma T(M)$. Hence, if f is harmonic, then $div\, S_f = 0$ and we say that f satisfies the conservation law for S_f.

Definition 4.7.1. The stress bienergy tensor $(S_2)_f \in \Gamma(\odot^2 T^*M)$ of a smooth map $f : M \to N$ is defined by

$$(S_2)_f(X,Y) = \frac{1}{2}||\tau(f)||^2(X,Y) + (d\tau(f), df)(X,Y)$$
$$- (\bar{D}_X\tau(f), df(Y)) - (\bar{D}_Y\tau(f), df(X)), \quad X, Y \in \Gamma(TM) \quad (4.113)$$

Let D, $\bar{D}$, $\tilde{D}$ be the connections of $T(M)$, $f^{-1}TN$ and $T^*(M) \otimes f^{-1}TN$, and $\bar{D}^*\bar{D} = \bar{D}_{e_i}\bar{D}_{e_i} - \bar{D}_{D_{e_i}e_i}$, where $\{e_i\}$ is a local orthonormal frame at some point in M. In (4.113), we have

$$(d\tau(f), df) = (d\tau(f)(e_i), df(e_i)) = (\bar{D}_{e_i}\tau(f), df(e_i)). \quad (4.114)$$

Theorem 4.7.2. *Let $f : M \to N$ be a smooth map between two Riemannian manifolds. Then*

$$div\,(S_2)_f(X) = (\bar{D}^*\bar{D}\tau(f) + R^N(df(e_i), \tau(f)df(e_i), df(X)) = (\tau_2(f), df(X)). \quad (4.115)$$

Hence, if $f : M \to N$ is a biharmonic map, then the conservation law for the stress bienergy tensor S_2 is satisfied, i.e., $div\,(S_2)_f = 0$.

Proof. According to the definition of the divergence, we have

$$div(S_2)_f(X) = -(\bar{D}_X\bar{D}_{e_i}\tau(f), df(e_i)) - (\bar{D}_{e_i}\tau(f), \bar{D}_X df(e_i))$$
$$+ (\bar{D}^*\bar{D}\tau(f), df(X)) + (\bar{D}_{e_i}\bar{D}_X\tau(f), df(e_i))$$
$$+ (\bar{D}_{e_i}\tau(f), (\tilde{D}e_i df)(X)) - (\bar{D}_{D_{e_i}X}\tau(f), df(e_i)), \quad (4.116)$$

where we use the fact that

$$(\tilde{D}_X df)(Y) = \bar{D}_X df(Y) - df(D_X Y), \ \ X, Y \in \Gamma(TM). \tag{4.117}$$

In (4.116), by adding and subtracting $(\bar{D}_{D_X e_i}\tau(f), df(e_i))$ we arrive at

$$\begin{aligned} div(S_2)_f(X) &= (\bar{D}^*\bar{D}\tau(f), df(X)) + (R^N(df(X), df(e_i))\tau(f), df(e_i)) \\ &+ (\bar{D}_{e_i}\tau(f), (\tilde{D}_{e_i}df)(X)) - (\bar{D}_{e_i}\tau(f), \bar{D}_X df(e_i)) - (\bar{D}_{D_X e_i}\tau(f), df(e_i)). \end{aligned} \tag{4.118}$$

By the symmetry of Riemannian curvature, we get

$$(R^N(df(X), df(e_i))\tau(f), df(e_i)) = (R^N(df(e_i), \tau(f))df(e_i), df(X)). \tag{4.119}$$

We have $(\tilde{D}_{e_i}df)(X) = (\tilde{D}_X df)(e_i)$ and $D_{e_i}e_j = \Gamma^k_{ij}e_k$. Writing $X = X^i e_i$, (4.117) shows that the third and fourth terms of the right-hand side of (4.118) can be expressed as

$$\begin{aligned} &(\bar{D}_{e_i}\tau(f), (\tilde{D}_{e_i}df)(X)) - (\bar{D}_{e_i}\tau(f), \bar{D}_X df(e_i)) \\ &= -(\bar{D}_{e_i}\tau(f), df(D_X e_i)) = (\bar{D}_{D_X e_i}\tau(f), df(e_i)). \end{aligned} \tag{4.120}$$

Substituting (4.119) and (4.120) into (4.118), we obtain (4.115). □

Let $f : M \to N$ be an isometric embedding with parallel mean curvature. Then $\bar{D}\tau(f)$ lies in f_*TM. By a calculation of Jiang [196],

$$\bar{D}^*\bar{D}\tau(f) = (\tau(f), R^N(df(e_k), df(e_j), df(e_k))df(e_j) - (\tau(f), \tilde{D}_{e_i}df(e_j))(\tilde{D}_{e_i}df)(e_j). \tag{4.121}$$

Therefore, the isometric embedding f with parallel mean curvature satisfies the conservation law for S_2.

Let $B(f) \in \Gamma(\odot^2 T^*M \otimes f^{-1}TN)$ be the second fundamental form of a smooth map $f : M \to N$. Then we have

$$B(f)(X, Y) = (\tilde{D}_X df)(Y) = (\tilde{D}_Y df)(X), \ X, Y \in \Gamma(TM). \tag{4.122}$$

Thus $(\tau(f), B(f))$ is a cross section of $\odot^2 T^*M$, which is called the second fundamental form of f along $\tau(f)$. Recall that the divergence to the stress energy tensor

$$\begin{aligned} div\, S_f(X) &= div(e(f)g - f^*h)(X) = (-)(\tilde{D}_{e_i} \cdot S_f)(e_i \cdot X) \\ &= (-)(\tau(f), df(X)), \ X \in \Gamma(TM), \end{aligned}$$

where there is a $-$ or $+$ sign convention in the above two equations. The map f satisfies the conservation law for S if $div\, S_f = 0$. Thus, if $f : M \to N$ is a harmonic map, then f satisfies the conservation law for S. The following theorem holds.

Theorem 4.7.3. *If $f : M \to N$ is a smooth map satisfying the conservation law for S, then the associated stress bienergy tensor is given by*

$$(S_2)_f = 2(\tau(f), B(f)) - \frac{1}{2}||\tau(f)||^2 g. \tag{4.123}$$

4.7.2 Applications

If $f : M \to N$ is a harmonic map, then $div(S_2)_f = 0$. Conversely, we have the following:

Theorem 4.7.4. *Suppose that M is compact and dim $M = m \neq 4$. If $f : M \to N$ satisfies $(S_2)_f = 0$, then f is harmonic.*

Proof. By (4.113), we have

$$trace(S_2)_f = \frac{m}{2}||\tau(f)||^2 + (m-2)(d\tau(f), df) = 0. \tag{4.124}$$

Let $\xi = (\tau(f), df(e_i))e_i$ be a tangent vector field on M. Then

$$-\, div\, \xi = (D_{e_k}\xi, e_k) = (d\tau(f), df) + ||\tau(f)||^2. \tag{4.125}$$

By the divergence theorem, (4.124) and (4.125) imply that

$$\frac{4-m}{2}\int_M ||\tau(f)||^2 dv = 0.$$

Hence, we can conclude that $\tau(f) = 0$. □

Proposition 4.7.5. *A non-minimal Riemannian immersion $f : (M^4, g) \to (N, h)$ satisfies $S_2 = 0$ if and only if f is pseudo-umbilical.*

Proof. For a Riemannian immersion, we have

$$S_2(X, Y) = -\frac{1}{2}||\tau(f)||^2(X, Y) + 2(\tau(f), B(X, Y)), \tag{4.126}$$

where $B = Ddf$ is the second fundamental form of f. Recall that a Riemannian immersion is pseudo-umbilical if and only if its shape operator satisfies

$$A_{\tau(f)} = \frac{1}{m}||\tau(f)||^2 I,$$

or equivalently,

$$(B(X,Y),\tau(f)) = \frac{1}{m}||\tau(f)||^2(X,Y).$$

We complete the proof by comparing with (4.126). □

Proposition 4.7.6. *Let $f : (M,g) \to (N,h)$ be a non-minimal pseudo-umbilical Riemannian immersion. (1) If $m = 4$, then $S_2 = 0$ and thus $DS_2 = 0$. (2) For $m \neq 4$, we have $DS_2 = 0$ if and only if $||\tau(f)||$ is constant.*

Proof. Since f is pseudo-umbilical, $(Ddf(X,Y)),\tau(f)) = \frac{1}{m}|\tau(f)|^2(X,Y)$. Hence,

$$S_2(X,Y) = \frac{4-m}{2m}|\tau(f)|^2(X,Y).$$ □

We next discuss how to use stress bienergy tensor to study the Gauss map. Let (M, g) be an m-dimensional Riemannian manifold and (N, h) be an $(m+p)$-dimensional compact and simply connected Riemannian manifold with constant sectional curvature c. Let Q denote the space of all m-dimensional totally geodesic subspaces of N. For an isometric embedding $f : M \hookrightarrow N$, based on [278], we define the Gauss map $\gamma : (M,g) \to (Q,k)$ as follows: for each point $x \in M$ corresponds to $f(x) \in N$ tangent to $f(M)$ which produces an m-dimensional totally geodesic submanifold $\gamma(x) \in Q$, where k is the canonical Riemannian metric on Q.

According to [18,278], the first fundamental form of γ is

$$\gamma^*k = (\tau(f), B(f)) - Ricc^M + c(m-1)g. \tag{4.127}$$

We can compute the stress energy tensor of the Gauss map γ as follows:

$$\begin{aligned} S_\gamma = e(\gamma)g - \gamma^*k = {} & \frac{||\tau(f)||^2 - R^M + cm(m-1)}{2}g \\ & - (\tau(f), B(f)) + Ricci^M - c(m-1)g, \end{aligned} \tag{4.128}$$

where $Ricc^M$ and R^M are the Ricci curvature and scalar curvature. Let us denote Einstein tensor field by

$$T = Ricc^M - \frac{R^M}{2}g. \tag{4.129}$$

Note that the isometric embedding f satisfies the conservation law of stress energy tensor, i.e., $div\, S_f = 0$. By Theorem 4.7.3, the stress bi-energy tensor has the form (4.123). Then we can write (4.128) as

$$S_\gamma = T - \frac{1}{2}(S_2)_f + \frac{||\tau(f)||^2 + 2c(m-1)(m-2)}{4}g. \tag{4.130}$$

Since the divergence of the Einstein tensor, $div\, T$, is equal to zero, we can write the divergence of $\frac{1}{4}||\tau(f)||^2$ as the 1-form

$$div(\frac{1}{4}(||\tau(f)||^2 g)(X) = -\frac{1}{2}(\bar{D}_X\tau(f), \tau(f)) = -\frac{1}{4}d||\tau(f)||^2(X), \quad X \in \Gamma(TM).$$

If we view $div S_\gamma$ and $div(S_2)_f$ as 1-forms, we obtain the following theorem.

Theorem 4.7.7 ([198]). *Let $f : M \to N$ be an isometric embedding from a Riemannian manifold M into a compact and connected manifold with parallel mean curvature, and $\gamma : (M, g) \to (Q, k)$ be its associated Gauss map. Then*

$$div\, S_\gamma + \frac{1}{2}div(S_2)_f + \frac{1}{4}d||\tau(f)||^2 = 0. \tag{4.131}$$

4.8 Maps into Lie Groups and Integrable Systems

We study biharmonic maps into compact Lie groups and integrable systems, which generalize the harmonic maps into Lie groups and integrable systems treated in Sect. 1.6. We discuss the formulations of bitension fields, biharmonic maps from real lines into Lie groups, biharmonic maps from open domains in $\mathbf{R}^2$ into Lie groups and the complexification of biharmonic map equations. All the theorems and results were obtained by Urakawa [392] in 2009.

4.8.1 Formulations of Bitension Fields

Let (M, g) be an m-dimensional compact Riemannian manifold, G be an n-dimensional compact Lie group with Lie algebra $\mathcal{G}$, and h the bi-invariant Riemannian metric on G corresponding to the $Ad(G)$-invariant inner product $(\, , \,)$ on $\mathcal{G}$. Let θ be the Maurer-Cartan form on G, i.e., the $\mathcal{G}$-valued left-invariant 1-form on G is defined by $\theta_y(Z_y) = Z$ $(y \in G,\ Z \in \mathcal{G})$. For every smooth map $f : (M, g) \to (G, h)$, we consider the $\mathcal{G}$-valued 1-form α on M given by $\alpha = f^*\theta$. We first have the following well-known fact.

Proposition 4.8.1 ([96]). *For any smooth map $f : (M, g) \to (G, h)$ from a compact Riemannian manifold into a compact Lie group,*

$$\theta(\tau(f)) = -\delta\alpha, \tag{4.132}$$

where $\alpha = f^*\theta$ and θ is the Maurer-Cartan form of G. Hence, $f : (M, g) \to (G, h)$ is harmonic if and only if $\delta\alpha = 0$.

Let D and D^h be the Levi-Civita connections of (M, g) and (G, h), respectively, and $\bar{D}$ be the induced connection on the bundle $f^{-1}TG$.

Lemma 4.8.2. *If $f : (M, g) \to (G, h)$ is a smooth map, then*

$$\theta(\bar{D}_\xi X) = \xi(\theta(X)) + \frac{1}{2}[\alpha(\xi), \theta(X)], \tag{4.133}$$

where $X \in \Gamma(f^{-1}TG)$ and $\xi \in \mathcal{X}(M)$.

Proof. Let $\{\xi_i\}$ be an orthonormal basis of $\mathcal{G}$ with respect to the inner product $(\, , \,)$. For each $X \in \Gamma(f^{-1}TG)$,

$$X(x) = \sum_{i=1}^{n} h_{f(x)}(X(x), \xi_{if(x)})\xi_{if(x)} \in T_{f(x)}G,$$

$$\theta(X)(x) = \sum_{i=1}^{n} h_{f(x)}(X(x), \xi_{if(x)})\xi_i \in \mathcal{G}, \tag{4.134}$$

for any $x \in M$. Thus, for $\xi \in \mathcal{X}(M)$,

$$\begin{aligned} \theta(\bar{D}_\xi X) &= \sum_{i=1}^{n} h(\bar{D}_\xi X, \xi_i)\xi_i = \sum_{i=1}^{n} \{\xi h(X, \xi_i)\xi_i - h(X, \bar{D}_\xi \xi_i)\}\xi_i \\ &= \xi(\theta(X)) - \sum_{i=1}^{n} h(X, \bar{D}_\xi \xi_i)\xi_i \end{aligned} \tag{4.135}$$

where a vector field $V \in \mathcal{X}(G)$ is viewed as an element in $\Gamma(f^{-1}TG)$ by the recipe $V(x) = V(f(x))$, $x \in M$ and we use the compatibility:

$$\xi h(V, W) = h(\bar{D}_\xi V, W) + h(V, \bar{D}_\xi W), \quad V, W \in \Gamma(f^{-1}TG)$$

for $\xi \in \mathcal{X}(M)$.

For $\xi \in \mathcal{X}(M)$, the differential of f, $f_*\xi \in \Gamma(f^{-1}TG)$ means $f_*\xi_x \in T_{f(x)}G$, $x \in M$. It can be expressed as

$$f_*\xi_x = \sum_{j=1}^{n} h(f_*\xi_x, \xi_{jf(x)})\xi_{jf(x)},$$

which implies

$$(\bar{D}_\xi \xi_i)_x = \sum_{j=1}^{n} h(f_*\xi_x, \xi_{jf(x)})(D^h_{\xi_j}\xi_i)_{f(x)}, \tag{4.136}$$

for $x \in M$. Note that the Levi-Civita connection D^h of (G, h) is given by Kobayashi and Nomizu [229]

$$D^h_{\xi_j}\xi_i = \frac{1}{2}[\xi_j, \xi_i] = \frac{1}{2}\sum_{k=1}^n C^k_{ji}\xi_k \tag{4.137}$$

where the structure constants C^k_{ji} of $\mathcal{G}$ are defined by $[\xi_j, \xi_i] = \sum_{k=1}^n C^k_{ji}\xi_k$ and satisfy

$$C^k_{ji} = ([\xi_j, \xi_i], \xi_k) = -(\xi_i, [\xi_j, \xi_k]) = -C^i_{jk}. \tag{4.138}$$

Therefore, by (4.137) and (4.138),

$$\begin{aligned}
\sum_{i=1}^n h(X, \bar{D}_\xi \xi_i)\xi_i &= \frac{1}{2}\sum_{i,j=1}^n h(X, \sum_{k=1}^n h(f_*\xi, \xi_j)C^k_{ji}\xi_k)\xi_i \\
&= -\frac{1}{2}\sum_{i,j,k=1}^n h(X, \xi_k)h(f_*\xi, \xi_j)C^i_{jk}\xi_i \\
&= -\frac{1}{2}\sum_{j,k=1}^n h(X, \xi_k)h(f_*\xi, \xi_j)[\xi_j, \xi_k] \\
&= -\frac{1}{2}\left[\sum_{j=1}^n h(f_*\xi, \xi_j)\xi_j, \sum_{k=1}^n h(X, \xi_k)\xi_k\right] \\
&= -\frac{1}{2}[\alpha(\xi), \theta(X)],
\end{aligned} \tag{4.139}$$

since

$$\begin{aligned}
\alpha(\xi) &= (f^*\theta)(\xi) = \theta(f_*\xi) \\
&= \theta\left(\sum_{j=1}^n h(f_*\xi, \xi_j)\xi_j\right) = \sum_{j=1}^n h(f_*\xi, \xi_j)\xi_j,
\end{aligned} \tag{4.140}$$

and

$$\theta(X) = \sum_{k=1}^n h(X, \xi_k)\theta(\xi_k) = \sum_{k=1}^n h(X, \xi_k)\xi_k. \tag{4.141}$$

Substituting (4.139) into (4.135), we conclude the result. □

Definition 4.8.3. For two $\mathcal{G}$-valued 1-forms α and β on M, we define the $\mathcal{G}$-valued symmetric 2-tensor $[\alpha, \beta]$ on M by

$$[\alpha,\beta](X,Y) = \frac{1}{2}\{[\alpha(X),\beta(Y)] - [\beta(X),\alpha(Y)]\},\ \ X,Y \in \mathcal{X}(M) \tag{4.142}$$

and its trace by

$$trace_g([\alpha,\beta]) = \sum_{i=1}^{m}[\alpha,\beta](e_i,e_i) \tag{4.143}$$

The $\mathcal{G}$-valued 2-form $[\alpha \wedge \beta]$ on M is given by

$$[\alpha\wedge\beta](X,Y) = \frac{1}{2}\{[\alpha(X),\beta(Y)] - [\alpha(Y),\beta(X)]\},\ \ X,Y \in \mathcal{X}(M). \tag{4.144}$$

Theorem 4.8.4. *For $f \in C^\infty(M,G)$, we have*

$$\theta(\tau_2(f)) = \theta(J(\tau(f))) = -\delta d\delta\alpha - trace_g([\alpha, d\delta\alpha]), \tag{4.145}$$

where $\alpha = f^*\theta$. Hence, $f : (M,\ g) \to (G,\ h)$ is biharmonic if and only if

$$\delta d\delta\alpha + trace_g([\alpha, d\delta\alpha]) = 0. \tag{4.146}$$

Proof. (1) We claim that for any $X \in \Gamma(f^{-1}TG)$,

$$\theta(\bar{\Delta}X) = \Delta_g\theta(X) - \sum_{i=}^{m}\Big\{\frac{1}{2}[e_i(\alpha(e_i)),\ \theta(X)] + [\alpha(e_i),\ e_i(\theta(X))]$$
$$+\frac{1}{4}[\alpha(e_i),\ [\alpha(e_i),\ \theta(X)]] - \frac{1}{2}[\alpha(D_{e_i}e_i),\ \theta(X)]\Big\}, \tag{4.147}$$

where $\{e_i\}_{i=1}^m$ is a local orthonormal frame on $(M,\ g)$,

$$\bar{\Delta}X = \bar{D}^*\bar{D}X = (-)\sum_{i=1}^{m}\{\bar{D}_{e_i}(\bar{\nabla}_{e_i}X) - \bar{D}_{D_{e_i}e_i}X\},$$

$\bar{D}$ is the induced connection on the induced bundle $f^{-1}TG$, and Δ_g is the Laplacian of $(M,\ g)$ acting on $C^\infty(M)$.

Indeed, applying Lemma 4.8.2 twice we have

$$\theta(\bar{\Delta}X) = -\sum_{i=1}^{m}\Big[\theta(\bar{D}_{e_i}(\bar{D}_{e_i})) - \theta(\bar{D}_{D_{e_i}e_i}X)\Big]$$
$$= -\sum_{i=1}^{m}\Big\{e_i(\theta(\bar{D}_{e_i}X) + \frac{1}{2}[\alpha(e_i),\ \theta(\bar{D}_{e_i}X)] - D_{e_i}e_i(\theta(X)) - \frac{1}{2}[\alpha(D_{e_i}e_i),\theta(X)]\Big\}$$

$$= -\sum_{i=1}^{m}\Big\{e_i(e_i(\theta(X)) + \frac{1}{2}[\alpha(e_i), \theta(X)]) + \frac{1}{2}[\alpha(e_i), e_i(\theta(X)) + \frac{1}{2}[\alpha(e_i), \theta(X)]]$$

$$- D_{e_i}e_i(\theta(X)) - \frac{1}{2}[\alpha(D_{e_i}e_i), \theta(X)]\Big\}$$

$$= -\sum_{i=1}^{m}\Big\{e_i(e_i(\theta(X))) - D_{e_i}e_i(\theta(X))\Big\} - \sum_{i=1}^{m}\Big\{\frac{1}{2}e_i([\alpha(e_i), \theta(X)])$$

$$+ \frac{1}{2}[\alpha(e_i), e_i(\theta(X))] + \frac{1}{4}\Big[\alpha(e_i), [\alpha(e_i), \theta(X)]\Big] - \frac{1}{2}[\alpha(D_{e_i}e_i), \theta(X)]\Big\}, \tag{4.148}$$

where used the fact that

$$e_i([\alpha(e_i), \theta(X)]) = [e_i(\alpha(e_i)), \theta(X)] + [\alpha(e_i), e_i(\theta(X))].$$

Hence, (4.147) holds, in view of the definition of $\triangle_g$.
(2) Consider the relation

$$-\sum_{i=1}^{m} R^h(X, f_*e_i)f_*e_i = -\sum_{i=1}^{m} R^h(L^{-1}_{f(x)*}X, L^{-1}_{f(x)*}f_*e_i)L^{-1}_{f(x)*}f_*e_i. \tag{4.149}$$

Using the identification $T_eG \ni Z_e \leftrightarrow Z \in \mathcal{G}$, we get

$$T_eG \ni L^{-1}_{f(x)*}f_*e_i \leftrightarrow \alpha(e_i) \in \mathcal{G}, \tag{4.150}$$

$$T_eG \ni L^{-1}_{f(x)*}X \leftrightarrow \theta(X) \in \mathcal{G}. \tag{4.151}$$

Indeed, (4.150) holds because

$$L^{-1}_{f(x)*}f_*e_i = \sum_{j=1}^{n} h(f_*e_i, \xi_{jf(x)})\xi_j$$

and

$$\alpha(e_i) = f^*\theta(e_i) = \theta(f_*e_i)$$
$$= \sum_{j=1}^{n} h(f_*e_i, \xi_{jf(x)})\theta(\xi_{jf(x)})$$
$$= \sum_{j=1}^{n} h(f_*e_i, \xi_{jf(x)})\xi_j,$$

we have (4.150). Similarly, we get (4.151).

Using the above identification, the curvature tensor of (G, h) is given by Kobayashi and Nomizu [229]

$$R^h(X,Y) = -\frac{1}{4}ad([X,Y]),\ \ X,Y \in \mathcal{G}.$$

Thus we obtain

$$\theta\left(-\sum_{i=1}^m R^h(X, f_*e_i)f_*e_i\right) = \frac{1}{4}\sum_{i=1}^m \Big[[\theta(X),\alpha(e_i)],\alpha(e_i)\Big] = \frac{1}{4}\sum_{i=1}^m \Big[\alpha(e_i),[\alpha(e_i),\theta(X)]\Big]. \tag{4.152}$$

(3) For $X \in \Gamma(f^{-1}TG)$, (4.147) and (4.152) yield

$$\begin{aligned}
\theta\left(\bar{\Delta}X - \sum_{i=1}^m R^h(X,, f_*e_i)f_*e_i\right) &= \Delta_g\theta(X) - \sum_{i=1}^m \Big\{\frac{1}{2}[e_i(\alpha(e_i)),\theta(X)] + [\alpha(e_i), e_i(\theta(X))] \\
&+ \frac{1}{4}[\alpha(e_i),[\alpha(e_i),\theta(X)]] - \frac{1}{2}[\alpha(D_{e_i}e_i,\theta(X)]\Big\} + \frac{1}{4}\sum_{i=1}^m \Big[\alpha(e_i),[\alpha(e_i),\theta(X)]\Big] \\
&= \Delta_g\theta(X) - \frac{1}{2}\sum_{i=1}^m[e_i(\alpha(e_i)),\theta(X)] + \sum_{i=1}^m[\alpha(e_i),e_i(\theta(X))] + \frac{1}{2}\sum_{i=1}^m[\alpha(D_{e_i}e_i),\theta(X)] \\
&= \Delta_g\theta(X) - \frac{1}{2}\left[\sum_{i=1}^m(e_i(\alpha(e_i)) - \alpha(D_{e_i}e_i)),\theta(X)\right] + \sum_{m=1}^m[\alpha(e_i),e_i(\theta(X))] \\
&= \Delta_g\theta(X) + \frac{1}{2}[\delta\alpha,\theta(X)] + \sum_{i=1}^m[\alpha(e_i),e_i(\theta(X))].
\end{aligned} \tag{4.153}$$

(4) For $X = \tau(f)$ in (4.153) and $\theta(\tau(f)) = -\delta\alpha$, we obtain

$$\begin{aligned}
\theta(J(\tau(f))) &= \Delta_g(\theta(\tau(f))) + \frac{1}{2}[\delta\alpha,\theta(\tau(f))] + \sum_{i=1}^m[\alpha(e_i),e_i(\theta(\tau(f)))] \\
&= -\Delta_g\delta\alpha - \frac{1}{2}[\delta\alpha,\delta\alpha] - \sum_{i=1}^m[\alpha(e_i),e_i(\delta\alpha)] \\
&= -\Delta_g\delta\alpha - \sum_{i=1}^m[\alpha(e_i),e_i(\delta\alpha)] \\
&= \Delta_g\delta\alpha - \sum_{i=1}^m[\alpha(e_i),d\delta\alpha(e_i)].
\end{aligned} \tag{4.154}$$

Hence, (4.145) follows from (4.154). □

4.8.2 Maps on the Real Line

We consider the simple case that $(M, g) = (\mathbf{R}, g_0)$ is the one-dimensional Euclidean space, and (G, h) is an n-dimensional compact Lie group with bi-invariant Riemannian metric h. Let $f : \mathbf{R} \ni t \mapsto f(t) \in (G, h)$ be a smooth curve in G. Then $\alpha = f^*\theta$ is a $\mathcal{G}$-valued 1-form on $\mathbf{R}$. Thus α can be expressed as

$$\alpha_t = F(t)dt, \tag{4.155}$$

where $F : \mathbf{R} \ni t \mapsto F(t) \in \mathcal{G}$ is the smooth function on $\mathbf{R}$ given by

$$F(t) = \alpha(\frac{\partial}{\partial t}) = f^*\theta(\frac{\partial}{\partial t}) = \theta(f_*(\frac{\partial}{\partial t})). \tag{4.156}$$

Since

$$f'(t) = f_*(\frac{\partial}{\partial t}) = \sum_{i=1}^{n} h_{f(x)}(f_*(\frac{\partial}{\partial t}), \xi_{if(t)})\xi_{if(t)}, \tag{4.157}$$

we get

$$F(t) = \sum_{i=1}^{n} h_{f(t)}(f_*(\frac{\partial}{\partial t}), \xi_{if(t)})\xi_i. \tag{4.158}$$

Therefore, we have the following correspondence:

$$\begin{aligned} T_eG \ni L_{f(t)*}^{-1} f'(t) &= \sum_{i=1}^{n} h_{f(t)}(f'(t), \xi_{if(t)})\xi_{ie} \\ &\leftrightarrow F(t) = \theta(f_* \frac{\partial}{\partial t}) \in \mathcal{G}. \end{aligned} \tag{4.159}$$

Since

$$\delta\alpha = -\{D^h_{e_i}(\alpha(e_i) - \alpha(D_{e_1}e_1)\} = -e_1(\alpha(e_1)) = -e_1(F(t)) = -F'(t), \tag{4.160}$$

we have

$$\delta\alpha = -F'(t). \tag{4.161}$$

Hence, $f : \mathbf{R} \to (G, h)$ is harmonic if and only if

$$\begin{aligned} &\delta\alpha = 0 \Longleftrightarrow F' = 0 \Longleftrightarrow \alpha = X \otimes dt \ (\exists X \in \mathcal{G}) \\ &\Longleftrightarrow f : \mathbf{R} \to (G, h) \text{ is geodesic.} \end{aligned} \tag{4.162}$$

Since

$$F(t) = \theta(f'(t)) = L^{-1}_{f(t)*} f'(t), \tag{4.163}$$

we get

$$f'(t) = L_{f(t)*} X, \tag{4.164}$$

for some $X \in \mathcal{G}$, which implies that

$$f(t) = x \exp(tX + Y)$$

for some $Y \in \mathcal{G}$. Thus any geodesic passing through $f(0) = x$ is given by

$$f(t) = x \exp(tX + Y), \ t \in \mathbf{R} \tag{4.165}$$

for some $X \in \mathcal{G}$ and $Y \in \mathcal{G}$.

Next, we need to determine a biharmonic curve $f : (\mathbf{R}, g_0) \to (G, h)$. It follows from (4.161) that

$$\delta d \delta \alpha = -\frac{\partial^2}{\partial t^2}(-F'(t)) = F^{(3)}(t), \tag{4.166}$$

and

$$trace_g[\alpha, d\delta\alpha] = [\alpha(\frac{\partial}{\partial t}), \ d\delta\alpha(\frac{\partial}{\partial t})] = [F(t), \ F''(t)]. \tag{4.167}$$

Hence, (4.164), (4.165), (4.146) show that $f : (\mathbf{R}, \ g_0) \to (G, \ h)$ is biharmonic if and only if

$$F^{(3)} - [F(t), \ F''(t)] = 0. \tag{4.168}$$

If we choose local coordinates in a neighborhood of $e \in G$ via

$$U \ni \exp\left(\sum_{i=1}^n x_i \xi_i\right) \mapsto (x_1, \cdots, x_n) \in \mathbf{R}^n,$$

then we have local coordinates in a neighborhood of each $x \in G$:

$$xU \ni x \exp\left(\sum_{i=1}^n x_i \xi_i\right) \mapsto (x_1, \cdots, x_n) \in \mathbf{R}^n.$$

Then any smooth curve $f : \mathbf{R} \ni t \mapsto f(t) \in G$ with $f(0) = x$ can be written locally as

$$f(t) = x \exp\left(\sum_{i=1}^{n} x_i(t)\xi_i\right), \ t \in \mathbf{R}.$$

Since

$$L^{-1}_{f(t)*} f'(t) = \sum_{i=1}^{n} h_{f(t)}(f'(t), \xi_i f(t))\xi_{ie} \in T_e G, \tag{4.169}$$

(4.163) becomes

$$F(t) = \sum_{i=1}^{n} h_{f(t)}(f'(t), \xi_{if(t)}\xi_i). \tag{4.170}$$

If we rewrite (4.169) as

$$L^{-1}_{f(t)*} f'(t) = \sum_{i=1}^{n} x_i'(t)\xi_i, \tag{4.171}$$

then we have

$$F(t) = \sum_{i=1}^{n} x_i{}'(t)\xi_i. \tag{4.172}$$

Hence, a smooth curve $f : \mathbf{R} \ni t \mapsto f(t) \in G$ is biharmonic if and only if

$$x_k^{(4)}(t) - \sum_{i,j=1}^{n} x_i{}'(t) x_j^{(3)}(t) C_{ij}^k = 0, \quad k = 1, \cdots, n, \tag{4.173}$$

where $[\xi_i, \xi_j] = \sum_{k=1}^{n} C_{ij}^k \xi_k$. If we set $y_k(t) = x_k{}'(t)\,(k = 1, \cdots, n)$, (4.168) reduces to the following system of third-order ODEs:

$$y_k^{(3)} - \sum_{i,j=1}^{n} y_i(t) y_j{}''(t) C_{ij}^k = 0, \quad k = 1, \cdots, n. \tag{4.174}$$

Theorem 4.8.5 ([392]). (i) *A C^∞ curve $f : \mathbf{R} \to (G, h)$ is harmonic if and only if $f(t) = x\,exp(tX + Y)$ for some $X, Y \in \mathcal{G}$.* (ii) *A C^∞ curve $f : \mathbf{R} \to (G, h)$ is biharmonic if and only if the system of ODEs (4.174) or (4.173) holds.*

4.8.3 Maps on Open Domains in $\mathbf{R}^2$

We consider a biharmonic map $f : \Omega \subset (\mathbf{R}^2, g) \to (G, h)$ from an open domain into a linear compact Lie group, i.e., G is a subgroup of the unitary group $U(n) \subset GL(n, \mathbf{C})$ of degree n with a bi-invariant Riemannian metric h. Let $\mathcal{G}$ be the Lie algebra of G, which thus is a Lie sub-algebra of the Lie algebra $u(n)$ of $U(n)$. The Riemannian metric g on $\mathbf{R}^2$ is a conformal metric given by $g = \lambda^2 g_0$ with a smooth positive function λ and $g_0 = dx \cdot dx + dy \cdot dy$, where (x, y) are the canonical coordinates in $\mathbf{R}^2$. All the theorems and results were obtained in [392].

Let $f : \Omega \ni (x, y) \mapsto f(x, y) = (f_{ij}(x, y)) \in U(n)$ be a smooth map. We consider

$$\frac{\partial f}{\partial x} = \left(\frac{\partial f_{ij}}{\partial x}\right), \ \frac{\partial f}{\partial y} = \left(\frac{\partial f_{ij}}{\partial y}\right).$$

Thus

$$A_x = f^{-1}\frac{\partial f}{\partial x}, \quad A_y = f^{-1}\frac{\partial f}{\partial y} \tag{4.175}$$

are $\mathcal{G}$-valued smooth functions on Ω. It is known that for two $\mathcal{G}$-valued 1-functions A_x and A_y on Ω, there exists a C^∞ map $f : \Omega \to G$ satisfying (4.175) if the integrability condition

$$\frac{\partial A_y}{\partial x} - \frac{\partial A_x}{\partial y} + [A_x, A_y] = 0 \tag{4.176}$$

is satisfied. The pull-back of the Maurer-Cartan form θ by f is

$$\begin{aligned}\alpha &= f^*\theta = f^{-1}df \\ &= f^{-1}\frac{\partial f}{\partial x}dx + f^{-1}\frac{\partial f}{\partial y}dy \\ &= A_x dx + A_y dy,\end{aligned} \tag{4.177}$$

and α is a $\mathcal{G}$-valued 1-form on Ω.

Proposition 4.8.6. *We have the following (well-known) fact:*

$$\begin{aligned}\delta\alpha &= -\lambda^{-2}\left\{\frac{\partial}{\partial x}(f^{-1}\frac{\partial f}{\partial x}) + \frac{\partial}{\partial x}(f^{-1}\frac{\partial f}{\partial y})\right\} \\ &= -\lambda^{-2}\left\{\frac{\partial A_x}{\partial x} + \frac{\partial A_y}{\partial y}\right\}.\end{aligned} \tag{4.178}$$

Hence, $f : (\Omega, g) \to (G, h)$ *is harmonic if and only if*

$$\delta\alpha = 0 \Longleftrightarrow \frac{\partial A_x}{\partial x} + \frac{\partial A_y}{\partial y} = 0. \tag{4.179}$$

We compute the Laplacian Δ_g of $(\mathbf{R}^2, g)$ with $g = \lambda^2 g_0$ as follows:

$$\Delta_g = -\sum_{i,j=1}^{2} g^{ij}\left(\frac{\partial^2}{\partial x^i \partial x^j} - \sum_{k=1}^{2}\Gamma_{ij}^k \frac{\partial}{\partial x^k}\right) = -\lambda^{-2}\left(\frac{\partial^2}{\partial x^2} + \frac{\partial^2}{\partial y^2}\right). \tag{4.180}$$

Therefore, we have

$$\begin{aligned}
\delta d\delta\alpha &= \Delta_g(\delta\alpha) = \lambda^{-2}\left(\frac{\partial^2}{\partial x^2} + \frac{\partial^2}{\partial y^2}\right)\left[\lambda^{-2}(\frac{\partial}{\partial x}(f^{-1}\frac{\partial f}{\partial x}) + \frac{\partial}{\partial y}(f^{-1}\frac{\partial f}{\partial y}))\right] \\
&= \lambda^{-2}\left(\frac{\partial^2}{\partial x^2} + \frac{\partial^2}{\partial y^2}\right)\left[\lambda^{-2}(\frac{\partial A_x}{\partial x} + \frac{\partial A_y}{\partial y})\right] \\
&= -\lambda^{-2}\left(\frac{\partial^2}{\partial x^2} + \frac{\partial^2}{\partial y^2}\right)(\delta\alpha).
\end{aligned} \tag{4.181}$$

On the other hand, choosing an orthonormal local frame $e_1 = \lambda^{-1}\frac{\partial}{\partial x}$, $e_2 = \lambda^{-1}\frac{\partial}{\partial y}$ in $(\mathbf{R}^2, g)$, we have

$$\begin{aligned}
trace_g([\alpha, d\delta\alpha]) &= [\alpha(e_1), d\delta\alpha(e_1)] + [\alpha(e_2), d\delta\alpha(e_2)] \\
&= -\lambda^{-2}[A_x, \frac{\partial}{\partial x}(\lambda^{-2}(\frac{\partial A_x}{\partial x} + \frac{\partial A_y}{\partial y}))] \\
&\quad -\lambda^{-2}[A_y, \frac{\partial}{\partial y}(\lambda^{-2}(\frac{\partial A_x}{\partial x} + \frac{\partial A_y}{\partial y}))] \\
&= \lambda^{-2}[A_x, \frac{\partial}{\partial x}(\delta\alpha)] + \lambda^{-2}[A_y, \frac{\partial}{\partial y}(\delta\alpha)].
\end{aligned} \tag{4.182}$$

Relations (4.181) and (4.182) yield

$$\begin{aligned}
\delta d\delta\alpha + trace_g([\alpha, d\delta\alpha]) &= -\lambda^2\left(\frac{\partial^2}{\partial x^2} + \frac{\partial^2}{\partial y^2}\right)(\delta\alpha) + \lambda^{-2}[A_x, \frac{\partial}{\partial x}(\delta\alpha)] + \lambda^{-2}[A_y, \frac{\partial}{\partial y}(\delta\alpha)] \\
&= -\lambda^2\left\{(\frac{\partial^2}{\partial x^2} + \frac{\partial^2}{\partial y^2})(\delta\alpha) - \frac{\partial}{\partial x}[A_x, \delta\alpha] - \frac{\partial}{\partial y}[A_y, \delta\alpha]\right\},
\end{aligned} \tag{4.183}$$

where the last two terms can be rewritten as

$$\begin{aligned}\frac{\partial}{\partial x}[A_x,\delta\alpha]+\frac{\partial}{\partial y}[A_y,\delta\alpha] &= [\frac{\partial}{\partial x}A_x,\delta\alpha]+[A_x,\frac{\partial}{\partial x}(\delta\alpha)]+[\frac{\partial}{\partial y}A_y,\delta\alpha]+[A_y,\frac{\partial}{\partial y}(\delta\alpha)]\\ &= [\frac{\partial}{\partial x}A_x+\frac{\partial}{\partial y}A_y,\delta\alpha]+[A_x,\frac{\partial}{\partial x}(\delta\alpha)]+[A_y,\frac{\partial}{\partial y}(\delta\alpha)]\\ &= [-\lambda^2\delta\alpha,\delta\alpha]+[A_x,\frac{\partial}{\partial x}(\delta\alpha)]+[A_y,\frac{\partial}{\partial y}(\delta\alpha)]\\ &= [A_x,\frac{\partial}{\partial x}(\delta\alpha)]+[A_y,\frac{\partial}{\partial y}\delta\alpha].\end{aligned}$$

Hence, we obtain the following theorem.

Theorem 4.8.7 ([392]). *Let $g=\lambda^{-2}g_0$ be a Riemannian metric conformal to the standard metric g_0 on an open domain $\Omega\subset\mathbf{R}^2$ with a C^∞ positive function λ on Ω, and let (G,h) be a compact linear Lie group with bi-invariant Riemannian metric h. We have*

(i) *$f:(\Omega,g)\to(G,h)$ is harmonic if and only if*

$$\delta\alpha=0\Longleftrightarrow\frac{\partial}{\partial x}A_x+\frac{\partial}{\partial y}A_y=0. \tag{4.184}$$

(ii) *$f:(\Omega,g)\to(G,h)$ is biharmonic if and only if*

$$\delta d\delta\alpha+\mathit{trace}_g([\alpha,d\delta\alpha])=0\Longleftrightarrow\left(\frac{\partial^2}{\partial x^2}+\frac{\partial^2}{\partial y^2}\right)(\delta\alpha)-\frac{\partial}{\partial x}[A_x,\delta\alpha]-\frac{\partial}{\partial y}[A_y,\delta\alpha]=0, \tag{4.185}$$

where $\delta\alpha$ is the co-differential of the $\mathcal{G}$-valued 1-form $\alpha=A_xdx+A_ydy$ ($A_x=f^{-1}\frac{\partial f}{\partial x}$, $A_y=f^{-1}\frac{\partial f}{\partial y}$) given by

$$\delta\alpha=-\lambda^{-2}\{\frac{\partial}{\partial x}A_x+\frac{\partial}{\partial y}A_y\}. \tag{4.186}$$

(iii) *Consider the two $\mathcal{G}$-valued 1-forms β and Θ on Ω defined by*

$$\beta=[A_x,\delta\alpha]dx+[A_y,\delta\alpha]dy, \tag{4.187}$$

$$\Theta=d\delta\alpha-\beta. \tag{4.188}$$

Hence, $f:(\Omega,g)\to(G,h)$ is biharmonic if and only if

$$\delta\Theta=0. \tag{4.189}$$

Proof. For (iii) we only need to observe (4.185) that is equivalent to

$$0=-\Delta_g(\delta\alpha)+\delta\beta=-\delta(d\delta\alpha-\beta)=-\delta\Theta \tag{4.190}$$

where

$$\begin{aligned}\Theta &= d\delta\alpha - \beta \\ &= \frac{\partial}{\partial x}(\delta\alpha)dx + \frac{\partial}{\partial y}(\delta\alpha)dy - [A_x, \delta\alpha]dx - [A_y, \delta\alpha]dy \\ &= \left\{\frac{\partial}{\partial x}(\delta\alpha) - [A_x, \delta\alpha]\right\} dx + \left\{\frac{\partial}{\partial y}(\delta\alpha) - [A_y, \delta\alpha]\right\} dy. \qquad (4.191)\end{aligned}$$

□

4.8.4 Complexification and Biharmonic Maps on Open Domains in $\mathbf{R}^2$

We use the complex coordinate $z = x + iy$ ($i = \sqrt{-1}$) in $\Omega \subset \mathbf{R}^2$ and set $A_z = \frac{1}{2}(A_x - iA_y)$ and $A_{\bar{z}} = \frac{1}{2}(A_x + iA_y)$. We know that

$$\begin{aligned}&\frac{\partial}{\partial \bar{z}}A_z + \frac{\partial}{\partial z}A_{\bar{z}} = \frac{1}{2}\left\{\frac{\partial}{\partial x}A_x + \frac{\partial}{\partial y}A_y\right\}, \\ &\frac{\partial}{\partial z}A_{\bar{z}} - \frac{\partial}{\partial \bar{z}}A_{\bar{z}} + [A_z, A_{\bar{z}}] = \frac{i}{2}\left\{\frac{\partial}{\partial x}A_y - \frac{\partial}{\partial y}A_x + [A_x, A_y]\right\}, \\ &\frac{\partial}{\partial \bar{z}}A_z + \frac{\partial}{\partial z}A_{\bar{z}} = \frac{1}{2}\left\{\frac{\partial}{\partial x}A_x + \frac{\partial}{\partial y}A_y\right\}.\end{aligned}$$

Also,

$$\begin{aligned}\alpha &= A_x dx + A_y dy = A_z dz + A_{\bar{z}} d\bar{z}, \\ \frac{\partial^2}{\partial x^2} &+ \frac{\partial^2}{\partial y^2} = 4\frac{\partial^2}{\partial z \partial \bar{z}}, \\ \delta\alpha &= -\lambda^{-2}\left(\frac{\partial}{\partial x}A_x + \frac{\partial}{\partial y}A_y\right) = -2\lambda^{-2}\left(\frac{\partial}{\partial \bar{z}}A_z + \frac{\partial}{\partial z}A_{\bar{z}}\right).\end{aligned}$$

Thus, (4.190) and (4.191) are equivalent to

$$\delta\bar{\Theta} = 0, \qquad (4.192)$$

where

$$\bar{\Theta} = \left\{\frac{\partial}{\partial z}(\delta\alpha) - [A_z, \delta\alpha]\right\} dz + \left\{\frac{\partial}{\partial \bar{z}}(\delta\alpha) - [A_{\bar{z}}, \delta\alpha]\right\} d\bar{z}. \qquad (4.193)$$

We want to determine all the biharmonic maps into a compact Lie group (G, h), where $g = \lambda^{-2} g_0$ with a positive smooth function λ on Ω and h is a bi-invariant Riemannian metric on G.

Firstly, we solve the harmonic map equations:

$$\begin{aligned} \frac{\partial}{\partial \bar{z}} B_z + \frac{\partial}{\partial z} B_{\bar{z}} &= 0, \\ -\frac{\partial}{\partial \bar{z}} B_z + \frac{\partial}{\partial z} B_{\bar{z}} + [B_z, B_{\bar{z}}] &= 0. \end{aligned} \tag{4.194}$$

Then there exists a harmonic map $f : (\Omega, g) \to (G, h)$ such that

$$\begin{aligned} f^{-1} \frac{\partial f}{\partial z} &= B_z, \\ f^{-1} \frac{\partial f}{\partial \bar{z}} &= B_{\bar{z}}. \end{aligned} \tag{4.195}$$

Secondly, for two $\mathcal{G}$-valued functions B_z and $B_{\bar{z}}$ on Ω satisfying (4.194), we solve for $\mathcal{G}$-valued functions A_z and $A_{\bar{z}}$ on Ω satisfying

$$\begin{aligned} \frac{\partial}{\partial z} \left(-2\lambda^{-2} \left(\frac{\partial A_z}{\partial \bar{z}} + \frac{\partial A_{\bar{z}}}{\partial z} \right) \right) - \left[A_z, -2\lambda^{-2} \left(\frac{\partial A_z}{\partial \bar{z}} + \frac{\partial A_{\bar{z}}}{\partial z} \right) \right] &= B_z, \\ \frac{\partial}{\partial \bar{z}} \left(-2\lambda^{-2} \left(\frac{\partial A_z}{\partial \bar{z}} + \frac{\partial A_{\bar{z}}}{\partial z} \right) \right) - \left[A_{\bar{z}}, -2\lambda^{-2} \left(\frac{\partial A_z}{\partial \bar{z}} + \frac{\partial A_{\bar{z}}}{\partial z} \right) \right] &= B_{\bar{z}}, \\ -\frac{\partial A_z}{\partial \bar{z}} + \frac{\partial A_{\bar{z}}}{\partial z} + [A_z, A_{\bar{z}}] &= 0. \end{aligned} \tag{4.196}$$

Lastly, for the above $\mathcal{G}$-valued functions A_z and $A_{\bar{z}}$ on Ω satisfying (4.196) and $a \in G$, there exists a smooth map $f : \Omega \to G$ such that

$$\begin{aligned} f(x_0, y_0) &= a, \\ f^{-1} \frac{\partial f}{\partial z} &= A_z, \\ f^{-1} \frac{\partial f}{\partial \bar{z}} &= A_{\bar{z}}. \end{aligned} \tag{4.197}$$

Then $f : (\Omega, g) \to (G, h)$ is a biharmonic map.

Theorem 4.8.8. *Every biharmonic map $f : (\Omega, g) \to (G, h)$ can be obtained in the described way. Here, $g = \lambda^2 g_0$, λ is a positive smooth function on Ω and g_0 is the canonical metric on $\mathbf{R}^2$.*

We next introduce a number of loop group notions for biharmonic maps. We consider a $\mathcal{G}^{\mathbf{C}}$-valued 1-form

$$B_\sigma = \frac{1}{2}(1-\sigma)B_z dz + \frac{1}{2}(1-\sigma^{-1})B_{\bar{z}} d\bar{z} \tag{4.198}$$

for a parameter $\sigma \in S^1$, such that

$$dB_\sigma + [B_\sigma \wedge B_\sigma] = 0, \quad \forall \sigma \in S^1. \tag{4.199}$$

Let α_ν^σ be a $\mathcal{G}^{\mathbf{C}}$-valued 1-form

$$\alpha_\nu^\sigma = \frac{1}{2}(1-\nu)A_z dz + \frac{1}{2}(1-\nu^{-1})A_{\bar{z}} d\bar{z} \tag{4.200}$$

satisfying

$$\begin{aligned} \frac{\partial}{\partial z}(\delta\alpha_\nu^\sigma) - [\frac{1}{2}(1-\nu)A_z, \delta\alpha_\nu^\sigma] &= B_z, \\ \frac{\partial}{\partial \bar{z}}(\delta\alpha_\nu^\sigma) - [\frac{1}{2}(1-\nu)A_{\bar{z}}, \delta\alpha_\nu^\sigma] &= B_{\bar{z}}, \\ d\alpha_\nu^\sigma + [\alpha_\nu^\sigma \wedge \alpha_\nu^\sigma] &= 0, \end{aligned} \tag{4.201}$$

for each $\nu, \sigma \in S^1$, where the co-differential $\delta\alpha_\nu^\sigma$ of α_ν^σ is given by

$$\delta\alpha_\nu^\sigma = -2\lambda^{-2}\left(\frac{1}{2}(1-\nu)\frac{\partial}{\partial \bar{z}}A_z + \frac{1}{2}(1-\nu^{-1})\frac{\partial}{\partial z}A_{\bar{z}}\right) \tag{4.202}$$

for $\nu \in S^1$. Thus, the map $f_\nu^\sigma : \Omega \to G$ satisfying $f_\nu^{\sigma *}\theta = \alpha_\nu^\sigma$ is a bi-harmonic map $(\Omega, g) \to (G, h)$, where $g = \lambda^{-2} g_0$ and λ is a positive smooth function on Ω.

Chapter 5
Biwave Maps

Biwave maps are biharmonic maps on Minkowski spaces which generalize wave maps. They have been first investigated by Chiang [75, 76] in 2009 and by Chiang and Wolak [84] later. In this chapter we study biwave maps, stability, equivariant biwave maps, biwave fields of inclusions of warped product manifolds, conservation law of stress bienergy tensors, and transversal biwave maps.

5.1 Maps into Manifolds

5.1.1 *Introduction*

The equations of biwave maps constitute a fourth-order hyperbolic system of PDEs, in contrast to the case of biharmonic maps whose equations constitute a fourth-order elliptic system of PDEs. All the theorems and results presented here were obtained in [75, 76, 84].

Bi-Yang-Mills fields, which generalize Yang-Mills fields, were first explored by Ichiyama, Inoguchi and Urakawa [191, 192] in 2009. The following connection between bi-Yang Mills fields and biwave equations motivates one to study biwave maps.

Let P be a principal fiber bundle over a manifold (M, g) with structure group G and canonical projection π, and $\mathcal{G}$ be the Lie algebra of G. A connection A can be locally considered as a $\mathcal{G}$-valued 1-form $A = A_\mu(x)dx^\mu$. The curvature of the connection A is given by the 2-form $F = F_{\mu\nu}dx^\mu dx^\nu$, with

$$F_{\mu\nu} = \partial_\mu A_\nu - \partial_\nu A_\mu + [A_\mu, A_\nu].$$

The bi-Yang-Mills Lagrangian is defined by

Y.-J. Chiang, *Developments of Harmonic Maps, Wave Maps and Yang-Mills Fields into Biharmonic Maps, Biwave Maps and Bi-Yang-Mills Fields*, Frontiers in Mathematics, DOI 10.1007/978-3-0348-0534-6_5,

$$L_2(A) = \frac{1}{2}\int_M ||\delta F||^2 dv_g, \tag{5.1}$$

where δ is the adjoint of the exterior differentiation operator d on the space of E-valued smooth forms on M ($E = End(P)$, the endormorphisms of P). Then the Euler-Lagrange equation describing the critical points of (5.1) has the form

$$(\delta d + F)\delta F = 0. \tag{5.2}$$

This is the bi-Yang-Mills system. In particular, letting $M = \mathbf{R}\times\mathbf{R}^2$ and $G = SO(2)$, the group of orthogonal transformations on $\mathbf{R}^2$, we have that $A_\mu(x)$ is a 2×2 skew-symmetric matrix A_μ^{ij}. The appropriate equivariant ansatz is

$$A_\mu^{ij}(x) = (\delta_\mu^i x^j - \delta_\mu^j x^i)h(t, |x|),$$

where $h : M \to \mathbf{R}$ is a spatially radial function. Setting $u = r^2 h$ and $r = |x|$, the bi-Yang-Mills system (5.2) becomes the following equation for $u(t, r)$:

$$u_{tttt} - u_{rrrr} - \frac{3}{r}u_{rrr} + \frac{2}{r^2}u_{rr} - \frac{2}{r^3}u_r = k(t, r),$$

which is a linear non-homogeneous biwave equation, where $k(t, r)$ is a function of t and r.

It is interesting to study biwave maps since their equations constitute a fourth-order hyperbolic system of partial differential equations, which generalizes the system for wave maps. This is the first attempt to study biwave maps and their relationship with wave maps.

5.1.2 Definition

If $f : \mathbf{R}^{1+m} \to N$ is a smooth map from a Minkowski space $\mathbf{R}^{1+m}$ into a Riemannian manifold N, then the bienergy functional is, by (4.1),

$$\begin{aligned} E_2(f) &= \frac{1}{2}\int_{R^{1+m}} ||(d + d^*)^2 f||^2 dt\, dx \\ &= \frac{1}{2}\int_{R^{1+m}} ||d^* df||^2 dt\, dx = \frac{1}{2}\int_{R^{1+m}} ||\tau_\Box(f)||^2 dt\, dx, \end{aligned} \tag{5.3}$$

where $\tau_\Box(f)$ is a wave field (i.e., tension field in the Minkowski space). The Euler-Lagrange equation describing the critical points of (5.3) is, by (4.3),

$$(\tau_2)_\Box(f) = J_f(\tau_\Box f) = \Delta\tau_\Box(f) + R'(df, df)\tau_\Box(f) = 0, \tag{5.4}$$

where R' is the Riemannian curvature of N.

Definition 5.1.1. A map $f : \mathbf{R}^{1+m} \to N$ from a Minkowski space into a Riemannian manifold is a biwave map if the biwave field (i.e., bitension field in the Minkowski space)

$$
\begin{aligned}
(\tau_2)_\Box(f)^\alpha &= J_f(\tau_\Box f)^\alpha = \Delta\tau_\Box(f)^\alpha + R'^\alpha(df,df)\tau_\Box(f) \\
&= \Box\tau_\Box(f)^\alpha + \Gamma'^\alpha_{\mu\gamma}\left(-\tau_\Box(f)^\mu_t\tau_\Box(f)^\gamma_t + \sum_{i=1}^m \tau_\Box(f)^\mu_i\tau_\Box(f)^\gamma_i\right) \\
&+ R'^\alpha_{\beta\gamma\mu}\left(-f^\beta_t f^\gamma_t + \sum_{i=1}^m f^\beta_i f^\gamma_i\right)\tau_\Box(f)^\mu = 0,
\end{aligned}
\tag{5.5}
$$

i.e., the wave field $\tau_\Box(f)$ is a Jacobi field in the Minkowski space.

If $\tau_\Box(f) = 0$, then $(\tau_2)_\Box(f) = 0$. Waves maps are obviously biwave maps, but biwave maps are not necessarily wave maps.

5.1.3 Examples and Theorems

Example 1. Let $u : \mathbf{R}^{1+m} \to \mathbf{R}$ be a function defined on a Minkowski space which satisfies the following conditions:

$$
\Box^2 u(t,x) = \Box(\Box u) = u_{tttt} - 2u_{ttxx} + u_{xxxx} = 0, \quad (t,x) \in (0,\infty) \times \mathbf{R}^m,
$$

$$
u = u_0,\ u_t = u_1,\ \Box u = \Box u_0,\ \frac{\partial}{\partial t}\Box u = \Box\frac{\partial u}{\partial t} = \Box u_1, \quad (t,x) \in \{t=0\} \times \mathbf{R}^m,
$$

where the initial data u_0 and u_1 are given. We want to derive a formula for u in terms of u_0 and u_1.

Recall that the concepts concern the solutions to the Cauchy problem: $(\star)$ $\Box u = g$, $(t,x) \in \mathbf{R} \times \mathbf{R}^m$, $u(0,x) = 0$, $\partial_t u(0,x) = u_1(x)$. Let R be the fundamental solution of $(\star)$, i.e., the solution of $\Box u = 0$, $(t,x) \in \mathbf{R} \times \mathbf{R}^m$, $u(0,x) = 0$, $\partial_t u(0,x) = \delta(x)$. Then for initial data $u(0,x) = 0$, $\partial_t u(0,x) = u_1(x)$ and $g = 0$, the solution to the Cauchy problem $(\star)$ is

$$
u(t,x) = \int_{\mathbf{R}^m} R(t,x-y)u_1(y)dy = (R(t) \star u_1)(x).
$$

Furthermore, $v = \partial_t u = \partial_t R \star u_1$ is a solution to $\Box v = 0, v(0) = u_1$, $\partial_t v(0) = 0$. Thus the solution $u^{(0)}$ to the homogeneous wave equation $\Box u^{(0)} = 0$ with initial data $u^{(0)}(0) = u_0$, $\partial_t u^{(0)}(0) = u_1$ is

$$
u^{(0)}(t) = \partial_t R(t) \star u_0 + R(t) \star u_1,
$$

and the solution to the non-homogeneous equation ($\star$) is given by Duhamel's principle

$$u(t) = u^{(0)}(t) - \int_0^t R(t-\tau) \star g(\tau)d\tau.$$

We first compute $\Box u(t,x)$ as in [140] as follows:

1. When $m = 1$, $\Box u(t,x) = \frac{1}{2}(\Box u_0(x+t) + \Box u_0(x-t)) + \frac{1}{2}\int_{x-t}^{x+t} \Box u_1(y)dy$, $x \in \mathbf{R}, t > 0$.
2. When $m = 2$,

$$\Box u(t,x) = \frac{1}{2}\int_{B(t,x)} \frac{t\Box u_0(y) + t^2 \Box u_1(y) + tD\Box u_0(y)(y-x)}{\sqrt{(t^2 - (y-x)^2)}} dy, \ x \in \mathbf{R}^2, \ t > 0.$$

3. When $m = 3$,

$$\Box u(t,x) = \frac{\partial}{\partial t}\left(t\int_{\partial B(t,x)} \Box u_0 dS\right) + t\int_{\partial B(t,x)} \Box u_1 dS, \ x \in \mathbf{R}^3, \ t > 0,$$

where dS is the surface measure on $\partial B(t,x)$.
4. When $m > 3$, (i) if m is odd,

$$\Box u(t,x) = \frac{1}{\gamma_m}\Big[\left(\frac{\partial}{\partial t}\right)\left(\frac{1}{t}\frac{\partial}{\partial t}\right)^{\frac{m-3}{2}}\left(t^{m-2}\int_{\partial B(t,x)} \Box u_0 dS\right)$$
$$+\left(\frac{1}{t}\frac{\partial}{\partial t}\right)^{\frac{m-3}{2}}\left(t^{m-2}\int_{\partial B(t,x)} \Box u_1 dS\right)\Big],$$

where $\gamma_m = 1\cdot 3\cdot 5\cdots(m-2)$, for $x \in \mathbf{R}^m$, $t > 0$. Since $\gamma_3 = 1$, this agrees for $m = 3$ with expression 3.
(ii) If m is even,

$$\Box u(t,x) = \frac{1}{\gamma_m}\Big[\left(\frac{\partial}{\partial t}\right)\left(\frac{1}{t}\frac{\partial}{\partial t}\right)^{\frac{m-2}{2}}\left(t^m\int_{B(t,x)} \frac{\Box u_0}{(t^2 - |y-x|^2)^{1/2}} dy\right)$$
$$+\left(\frac{1}{t}\frac{\partial}{\partial t}\right)^{\frac{m-2}{2}}\left(t^m\int_{B(t,x)} \frac{\Box u_1}{(t^2 - |y-x|^2)^{1/2}} dy\right)\Big],$$

where $\gamma_m = 2\cdot 4\cdots(m-2)\cdot m$, for $x \in \mathbf{R}^m$, $t > 0$. Since $\gamma_2 = 2$, this agrees for $m = 2$ with expression 2.

After we compute $\Box u(t,x)$ out, we discuss $\Box u$ as follows:

(1) If $\Box u = 0$, then a wave function u is obviously a biwave function.
(2) If $\Box u = constant \neq 0$, then u is a biwave function with constant wave.
(3) If $\Box u = g$ is neither 0 nor constant, then u is a biwave non-wave function. We can compute $u(t,x)$ as in [140, 332] as follows:

(a) When $m = 1$,

$$u(t,x) = \frac{1}{2}(u_0(x+t) + u_0(x-t)) + \frac{1}{2}\int_{x-t}^{x+t} u_1(y)dy$$
$$+\frac{1}{2}\int_0^t \int_{x-s}^{x+s} g(t-s,y)dy\,ds.$$

(b) When $m = 2$, $u(t,x) = \frac{1}{2\pi}\int_{B(t,x)} \frac{u_1(y)}{\sqrt{t^2-|x-y|^2}}dy^1dy^2$, and thus the fundamental solution is given by

$$R(t,x) = \frac{1}{2\pi}\frac{1}{\sqrt{t^2-|x|^2}}\chi_{B(t,x)},$$

where χ_A is the characteristic function of the set A.

(c) When $m = 3$,

$$u(t,x) = \frac{d}{dt}\left(\frac{t}{4\pi}\int_{S^2} u_0(x+t\xi)d\omega\right) + \frac{t}{4\pi}\int_{S^2} u_1(x+t\xi)d\omega$$
$$+\int_0^t \frac{s}{4\pi}\int_{S^2} g(t-s,x+s\xi)d\omega\,ds,$$

where $d\omega$ is the surface measure of S^2.

(d) When $m > 3$, (i) if m is odd,

$$u(t) = \partial_t R(t)\star u_0 + R(t)\star u_1 - \int_0^t R(t-\tau)\star g(\tau)d\tau,$$

where $R(t,x) = A_m(\frac{1}{t}\partial_t)^{\frac{m-3}{2}}\frac{1}{t}\delta(|x|-t)$, $A_m = \frac{1}{\omega_{m-1}(m-2)(m-4)\cdots 3\cdot 1}$, ω_{m-1} is the area of the unit sphere S^{m-1}.

(ii) If m is even,

$$u(t) = \partial_t R(t)\star u_0 + R(t)\star u_1 - \int_0^t R(t-\tau)\star g(\tau)d\tau,$$

where $R(t,x) = A_m\left(\frac{1}{t}\partial_t\right)^{\frac{m-2}{2}}\frac{1}{\sqrt{t^2-|x|^2}}\chi_{B(t,x)}$, and $A_m = \frac{2}{\omega_m(m-1)(m-3)\cdots 3\cdot 1}$.

Let $f : \mathbf{R}^{1+m} \to N_1$ be a smooth map from a Minkowski space $\mathbf{R}^{1+m}$ into a Riemannian manifold N_1 and $f_1 : N_1 \to N_2$ be a smooth map from N_1 into another Riemannian manifold N_2. Then the composition $f_1 \circ f : \mathbf{R}^{1+m} \to N_2$ is a smooth map. Since $\mathbf{R}^{1+m}$ is a semi-Riemannian manifold (i.e., a pseudo-Riemannian manifold), we can define a Levi-Civita connection on $\mathbf{R}^{1+m}$ (see O'Neill [279]). Let $D, D', \bar{D}, \bar{D}', \bar{D}'', \hat{D}, \hat{D}', \hat{D}''$ be the connections on $\mathbf{R}^{1+m}, TN_1, f^{-1}TN_1,$

$f_1^{-1}TN_2, (f_1 \circ f)^{-1}TN_2,\ T^*\mathbf{R}^{1+m} \otimes f^{-1}TN_1, T^*N_1 \otimes f_1^{-1}TN_2,\ T^*\mathbf{R}^{1+m} \otimes (f_1 \circ f)^{-1}TN_2$, respectively, and let $R^{N_2}(\cdot,\cdot)$ and $R^{f_1^{-1}TN_2}(\cdot,\cdot)$ be the curvatures on TN_2 and $f^{-1}TN_2$, respectively. We first have the following formulas:

$$\bar{D}''_X d(f_1 \circ f)(Y) = (\hat{D}'_{df(X)} df_1)df(Y) + df_1 \circ \bar{D}_X df(Y), \tag{5.6}$$

for $X, Y \in \mathbf{R}^{1+m}$ and

$$R^{N_2}(df_1(X'), df_1(Y'))df_1(Z') = R^{f_1^{-1}TN_2}(X', Y')df_1(Z'), \tag{5.7}$$

for $X',\ Y',\ Z' \in \Gamma(TN_1)$.

Theorem 5.1.2 ([75]). *If $f : \mathbf{R}^{1+m} \to N_1$ is a biwave map and $f_1 : N_1 \to N_2$ is a totally geodesic map between two Riemannian manifolds N_1 and N_2, then the composition $f_1 \circ f : \mathbf{R}^{1+m} \to N_2$ is a biwave map and*

$$(\tau_2)_\square(f_1 \circ f) = df_1 \circ (\tau_2)_\square(f).$$

Proof. Let $x^0 = t, x^1, \cdots, x^m$ be the coordinates of a point p in $\mathbf{R}^{1+m}$ and $e_0 = \frac{\partial}{\partial t}, e_1 = (1, 0, \cdots, 0), e_2 = (0, 1, 0, \cdots, 0), \cdots, e_m = (0, \cdots, 0, 1)$ be the frame at p. We know from [196] that $\bar{D}''^*\bar{D}'' = \bar{D}''_{e_k}\bar{D}''_{e_k} - \bar{D}''_{D_{e_k}e_k}$. Since f_1 is totally geodesic, by applying the chain rule of the wave field to $f_1 \circ f$ as in [129] we have $\tau_\square(f_1 \circ f) = df_1 \circ \tau_\square(f)$. Then we arrive at

$$\begin{aligned}\bar{D}''^*\bar{D}''\tau_\square(f_1 \circ f) &= \bar{D}''^*\bar{D}''(df_1 \circ \tau_\square(f)) \\ &= \bar{D}''_{e_k}\bar{D}''_{e_k}(df_1 \circ \tau_\square(f)) - \bar{D}''_{D_{e_k}e_k}(df_1 \circ \tau_\square(f)).\end{aligned} \tag{5.8}$$

Recalling that $\tau_\square(f) = \hat{D}_{e_j} df(e_j)$, we derive from (5.6) that

$$\begin{aligned}&\bar{D}''_{e_k}(df_1 \circ \tau_\square(f)) = \bar{D}''_{e_k}(df_1 \circ \hat{D}_{e_j} df(e_j)) \\ &= (\hat{D}'_{\hat{D}_{e_j}df(e_k)} df_1)(\hat{D}_{e_j} df(e_j)) + df_1 \circ \bar{D}_{e_k}(\hat{D}_{e_j} df(e_j)) = df_1 \circ \bar{D}_{e_k}\tau_\square(f),\end{aligned}$$

because f_1 is totally geodesic. Therefore,

$$\bar{D}''_{e_k}\bar{D}''_{e_k}(df_1 \circ \tau_\square(f)) = \bar{D}''_{e_k}(df_1 \circ \bar{D}_{e_k}\tau(f)) = df_1 \circ \bar{D}_{e_k}\bar{D}_{e_k}\tau_\square(f), \tag{5.9}$$

$$\bar{D}''_{D_{e_k}e_k}(df_1 \circ \tau(f)) = df_1 \circ \bar{D}_{D_{e_k}e_k}\tau_\square(f). \tag{5.10}$$

Substituting (5.9), (5.10) into (5.8) yields

$$\bar{D}''^*\bar{D}''\tau_\square(f_1 \circ f) = df_1 \circ \bar{D}^*\bar{D}\tau_\square(f), \tag{5.11}$$

where $\bar{D}^*\bar{D} = \bar{D}_{e_k}\bar{D}_{e_k} - \bar{D}_{D_{e_k}e_k}$.

On the other hand, we have, by (5.7),

$$R^{N_2}(d(f_1 \circ f)(e_i), \tau_\Box(f_1 \circ f))d(f_1 \circ f)(e_i) = R^{f_1^{-1}TN_2}(df(e_i), \tau_\Box(f))df_1(df(e_i)) \\ = df_1 \circ R^{N_1}(df(e_i), \tau_\Box(f))df(e_i). \tag{5.12}$$

It follows from (5.11) and (5.12) that

$$\bar{D}''^{*}\bar{D}''(f_1 \circ f) + R^{N_2}(d(f_1 \circ f)(e_i), \tau_\Box(f_1 \circ f))d(f_1 \circ f)(e_i) \\ = df_1 \circ [\bar{D}^{*}\bar{D}\tau_\Box(f) + R^{N_1}(df(e_i), \tau_\Box(f))df(e_i)], \tag{5.13}$$

i.e., $(\tau_2)_\Box(f_1 \circ f) = df_1 \circ (\tau_2)_\Box(f)$. Hence, if f is a biwave map and f_1 is totally geodesic, then $f_1 \circ f$ is a biwave map. □

Example 2. Let N_1 be a submanifold of N. Are the biwave maps into N_1 also biwave maps into N? The answer is affirmative iff N_1 is a totally geodesic submanifold of N, i.e., N_1 geodesics are N geodesics. N_1 is a geodesic $\gamma(t) = (\gamma^1, \cdots, \gamma^n) : \mathbf{R} \to N \subset \mathbf{R}^n$ with $|\dot{\gamma}(t)| = 1$ iff $\dot{\gamma}$ is parallel, i.e., $D_{\frac{\partial}{\partial t}}\dot{\gamma} = 0$ iff $\ddot{\gamma} \perp T_\gamma N$. For a map $v : \mathbf{R}^{1+m} \to \mathbf{R}$, let $f = \gamma \circ v = (f^1, \cdots, f^n) : \mathbf{R}^{1+m} \to N \subset \mathbf{R}^n$, we have, by (5.13),

$$(\tau_2)_\Box(f) = d\gamma \circ (\tau_2)_\Box(v) = d\gamma \circ \Box^2 v,$$

since γ is a geodesic. Hence, $f = \gamma \circ v$ is a biwave map if and only if v solves the fourth-order homogeneous linear biwave equation $\Box^2 v = 0$. It follows from Theorem 5.1.2 that there are many biwave maps $f : \mathbf{R}^{1+m} \to N$, provided by geodesics of N.

We also can construct examples of biwave non-wave maps from some wave maps with constant energy using Theorem 5.1.3. Let

$$S^n(1/\sqrt{2}) = S^n(1/\sqrt{2}) \times \{1/\sqrt{2}\} = \{(x_1, x_2, \cdots, x_{n+1}, 1/\sqrt{2}) | x_1^2 + \cdots + x_{n+1}^2 = 1/2\}$$

be a hypersphere of $S^{n+1}(1)$. Then $S^n(1/\sqrt{2})$ is a biharmonic non-minimal submanifold of $S^{n+1}(1)$, by Theorem 4.2.2 and Example 1 in Sect. 4.2. Let $\zeta = (x_1, \cdots, x_{n+1}, -1/\sqrt{2})$ be a unit section of the normal bundle of $S^n(1/\sqrt{2})$ in $S^{n+1}(1)$. The second fundamental form of the inclusion $i : S^n(1/\sqrt{2}) \to S^{n+1}(1)$ is $B(X, Y) = Ddi(X, Y) = -(X, Y)\zeta$. The tension field of i is $\tau(i) = -n\zeta$. Then we can compute the bitension field $\tau_2(i) = 0$. Let $\Omega \subset \mathbf{R}^{1+m}$ be a compact space-time domain in Minkowski space.

Theorem 5.1.3. *Let $h : \Omega \to S^n(1/\sqrt{2})$ be a non-constant wave map and $i : S^n(1/\sqrt{2}) \to S^{n+1}(1)$ be an inclusion. Then $f = i \circ h : R^{1+m} \to S^{n+1}(1)$ is a biwave non-wave map if and only if h has constant energy density $e(h) = \frac{1}{2}|dh|^2$.*

Proof. Let $x^0 = t, x^1, \cdots, x^m$ be the coordinates of an arbitrary point in $\Omega \subset \mathbf{R}^{1+m}$ and $e_0 = \frac{\partial}{\partial t}, e_1 = (1, 0, \cdots, 0), e_2 = (0, 1, 0, \cdots, 0), \cdots, e_m = (0, \cdots, 0, 1)$ be the local frame at the point. Recall that ζ is a unit section of the normal bundle. By applying the chain rule of the wave field to $f = i \circ h$, we have

$$\tau_\Box(f) = di(\tau_\Box(h)) + trace\, Ddi(dh, dh) = -2e(h)\zeta,$$

since h is a wave map. By straightforward calculation we derive

$$\begin{aligned} D^* D\tau_\Box(f) &= -D^f_{e_i} D^f_{e_i} \tau_\Box(f) = -D^f_{e_i} D^f_{e_i}(-2e(h)\zeta) \\ &= 2(e_i e_i e(h))\zeta - 2e(h)(dh(e_i), dh(e_i))\zeta + 4df[(e_i e(h))e_i] + 2e(h)Ddh(e_i, e_i) \end{aligned}$$

and

$$R^{S^{n+1}}(df(e_i), \tau_\Box(f))df(e_i) = -(dh(e_i), dh(e_i))\tau_\Box(f) = 2(dh(e_i), dh(e_i))e(h)\zeta.$$

Therefore,

$$\tau_{2\Box}(f) = -2(\Delta e(h))\zeta + 4df(grad\, e(h)).$$

Suppose that $f = i \circ h : \Omega \to S^n(1/\sqrt{2}) \times \{1/\sqrt{2}\} \to S^{n+1}(1)$ is a biwave non-wave map ($\tau_\Box(f) \neq 0$). As the ζ-part of $\tau_{2\Box}(f)$, $\Delta e(h)$ vanishes, which implies that $e(h)$ is constant since Ω is compact. The converse is obvious. □

Let $x^0 = t, x^1, \cdots, x^m$ be the coordinates of an arbitrary point in a compact space-time domain $\Omega \subset \mathbf{R}^{1+m}$ and $e_0 = \frac{\partial}{\partial t}$, $e_1 = (1, 0, \cdots, 0)$, $e_2 = (0, 1, 0, \cdots, 0), \cdots, e_m = (0, \cdots, 0, 1)$ be the frame at the point. Suppose that $f : \Omega \to N$ is a biwave map into a Riemannian manifold N such that the compact supports of $\frac{\partial f}{\partial x_i}$ and $D_{e_i} \frac{\partial f}{\partial x_i}$ are contained in the interior of Ω.

Theorem 5.1.4. *If $f : \Omega \to N$ is a biwave map as above and satisfies*

$$- |\tau_\Box f|^2_t + \sum_{i=1}^m |\tau_\Box f|^2_{x^i} - R'^{\alpha}_{\beta\gamma\mu}(-f^\beta_t f^\gamma_t + \sum_{i=1}^m f^\beta_i f^\gamma_i)\tau_\Box(f)^\mu \geq 0 \quad (5.14)$$

then f is a wave map.

Proof. Since f is a biwave map, (5.4) yields

$$(\tau_2)_\Box(f) = \Delta\tau_\Box(f) + R'(df, df)\tau_\Box(f).$$

Recall that $x^0 = t, x^1, \cdots, x^m$ are the coordinates of a point in $\Omega \subset \mathbf{R}^{1+m}$ and $e_0 = \frac{\partial}{\partial t}, e_1 = (1, 0, \cdots, 0), e_2 = (0, 1, 0, \cdots, 0), \cdots, e_m = (0, \cdots, 0, 1)$. We compute

$$\frac{1}{2}\Delta||\tau_\Box(f)||^2 = (D_{e_i}\tau_\Box(f), D_{e_i}\tau_\Box(f)) + (D^* D\tau_\Box(f), \tau_\Box(f)) \quad (5.15)$$

$$= \sum_{i=0}^{m}(D_{e_i}\tau_\Box(f), D_{e_i}\tau_\Box(f)) - (R'^{\alpha}_{\beta\gamma\mu}(-f_t^\beta f_t^\gamma + \sum_{i=1}^{m} f_i^\beta f_i^\gamma)\tau_\Box(f)^\mu, \tau_\Box(f))$$

$$= -|\tau_\Box f|_t^2 + \sum_{i=1}^{m}|\tau_\Box f|_{x^i}^2 - (R'^{\alpha}_{\beta\gamma\mu}(-f_t^\beta f_t^\gamma + \sum_{i=1}^{m} f_i^\beta f_i^\gamma)\tau_\Box(f)^\mu, \tau_\Box(f)) \geq 0.$$

Applying Bochner's technique to the above equation, and using (5.14) and the assumption that the compact supports of $\frac{\partial f}{\partial x_i}$ and $D_{e_i}\frac{\partial f}{\partial x_i}$ are contained in the interior of Ω, we deduce that $||\tau_\Box(f)||^2$ is constant, i.e., $d\tau_\Box(f) = 0$. If we use the identity

$$\int_\Omega div(df, \tau_\Box(f))dz = \int_\Omega (|\tau_\Box(f)|^2 + (df, d\tau_\Box(f)))dz,\ z = (t, x),$$

the fact that $d\tau_\Box(f) = 0$, and the divergence theorem, we get $\tau_\Box(f) = 0$. □

The above theorem is different than Theorem 4.1.2 obtained by Jiang [196] which asserts that *if $f : M \to N$ is a biharmonic map from a compact Riemannian manifold M into a Riemannian manifold N with non-positive curvature, then f is harmonic.*

5.2 Stability

5.2.1 Definition and Properties

Let $x^0 = t, x^1, \cdots, x^m$ be the coordinates of an arbitrary point in a compact space-time domain $\Omega \subset \mathbf{R}^{1+m}$ and $e_0 = \frac{\partial}{\partial t}, e_1 = (1, 0, \cdots, 0), \cdots, e_m = (0, \cdots, 0, 1)$ be the local frame at the point. Suppose that $f : \Omega \to N$ is a biwave map from Ω into a Riemannian manifold N such that the compact supports of $\frac{\partial f}{\partial x_i}$ and $D_{e_i}\frac{\partial f}{\partial x_i}$ are contained in the interior of Ω. Let $V \in \Gamma(f^{-1}TN)$ be a vector field such that $\frac{\partial f}{\partial t}|_{t=0} = V$. If we apply the formula for the second variation of a biharmonic map in Theorem 4.3.13 to a biwave map, we obtain:

Lemma 5.2.1. *If $f : \Omega \to N$ is a biwave map from a compact domain into a Riemannian manifold, then*

$$\frac{1}{2}\frac{d^2}{dt^2}E_2(f)\Big|_{t=0} = \int_\Omega ||\Delta V + R^N(df(e_i), V)df(e_i)||^2 dz$$
$$+ \int_\Omega < V, (D'_{df(e_i)}R^N)(f(e_i), \tau_\Box(f))V + (D'_{\tau_\Box(f)}R^N)(df(e_i), V)df(e_i)$$
$$+ R^N(\tau_\Box(f), V)\tau_\Box(f) + 2R^N(df(e_i), V)D_{e_i}\tau_\Box(f) + 2R^N(df(e_i), \tau_\Box(f))D_{e_i}V > dz \tag{5.16}$$

where $z = (t,x) \in \mathbf{R} \times \mathbf{R}^m$, $\triangle = \bar{D}^*\bar{D} = \bar{D}_{e_i}\bar{D}_{e_i} - \bar{D}_{D_{e_i}e_i}$, D, $\bar{D}$ *are the connections of* $\mathbf{R}^{1+m}$ *and* $f_t^{-1}TN$, D' *is the Riemannian connection on* TN *and* V *is the vector field along* f.

Definition 5.2.2. Let $f : R^{1+m} \to N$ be a biwave map. If $\frac{d^2}{dt^2}E_2(f)\big|_{t=0} \geq 0$, then f is said to be *stable*.

If we consider a wave map f, i.e., $\tau_\Box(f) = 0$, as a biwave map, then by (5.16) we have $\frac{d^2}{dt^2}E_2(f)\big|_{t=0} \geq 0$ and f is automatically stable.

Definition 5.2.3. Let $f : (\mathbf{R}^{1+m}, g) \to (N,h)$ be a smooth map from a Minkowski space into a Riemannian manifold (N,h). The *stress energy* is defined by $S(f) = e(f)g - f^*h$, where $e(f) = \frac{1}{2}|df|^2$ is the energy function. The map f satisfies the *conservation law* if $div\, S(f) = 0$.

Proposition 5.2.4. *Let* $f : \mathbf{R}^{1+m} \to (N,h)$ *be a smooth map from a Minkowski space into a Riemannian manifold* (N,h). *Then*

$$div\, S(f)(X) = - < \tau_\Box(f), df(X) >, \quad X \in \mathbf{R}^{1+m}. \tag{5.17}$$

Proof. Let $x^0 = t, x^1, \cdots, x^m$ be the coordinates of a point in $\mathbf{R}^{m,1}$, $e_0 = \frac{\partial}{\partial t}$, $e_1 = (1,0,\cdots,0), \cdots, e_m = (0,\cdots,0,1)$ and $g = \begin{bmatrix} -1 & 0 \\ 0 & I \end{bmatrix}$, where I is the $m \times m$ identity matrix. We compute

$$div\, S(f)(X) = D_{e_i}S(f)(e_i,X) = D_{e_i}\left((\frac{1}{2}|df|^2 \begin{bmatrix} -1 & 0 \\ 0 & I \end{bmatrix} - f^*h)(e_i,X) \right)$$

$$= D_{e_i}\left(\frac{1}{2}|df|^2 \begin{bmatrix} -1 & 0 \\ 0 & I \end{bmatrix} \right)(e_i,X) - (D_{e_i}f^*h)(e_i,X)$$

$$= (-(D\frac{\partial f}{\partial t}, \frac{\partial f}{\partial t})(-1))(e_0,X) + (D\frac{\partial f}{\partial x_i}, \frac{\partial f}{\partial x_i})I(e_i,X) - D_{e_i}(f_*e_i, f_*X)$$

$$= (D\frac{\partial f}{\partial t}, \frac{\partial f}{\partial t})(e_0,X) + (D\frac{\partial f}{\partial x_i}, \frac{\partial f}{\partial x_i})(e_i,X) - (D_{e_i}f_*e_i, f_*X) - (f_*e_i, D_{e_i}f_*X)$$

$$= ((D_X df)e_i, f_*e_i) - (\tau_\Box(f), f_*X) - (f_*e_i, D_{e_i}f_*X), \tag{5.18}$$

where the first term and the third term cancel out and $D_{e_i}f_*e_i = \tau_\Box(f)$. □

Theorem 5.2.5. *Let* $\Omega \subset \mathbf{R}^{1+m}$ *be a compact domain and* (N,h) *be a Riemannian manifold with constant sectional curvature* $K > 0$. *If* $f : \Omega \to N$ *is a stable biwave map satisfying the conservation law, then* f *is a wave map.*

Proof. Because N has constant sectional curvature, the second term in (5.16) vanishes and (5.16) becomes

$$\frac{1}{2}\frac{d^2}{dt^2}E_2(f_t)\Big|_{t=0} = \int_\Omega ||\Delta V + R^N(df(e_i),V)df(e_i)||^2 dz + \int_\Omega < V, R^N(\tau_\Box(f),V)\tau_\Box(f)$$
$$+ 2R^N(df(e_i),V)\bar{D}_{e_i}\tau_\Box(f) + 2R^N(df(e_i),\tau_\Box(f))D_{e_i}V > dz. \tag{5.19}$$

In particular, let $V = \tau_\Box(f)$. Since f is a biwave map and N has constant sectional curvature $K > 0$, (5.19) reduces to

$$\frac{1}{2}\frac{d^2}{dt^2}E_2(f)\Big|_{t=0} = 4\int_\Omega < R^N(df(e_i), \tau_\Box(f))D_{e_i}\tau_\Box(f), \tau_\Box(f) > dz$$

$$= 4K\int_\Omega [< df(e_i), D_{e_i}\tau_\Box(f) > ||\tau_\Box(f)||^2 - < df(e_i), \tau_\Box(f) >< \tau_\Box(f), D_{e_i}\tau_\Box(f) >]dz. \tag{5.20}$$

Since f satisfies the conservation law, by Definition 5.2.3, Proposition 5.2.4 and (5.17) we have

$$< df(e_i), \tau_\Box(f) > = 0,$$
$$< df(e_i), D_{e_i}\tau_\Box(f) > = - < D_{e_i}df(e_i), \tau_\Box(f) > = -||\tau_\Box(f)||^2. \tag{5.21}$$

Substituting (5.21) into (5.20) and using the stability of f, we obtain

$$\frac{1}{2}\frac{d^2}{dt^2}E_2(f_t)\Big|_{t=0} = -4K\int_\Omega ||\tau_\Box f||^4 dz \geq 0,$$

which implies that $\tau_\Box(f) = 0$. Hence, $f : \Omega \to N$ is a wave map. □

5.2.2 *An Example of Unstable Biwave Map*

If we apply the Hessian of the bienergy of a biharmonic map [196] to a biwave map $f : \Omega \to S^n(1)$, then we have the following:

Lemma 5.2.6. *Let $f : \Omega \to S^{n+1}(1)$ be a biwave map. The Hessian of the bienergy functional E_2 of f is*

$$H(E_2)_f(X,Y) = \int_\Omega (I_f(X), Y)dz, \quad for\, X, Y \in \Gamma(f^{-1}TS^{n+1}(1)),$$
$$X, Y \in \Gamma(f^{-1}TS^{n+1}(1)),$$

where

$$\begin{aligned} I_f(X) &= \Delta^f(\Delta^f X) + \Delta^f(trace(X, df\cdot)df\cdot - |df|^2 X) + 2(d\tau_\Box(f), df)X \\ &+ |\tau_\Box(f)|^2 X - 2\,trace(X, d\tau_\Box(f)\cdot)df - 2\,trace(\tau_\Box(f), dX\cdot)df\cdot \\ &- (\tau_\Box(f), X)\tau_\Box(f) + trace(df\cdot, \Delta^f X)df\cdot + trace(df, trace(X, df\cdot)df\cdot)df\cdot \\ &- 2|df|^2\, trace(df\cdot, X)df\cdot + 2(dX, df)\tau_\Box(f) - |df|^2\Delta^f X + |df|^4 X. \end{aligned}$$

Theorem 5.2.7. *Let $h : \Omega \to S^n(1/\sqrt{2})$ be a wave map on a compact domain with constant energy and $i : S^n(1/\sqrt{2}) \to S^{n+1}(1)$ be the inclusion map. Then $f = i \circ h : \Omega \to S^{n+1}(1)$ is an unstable biwave map.*

Proof. We obtain the following identities from Theorem 5.1.3:

$$|df|^2 = 2e(h), \quad trace(\zeta, df\cdot)df\cdot = 0, \; (d\tau_{\Box}(f), df)\zeta = -4(e(h))^2\zeta,$$

$$|\tau_{\Box}(f)|^2 = 4(e(h))^2, \quad trace(\zeta, d\tau_{\Box}(f))df\cdot = 0, \quad trace(\tau_{\Box}(f), d\zeta\cdot)df = 0,$$

$$(\tau_{\Box}(f), \zeta)\tau_{\Box}(f) = 4(e(h))^2\zeta, \quad trace(df, \Delta^f\zeta)df\cdot = (\Delta^f\zeta)^T$$

$$(d\zeta, df)\tau_{\Box}(f) = -4(e(h))^2\zeta.$$

Lemma 5.2.6 and the above identities yield

$$(I_f(\zeta), \zeta) = \int_{\Omega} (|\Delta^f\zeta|^2 - 4e(h)(\Delta^f\zeta, \zeta) - 12(e(h))^2)dz,$$

which is strictly negative, where $\Delta^f\zeta = 2e(h)\zeta$. Hence, f is an unstable biwave map. □

5.3 Equivariant Biwave Maps

We first review warped products of two semi-Riemannian manifolds or Riemannian manifolds. Then we utilize warped products to study equivariant biwave maps into various spaces by applying eigenmaps between spheres based on [76].

5.3.1 Warped Product

Let M and N be two semi-Riemannian or Riemannian manifolds of dimensions m and n equipped with metrics g and h, respectively, and let $\eta \in C^\infty(M)$ be a positive function. On the product manifold $M \times N$, let $\pi : M \times N \to M$ and $\sigma : M \times N \to N$ be its projections, respectively.

Definition 5.3.1 ([279]). The *warped product* $M \times_\eta N$ is the product manifold $M \times N$, furnished with the metric tensor

$$G_\eta(X, Y) = \pi^*(g) + (\eta \circ \pi)^2\sigma^*(h(X, Y)) = g(d\pi(X), d\pi(Y)) + (\eta \circ \pi)^2 h(d\sigma(X), d\sigma(Y)), \tag{5.22}$$

for $X, Y \in T_{(x,y)}(M \times N)$. The function η is called the *warping function* of the warped product.

Let $X, Y \in \Gamma(T(M \times N))$, $X = (X_1, X_2)$, $Y = (Y_1, Y_2)$, where $X_1, Y_1 \in \Gamma(TM)$ and $X_2, Y_2 \in \Gamma(TN)$. Denote by D and R the Levi-Civita connection and the curvature tensor field on the product manifold $M \times N$ with respect to the metric G. Then the Levi-Civita connection $\tilde{D}$ on the warped product manifold $M \times_\eta N$ with respect to G_η is given by

$$\begin{aligned}\tilde{D}_X Y &= D_X Y + \frac{1}{2\eta^2} X_1(\eta^2)(0, Y_2)\\ &+ \frac{1}{2\eta^2} Y_1(\eta^2)(0, X_2) - \frac{1}{2} h(X_2, Y_2)(grad\, \eta^2, 0). \end{aligned} \tag{5.23}$$

The curvature tensor field $\tilde{R}(X, Y)$ on $M \times_\eta N$ with respect to G_η is given by

$$\begin{aligned}\tilde{R}(X, Y) = R(X, Y) &+ \frac{1}{2\eta^2}\Big\{\Big(D^M_{Y_1} grad\, \eta^2 - \frac{1}{2\eta^2} Y_1(\eta^2) grad\, \eta^2, 0\Big) \wedge_{G_\eta} (0, X_2)\\ &- \Big(D^M_{X_1} grad\, \eta^2 - \frac{1}{2\eta^2} X_1(\eta^2) grad\, \eta^2, 0\Big) \wedge_{G_\eta} (0, Y_2)\\ &- \frac{1}{2\eta^2} |grad\, \eta^2|^2 (0, X_2) \wedge_{G_\eta} (0, Y_2)\Big\}, \end{aligned} \tag{5.24}$$

where

$$(X \wedge_{G_\eta} Y)Z = G_\eta(Z, Y)X - G_\eta(Z, X)Y,$$

for $X, Y, Z \in \Gamma(T(M \times N))$. Please refer to [31,279] for more details about warped products.

5.3.2 *Formulation*

Let M be the Minkowski space $\mathbf{R}^{1+m} = \mathbf{R} \times \mathbf{R}^m$ with spatial polar coordinates

$$(t, r, w) \in \mathbf{R} \times \mathbf{R}^+ \times S^{m-1}, \; r = |x|, \; w^i = \frac{x^i}{r}, \quad i = 1, \cdots, m.$$

In these coordinates, the metric g on M takes the form

$$-dt^2 + dr^2 + r^2 dw^2,$$

where dw^2 is the standard metric on $S^{m-1} \hookrightarrow \mathbf{R}^m$. Let N be a smooth, n-dimensional, rotationally symmetric, warped product manifold defined by

$$N = [0, R^*) \times_h S^{n-1},$$

where $R^* \in \mathbf{R}^+ \cup \{\infty\}$ and $h : \mathbf{R} \to \mathbf{R}$ is a smooth function. On N we have the polar coordinates $(\phi, \chi) \in [0, R^*) \times S^{n-1}$. In these coordinates, the metric on N takes the form

$$d\phi^2 + h^2(\phi)d\chi^2,$$

where $d\chi^2$ is the standard metric of $S^{n-1} \hookrightarrow \mathbf{R}^n$. We also can define the normal coordinates $(f^1, \cdots, f^n)$ on N by letting $f^i = \phi \cdot \chi^i$, $i = 1, \cdots, n$, and then $(\phi, \chi) = (|f|, f/|f|)$. In this setting N can be identified with the ball $B_{R^*}(0)$ in $\mathbf{R}^n$.

Let $f : M^m \to N^n$ be a map between two rotationally symmetric manifolds (i.e., $SO(m)$ and $SO(n)$ act on M and N as isometries). Then f is an *equivariant* map if the orbit of any point $p \in M$ is mapped into the orbit of $f(p) \in N$.

For a map $f : \mathbf{R}^{1+m} = \mathbf{R} \times \mathbf{R}^+ \times_r S^{m-1} \to N = [0, R^*) \times_h S^{n-1}$, the Cauchy problem takes the form

$$\partial^i \partial_i f^\alpha + \Gamma^\alpha_{\beta\gamma} \partial_i f^\beta \partial^i f^\gamma = 0, \tag{5.25}$$

$$f(0, x) = f_0, \ \partial_t f(0, x) = f_1. \tag{5.26}$$

We assume that the initial data (f_0, f_1) are equivariant in the sense that there exist functions $\phi_0, \phi_1 : \mathbf{R} \to \mathbf{R}$ and a map $\chi : S^{m-1} \to S^{n-1}$ such that for $x = (r, w) \in \mathbf{R}^m$,

$$f_0^i(x) = \phi_0(r) \cdot \chi^i(w), \ \ f_1^i(x) = \phi_1(r) \cdot \chi^i(w), \ \ i = 1, 2, \cdots, n. \tag{5.27}$$

By the conservation law for (5.25), we know that any solution to the equation will also be equivariant for $t > 0$, i.e., there is a radial function ϕ such that

$$f^i(t, x) = \phi(t, r) \cdot \chi^i(w). \tag{5.28}$$

Definition 5.3.2 ([128]). A map $\chi : S^{m-1} \to S^{n-1}$ is an *eigenmap* with eigenvalue λ if χ is harmonic and $|d\chi|^2 = \lambda$ (note that the energy density $e(\chi) = \frac{1}{2}|d\chi|^2 = \lambda/2$ is constant).

Proposition 5.3.3. *Let $\chi : S^{m-1} \to S^{n-1}$ be an eigenmap of eigenvalue λ. The map $f = \phi \cdot \chi : \mathbf{R} \times \mathbf{R}^m \to N$ into a rotationally symmetric manifold is an equivariant wave map iff ϕ is a solution of*

$$-\phi_{tt} + \phi_{rr} + \frac{m-1}{r}\phi_r - \frac{\lambda}{r^2}h(\phi)h'(\phi) = 0. \tag{5.29}$$

Note that (5.26) reduces to $\phi(0, r) = \phi_0(r)$, $\phi_t(0, r) = \phi_1(r)$.

Proof. We first consider $\psi(r, w) = \phi(r) \times \chi(w) = \phi \cdot \chi : (0, \infty) \times_r S^{m-1} \to N = [0, R^*) \times_h S^{n-1}$. Take $\frac{\partial}{\partial r}, \frac{\partial}{\partial \rho} \in \Gamma(T(0, \infty))$, and let $\{\xi_i\}_{i=1}^{m-1}$ be a local orthonormal frame on S^{m-1}. Then

$$Dd\psi((\frac{\partial}{\partial r},0),(\frac{\partial}{\partial r},0)) = D_{\partial/\partial r}d\phi(\partial/\partial r) - d\phi(D_{\partial/\partial r}\partial/\partial r,0)$$
$$= \phi_{rr}(\partial/\partial\rho,0)\circ\psi, \tag{5.30}$$

and

$$Dd\psi((0,\xi_i),(0,\xi_i)) = D_{(0,\xi_i)}(0,d\chi(\xi_i)) - d\psi(\tilde{D}_{(0,\xi_i)}(0,\xi_i))$$
$$= (0, D_{\xi_i}d\chi(\xi_i)) - \frac{1}{2}(d\chi(\xi_i),d\chi(\xi_i))(grad\, h^2,0)\circ\psi$$
$$- d\psi((0,D_{\xi_i}\xi_i) - \frac{1}{2}(grad\, r^2,0))$$
$$= (r\phi_r - (d\chi(\xi_i),d\chi(\xi_i))(hh')\circ\phi)(\partial/\partial\rho,0)\circ\psi + (0, Dd\chi(\xi_i,\xi_i)), \tag{5.31}$$

by (5.23). Since $\{(\frac{\partial}{\partial r},0),\frac{1}{r}(0,\xi_i)\}_{i=1}^{m-1}$ is a local orthonormal frame on $(0,\infty)\times_r S^{m-1}$, by adding (5.30) and (5.31) we arrive at

$$\tau(\psi)_{(r,w)} = trace\, Dd\psi = H(r,w)(\frac{\partial}{\partial\rho},0) + \frac{1}{r^2}(0,\tau(\chi)_w), \tag{5.32}$$

where $H:(0,\infty)\times_r S^{m-1}\to\mathbf{R}$ is given by

$$H(r,w) = \phi_{rr} + \frac{m-1}{r}\phi_r - 2\frac{e(\chi)(w)}{r^2}h(\phi)h'(\phi). \tag{5.33}$$

Since χ is an eigenmap, (5.32) reduces to $\tau(\psi) = H(\frac{\partial}{\partial\rho},0)$.

By applying (5.32) and (5.33) to the equivariant wave map $f(t,x) = \phi(t,r)\cdot\chi(w):\mathbf{R}^{1+m} = \mathbf{R}\times\mathbf{R}^+\times_r S^{m-1}\to N = [0,R^*)\times_h S^{n-1}$ satisfying $trace_g\, Ddf = 0$, we obtain (5.29). □

Theorem 5.3.4. *Let $\chi: S^{m-1}\to S^{n-1}$ be an eigenmap of eigenvalue λ. The map $f=\phi\cdot\chi:\mathbf{R}\times\mathbf{R}^m\to N$ into a rotationally symmetric manifold is an equivariant biwave map if and only if ϕ is a solution of*

$$\phi_{tttt}-\phi_{ttr}-\frac{m-1}{r}\phi_{ttr}-H_{tt}+H_{rr}+\frac{m-1}{r}H_r-\frac{\lambda}{r^2}(h'^2(\phi)+h(\phi)h''(\phi))(H-\phi_{tt}) = 0, \tag{5.34}$$

where $H = \phi_{rr} + \frac{m-1}{r}\phi_r - \frac{\lambda}{r^2}h(\phi)h'(\phi)$.

Proof. As above, let $\psi(r,w) = \phi\cdot\chi:(0,\infty)\times_r S^{m-1}\to N = [0,R^*)\times_h S^{n-1}$. We continue to use the same notations as in the proof of Proposition 5.3.3. We know from [196] that $\Delta = (-)DD - D_{\tilde{D}}$. By substituting $\tau(\psi) = H(\frac{\partial}{\partial\rho},0)$, we first calculate

$$D_{(\partial/\partial r,0)}\tau(\psi) = \frac{\partial H}{\partial r}(\partial/\partial\rho,0)\circ\psi,$$

and

$$D_{(\partial/\partial r,0)}D_{(\partial/\partial r,0)}\tau(\psi) = \frac{\partial^2 H}{\partial r^2}(\partial/\partial\rho,0)\circ\psi. \tag{5.35}$$

Further,

$$\begin{aligned} D_{(0,\xi_i)}\tau(\psi) &= \xi_i(H)(\partial/\partial\rho,0)\circ\psi + HD_{(0,\xi_i)}(\partial/\partial\rho,0)\circ\psi \\ &= \xi_i(H)(\partial/\partial\rho,0)\circ\psi + \frac{h'(\phi)}{h(\phi)}H(0,d\chi(\xi_i)), \end{aligned} \tag{5.36}$$

thus

$$\begin{aligned} D_{(0,\xi_i)}D_{(0,\xi_i)}\tau(\psi) &= \Big(\xi_i(\xi_i(H)) - Hh'^2(\phi)(d\chi(\xi_i),d\chi(\xi_i))\Big)(\partial/\partial\rho,0)\circ\psi \\ &\quad + 2\frac{h'(\phi)}{h(\phi)}\xi_i(H)(0,d\chi(\xi_i)) + \frac{h'(\phi)}{h(\phi)}H(0,D_{\xi_i}d\chi(\xi_i)), \end{aligned} \tag{5.37}$$

and

$$D_{\tilde{D}_{(0,\xi)}(0,\xi_i)}\tau(\psi) = \left((D_{\xi_i}\xi_i)(H) - r\frac{\partial H}{\partial r}\right)(\partial/\partial\rho,0)\circ\psi + H\frac{h'(\phi)}{h(\phi)}(0,d\chi(D_{\xi_i}\xi_i)). \tag{5.38}$$

Since χ is an eigenmap, $\tau(\chi) = 0$ and the energy density $e(\chi) = \lambda/2$ is constant. We have

$$trace_{G_r}\tilde{R}'(d\psi,\tau(\psi))d\psi = 2H\frac{h(\phi)h''(\phi)}{r^2}e(\chi)(\partial/\partial\rho,0)\circ\psi, \tag{5.39}$$

by (5.24). Note that $\tau_2(\psi) = (-)(\Delta\tau(\psi) + trace_{G_r}\tilde{R}(d\psi,\tau(\psi))d\psi)$ (there is a $-$ or $+$ sign convention in the bitension field). By combining the above equations, the bitension field of ψ (using the $-$ sign here) is given by

$$\begin{aligned} \tau_2(\psi) &= \Big\{\frac{\partial^2 H}{\partial r^2} + \frac{m-1}{r}\frac{\partial H}{\partial r} - \frac{2e(\chi)}{r^2}(h'^2(\phi) + h(\phi)h''(\phi))H \\ &\quad + \frac{2h(\phi)h'(\phi)}{r^4}\Delta e(\chi)\Big\}(\partial/\partial\rho,0)\circ\psi - \frac{4}{r^4}h'^2(\phi)(0,d\chi(grad\, e(\chi))) \\ &= \frac{\partial^2 H}{\partial r^2} + \frac{m-1}{r}\frac{\partial H}{\partial r} - \frac{2e(\chi)}{r^2}(h'^2(\phi) + h(\phi)h''(\phi))H, \end{aligned} \tag{5.40}$$

where the last two terms in the first equality vanish since χ is an eigenmap.

If $f(t,x) = \phi(t,r)\cdot\chi(w) : \mathbf{R}^{1+m} \to N$ is an equivariant map, then by Proposition 5.3.3, $\tau_\Box(f) = -\phi_{tt} + H$. We can compute $\Delta\tau_\Box(f) = \Delta(-\phi_{tt} + H)$ similarly to (5.35)–(5.38) and compute $trace_g\tilde{R}'(df,\tau_\Box(f))df$ similarly to (5.39). Hence, we can derive (5.34) for the equivariant biwave map $f(t,x) = \phi(t,r)\cdot\chi(w) : \mathbf{R}^{1+m} = \mathbf{R}\times\mathbf{R}^+ \times_r S^{m-1} \to N = [0,R^*) \times_h S^{n-1}$ satisfying $\tau_{2\Box}(f) = (-)(\Delta\tau_\Box(f) + trace_g\tilde{R}(df,\tau_\Box(f))df) = 0$, similarly to (5.40). □

If f is an equivariant wave map, i.e., $\tau_{\Box}(f) = -\phi_{tt} + H = 0$, (5.34) can be written as

$$\phi_{tttt} - \phi_{ttrr} - \frac{m-1}{r}\phi_{ttr} - H_{tt} + H_{rr} + \frac{m-1}{r}H_r - \frac{\lambda}{r^2}(h'^2(\phi) + h(\phi)h''(\phi))(H - \phi_{tt})$$
$$= \left(-\frac{\partial^2}{\partial t^2} + \frac{\partial^2}{\partial r^2} + \frac{m-1}{r}\frac{\partial}{\partial r}\right)(H - \phi_{tt}) - \frac{\lambda}{r^2}(h'^2(\phi) + h(\phi)h''(\phi))(H - \phi_{tt}) = 0, \tag{5.41}$$

then f is an equivariant biwave map.

When the target is $\mathbf{R}^n - \{0\} = (0,\infty) \times_r S^{n-1}$, $f : \mathbf{R}^{1+m} \to \mathbf{R}^n - \{0\}$ is an equivariant map if there exist a map $\chi : S^{m-1} \to S^{n-1}$ and a function $\phi : (0,\infty) \to (0,\infty)$ such that for $y \in \mathbf{R}^m - \{0\}$, $\psi(y) = \phi(|y|)\chi(y/|y|)$. Thus the function $H : (0,\infty) \times_r S^{m-1} \to \mathbf{R}$ becomes

$$H(t,w) = \phi_{rr} + \frac{m-1}{r}\phi_r - \frac{2e(\chi)(w)}{r^2}\phi. \tag{5.42}$$

We can apply Proposition 5.3.3 and Theorem 5.3.4 to f as follows.

Corollary 5.3.5. *Let $\chi : S^{m-1} \to S^{n-1}$ be an eigenmap with eigenvalue λ.*

(1) $f = \phi\cdot\chi : \mathbf{R}^{1+m} \to \mathbf{R}^n - \{0\}$ *is an equivariant wave map iff ϕ is a solution of*

$$-\phi_{tt} + \phi_{rr} + \frac{m-1}{r}\phi_r - \frac{\lambda}{r^2}\phi = 0. \tag{5.43}$$

(2) $f = \phi\cdot\chi : \mathbf{R}^{1+m} \to \mathbf{R}^n - \{0\}$ *is an equivariant biwave map iff ϕ is a solution of*

$$\phi_{tttt} - H_{tt} - \phi_{ttrr} - \frac{m-1}{r}\phi_{ttr} + H_{rr} + \frac{m-1}{r}H_r - \frac{\lambda}{r^2}(H - \phi_{tt}) = 0, \tag{5.44}$$

where $H = \phi_{rr} + \frac{m-1}{r}\phi_r - \frac{\lambda}{r^2}\phi$.

Remark that $\chi : S^{m-1} \to S^{n-1}$ is a harmonic polynomial map, i.e. the restriction of a map from $\mathbf{R}^m$ to $\mathbf{R}^n$, each component of which is a harmonic homogeneous polynomial of a certain degree $k > 0$. Such a map has a constant density with $\lambda = 2e(\chi) = |d_w\chi|^2 = k(k+m-2)$ [128]. In particular, when $m = n = 2$, $k = 1$, (5.43) was solved in Chap. 8 of [332]. In general, (5.43) may be solved by Evans [140] and Shatah and Struwe [332], and (5.44) may be solved by solving (5.43) twice.

When the target is $S^n - \{\pm p\} = (0,\pi) \times_{\sin\rho} S^{n-1}$, $f : \mathbf{R}^{1+m} \to S^n - \{\pm p\}$ is an equivariant map if there are a map $\chi : S^{m-1} \to S^{n-1}$ and a function $\phi : (0,\infty) \to (0,\pi)$ such that for $y \in (0,\infty) \times_r S^{m-1}$,

$$\psi(y) = (\cos\phi(|y|), \chi(y/|y|)\sin\phi(|y|)). \tag{5.45}$$

Thus $H : (0,\infty)\times_r S^{m-1} \to \mathbf{R}$ becomes

$$H(r,x) = \phi_{rr} + \frac{m-1}{r}\phi_r - \frac{1}{r^2}e(\chi)\sin(2\phi(r)). \tag{5.46}$$

Corollary 5.3.6. *Let $\chi : S^{m-1} \to S^{n-1}$ be an eigenmap with eigenvalue λ.*

(1) *$f = \phi\cdot\chi : \mathbf{R}^{1+m} \to S^n - \{\pm p\}$ is an equivariant wave map iff ϕ is a solution of*

$$-\phi_{tt} + \phi_{rr} + \frac{m-1}{r}\phi_r - \frac{\lambda}{2r^2}\sin(2\phi(r)) = 0. \tag{5.47}$$

(2) *$f = \phi\cdot\chi : \mathbf{R}^{1+m} \to S^n - \{\pm p\}$ is an equivariant biwave map iff ϕ is a solution of*

$$\phi_{tttt} - H_{tt} - \phi_{ttrr} - \frac{m-1}{r}\phi_{ttr} + H_{rr} + \frac{m-1}{r}H_r - \frac{\lambda}{2r^2}\cos(2\phi)(H-\phi_{tt}) = 0, \tag{5.48}$$

where $H = \phi_{rr} + \frac{m-1}{r}\phi_r - \frac{\lambda}{2r^2}\sin(2\phi(r)) = 0$.

Let $\mathbf{R}^{n+1}$ be equipped with the standard Lorentzian inner product

$$(x,y)_0 = x^1y^1 + \cdots + x^ny^n - x^{n+1}y^{n+1}, \quad x,y \in \mathbf{R}^{n+1}$$

The hyperboloid model is the upper sheet of the hyperboloid

$$\mathbf{H}^n = H^n_+ = \{(x^1,\cdots,x^n,x^{n+1}) \in \mathbf{R}^{n+1} \mid (x^1)^2+\cdots+(x^n)^2-(x^{n+1})^2 = -1,\ x^{n+1} \geq 1\},$$

equipped with the Riemannian metric given by the restriction to H^n_+ of $(\cdot,\cdot)_0$. Setting $p = (0,\cdots,0,1)$, $\mathbf{H}^n - \{p\}$ is diffeomorphic to $(0,\infty)\times S^{n-1}$ by

$$(x^1,\cdots,x^n,x^{n+1}) \mapsto (\rho, \frac{(x^1,\cdots,x^n)}{\sinh\rho}),$$

where $x^{n+1} = \cosh\rho$. In these coordinates, the Riemannian metric on $\mathbf{H}^n - \{p\}$ takes the form

$$d\rho^2 + \sinh^2\rho\, dw_{S^{n-1}}.$$

Thus $\mathbf{H}^n - \{p\}$ can be considered as the warped product

$$\mathbf{H}^n - \{p\} = (0,\infty)\times_{\sinh\rho} S^{n-1}.$$

The map $f : \mathbf{R}^{1+m} \to \mathbf{H}^n - \{p\}$ is an equivariant wave map if there exist a map $\chi : S^{m-1} \to S^{n-1}$ and a function $\phi : (0,\infty) \to (0,\infty)$ such that for $y \in \mathbf{R}^m - \{0\}$,

$$\psi(y) = (\cosh\phi(|y|), \chi(y/|y|)\sinh\phi(|y|)).$$

Corollary 5.3.7. *Let $\chi : S^{m-1} \to S^{n-1}$ be an eigenmap with eigenvalue λ.*

(1) $f : \mathbf{R}^{1+m} \to \mathbf{H}^n - \{p\}$ *is an equivariant wave map iff ϕ is a solution of*

$$-\phi_{tt} + \phi_{rr} + \frac{m-1}{r} - \frac{\lambda}{2r^2}\sinh(2\phi) = 0. \tag{5.49}$$

(2) $f : \mathbf{R}^{1+m} \to \mathbf{H}^n - \{p\}$ *is an equivariant biwave map iff ϕ is a solution of*

$$\phi_{tttt} - H_{tt} - \phi_{ttrr} - \frac{m-1}{2}\phi_{ttr} + H_{rr} + \frac{m-1}{r}H_r - \frac{\lambda}{2r^2}\cosh(2\phi)(H - \phi_{tt}) = 0, \tag{5.50}$$

where $H = \phi_{rr} + \frac{m-1}{r}\phi_r - \frac{\lambda}{2r^2}\sinh(2\phi) = 0.$

5.4 Biwave Fields of Inclusions and Examples

5.4.1 *Biwave Fields of Inclusions*

We compute the biwave fields of inclusions of warped product manifolds, and then construct some examples of biwave maps from the inclusions and projections of warped product manifolds based on [76]. We now consider the warped product $N \times_\eta \mathbf{R}^{1+m}$ of an n-dimensional Riemannian manifold (N, h) and the Minkowski space $(\mathbf{R}^{1+m}, g)$ (which is viewed as a semi-Riemannian manifold with metric $g = diag(-1, 1, \cdots, 1)$). Let

$$i_a : (\mathbf{R}^{1+m}, g) \to (N \times_\eta \mathbf{R}^{1+m}, G_\eta)$$
$$y \mapsto (a, y), \quad a \in N$$

be the inclusion of $\mathbf{R}^{1+m}$ at the a-level in $N \times_\eta \mathbf{R}^{1+m}$.

Theorem 5.4.1. *The biwave field of the inclusion $i_a : \mathbf{R}^{1+m} \to N \times_\eta \mathbf{R}^{1+m}$ is given in terms of η as follows:*

$$\tau_{\square^2}(i_a) = \frac{(m-1)^2}{8}(grad(|grad\,\eta^2|^2), 0) \circ i_a. \tag{5.51}$$

Proof. Let $x_0 = t, x_1, , \cdots, x_m$ be the coordinates of any point in $\mathbf{R}^{1+m}$ and $e_0 = \frac{\partial}{\partial t}, e_1 = (1, 0, \cdots, 0), e_m = (0, \cdots, 0, 1)$ be the local frame of the point in $\mathbf{R}^{1+m}$. Denote by $D^1 = D^N$, $D^2 = D^{\mathbf{R}^{1+m}}$ the Levi-Civita connections of N and $\mathbf{R}^{1+m}$. We first calculate the wave field by applying (5.23) as follows:

$$\tau_{\Box}(i_a) = trace_g Ddi_a = \sum_{i=0}^{m}\{D_{e_i}di_a(e_i) - di_a(D^2_{e_i}e_i)\}$$

$$= \sum_{i=0}^{m}\{\tilde{D}_{(0,e_i)}(0,e_i) - (0, D^2_{e_i}e_i)\} \circ i_a = -\frac{m-1}{2}(grad\,\eta^2, 0) \circ i_a. \quad (5.52)$$

Remark that for $m \neq 1$ we have that i_a is a wave map iff $(grad\,\eta^2)_a = 0$.

We next calculate the biwave field of the inclusion i_a. Recall from [196] that $\Delta = (-)D_{e_i}D_{e_i} - D_{D^2_{e_i}e_i}$. We first have

$$D_{e_i}\tau_{\Box}(i_a) = -\frac{m-1}{2}D_{e_i}(grad\,\eta^2, 0) \circ i_a = -\frac{m-1}{2}(\tilde{D}_{(0,e_i)}(grad\,\eta^2, 0)) \circ i_a$$
$$= (-\frac{m-1}{4\eta^2}|grad\,\eta^2|^2(0, e_i)) \circ i_a.$$

Then

$$D_{e_i}D_{e_i}\tau_{\Box}(i_a) = -\frac{m-1}{2}\{\frac{1}{2\eta^2}|grad\,\eta^2|^2((0, D^2_{e_i}e_i) - \frac{1}{2}(grad\,\eta^2, 0))\} \circ i_a, \quad (5.53)$$

and

$$\nabla_{D^2_{e_i}e_i}\tau_{\Box}(i_a) = -\frac{m-1}{2}(\frac{1}{2\eta^2}|grad\,\eta^2|^2(0, \nabla^2_{e_i}e_i)) \circ i_a. \quad (5.54)$$

By adding (5.53) and (5.54), we obtain

$$\Delta\tau_{\Box}(i_a) = \frac{(m-1)^2}{8\eta^2}|grad\,\eta^2|^2(grad\,\eta^2, 0) \circ i_a. \quad (5.55)$$

By applying (5.24), we arrive at

$$trace_g\tilde{R}(i_a, \tau_{\Box}(i_a))di_a = \frac{(m-1)^2}{4}\{-(D^1_{grad\,\eta^2}grad\,\eta^2, 0) + \frac{1}{2\eta^2}|grad\,\eta^2|^2(grad\,\eta^2, 0)\} \circ i_a$$
$$= \frac{(m-1)^2}{8}\{-(grad(|grad\,\eta^2|^2), 0) + \frac{1}{\eta^2}|grad\,\eta^2|^2(grad\,\eta^2, 0)\} \circ i_a.$$

Note that $\tau_{2\Box}(i_a) = (-)(\Delta\tau_{\Box}(i_a) + trace_g\tilde{R}(di_a, \tau_{\Box}(i_a))di_a)$ (using the $-$ sign here), and we conclude the result. □

A function $\mathcal{A} \in C^\infty(N)$ on a Riemannian manifold (N, h) is an *affine function* if $\mathcal{A}\circ\gamma : I \subset \mathbf{R} \to \mathbf{R}$ is an affine function for any geodesic γ on N. For $\mathcal{A} \in C^\infty(N)$, let $Hess\,\mathcal{A}(X, Y) = X(Y\mathcal{A}) - (D_X Y)\mathcal{A}$ be the Hessian of $\mathcal{A}$ for $X, Y \in \Gamma(TN)$.

Lemma 5.4.2 ([31]). *Let (N, h) be a Riemannian manifold and $\mathcal{A} \in C^\infty(N)$. Then the following statements are equivalent:*

(a) *$\mathcal{A}$ is an affine function;*
(b) *grad $\mathcal{A}$ is a parallel vector field;*
(c) *the Hessian of $\mathcal{A}$ vanishes identically;*
(d) *grad $\mathcal{A}$ is a Killing field.*

Corollary 5.4.3. *Assume that $m \neq 1$. We have the following:*

(1) *The inclusion $i_a : \mathbf{R}^{1+m} \to N \times_\eta \mathbf{R}^{1+m}$ $(a \in N)$ is a non-trivial biwave map iff a is not a critical point of η^2, but it is a critical point of $|grad\, \eta^2|^2$.*
(2) *Every inclusion $i_x : \mathbf{R}^{1+m} \to N \times_\eta \mathbf{R}^{1+m}$ $(x \in N)$ is a non-trivial biwave map iff grad η^2 is a non-zero constant norm vector field.*
(3) *Let (N, h) be a Riemannian manifold with a positive non-trivial affine function η^2. Then $i_x : \mathbf{R}^{1+m} \to N \times_\eta \mathbf{R}^{1+m}$ $(x \in N)$ is a non-trivial biwave map.*

Proof. (1) follows from Theorem 5.4.1. (2) and (3) follow from Theorem 5.4.1 and Lemma 5.4.2 by setting $\eta^2 = \mathcal{A}$.

Viewing $\mathbf{R}^{1+m}$ as a semi-Riemannain manifold, let $p : M = N \times_\eta \mathbf{R}^{1+m} \to N$ be a semi-Riemannian submersion [279] which maps horizontal geodesics in $M = N \times_\eta \mathbf{R}^{1+m}$ to geodesics in N. □

Corollary 5.4.4. *Assume $m \neq 1$ and let $p : N \times_\eta \mathbf{R}^{1+m} \to N$, be a submersion. We have the following:* (1) *it $f = p \circ i_a : \mathbf{R}^{1+m} \to N$ is a biwave map iff a is not a critical point of η^2, but it is a critical point of $|grad\, \eta^2|^2$.* (2) *$f = p \circ i_x : \mathbf{R}^{1+m} \to N$ $(x \in N)$ is a biwave map iff grad η^2 is a non-zero constant norm vector field.* (3) *Let (N, h) be a Riemannian manifold with a positive non-trivial affine function η^2. Then $f = p \circ i_x : \mathbf{R}^{1+m} \to N$ $(x \in N)$ is a non-trivial biwave map.*

Proof. It follows from Theorem 5.4.1, Corollary 5.4.3 and Theorem 5.1.2. □

5.4.2 Examples

Example 1. (1) Let N be compact and let $\eta \in C^\infty(N)$ be a non-constant function. Then there exists $a \in N$ a maximum point for $|grad\, \eta^2|^2$. By Theorem 5.4.1, the inclusion $i_a : \mathbf{R}^{1+m} \to N \times_\eta \mathbf{R}^{1+m}$ $(m \neq 1)$ is a non-trivial biwave map. (2) Let $N = \mathbf{R}^n_+ = \{(x^1, \cdots, x^n) \in \mathbf{R}^n \,|\, x^i > 0,\ i = 1, \cdots, n\}$. By Corollary 5.4.3 (2) and [31], the inclusions $i_x : \mathbf{R}^{1+m} \to \mathbf{R}^n_+ \times_\eta \mathbf{R}^{1+m}$ $(m \neq 1,\ x \in \mathbf{R}^n_+)$ are non-trivial biwave maps iff there exists $a \in \mathbf{R}^n_+$ and $c \in \mathbf{R}_+$ such that $\eta^2(x) = (a, x) + c,\ \forall x \in \mathbf{R}^n_+$. (3) If $N = \mathbf{R}^n - \{0\}$ and $\eta(x) = \sqrt{|x|}$ for $x \in N$, then $grad\, \eta^2(x) = \frac{x}{|x|}$ and $|grad\, \eta^2|$ is constant. By Corollary 5.4.3 (2), $i_x : \mathbf{R}^{1+m} \to \mathbf{R}^n - \{0\} \times_\eta \mathbf{R}^{1+m}$ $(m \neq 1,\ x \in N)$ is a non-trivial biwave map.

Example 2. (1) Let $p : N \times \mathbf{R}^{1+m} \to N$ be a submersion [279] which maps horizontal geodesics in $N \times \mathbf{R}^{1+m}$ to geodesics in N. If N is compact and $\eta \in C^\infty(N)$ is a non-constant function, then there exists $a \in N$ a maximum point for $|grad\, \eta^2|^2$ and $f = p \circ i_a : \mathbf{R}^{1+m} \to N$ $(m \neq 1)$ is a non-trivial biwave map.

(2) Following Example 1(2), $f = p \circ i_x : \mathbf{R}^{1+m} \to \mathbf{R}^n_+$ $(m \neq 1)$ is a non-trivial biwave map. (3) Following Example 1(3) and taking $N = \mathbf{R}^n - \{0\} \approx S^{n-1}$, $f = p \circ i_x : \mathbf{R}^{1+m} \to S^{n-1} (m \neq 1)$ is a non-trivial biwave map.

5.5 Stress Bienergy Tensor

5.5.1 Definition

In Hilbert's paper [181], the stress-energy tensor associated to a variational problem is a symmetric 2-covariant tensor conserved at critical points, i.e., $div\, S = 0$. Let $f : (\mathbf{R}^{1+m}, g) \to (N, h)$ be a smooth map from a Minkowski space to a Riemannian manifold N. Recall from Definition 5.2.2 that the stress-energy tensor of f is defined by $S_f = e(f)g - f^*h$, where $e(f) = \frac{1}{2}||df||^2$ is the energy density and $g = \begin{bmatrix} -1 & 0 \\ 0 & I \end{bmatrix}$, I is the $m \times m$ identity matrix. The map f is said to satisfy the *conservation law* for S if $div\, S_f = 0$. Let us review Proposition 5.2.4 as follows:

Let $f : \mathbf{R}^{1+m} \to N$ be a smooth map from a Minkowski space into a Riemannian manifold N. Then

$$divS_f(X) = -(\tau_\Box(f), df(X)), \quad X \in \mathbf{R}^{1+m}. \tag{5.56}$$

Hence, if $f : \mathbf{R}^{1+m} \to N$ is a wave map, then f satisfies the conservation law for the stress-energy tensor S.

The conservation laws of biharmonic maps were first studied by Jiang [198], and we can apply the notions and techniques used therein to study the conservation laws of biwave maps. All the following theorems and results are obtained in [76].

Definition 5.5.1. Let $f : \mathbf{R}^{1+m} \to N$ be a smooth map from a Minkowski space into a Riemannian manifold N. The stress bienergy tensor of $f : \mathbf{R}^{1+m} \to N$ is defined by

$$\begin{aligned} S_2(X, Y) = {} & \frac{1}{2}|\tau_\Box(f)|^2(X, Y) + (df, \nabla(\tau_\Box(f)))(X, Y) \\ & - (df(X), \nabla_Y \tau_\Box(f)) - (df(Y), \nabla_X \tau_\Box(f)), \end{aligned} \tag{5.57}$$

for $X, Y \in \mathbf{R}^{1+m}$.

Theorem 5.5.2. *Let $f : R^{1+m} \to N$ be a smooth map from a Minkowski space into a Riemannian manifold N. Then*

$$div\, S_2(Y) = (-)(\tau_{2\Box}(f), df(Y)). \tag{5.58}$$

Consequently, if $f : \mathbf{R}^{1+m} \to N$ is a biwave map, then f satisfies the conservation law for the stress bienergy tensor S_2.

Proof. Given the map $f : \mathbf{R}^{1+m} \to N$, set $S_2 = Q_1 + Q_2$, where Q_1 and Q_2 are the $(0, 2)$-tensors defined by

$$Q_1(X,Y) = \frac{1}{2}|\tau_\Box f|^2(X,Y) + (df, \nabla\tau_\Box(f))(X,Y),$$

$$Q_2(X,Y) = -(df(X), \nabla_Y \tau_\Box(f)) - (df, \nabla_X \tau_\Box(f)).$$

Let $x_0 = t, x_1, x_2, \cdots, x_m$ be the coordinates at any point in $\mathbf{R}^{1+m}$, and $e_0 = \frac{\partial}{\partial t}, e_1 = (1,0,\cdots,0),\ e_2 = (0,1,0,\cdots,0), \cdots, e_m = (0,\cdots,0,1)$ be the local frame at the point. Write $Y = Y^i e_i$, and compute

$$\begin{aligned}
&div\, Q_1(Y) = \sum_i (\nabla_{e_i} Q_1)(e_i, Y) = \sum_i (e_i(Q_1(e_i,Y) - Q_1(e_i, \nabla_{e_i} Y)))\\
&= \sum_i \Big[e_i(\frac{1}{2}|\tau_\Box(f)|^2 Y^i + \sum_k (df(e_k, \nabla_{e_k}\tau_\Box(f))Y^i)\\
&- \frac{1}{2}|\tau_\Box(f)|^2 Y^i e_i - \sum_k (df(e_k), \nabla_{e_k}\tau_\Box(f))Y^i e_i)\Big]\\
&= (\nabla_Y \tau_\Box(f), \tau_\Box(f)) + \sum_i (df(Y, e_i), \nabla_{e_i}\tau_\Box(f)) + \sum_i (df(e_i), \nabla_Y \nabla_{e_i}\tau_\Box(f))\\
&= (\nabla_Y \tau_\Box(f), \tau_\Box(f)) + trace(\nabla df(Y,\cdot), \nabla.\tau_\Box(f)) + trace(df(\cdot), \nabla^2\tau_\Box(f)(Y,\cdot)).
\end{aligned} \tag{5.59}$$

We then compute

$$\begin{aligned}
&div\, Q_2(Y) = \sum_i (\nabla_{e_i} Q_2)(e_i, Y) = \sum_i (e_i(Q_2(e_i,Y) - Q_2(e_i, \nabla_{e_i} Y))\\
&= -(\nabla_Y \tau_\Box(f), \tau_\Box(f)) - \sum_i (\nabla df(Y, e_i), \nabla_{e_i}\tau_\Box(f))\\
&- \sum_i \Big(df(e_i), \nabla_{e_i}\nabla_Y \tau_\Box(f) - \nabla_{\nabla_{e_i}}\tau_\Box(f)\Big) + (df(Y), \Delta\tau_\Box(f))\\
&= -(\nabla_Y \tau_\Box(f), \tau_\Box(f)) - trace(\nabla df(Y,\cdot), \nabla.\tau_\Box(f))\\
&- trace(df(\cdot), \nabla^2\tau_\Box(f)(\cdot, Y)) + (df(Y), \Delta\tau_\Box(f)).
\end{aligned} \tag{5.60}$$

Adding (5.59) and (5.60), we arrive at

$$\begin{aligned}
divS_2(Y) &= (df(Y), \Delta\tau_\Box(f)) + \sum_i (df(e_i), R(Y, e_i)\tau_\Box(f))\\
&= (-)(\tau_{2\Box}(f), df(Y)).
\end{aligned} \tag{5.61}$$

This completes the proof. □

5.5.2 Applications

Let $\Omega \subset \mathbf{R}^{1+m}$ be a domain. Viewing $\mathbf{R}^{1+m}$ as a semi-Riemannian manifold, by O'Neill [279] we can consider a submersion $f : \mathbf{R}^{1+m} \to N$.

Proposition 5.5.3. *Let $f : \Omega \subset \mathbf{R}^{1+m} \to (N,h)$ be a submersion such that $\tau_\square(f)$ is basic, i.e., $\tau_\square(f) = W \circ f$ for $W \in \Gamma(TN)$. Assume that W is Killing and $|W|^2 = c^2$ is non-zero constant. Then* (1) *f is a non-trivial biwave map if Ω is non-compact;* (2) *f is a wave map if Ω is compact.*

Proof. Since $\tau_\square(f)$ is basic,

$$S_2(X,Y) = \{\frac{c^2}{2} + (df, \nabla\tau_\square(f))\}(X,Y) - (df(X), \nabla_Y\tau_\square(f)) - (df(Y), \nabla_X\tau_\square(f)). \tag{5.62}$$

Choose a point p in $\mathbf{R}^{1+m}$ with the frame $\{e_i\}_{i=0}^m$ such that $\{e_j\}_{j=0}^n$ are in $T_p^H\mathbf{R}^{1+m} = (\mathbf{R}^{1+m})^H = (\mathbf{R}^{1+m})^{V\perp}$ and $\{e_k\}_{k=n+1}^m$ are in $T_p^V\mathbf{R}^{1+m} = (\mathbf{R}^{1+m})^V = \ker df(p)$. Because W is Killing,

$$\begin{aligned}(df, \nabla\tau_\square(f))(p) &= \sum_j (df_p(e_j), \nabla_{e_j}\tau_\square(f)) + \sum_k (df_p(e_k), \nabla_{e_k}\tau_\square(f)) \\ &= \sum_j (df_p(e_j), \nabla^N_{df_p(e_j)}W) = 0.\end{aligned} \tag{5.63}$$

Then

$$S_2(p)(X,Y) = \frac{c^2}{2}(X,Y) - ((df_p(X), \nabla^N_{df_p(Y)}W) + (df_p(Y), \nabla^N_{df_p(X)}W)) = \frac{c^2}{2}(X,Y)$$

for $X, Y \in \mathbf{R}^{1+m}$. Hence, if Ω is not compact, $S_2 = \frac{c^2}{2}g$, $g = diag(-1, 1, \cdots, 1)$ is divergence free and f is a non-trivial biwave map since $c \neq 0$. If Ω is compact, using the relation

$$div(df \cdot \tau_\square(f)) = |\tau_\square(f)|^2 + (df, \nabla\tau_\square(f)),$$

and integrating it over Ω by applying the divergence theorem, we obtain that $\tau_\square(f) = 0$ by (5.63). Hence, f is a wave map. □

If $f : \mathbf{R}^{1+m} \to N$ is a wave map, then $S_2(f) = 0$. However, $S_2(f) = 0$ does not imply that f is a wave map.

Proposition 5.5.4. *Let $f : (\mathbf{R}^{1+m}, g) \to (N,h)$, $m \neq 3$. Then $S_2 = 0$ if and only if*

$$\frac{1}{m-3}|\tau_\square(f)|^2(X,Y) + (\nabla_X\tau_\square(f), df(Y)) + (\nabla_Y\tau_\square(f), df(X)) = 0, \ \text{for } X, Y \in \mathbf{R}^{1+m}. \tag{5.64}$$

Proof. Suppose that $S_2 = 0$, which implies trace $S_2 = 0$. Therefore,

$$(\nabla \tau_{\square}(f), df) = -\frac{m-1}{2(m-3)}|\tau_{\square}(f)|^2 \quad (m \neq 3). \tag{5.65}$$

Substituting this expression into the definition of S_2, we get

$$0 = S_2(X,Y) = -\frac{1}{m-3}|\tau_{\square}(f)|^2(X,Y) - (\nabla_X \tau_{\square}(Y), df(Y)) - (\nabla_Y \tau_{\square}(f), df(X)). \tag{5.66}$$

The converse is similar. □

Corollary 5.5.5. *If $f : \mathbf{R}^{1+m} \to (N,h)$ $(m \neq 3)$ with $S_2 = 0$ and rank $f \leq m$, then f is a wave map.*

Proof. Choose a point $p \in \mathbf{R}^{1+m}$ such that *rank* $f \leq m$. Then there exists a vector $X_p \in \ker df$. It follows from Proposition 5.5.4 that $\tau_{\square}(f)(p) = 0$ for $X = Y = X_p$. □

Theorem 5.5.6. *If $f : \Omega \subset \mathbf{R}^{1+m} \to (N,h)$ $(m \neq 5)$ is a map from a compact domain into a Riemannian manifold with $S_2 = 0$, then f is a wave map.*

Proof. Since *trace* $S_2 = 0$, we have

$$0 = trace\ S_2 = \frac{m-1}{2}|\tau_{\square}(f)|^2 + (m-1)(\nabla \tau_{\square}(f), df) - 2(\nabla \tau_{\square}(f), df). \tag{5.67}$$

Integrating (5.67) over Ω and applying the divergence theorem, we have

$$0 = \frac{5-m}{2}\int_{\Omega} |\tau_{\square}(f)|^2 dv. \tag{5.68}$$

Hence, f is a wave map if $m \neq 5$. □

5.6 Well-Posedness Problem

There are interesting and difficult problems involving local well-posedness, global well-posedness and global regularity of biwave maps into Riemannian manifolds or Lie groups (or Riemannian symmetric spaces) that require future exploration.

Let $\Omega \subset\subset \mathbf{R}^{1+m}$ be a space-time domain and $(N^n, h) \subset \mathbf{R}^k$ be a compact Riemannian manifold without boundary. The Sobolev space $W^{2,2}(\Omega, N)$ is defined by

$$W^{2,2}(\Omega, N) = \{g \in W^{2,2}(\Omega, R^k) |\, g(x) \in N, \text{ for a.e. } x \in \Omega\}.$$

We consider the second-order energy functional on $W^{2,2}(\Omega, N)$, defined by

$$E_2(f) = \frac{1}{2}\int_\Omega |\Box f|^2 dz, \ z = (t,\, x), \tag{5.69}$$

where $\Box f = -f_{tt} + \triangle_x$, $\triangle_x = \sum_{i=1}^m \frac{\partial^2 f}{\partial x^{i\,2}}$. Since $N \subset \mathbf{R}^k$ is compact, there exists a tubular neighborhood of sufficient small uniform width $\delta > 0$ and a smooth nearest neighbor projection $\Pi : U_\delta(N) \to N$. For $y \in N$, $P(y) = \nabla\Pi : \mathbf{R}^k \to T_y N$ is the orthonormal projection, and $P^\perp(y) = Id - P(y) : \mathbf{R}^k \to (T_y N)^\perp$ is the orthonormal projection to the normal space of N at y.

Definition 5.6.1. $f \in W^{2,2}(\Omega, N)$ is a (weakly) biwave map if it is a critical point of the bienergy functional $E_2(\cdot)$ over $W^{2,2}(\Omega, N)$.

Theorem 5.6.2. *Any biwave map $f \in W^{2,2}(\Omega, N)$ satisfies*

$$\Box^2 f \perp T_f N \tag{5.70}$$

in the sense of distributions.

Proof. For any $\phi \in C_0^\infty(\Omega, \mathbf{R}^k)$, set $f_s = \Pi(f + s\phi)$, $0 \le s < 1$; then $\frac{df_s}{ds}\big|_{s=0} = P(f)\phi \in T_f(N)$. Therefore, we have

$$0 = \frac{d}{ds}E_2(f)\Big|_{s=0} = \frac{d}{ds}\frac{1}{2}\int_\Omega (\Box\Pi(f + s\phi),\, \Box\Pi(f+s\phi))dz\Big|_{s=0}$$
$$= \int_\Omega (\Box f,\, \Box P(f)\cdot\phi)dz = -\int_\Omega (\Box^2 f,\, P(f)\cdot\phi) = -\int_\Omega (P(f)\cdot\Box^2 f,\, \phi).$$

Integrating by parts in the above equation and observing that $\phi \in C_0^\infty(\Omega, \mathbf{R}^k)$ is arbitrary, we deduce that $P(f)\cdot\Box^2 f = 0$, which implies $\Box^2(f) \perp T_f N$. □

Let $f : \mathbf{R}^{1+m} \to N \subset \mathbf{R}^k$ be a biwave map such that $(f, f_t)|_{t=0} = (f_0, f_1) : \mathbf{R}^m \to TN$, i.e., $f_0(x) \in N \subset \mathbf{R}^k$ and $f_1(x) \in T_{f_0}N \subset \mathbf{R}^k$ for almost every $x \in \mathbf{R}^m$. We then have $\Box f_0 = \Box f$, $\frac{\partial}{\partial t}\Box f = \Box f_1$. For simplicity, we assume that N is compact. The biwave field is

$$(\tau_2)_\Box(f) = \Box\tau_\Box(f) + R'(df, df)\tau_\Box(f), \tag{5.71}$$

where R' is the Riemannian curvature of N. Suppose that we choose normal coordinates on N, then (5.71) becomes

$$\Box\Box f + R'(df, df)\Box f = 0. \tag{5.72}$$

To see how biwave maps are related to the local well-posedness in H^s for $s < \frac{m}{4}+2$, we begin by observing that, by Theorem 5.6.2,

$$\Box\Box f = -R'(df, df)\Box f \perp T_f N.$$

Proposition 5.6.3. *Solutions of the biwave equation satisfy the energy conservation*

$$0 = < \Box^2 f, f_t >= \int_{\mathbf{R}^m} \Box^2 f \cdot f_t \, dx$$
$$= \frac{d}{dt} \int_{\mathbf{R}^m} \{ f_t f_{ttt} - \frac{1}{2}(f_{tt})^2 + (f_{tx})^2 + \frac{1}{2}(f_{xx})^2 \} dx. \tag{5.73}$$

Proof. We compute

$$\frac{d}{dt} \int_{\mathbf{R}^m} \{ f_t f_{ttt} - \frac{1}{2}(f_{tt})^2 + (f_{tx})^2 + \frac{1}{2}(f_{xx})^2 \} \, dx$$
$$= \int_{\mathbf{R}^m} (f_{tt} f_{ttt} + f_t f_{tttt} - f_{tt} f_{ttt} + 2 f_{tx} f_{ttx} + f_{xx} f_{xxt}) \, dx$$
$$= \int_{\mathbf{R}^m} (f_t f_{tttt} - 2 f_t f_{ttxx} + f_{xxxx} f_t) \, dx$$
$$= \int_{\mathbf{R}^m} (f_{tttt} - 2 f_{ttxx} + f_{xxxx}) f_t \, dx$$
$$= \int_{\mathbf{R}^m} \Box^2 f \cdot f_t \, dx = 0. \quad \Box$$

Thus we have the energy identity

$$E_2(f(t)) = \frac{1}{2} ||Df(t)||^2_{L^4(\mathbf{R}^m)} = \frac{1}{2} ||Df(0)||^2_{L^4(\mathbf{R}^m)}, \tag{5.74}$$

where $Df = (\partial_t f, \nabla f)$ is the vector of space-time derivatives of f. By applying the first spatial derivatives ∂ to (5.72) for ∂f, we derive

$$\partial \Box^2 \partial f = -\partial (R'(d\partial f, d\partial f) \Box (\partial f))$$
$$= -2dR'(d\partial f, d\partial f)(\partial d\partial f, d\partial f) \Box (\partial f) - R'(d\partial, d\partial f) \partial \Box (\partial f), \tag{5.75}$$

which implies

$$(\partial f_t, \Box^2 \partial f) = -(f_t, \partial \Box^2 (\partial f))$$
$$= (f_t, 2dR'(d\partial f, d\partial f)(\partial d\partial f, d\partial f) \Box (\partial f)), \tag{5.76}$$

by the orthogonality $(f_t, R'(\cdot, \cdot)\cdot) = 0$. It follows from (5.75) and (5.76) that

$$\frac{dt}{dt} ||D^2 f(t)||^2_{L^4(\mathbf{R}^m)} = \frac{d}{dt} E_2(\partial f(t)) = \int_{\mathbf{R}^m} (\Box^2 (\partial f), \partial f_t) dx$$
$$= \int_{\mathbf{R}^m} (f_t, 2dR'(d\partial f, d\partial f)(\partial d\partial f, d\partial f) \Box \partial f) dx$$

$$\leq const \cdot |dR'|_{L^\infty} \int_{\mathbf{R}^m} |D^3 f(t)|\,|D^2 f(t)|\,|Df(t)|\,|\Box\partial f|dx$$
$$\leq const \cdot |dR'|_{L^\infty} \int_{\mathbf{R}^m} |D^3 f(t)|^2 |D^2 f(t)||Df(t)|dx. \tag{5.77}$$

Obtaining well-posedness results for biwave maps is an unfinished difficult task which requires a lot of hard work.

5.7 Transversal Biwave Maps

Transversal biwave maps, whose equations are a fourth-order system of hyperbolic PDEs, are different from the transversally biharmonic maps, whose equations are a fourth-order system of elliptic PDEs. In this section we discuss transversal biwave maps and their properties based on Chiang and Wolak [84].

5.7.1 *Definition and Examples*

We follow the notions and notations of transversal wave maps in Sect. 2.8. Let $\mathbf{R}^{1+m}$ be a $m+1$ dimensional Minkowski space $\mathbf{R} \times \mathbf{R}^m$ with the metric $(\eta_{ab}) = diag(-1, 1, \cdots, 1)$ and the coordinates $x_0 = t, x_1, x_2, \cdots, x_m$, foliated by planes parallel to $\{0\} \times \mathbf{R}^p \subset \mathbf{R} \times \mathbf{R}^m$ $(p + q = m)$. Then $(\mathbf{R}^{1+m}, \mathcal{H}^p)$ is a transversally Minkowski foliation defined by the global submersion $\iota \times \phi \colon \mathbf{R} \times \mathbf{R}^m \to \mathbf{R} \times \mathbf{R}^q$; $\mathbf{R} \times \mathbf{R}^q$ can be considered as its complete transverse manifold. Let $\mathcal{F}$ be a Riemannian foliation for a Riemannian metric g of an n-dimensional Riemannian manifold M, which induces a Riemannian metric $\bar{g}$ on a q_1 $(p_1 + q_1 = n)$ dimensional transverse manifold $N = \coprod_i \bar{U}_i$. Let $f : (\mathbf{R}^{1+m}, \mathcal{H}) \to (M, \mathcal{F})$ be a smooth foliated map from a foliated Minkowski space into a foliated Riemannian manifold. Form $V_i = f^{-1}(U_i) \subset \mathbf{R}^{1+m}$ for each i. Let $\bar{V}_i$ be the quotient of V_i; it is an open subset of $\mathbf{R}^{1+q}$ for each i. Then f induces a map $\bar{f} = \coprod_i \bar{f}_i : \coprod_i \bar{V}_i \to \coprod_i \bar{U}_i$ with $\bar{f}_i : \bar{V}_i \to \bar{U}_i$ (for convenience, we drop the subscript i from $\bar{f}_i$ if there is no confusion) such that the Diagram 2.8.1 commutes, i.e., $\bar{f} \circ (\iota \times \phi) = \phi_1 \circ f$, where $\iota \times \phi : V_i \to \bar{V}_i$ is a submersion defined by the foliation $\mathcal{H}$ on the open subset V_i, $\phi_1 : U_i \to \bar{U}_i$ is a Riemannian submersion defining the foliation $\mathcal{F}$ on the open set U_i and $\iota(t) = t$. By taking a smaller V_i we can assume that $V_i = T_i \times W_i \subset \mathbf{R} \times \mathbf{R}^m$ and $\bar{V}_i = T_i \times \bar{W}_i \subset \mathbf{R} \times \mathbf{R}^q$, T_i is an open subset of $\mathbf{R}$ and $\bar{W}_i$ is an open subset of $\mathbf{R}^q$.

A transversal biwave map $f : \mathbf{R}^{1+m} \to (M, \mathcal{F})$ is a transversally biharmonic map on the Minkowski space $\mathbf{R}^{1+m}$ with the transversal bienergy functional

$$E_2(\bar{f}) = \frac{1}{2}\int_{\coprod \bar{V}_i} \tau_{\Box}(\bar{f}) dt\, dx$$

$$= \frac{1}{2}\int_{\coprod \bar{V}_i} \left[\Box \bar{f}^k + \bar{\Gamma}^k_{rs} \left(-\bar{f}^r_t \bar{f}^s_t + \sum_{a=1}^{q} \bar{f}^r_a \bar{f}^s_a \right) \right] dt\, dx, \qquad (5.78)$$

where $\Box = -\frac{\partial^2}{\partial t^2} + \sum_{a=1}^{q} \frac{\partial^2}{\partial x_a^2}$ is the wave operator and $\bar{\Gamma}^k_{rs}$ are the Christoffel symbols of $\bar{U}_i$ for each i. The Euler-Lagrange equation describing the critical points of (5.78) gives the following definition.

Definition 5.7.1. $f : \mathbf{R}^{1+m} \to (M, \mathcal{F})$ is a transversal biwave map iff

$$(\tau_2)_{\Box}(\bar{f}) = J_{\bar{f}}(\tau_{\Box}\bar{f}) = \Delta\tau_{\Box}(\bar{f}) + \bar{R}'(d\bar{f}, d\bar{f})\tau_{\Box}(\bar{f})$$

$$= \Box\tau_{\Box}(\bar{f})^k + \Gamma'^k_{rs} \left(-\tau_{\Box}(\bar{f})^r_t \tau_{\Box}(\bar{f})^s + \sum_{a=1}^{q} \tau_{\Box}(\bar{f})^{\mu}_a \tau_{\Box}(\bar{f})^{\gamma}_a \right)$$

$$+ \bar{R}'^k_{rsl} \left(-\bar{f}^r_t \bar{f}^s_t + \sum_{a=1}^{q} \bar{f}^r_a \bar{f}^s_a \right) \tau_{\Box}(\bar{f})^l = 0 \qquad (5.79)$$

for each $\bar{f}_i : \bar{V}_i \to \bar{U}_i$, where $\bar{R}'$ is the Riemannian curvature on each $\bar{U}_i$.

Since Diagram 2.8.1 commutes, the definition of a transversal biwave map does not depend on the choice of the local Riemannian submersion defining the Riemannian foliation.

Example 1. Let $u : \mathbf{R}^{1+m} \to \mathbf{R}$ be a transversal biwave function, which satisfies $\Box^2 u(t,x) = \Box(\Box u) = 0$ with initial data $u_0 = u$, $u_1 = \frac{\partial u}{\partial t}$. We have $\Box u_0 = \Box u$ and $\frac{\partial}{\partial t}\Box u = \Box \frac{\partial u}{\partial t} = \Box u_1$. The transversal biwave function u induces $\bar{u} : \bar{V} \subset R^{1+q} \to R$ locally satisfying

$$\Box^2 \bar{u}(t,x) = \bar{u}_{tttt} - 2\bar{u}_{ttxx} + \bar{u}_{xxxx} = 0, \quad (t,x) \in (0,\infty) \times \mathbf{R}^q,$$

$$\bar{u}_0 = \bar{u},\ \bar{u}_1 = \frac{\partial \bar{u}}{\partial t}, \Box\bar{u}_0 = \Box\bar{u},\ \frac{\partial}{\partial t}\Box\bar{u} = \Box\bar{u}_1, \quad (t,x) \in \{t = 0\} \times \mathbf{R}^q,$$

where the initial data $\bar{u}_0$, $\bar{u}_1$ are given. This is a fourth-order homogeneous linear equation with constant coefficients. Similarly to Example 1 of Sect. 5.1, $\bar{u}(t,x)$ can be solved in each $\bar{V} \subset \mathbf{R}^{1+q}$.

Let $(M_1, \mathcal{F}_1, g_1)$ and $(M_2, \mathcal{F}_2, g_2)$ be two Riemannian manifolds with Riemannian foliations. Suppose that $f_1 : (M_1, \mathcal{F}_1) \to (M_2, \mathcal{F}_2)$ is a smooth foliated leaf-preserving map, i.e., $df_1(T\mathcal{F}_1) \subset T\mathcal{F}_2$. Let $U_i \subset M_i$ be open subsets and let $\phi_i : (U_i, g_i) \to (\bar{U}_i, \bar{g}_i)$ be Riemannian submersions on U_i, which define locally the Riemannian foliations $\mathcal{F}_i$ for $i = 1, 2$. Suppose that $f_1(U_1) \subset U_2$. Based on the notions discussed in Sect. 1.9, there is a closed relationship between the transversally second fundamental form of f_1 and the second fundamental form of the induced maps $\bar{f}_1$, obtained by using the local submersions defining the foliations $\mathcal{F}_1$ and $\mathcal{F}_2$. It follows from Sect. 1.9 that

$$d\phi_2 S_b(f_1)_x = S(\bar{f}_1)_{\phi_1(x)}$$

holds for each of the foliation defining local submersions $\phi_i : U_i \to \bar{U}_i$ such that $f_1(U_1) \subset U_2$.

Definition 5.7.2. $f_1 : (M_1, \mathcal{F}_1) \to (M_2, \mathcal{F}_2)$ is a *transversally totally geodesic map* if $S(\bar{f}_1)_{\phi_1(x)} = \nabla d(\bar{f}_1)_{\phi_1(x)} = 0$ in each $\bar{U}_1$, where ∇ is the connection on $T^*\bar{U}_1 \otimes \bar{f}_1^{-1} T\bar{U}_2$.

Theorem 5.7.3. *If $f : (\mathbf{R}^{1+m}, \mathcal{H}) \to (M_1, \mathcal{F}_1)$ is a transversal biwave map and $f_1 : (M_1, \mathcal{F}_1) \to (M_2, \mathcal{F}_2)$ is a transversally totally geodesic map between two foliated Riemannian manifolds $(M_1, \mathcal{F}_1)$ and $(M_2, \mathcal{F}_2)$, then the composition $f_1 \circ f : (\mathbf{R}^{1+m}, \mathcal{H}) \to (M_2, \mathcal{F}_2)$ is a transversal biwave map. That is,*

$$(\tau_2)_\square(\bar{f}_1 \circ \bar{f}|_{\bar{U}_1}) = d\bar{f}_1 \circ (\tau_2)_\square(\bar{f}|_{\bar{U}_1}) \tag{5.80}$$

for each i, where $N_1 = \coprod (\bar{U}_1)_i$ is the transverse manifold of $(M_1, \mathcal{F}_1)$.

The proof of the above theorem is similar to the proof of Theorem 5.1.2, combining the concepts of foliated Riemannian manifolds (cf. [84]).

Example 2. Let $(M_1, \mathcal{F}_1)$ be a foliated submanifold of $(M_2, \mathcal{F}_2)$ such that the traces of leaves of $\mathcal{F}_2$ on M_1 are leaves of $\mathcal{F}_1$. This condition implies that for suitable choices of foliation cycles, the transverse manifold N_1 is a submanifold of the transverse manifold N_2. Are the transversal biwave maps into $(M_1, \mathcal{F}_1)$ also transversal biwave maps into $(M_2, \mathcal{F}_2)$? By Theorem 5.7.3 the answer is affirmative if $(M_1, \mathcal{F}_1)$ is a transversally totally geodesic foliated submanifold of $(M_2, \mathcal{F}_2)$, i.e., $N_1 = \coprod (\bar{U}_1)_i$ is a totally geodesic submanifold of N_2, that is, geodesics in N_1 are also geodesics in N_2. If γ is a transversally geodesic of $(M_1, \mathcal{F}_1)$, i.e., $\bar{\gamma} = \phi \circ \gamma : \mathbf{R} \to U_1 \to \bar{U}_1$ is a geodesic in N_1, then $\bar{\gamma}$ is also a N_2 geodesic, $\bar{\gamma}$ has dimension one and has no curvature. For a map $v : \mathbf{R}^{1+m} \to \mathbf{R}$, let $u = \gamma \circ v : \mathbf{R}^{1+m} \to \mathbf{R} \to U_1$, which induces $\bar{u} = \bar{\gamma} \circ \bar{v} : \bar{V}_1 \to \mathbf{R} \to \bar{U}_1$. By (5.80), we have

$$(\tau_2)_\square(\bar{f}) = d\gamma \circ (\tau_2)_\square(\bar{v}) = d\gamma \circ \square^2 \bar{v}, \tag{5.81}$$

since $\bar{\gamma}$ is a geodesic. Therefore, u is a transversal biwave map iff $\bar{v}$ solves the homogeneous linear biwave equation $\square^2 \bar{v} = 0$. Hence, with respect to the arclength parameterization, the transversal biwave map equation into $\bar{\gamma}$ is equivalent to the linear biwave equation, by (5.81). Then for any target foliated manifold $(M_2, \mathcal{F}_2)$ we can provide many transversal biwave maps associated to the transversal geodesics of $(M_2, \mathcal{F}_2)$.

There are biwave maps which are not transversal biwave maps. We construct such an example using a warped product of two manifolds as in Example 3. Recall that by B. O'Neill [279] a warped product can be defined on semi-Riemannian manifolds (i.e. pseudo-Riemannian manifolds) or Riemannian manifolds. Let (B, g), (F, h) be semi-Riemannian manifolds or Riemannian manifolds and $\alpha : B \to \mathbf{R}$ be a smooth map. On the product manifold $B \times F$ we define a metric tensor $k = g \oplus e^{2\alpha} h$. Let ∇^g and ∇^h be the Levi-Civita connections on (B, g) and (F, h), respectively. Recall that the Levi-Civita connection ∇^k on $B \times F$ can be related to those of B and F as follows:

$\nabla^k_X Y = \nabla^g_X Y$, where X and Y are vector fields on B.
$\nabla^k_X V = \nabla^k_V X = X(\alpha)V$, where V is a vector field on F.
$\nabla^k_V W = -h(V, W) grad_g \alpha + \nabla^h_V(W)$, where V and W are vector fields on F.

Example 3. Let $f : B_1 \times F_1 \to B_2 \times F_2$ be a smooth map preserving the leaves such that $f(t, x, y) = (f_1(t, x), f_2(t, x, y))$, where $B_1 = \mathbf{R} \times R = \mathbf{R}^{1+1}$, $F_1 = \mathbf{R}$, $B_2 = F_2 = \mathbf{R}$, $\alpha_1(x) = 0$, $\alpha_2(x) = x$, $f_1(t, x) = t + \frac{4}{3}x^4$, $f_2(t, x, y) = 2x^2$. By (1.58) and (1.59), we have

$$\tau_\Box(f) = \tau_\Box(f_1) + \tau_\Box(f_2|_{B_1}) + \tau_\Box(f_2|_{F_1}) - ||df_2||^2 (grad_{g_2}\alpha_2) \circ f_1$$
$$= 16x^2 + 4 - 16x^2 = 4 \neq 0, \tag{5.82}$$

where the third term vanishes. Then $(\tau_2)_\Box(f) = 0$. But, $(\tau_2)_\Box(f_1) = 32 \neq 0$. Therefore, f is a biwave map, but it is not a transversal biwave map.

The following example shows that there are transversal biwave maps which are not biwave maps either.

Example 4. Let (B_1, g_1), (B_2, g_2), (F_1, h_1), and (F_2, h_2) be Riemannian manifolds. Consider the foliations on the Riemannian manifolds $B_1 \times F_1$ and $B_2 \times F_2$ given by the projections on the first component $\pi_1 : B_1 \times F_1 \to B_1$ and $\pi_2 : B_2 \times F_2 \to B_2$, respectively. The projections π_1 and π_2 are Riemannian submersions, and the foliations defined by them are Riemannian. Let $h : B_1 \times F_1 \to B_2 \times F_2$ be a smooth leaf-preserving map. Then h must be of the form $h(x, y) = (h_1(x), h_2(x, y))$, $x \in B_1$, $y \in F_1$, where $h_1 : B_1 \to B_2$, $h_2 : B_1 \times F_1 \to F_2$ are smooth. For the product Riemannian metrics on $B_1 \times F_1$ and $B_2 \times F_2$, the connection of dh is equal to

$$\nabla d(h) = (\nabla d(h_1), \nabla d(h_2|_{B_1}) + \nabla d(h_2|_{F_1})), \tag{5.83}$$

where $\nabla d(h_1)$ is the connection of dh_1 at x of the map $h_1 : B_1 \to B_2$, $\nabla d(h_2|_{B_1})$ is the connection of dh_2 at x of the map $x \mapsto h_2(x, y)$ with y fixed, and $\nabla d(h_2|_{F_1})$ is the connection of dh_2 at y of the map $y \mapsto h_2(x, y)$ with x fixed. On the one hand, by (5.83) the property "totally geodesic" of $h = (h_1, h_2)$ is equivalent to h_1 is totally geodesic and $\nabla d(h_2|_{B_1}) + \nabla d(h_2|_{F_1}) = 0$, i.e., the vertical and horizontal contributions to the totally geodesic annihilate each other. On the other hand, if h_1 is totally geodesic and $h_2|_{B_1}$, $h_2|_{F_1}$ are totally geodesic for $x \in B_1, y \in F_1$, then h is totally geodesic. Therefore, it follows that there are maps h which are transversally totally geodesic, but not totally geodesic. Hence, by Theorem 5.7.3 there are transversal biwave maps that are not biwave maps.

Let Ω be a compact domain in $\mathbf{R}^{1+m}$. We can consider $(\Omega, \mathcal{H}|_\Omega)$ as a compact foliated domain in $(\mathbf{R}^{1+m}, \mathcal{H})$. Let $f : (\Omega, \mathcal{H}|_\Omega) \subset (\mathbf{R}^{1+m}, \mathcal{H}) \to (M, \mathcal{F})$ is a transversal biwave map from a compact foliated space-time domain into a Riemannian manifold which induces $\bar{f} : \bar{V}_i \to \bar{U}_i$ for each i, where for simplicity we still denote $V_i = V_i \cap \Omega$, $V_i = f^{-1}U_i$, $\bar{V}_i$ is the quotient of V_i and it is an open subset of $\mathbf{R}^{1+q}$ for each i. Suppose that the compact supports of $\frac{\partial \bar{f}_s}{\partial s}$ and $\nabla_{e_k} \frac{\partial \bar{f}_s}{\partial s}$ $(0 \le k \le q)$ are contained in the interior of $\bar{V}_i$ for each i.

Theorem 5.7.4. *If $f : (\Omega, \mathcal{H}|_\Omega) \to (M, \mathcal{F})$ is a transversal biwave map from a compact foliated space-time domain into a foliated Riemannian manifold such that*

$$-|\tau_\Box f|_t^2 + \sum_{i=1}^{m} |\tau_\Box f|_{x^i}^2 - R'^{\alpha}_{\beta\gamma\mu} \left(-f_t^\beta f_t^\gamma + \sum_{i=1}^{m} f_i^\beta f_i^\gamma \right) \tau_\Box(f)^\mu \ge 0 \quad (5.84)$$

then f is a transversal wave map.

Proof. Since $f : (\Omega, \mathcal{H}|_\Omega) \subset (\mathbf{R}^{1+m}, \mathcal{H}) \to (M, \mathcal{F})$ is a transversal biwave map which induces $\bar{f} : \bar{V}_i \to \bar{U}_i$ for each i, we have

$$(\tau_2)_\Box(\bar{f}) = \Delta \tau_\Box(\bar{f}) + R'(d\bar{f}, d\bar{f})\tau_\Box(\bar{f}) = 0,$$

where $\Delta = \nabla^*\nabla$ and ∇ is the connection on $T^*\bar{V} \otimes f^{-1}T\bar{U}_i$ by Jiang [196]. Let $x_0 = t, x_1, \cdots, x_q$ be the coordinates of a point p in $\bar{V}_i$ and $e_0 = \frac{\partial}{\partial t}, e_1 = (1, 0, \cdots, 0), e_2 = (0, 1, 0, \cdots, 0), \cdots, e_q = (0, \cdots, 0, 1)$ be the frame at the point. We compute

$$\begin{aligned}
\frac{1}{2}\Delta ||\tau_\Box(\bar{f})||^2 &= (\nabla_{e_i}\tau_\Box(\bar{f}), \nabla_{e_i}\tau_\Box(\bar{f})) + (\nabla_{e_i}^*\nabla_{e_i}\tau_\Box(\bar{f}), \tau_\Box(\bar{f})) \\
&= \sum_{i=0}^{q} (\nabla_{e_i}\tau_\Box(\bar{f}), \nabla_{e_i}\tau_\Box(\bar{f})) - (R'^{\alpha}_{\beta\gamma\mu}(-\bar{f}_t^\beta \bar{f}_t^\gamma + \sum_{i=1}^{q} f_i^\beta \bar{f}_i^\gamma)\tau_\Box(\bar{f})^\mu, \tau_\Box(\bar{f})) \\
&= -|\tau_\Box \bar{f}|_t^2 + \sum_{i=1}^{m} |\tau_\Box \bar{f}|_{x^i}^2 - (R'^{\alpha}_{\beta\gamma\mu}(-\bar{f}_t^\beta \bar{f}_t^\gamma + \sum_{i=1}^{m} \bar{f}_i^\beta \bar{f}_i^\gamma)\tau_\Box(\bar{f})^\mu, \tau_\Box(\bar{f})).
\end{aligned} \quad (5.85)$$

By applying the Bochner's techniques from (5.85) and the assumption that the compact supports of $\frac{\partial \bar{f}_s}{\partial s}$ and $\nabla \frac{\partial f_s}{\partial s}$ are contained in the interior of $\bar{V}_i$, we deduce that $||\tau_\Box(f)||^2$ is constant, i.e., $d\tau_\Box(f) = 0$. If we use the identity

$$\int_{\amalg \bar{V}_i} div(d\bar{f}, \tau_\Box(\bar{f}))dz = \int_{\amalg \bar{V}_i} (|\tau_\Box(\bar{f})|^2 + (d\bar{f}, d\tau_\Box(\bar{f})))dz, \; z = (t, x),$$

and the fact $d\tau_{\Box}(\bar{f}) = 0$, and then apply the divergence theorem, we conclude that $\tau_{\Box}(\bar{f}) = 0$ for each i. Hence, f is a transversal wave map. □

Note that the above theorem is different from Theorem 4.6.2.

5.7.2 *Transversal Conservation Law*

Suppose that $f : (\Omega, \mathcal{H}) \to (M, \mathcal{F})$ is a transversal biwave map from a compact foliated space-time domain in a foliated Minkowski space into a foliated Riemannian manifold $(M, \mathcal{F})$ with a transverse manifold $N = \coprod \bar{U}_i$, which induces $\bar{f} : \bar{V}_i \to \bar{U}_i$ for each i such that Diagram 2.8.1 commutes, where we still denote $V_i = V_i \cap \Omega, V_i = f^{-1}(U_i)$, $\bar{V}_i$ is an open subset of $\mathbf{R}^{1+q}$ for each i. Let $x_0 = t, x_1, x_2, \cdots, x_q$ be the coordinates at a point in $\bar{V}_i$ for each i and let $e_0 = \frac{\partial}{\partial t}, e_1 = (1, 0, \cdots, 0)$, $e_2 = (0, 1, 0, \cdots, 0), \cdots, e_q = (0, \cdots, 0, 1)$ be the frame at the point. Suppose that the compact supports of $\frac{\partial \bar{f}}{\partial s}$ and $\nabla_{e_k} \frac{\partial \bar{f}}{\partial s}$ are contained in the interior of $\bar{V}_i$ for each i. Let $\xi_i \in \Gamma(\bar{f}^{-1} T\bar{U}_i)$ be a vector field with $\frac{\partial f}{\partial s}|_{s=0} = \xi_i$ for each i. By Jiang [196] and using concepts of foliated Riemannian manifolds, we have the following:

Lemma 5.7.5. *If $f : (\Omega, \mathcal{H}) \to (M, \mathcal{F})$ is a transversal biwave map, then*

$$\begin{aligned}\frac{d^2}{ds^2} E_2(\bar{f}_s)\Big|_{s=0} &= 2\int_{\coprod \bar{V}_i} ||\Delta\xi_i + R^{\bar{U}_i}(d\bar{f}(e_k), \xi_i) d\bar{f}(e_k)||^2 dz \\ &+ 2\int_{\coprod \bar{V}_i} < \xi_i, (\nabla'_{d\bar{f}(e_k)}) R^{\bar{U}_i}(f(e_k), \tau_{\Box}(\bar{f}))\xi_i + (\nabla'_{\tau_{\Box}(\bar{f})} R^{\bar{U}_i})(d\bar{f}(e_k), \xi_i) d\bar{f}(e_k) \\ &+ R^{\bar{U}_i}(\tau_{\Box}(\bar{f}), \xi_i)\tau(\bar{f}) + 2R^{\bar{U}_i}(d\bar{f}(e_k), \xi_i)\bar{\nabla}_{e_k}\tau_{\Box}(f) \\ &+ 2R^{\bar{U}_i}(d\bar{f}(e_k), \tau_{\Box}(\bar{f}))\bar{\nabla}_{e_k}\xi_i > dz \end{aligned} \tag{5.86}$$

where $z = (t, x) \in \bar{V}_i \subset \mathbf{R}^{1+q}$, ∇' is the Riemannian connection on $T\bar{U}_i$, for each i and ξ_i is the vector field along $\{f_s\}$.

Definition 5.7.6. Let $f : (\mathbf{R}^{1+m}, \mathcal{H}) \to (M, \mathcal{F})$ be a transversal biwave map from a foliated Minkowski space to a foliated Riemannian manifold. If $\frac{d^2}{ds^2} E_2(\bar{f}_s)\Big|_{s=0} \geq 0$, then f is a *stable* transversal biwave map.

If we consider a transversal wave map as a transversal biwave map, then by (5.86) we have $\frac{d^2}{ds^2} E_2(\bar{f}_s)\big|_{s=0} \geq 0$ and f is automatically stable.

Definition 5.7.7. Let $f : (\mathbf{R}^{1+m}, \mathcal{H}) \to (M, \mathcal{F})$ be a smooth map from a foliated Minkowski space to a foliated Riemannian manifold, which induces $\bar{f}$: $(\mathbf{R}^{1+q}, \eta) \to (N = \coprod \bar{U}_i, \bar{g})$. The *transversal stress energy* is defined by $S_{\bar{f}} = e(\bar{f})\eta - \bar{f}^*\bar{g}$, where $e(\bar{f}) = \frac{1}{2}||d\bar{f}||^2$ for $\bar{f}$ in each $\bar{U}_i$. We say that f satisfies *transversal conservation law* if $div\, S_{\bar{f}} = 0$ for $\bar{f}$ in each $\bar{V}_i$.

By Chiang [75] for $\forall \bar{X} \in \Gamma(T\bar{V}_i)$ we have

$$(div\, S_{\bar{f}})(\bar{X}) = -(\tau(\bar{f}), d\bar{f}(\bar{X})) \tag{5.87}$$

in each $\bar{V}_i$. Hence, if $f : (\mathbf{R}^{m,1}, \mathcal{H}) \to (M, \mathcal{F})$ is a transversal wave map, then f satisfies the transversal conservation law for S.

Definition 5.7.8. Let $f : (\mathbf{R}^{1+m}, \mathcal{H}) \to (M, \mathcal{F})$ be a smooth map from a foliated Minkowski space into a foliated Riemannian manifold, which induces $\bar{f} : (\mathbf{R}^{1+q}, \eta) \to (N = \coprod \bar{U}_i, \bar{g})$. The transversal stress bienergy tensor of f is defined by

$$\begin{aligned} S_2(\bar{X}, \bar{Y}) &= \frac{1}{2}|\tau_{\Box}(\bar{f})|^2(\bar{X}, \bar{Y}) + (d\bar{f}, \nabla(\tau_{\Box}(\bar{f})))(\bar{X}, \bar{Y}) \\ &\quad - (d\bar{f}(\bar{X}), \nabla_{\bar{Y}}\tau_{\Box}(\bar{f})) - (d\bar{f}(\bar{Y}), \nabla_{\bar{X}}\tau_{\Box}(\bar{f})), \end{aligned}$$

for each $\bar{f} : \bar{V}_i \to \bar{U}_i$ locally, where $\bar{X}, \bar{Y}$ are vector fields on $\bar{V}_i$.

Theorem 5.7.9. *Let* $f : (\mathbf{R}^{1+m}, \mathcal{H}) \to (M, \mathcal{F})$ *be a smooth map from a foliated Minkowski space into a foliated Riemannian manifold, which induces* $\bar{f} : (\mathbf{R}^{1+q}, \eta) \to (N = \amalg \bar{U}_i, \bar{g})$. *Then we have*

$$div\, S_2(\bar{Y}) = (-)(\tau_{2\Box}(\bar{f}), d\bar{f}(\bar{Y})),$$

for each $\bar{f} : \bar{V}_i \to \bar{U}_i$ *locally. Consequently, if* $f : \mathbf{R}^{1+m} \to N$ *is a transversal biwave map, then f satisfies the transversal conservation law for the stress bienergy tensor* S_2.

The proof is similar to the proof of Theorem 5.5.3, using concepts of foliations, see [84].

Theorem 5.7.10. *There does not exist a non-trivial stable transversal biwave map* $f : (\Omega, \mathcal{H}) \to (M, \mathcal{F})$ *from a compact foliated domain into a foliated Riemannian manifold with constant transverse sectional curvature* $K > 0$ *satisfying the transversal conservation law for S.*

The proof is analogous to the proof of Theorem 5.2.5, using concepts of foliations, see [84].

Chapter 6
Bi-Yang-Mills Fields

Bi-Yang-Mills fields are the critical points of the bi-Yang-Mills functionals of connections whose curvature tensors satisfy a certain condition, which generalize the Yang-Mills fields studied by Bourguignon and Lawson [38, 39]. Bi-Yang-Mills fields were first explored by Ichiyama, Inoguchi and Urakawa [191, 192] in 2009. In this chapter, we study first and second variations of bi-Yang-Mills functionals and the isolation phenomena of bi-Yang-Mills fields, based on [191, 192].

6.1 First Variation

We follow the notions and notations of Yang-Mills fields in Sect. 3.1, and recall some basic concepts as follows. Let (E, h) be a real vector bundle of rank N with an inner product over an m-dimensional compact Riemannian manifold (M, g). Let $\mathcal{C}(E, h)$ be the space of all C^∞ connections of E satisfying the compatibility condition

$$X(s,t)_h = (D_X s, t)_h + (s, D_X t)_h, \quad s, t \in \Gamma(E),$$

where X is a vector field in $\mathcal{X}(M)$ and $\Gamma(E)$ is the space of all C^∞ sections of E. For $D \in \mathcal{C}(E, h)$, let R^D be its curvature tensor, defined by

$$R(X, Y)s = D_X(D_Y s) - D_Y(D_X s) - D_{[X,Y]}s,$$

for all $X, Y \in \mathcal{X}(M)$, $s \in \Gamma(E)$. Let $F = End(E, h)$ be the bundle of endomorphisms of E which are skew-symmetric with respect to the inner product h on E. We define an inner product $< \cdot, \cdot >$ on F by

Y.-J. Chiang, *Developments of Harmonic Maps, Wave Maps and Yang-Mills Fields into Biharmonic Maps, Biwave Maps and Bi-Yang-Mills Fields*, Frontiers in Mathematics, DOI 10.1007/978-3-0348-0534-6_6, © Springer Basel 2013

$$< \phi, \psi >= \sum_{i=1}^{N} (\phi u_i, \psi u_i)_h, \ \phi, \psi \in F_x,$$

where $\{u_i\}_{i=1}^N$ is an orthonormal basis of E_x with respect to h and $x \in M$. Let $\Omega^k(F) = \Gamma(\wedge^k T^*M \otimes F)$ be the space of F-valued k-forms on M, which admits the global inner product $(\cdot, \cdot)$ given by

$$(\alpha, \beta) = \int_M < \alpha, \beta > dv_g,$$

where the pointwise inner product $< \alpha, \beta >$ is given by

$$< \alpha, \beta >= \sum_{i_1 < \cdots < i_k} < \alpha(e_{i_1}, \cdots, e_{i_k}), \beta(e_{i_1}, \cdots, e_{i_k}) >$$

Here, $\{e_i\}_{i=1}^m$ is a locally defined orthonormal frame on (M, g).

For every $D \in \mathcal{C}(E, h)$, let $d^D : \Omega^k(F) \to \Omega^{k+1}(F)$ be the exterior differentiation with respect to D. The adjoint operator $\delta^D : \Omega^{k+1}(F) \to \Omega^k(F)$ is given by

$$\delta^D \alpha = (-1)^{k+1} * d^D * \alpha, \ \ \alpha \in \Omega^{k+1}(F),$$

where $* : \Omega^k(F) \to \Omega^{m-k}(F)$ is the extension of the usual Hodge star operator on (M, g). Then we have

$$(d^D \alpha, \beta) = (\alpha, \delta^D \beta), \ \alpha \in \Omega^k(F), \ \beta \in \Omega^{k+1}(F).$$

Let

$$\mathcal{YM}_2(D) = \frac{1}{2} \int_M ||\delta^D R^D||^2 dv, \ \ D \in \mathcal{C}(E, h), \tag{6.1}$$

be the bi-Yang-Mills functional of a compact Riemannian manifold (M, g).

Definition 6.1.1. $D \in \mathcal{C}(E, h)$ is a bi-Yang-Mills connection, or the associated curvature R^D is a bi-Yang-Mills field, if for each smooth one-parameter family D^t $(-\epsilon < t < \epsilon)$ with $D^0 = D$,

$$\left.\frac{d}{dt}\right|_{t=0} \mathcal{YM}_2(D^t) = 0.$$

Remark that the first variation of bi-Yang-Mills functional is exactly the same as the second variation of the Yang-Mills functional.

Theorem 6.1.2 ([39, 191]). *Let* $\alpha = \left.\frac{d}{dt}\right|_{t=0} D^t \in \Omega^1(F)$ $(F = End(E, h))$. *Then we have*

$$\frac{d}{dt}\Big|_{t=0} \mathcal{YM}_2(D^t) = \int_M < \alpha, (\delta^D d^D + \mathcal{R}^D)(\delta^D R^D) > dv, \tag{6.2}$$

where $\mathcal{R}^D(\beta)(X) = \sum_{i=1}^m [R^D(e_i, X), \beta(e_i)] \in \Omega^1(F)$, $\beta \in \Omega^1(F)$, $X \in \mathcal{X}(M)$ *and* $\{e_i\}_{i=1}^m$ *is a local orthonormal frame on* M *with respect to* g*. Hence,* D *is a bi-Yang-Mills connection if and only if*

$$(\delta^D d^D + \mathcal{R}^D)(\delta^D R^D) = 0. \tag{6.3}$$

Proof. First, computing the derivative we have

$$\begin{aligned} \frac{d}{dt}\Big|_{t=0} \mathcal{YM}_2(D^t) &= \int_M < \frac{d}{dt}\Big|_{t=0} \delta^{D^t} R^{D^t}, \delta^D R^D > dv \\ &= \int_M < \frac{d}{dt}\Big|_{t=0} (\delta^{D^t}) R^D + \delta^D (\frac{d}{dt}\Big|_{t=0} R^{D^t}), \delta^D R^D > dv. \end{aligned} \tag{6.4}$$

Since $\frac{d}{dt}\Big|_{t=0} R^{D^t} = d^D\alpha$, the second term in the right-hand side of (6.4) becomes

$$\begin{aligned} \int_M < \delta^D (\frac{d}{dt}\Big|_{t=0} R^{D^t}), \delta^D R^D > dv &= \int_M < \delta^D d^D \alpha, \delta^D R^D > dv \\ &= \int_M < \alpha, \delta^D d^D (\delta^D R^D) > dv. \end{aligned} \tag{6.5}$$

The first term in the right-hand side of (6.4) is equal to

$$\begin{aligned} \int_M < \frac{d}{dt}\Big|_{t=0} (\delta^{D^t}) R^D, \delta^D R^D > dv &= \int_M < R^D, [\alpha \wedge \delta^D R^D] > dv \\ &= \int_M < \alpha, \mathcal{R}^D (\delta^D R^D) > dv. \end{aligned} \tag{6.6}$$

To check the first equality in (6.6), we need to show that for all $\phi \in \Omega^2(F)$ and $\beta \in \Omega^1(F)$

$$\begin{aligned} \int_M < (\frac{d}{dt}\Big|_{t=0} \delta^{D^t}) \phi, \beta > dv &= \int_M < \phi, \frac{d}{dt}\Big|_{t=0} (d^{D^t} \beta) > dv, \\ \frac{d}{dt}\Big|_{t=0} (d^{D^t} \beta)(X, Y) &= [\alpha \wedge \beta](X, Y), \end{aligned} \tag{6.7}$$

where $[\alpha \wedge \beta](X, Y) = [\alpha(X), \beta(Y)] - [\alpha(Y), \beta(X)]$ (cf. [39]). By definition, we have

$$(d^{D^t} \beta)(X, Y) = D^t_X(\beta(Y)) - D^t_Y(\beta(X)) - \beta([X, Y]).$$

Therefore,

$$\left(\frac{d}{dt}\Big|_{t=0} d^{D^t}\beta\right)(X,Y) = \frac{d}{dt}\Big|_{t=0} D^t_X(\beta(Y)) - \frac{d}{dt}\Big|_{t=0} D^t_Y(\beta(X))$$
$$= [\alpha(X), \beta(Y)] - [\alpha(Y), \beta(X)] = [\alpha \wedge \beta](X,Y),$$

provided we can show that $\frac{d}{dt}\big|_{t=0} D^t_X(\beta(Y)) = [\alpha(X), \beta(Y)]$. We set $D^t = D + \alpha^t$ with $\alpha^t \in \Omega^1(F)$ and $\alpha^0 = 0$. Since $\beta(Y) \in \Gamma(F)$ with $F = End(E,h)$, we have

$$D^t_X(\beta(Y))(u) = D^t_X(\beta(Y)u) - \beta(Y)(D^t_X u)$$
$$= (D_X + \alpha^t(X))(\beta(Y)u) - \beta(Y)(D_X u + \alpha^t(X)u)$$

for $u \in \Gamma(E)$, whence

$$\frac{d}{dt}\Big|_{t=0} D^t_X(\beta(Y))(u) = \frac{d}{dt}\Big|_{t=0} \alpha^t(X)(\beta(Y)u) - \beta(Y)\left(\frac{d}{dt}\Big|_{t=0} \alpha^t(X)u\right)$$
$$= \alpha(X)(\beta(Y)u) - \beta(Y)(\alpha(X)u) = [\alpha(X), \beta(Y)]u$$

as desired.

For the second equality of (6.6), we need to show that

$$< R^D, [\alpha \wedge \beta] > = < \mathcal{R}^D(\beta), \alpha >$$

for all $\beta \in \Omega(F)$. This follows from

$$\sum_{j<i} < R^D(e_j, e_i), [\alpha \wedge \beta](e_j, e_i) > = \sum_{j<i} < R^D(e_j, e_i), [\alpha(e_j), \beta(e_i)] - [\alpha(e_i), \beta(e_j)] >$$
$$= \sum_{j,i=1}^{m} < R^D(e_j, e_i), [\alpha(e_j), \beta(e_i)] > = \sum_{j=1}^{m} < \sum_{i=1}^{m} [\beta(e_i), R^D(e_j, e_i)], \alpha(e_j) > = < \mathcal{R}^D(\beta), \alpha > .$$

This completes the proof of Theorem 6.1.2. □

6.2 Second Variation

We next calculate the second variation of the bi-Yang-Mills functional following the work of Ichiyama, Inoguchi and Urakawa [191, 192].

Theorem 6.2.1 ([191]). *Let $D \in \mathcal{C}(E,h)$ be a bi-Yang-Mills connection. Then for each smooth one-parameter family D^t ($|t| < \epsilon$) with $D^0 = D$ we have*

$$\frac{d^2}{dt^2}\Big|_{t=0} \mathcal{YM}_2(D^t)$$
$$= \int_M < (\delta^D d^D + \mathcal{R}^D)^2(\alpha) + 2\delta^D(\alpha \wedge \delta^D R^D) + \mathcal{R}(d^D \delta^D R^D)(\alpha), \alpha > dv. \tag{6.8}$$

In order to prove the theorem, we require the following lemmas (proofs are given in [191]).

Lemma 6.2.2. *If D is a connection in $\mathcal{C}(E,h)$ and D^t ($|t| < \epsilon$) is a smooth one-parameter family in $\mathcal{C}(E,h)$ with $D^0 = D$, then*

$$\frac{d}{dt}\Big|_{t=0} d^{D^t}\beta = [\alpha \wedge \beta], \quad \frac{d^2}{dt^2}\Big|_{t=0} d^{D^t}\beta = [\gamma \wedge \beta], \tag{6.9}$$

where $\alpha = \frac{d}{dt}\big|_{t=0} D^t \in \Omega^1(F)$ and $\gamma = \frac{d^2}{dt^2}\big|_{t=0} D^t \in \Omega^1(F)$.

Lemma 6.2.3. *We have*

$$< \phi, [\beta_1 \wedge \beta_2] > = < \mathcal{R}(\phi)(\beta_2), \beta_1 > = < \beta_2, \mathcal{R}(\phi)(\beta_1) >, \tag{6.10}$$

for all $\beta_1, \beta_2 \in \Omega^1(F)$ and $\phi \in \Omega^2(F)$.

Lemma 6.2.4. *For $\alpha = \frac{d}{dt}\big|_{t=0} D^t$ and $\gamma = \frac{d^2}{dt^2}\big|_{t=0} D^t$, we have*

$$\frac{d}{dt}\Big|_{t=0} R^{D^t} = d^D\alpha, \quad \frac{d^2}{dt^2}\Big|_{t=0} R^{D^t} = d^D\gamma + [\alpha \wedge \alpha]. \tag{6.11}$$

Lemma 6.2.5. *We have*

$$\frac{d}{dt}\Big|_{t=0} \delta^{D^t}\phi = \mathcal{R}(\phi)(\alpha), \; \phi \in \Omega^2(F), \tag{6.12}$$

where

$$\mathcal{R}(\phi)(\alpha)(X) = \sum_{i=1}^m [\phi(e_i, X), \alpha(e_i)], \; X \in \mathcal{X}(M). \tag{6.13}$$

(Note that $R^D = R(R^D)$ if we take $\phi = R^D$.) In particular,

$$\frac{d}{dt}\Big|_{t=0} \delta^{D^t} R^D = \mathcal{R}^D(\alpha). \tag{6.14}$$

Moreover,

$$\frac{d^2}{dt^2}\bigg|_{t=0} \delta^{D^t} R^D = \mathcal{R}^D(\gamma), \tag{6.15}$$

where $\alpha = \frac{d}{dt}\big|_{t=0} D^t$ *and* $\gamma = \frac{d^2}{dt^2}\big|_{t=0} D^t$.

Lemma 6.2.6. *We have*

$$\frac{d}{dt}\bigg|_{t=0} \delta^{D^t} R^{D^t} = \mathcal{R}^D(\alpha) + \delta^D d^D \alpha, \tag{6.16}$$

$$\frac{d^2}{dt^2}\bigg|_{t=0} \delta^{D^t} R^{D^t} = \mathcal{R}^D(\gamma) + 2\mathcal{R}(d^D\alpha)(\alpha) + \delta^D d^D \gamma + \delta^D[\alpha \wedge \alpha], \tag{6.17}$$

where $\mathcal{R}(\phi)$ *is given by (6.13),* $\alpha = \frac{d}{dt}\big|_{t=0} D^t$ *and* $\gamma = \frac{d^2}{dt^2}\big|_{t=0} D^t$.

Proof (Proof of Theorem 6.2.1). We compute the second derivative of the integrand of the bi-Yang-Mills functional $\mathcal{YM}_2(D^t)$ at $t = 0$:

$$\begin{aligned}\frac{1}{2}\frac{d^2}{dt^2}\bigg|_{t=0} ||\delta^{D^t} R^{D^t}||^2 &= < \frac{d^2}{dt^2}\bigg|_{t=0} \delta^{D^t} R^{D^t}, \delta^D R^D > + < \frac{d}{dt}\bigg|_{t=0} \delta^{D^t}, \frac{d}{dt}\bigg|_{t=0} \delta^{D^t} > \\ &= < \mathcal{R}^D(\gamma) + 2\mathcal{R}(d^D\alpha)(\alpha) + \delta^D d^D \gamma + \delta^D[\alpha \wedge \alpha], \delta^D R^D > \\ &+ < \mathcal{R}^D(\alpha) + \delta^D d^D(\alpha), \mathcal{R}^D(\alpha) + \delta^D d^D \alpha >, \end{aligned} \tag{6.18}$$

by (6.16) and (6.17) in Lemma 6.2.6. If we integrate (6.18) over M, we obtain

$$\begin{aligned}\frac{d^2}{dt^2}\bigg|_{t=0} \mathcal{YM}_2(D^t) &= (\mathcal{R}^D(\gamma) + 2\mathcal{R}(d^D\alpha)(\alpha) + \delta^D d^\nabla \gamma + \delta^D[\alpha \wedge \alpha], \delta^D R^D) \\ &+ (\mathcal{R}^D(\alpha) + \delta^D d^D(\alpha), \mathcal{R}^D(\alpha) + \delta^D d^\nabla \alpha) \\ &= (2\mathcal{R}(d^D\alpha)(\alpha) + \delta^D[\alpha \wedge \alpha], \delta^D R^D) \\ &+ (\mathcal{R}^D(\alpha) + \delta^D d^D(\alpha), \mathcal{R}^D(\alpha) + \delta^D d^D \alpha), \end{aligned} \tag{6.19}$$

since $(\mathcal{R}^D(\gamma) + \delta^D d^D \gamma, \delta^D R^\nabla) = (\gamma, (\mathcal{R}^D + \delta^D d^D)(\delta^D R^D)) = 0$ and D is a bi-Yang-Mills field, i.e., $(\mathcal{R}^D + \delta^D d^D)(\delta^D R^D) = 0$.

Moreover, for the first term in the right-hand side of (6.19), we have

$$\begin{aligned}(2\mathcal{R}(d^D\alpha)(\alpha) + \delta^D[\alpha \wedge \alpha], \delta^D R^D) &= 2(\mathcal{R}(d^D\alpha)(\alpha), \delta^D R^D) + ([\alpha \wedge \alpha], d^D\delta^D R^D) \\ &= 2(d^D\alpha, [\alpha \wedge \delta^D R^D]) + (\alpha, \mathcal{R}(d^D\delta^D R^D)(\alpha)) \\ &= (\alpha, 2\delta^D[\alpha \wedge \delta^D R^D] + \mathcal{R}(d^D\delta^D R^D)(\alpha)), \end{aligned} \tag{6.20}$$

by Lemma 6.2.3. The second term in the right-hand side of (6.19) becomes

$$(\alpha, (\mathcal{R}^D + \delta^D d^D)^2(\alpha)), \tag{6.21}$$

because $\mathcal{R}^D + \delta^D d^D$ is self-adjoint with respect to the global inner product (,). Now (6.8) follows from (6.19)–(6.21). □

6.3 Isolation Phenomena

Ichiyama, Inoguchi and Urakawa [191, 192] have recently studied the isolation phenomena of bi-Yang-Mills fields, which generalize the isolation phenomena of Yang-Mills fields analyzed by Bourguignon and Lawson [39]. In this section, we discuss such phenomena based on [191, 192]. We first rewrite Theorem 6.2.1 as follows.

Proposition 6.3.1. *Let $D \in \mathcal{C}(E,h)$ be a bi-Yang-Mills connection and D^t ($|t| < \epsilon$) be a smooth one-parameter family in $\mathcal{C}(E,h)$ with $D^0 = D$. Then we have*

$$\frac{d^2}{dt^2}\Big|_{t=0} \mathcal{YM}_2(D^t) = \int_M < S_2^D(\alpha), \alpha > dv, \tag{6.22}$$

where

$$S_2^D(\alpha) = S^D(S^D(\alpha)) + 2\delta^D(\alpha \wedge \delta^D R^D) + \mathcal{R}(d^D \delta^D R^D)(\alpha) \tag{6.23}$$

is a fourth-order self-adjoint elliptic differential operator acting on $\Omega^1(F)$.

Recall that $S^D(\alpha) = (d^D\delta^D + \delta^D d^D)(\alpha) + \mathcal{R}^D(\alpha)$ is a second-order self-adjoint elliptic differential operator acting on $\Omega^1(F)$ by (3.16).

Let $D \in \mathcal{C}(E,h)$ is a bi-Yang-Mills connection, and let E^2_λ be the eigenspace of S_2^D on $\Omega^1(F)$ with eigenvalue λ. Since S_2^D is a self-adjoint elliptic differential operator, it keeps *Ker*(δ^D) invariant. The restriction of S_2^D to *Ker*(δ^D) has a discrete spectrum consisting of distinct eigenvalues $\lambda_1^2 < \lambda_2^2 < \cdots < \lambda_i^2 < \cdots \to \infty$ corresponding to finite-dimensional eigenspaces $E^2_{\lambda_i}$. Then the *index* and *nullity* of D are defined by

$$index_2(D) = dim(\bigoplus_{\lambda<0} E^2_\lambda), \quad nullity_2(D) = dim(E^2_0)$$

Definition 6.3.2. A bi-Yang-Mills connection D or a bi-Yang-Mills field R^D is *stable* if $i_2(D) = n_2(D) = 0$, i.e., $\frac{d^2}{dt^2}\big|_{t=0}\mathcal{YM}_2(D^t) > 0$.

Note that the weak stability $\frac{d^2}{dt^2}\big|_{t=0}\mathcal{YM}_2(D^t) \geq 0$ is equivalent to the condition $i_2(D) = 0$. In particular, stability (i.e. $i_2(D) = n_2(D) = 0$) implies weak stability.

Proposition 6.3.3. *If D is a Yang-Mills connection, then D is a weakly stable bi-Yang-Mills connection.*

Proof. If D is a Yang-Mills field, then

$$\frac{d^2}{dt^2}\bigg|_{t=0} \mathcal{YM}_2(D^t) = \int_M < S_2^D(\alpha), \alpha > dv$$

$$= \int_M < S^D(S^D(\alpha)), \alpha > dv = \int_M ||S^D(\alpha)||^2 dv \geq 0,$$

since $S^D = (\delta^D d^D + \mathcal{R}^D)(\alpha)$ is the second-order self-adjoint elliptic differential operator and $\delta^D R^D = 0$ by (3.16). Then D is a weakly stable bi-Yang-Mills field (i.e., $i_2(D) = 0$). □

Theorem 6.3.4 (Bounded Isolation Phenomena [191]). *Let (M, g) be a compact Riemannian manifold whose Ricci curvature is bounded below by a positive constant $c > 0$ (i.e., $Ric(M) \geq c \cdot id$). Suppose that $D \in \mathcal{C}(E, h)$ is a bi-Yang-Mills connection with $||R^D|| < c/2$ point-wisely everywhere on M. Then D is a Yang-Mills connection.*

We require the following lemma to prove Theorem 6.3.4.

Lemma 6.3.5. *If $D \in \mathcal{C}(E, h)$ is a bi-Yang-Mills connection, then*

$$\frac{1}{2}\Delta||\delta^D R^D||^2 = < 2\mathcal{R}^D(\delta^D R^D) + d^D R^D \circ Ric,\ \delta^D R^D > + \sum_{i=1}^m ||D_{e_i}(\delta^D R^D)||^2, \tag{6.24}$$

where $\Delta f = \sum_{i=1}^m (e_i^2 - D_{e_i}(e_i)) f$ is the Laplacian acting on smooth functions f on M and for all $\alpha \in \Omega^1(F)$,

$$(\alpha \circ Ric)(X) = \alpha(Ric(X)),\ X \in \mathcal{X}(M). \tag{6.25}$$

Here Ric is the Ricci transform of (M, g).

Proof (Proof of Theorem 6.3.4). If we integrate (6.24) over M and apply Green's divergence theorem we get

$$2\int_M < \mathcal{R}^D(\delta^D R^D), \delta^D R^D > dv + \int_M < \delta^D R^D \circ Ric, \delta^D R^D > dv$$

$$+ \int_M \sum_{i=1}^m < D_{e_i}(\delta^D R^D), D_{e_i}(\delta^D R^D) > dv = 0. \tag{6.26}$$

Remark that

$$| < \mathcal{R}^D(\alpha), \alpha > | \leq ||R^D||||\alpha||^2,\ \ \alpha \in \Omega^1(F). \tag{6.27}$$

By Lemma 6.2.3 and Schwartz's inequality, we obtain

$$| < \mathcal{R}^D(\alpha), \alpha > | = | < R^D, [\alpha \wedge \alpha] > | \leq ||R^D|| \, ||[\alpha \wedge \alpha]|| \leq ||R^D|| \, ||\alpha||^2, \tag{6.28}$$

where the last inequality holds because

$$||\alpha \wedge \alpha||^2 = \sum_{i<j} ||[\alpha \wedge \alpha](e_i, e_j)||^2 = \frac{1}{2} \sum_{i,j=1}^{m} ||[\alpha \wedge \alpha](e_i, e_j)||^2$$
$$\leq \frac{1}{2} \sum_{i,j=1}^{m} 2||\alpha(e_i)||^2||\alpha(e_j)||^2 = \left(\sum_{i=1}^{m} ||\alpha(e_i)||^2 \right)^2$$

(Lemma (2.30) in [39], p. 197), which is $||\alpha||^4$.

Moreover, by the hypothesis of the Ricci curvature of (M, g), we get

$$< \delta^D R^D \circ Ric, \delta^D R^D > \geq c||\delta^D R^D||^2. \tag{6.29}$$

We can choose an orthonormal basis $\{e_i\}_{i=1}^m$ of $(T_x M, g_x)$ such that $Ric(e_i) = \mu e_i$ for $i = 1, \cdots, m$, where $\mu_i \geq c$ for each i. Then

$$< \delta^D R^D \circ Ric, \delta^D R^D > = \sum_{i=1}^{m} < \delta^D R^D(Ric(e_i)), \delta^D R^D(e_i) >$$
$$= \sum_{i=1}^{m} \mu_i ||\delta^D R^D(e_i)||^2 \geq c||\delta^D R^D||^2.$$

Since $||R^D|| < \frac{c}{2}$ at each point of M, we obtain

$$< 2\mathcal{R}^D(\delta^D R^D) + \delta^D R^D \circ Ric, \ \delta^D R^D > \geq 0, \tag{6.30}$$

where the equality holds iff $\delta^D R^D = 0$. By (6.27) and (6.29), we arrive at

$$(2\mathcal{R}^D(\delta^D R^D) + \delta^D R^D \circ Ric, \ \delta^D R^D) \geq (-2||R^D|| + c)||\delta^D R^D||^2 \geq 0,$$

where the equality holds iff $||\delta^D R^D|| = 0$, by the hypothesis $||\delta^D R^D|| < \frac{c}{2}$.

By (6.30), the sum of the first and second terms, and the third term in the left-hand side of (6.26) are greater than or equal to zero. Therefore, (6.26) implies that the sum of the first and second terms of (6.26) is zero, and by (6.30) we have $\delta^D R^D = 0$ on M. □

Remarks. (a) If $||R^D|| = \frac{c}{2}$, then we have $D_X(\delta^D R^D) = 0$ for all $X \in \mathcal{X}(M)$. (b) If $(M, g) = (S^m, can)$ the unit m-sphere with canonical metric, then $c = m - 1$.

Theorem 6.3.6 (**L^2-isolation phenomena [192]**). *Let (M, g) be a four-dimensional compact Riemannian manifold whose Ricci curvature is bounded below by a positive constant $c > 0$ (i.e., $Ric \geq c \cdot id$). Suppose that $D \in \mathcal{C}(E, h)$ is a bi-Yang-Mills connection satisfying*

$$||R^D||_{L^2} < \frac{1}{2} min \left\{ \frac{\sqrt{c_1}}{18}, \frac{c}{2} Vol(M, g)^{1/2} \right\}. \tag{6.31}$$

Then D is a Yang-Mills connection. Here, c_1 is the isoparametric constant of (M, g), defined by

$$c_1 = \inf_{W \subset M} \frac{Vol_3(W)^4}{(min\{Vol(M_1), Vol(M_2)\})^3}, \tag{6.32}$$

where $W \subset M$ runs over all the hypersurfaces in M and $Vol_3(W)$ is the three-dimensional volume of W with respect to the Riemannian metric on W induced by g and the complement of W in M has a disjoint union of M_1 and M_2.

Proof. For a bi-Yang-Mills connection $D \in \mathcal{C}(E, h)$, (6.26) can be estimated by (6.27) and (6.29) as follows:

$$\begin{aligned} 0 = 2 \int_M &< \mathcal{R}^D(\delta^D R^D), \delta^D R^D > dv + \int_M < \delta^D R^D \circ Ric, \delta^D R^D > dv \\ &+ \int_M ||D(\delta^D R^D)||^2 dv \geq \int_M ||D(\delta^D R^D)||^2 dv + c \int_M ||\delta^D R^D||^2 dv \\ &- 2 \int_M ||R^D|| ||\delta^D R^D||^2 dv \\ &\geq ||D(\delta^D R^D)||^2_{L^2} + c||\delta^D R^D||^2_{L^2} - 2||R^D||^2_{L^2} ||\delta^D R^D||^2_{L^4}, \end{aligned} \tag{6.33}$$

by the Schwartz inequality.

We will use the Sobolev inequality for a four-dimensional Riemannian manifold (M, g) (see [265]):

$$||Df||^2_{L^2} \geq \frac{\sqrt{c_1}}{18} ||f||^2_{L^4} - \frac{1}{9} \left(\frac{c_1}{Vol(M, g)} \right)^{1/2} ||f||^2_{L^2}, \quad f \in W^{1,2}(M), \tag{6.34}$$

where $W^{1,2}(M)$ is the Sobolev space of (M, g). Applying (6.34) to the first term of (6.33), we get

$$\begin{aligned} \text{r.h.s. of } (6.33) &\geq \frac{\sqrt{c_1}}{18} ||\delta^D R^D||^2_{L^4} - \frac{1}{9} \left(\frac{c_1}{Vol(M, g)} \right)^{1/2} ||\delta^D R^D||^2_{L^2} \\ &+ c||\delta^D R^D||^2_{L^2} - 2||R^D||_{L^2} ||\delta^D R^D||^2_{L^4} \end{aligned}$$

$$= \left(\frac{\sqrt{c_1}}{18} - 2||R^D||_{L^2}\right) ||\delta^D R^D||^2_{L^4} + \left[c - \frac{1}{9}\left(\frac{c_1}{Vol(M,g)}\right)^{1/2}\right] ||\delta^D R^D||^2_{L^2}. \tag{6.35}$$

Because $||\delta^D R^D||^2_{L^2} \geq 0$ in (6.33), we also get

$$\text{r.h.s. of (6.33)} \geq c||\delta^D R^D||^2_{L^2} - 2||R^D||_{L^2}||\delta^D R^D||^2_{L^4}. \tag{6.36}$$

We consider two cases separately.

Case (a): $||\delta^D R^D||^2{}_{L^2} \geq \frac{Vol(M,g)^{1/2}}{2}||\delta^D R^D||^2{}_{L^4}$. If $||\delta^D R^D||_{L^4} > 0$, then

$$\text{r.h.s. of (6.36)} > c||\delta^D R^D||^2{}_{L^2} - \frac{c}{2}Vol(M,g)^{1/2}||\delta^D R^D||^2{}_{L^4}$$

$$= c\left(||\delta^D R^D||^2{}_{L^2} - \frac{Vol(M,g)^{1/2}}{2}||\delta^D R^D||^2{}_{L^4}\right) \geq 0$$

which is a contradiction, here the first inequality follows from $2||R^D||_{L^2} < \frac{c}{2}vol(M,g)^{1/2}$. Hence, we obtain $||\delta^D R^R||_{L^4} = 0$, i.e., $\delta^D R^D = 0$.

Case (b): $||\delta^D R^D|^2_{L^2} \leq \frac{Vol(M,g)^{1/2}}{2}||\delta^D R^D||^2_{L^4}$. If $||\delta^D R^D||_{L^2} > 0$, then

$$\begin{aligned}
\text{r.h.s. of (6.35)} &= \left(\frac{\sqrt{c_1}}{18} - 2||R^D||_{L^2}\right) ||d^D R^D||^2{}_{L^4} \\
&\quad + \left(c - \frac{1}{9}\left(\frac{c_1}{Vol(M,g)}\right)^{1/2}\right) ||\delta^D R^D||^2{}_{L^2} \\
&\geq \left(\frac{\sqrt{c_1}}{18} - 2||R^D||_{L^2}||\right) 2Vol(M,g)^{-1/2}||\delta^D R^D||^2_{L^2} \\
&\quad + \left(c - \frac{1}{9}\left(\frac{c_1}{Vol(M,g)}\right)^{1/2}\right) ||\delta^D R^D||^2_{L^2} \ \left(\text{since } \frac{\sqrt{c_1}}{18} - 2||R^D||_{L^2} \geq 0\right) \\
&= \left\{\frac{\sqrt{c_1}}{9}Vol(M,g)^{-1/2} - 2||R^D||_{L^2} \cdot 2Vol(M,g)^{-1/2} \right. \\
&\quad \left. + c - \frac{1}{9}\left(\frac{c_1}{Vol(M,g)}\right)^{1/2}\right\} ||\delta^D R^D||^2{}_{L^2} \\
&= (c - 2)||R^D||_{L^2} \cdot 2Vol(M,g)^{-1/2})||\delta^D R^D||^2{}_{L^2} > 0,
\end{aligned}$$

which is a contradiction too. Hence, $||\delta^D R^D||_{L^2} = 0$, i.e., $\delta^D R^D = 0$. □

Following the notions of weakly Yang-Mills connections considered in Chap. 3, we may similarly define weakly bi-Yang-Mills connections. Then we can study weak and strong compactness for bi-Yang-Mills connections. Moreover, as a future project, we may investigate monotonicity, blow-up loci, and removable singularities for bi-Yang-Mills connections, etc.

Chapter 7
Exponentially Harmonic Maps

Exponentially harmonic maps were first introduced by Eells and Lemaire [125] in 1990. Afterwards, Hong and Yang [188] also studied exponentially harmonic maps in 1993. In this chapter, we first compute the first and second variations of exponential energy functionals explicitly using tensor techniques and obtain the stability of exponentially harmonic maps in Theorems 7.1.3 and 7.1.4, which are based on the work of Chiang and Yang [88], published in 2007. We then discuss the regularity of exponentially harmonic functions, which is based on Eells and Duc [106].

7.1 First and Second Variations

An exponentially harmonic map $f : M \to N$ from an m-dimensional Riemannian manifold (M^m, g_{ij}) into an n-dimensional Riemannian manifold $(N^n, h_{\alpha\beta})$ is a critical point of the exponential energy functional

$$E(f) = \int_M e^{|df|^2} dv = \int_M e^{h_{\alpha\beta} f_i^\alpha f_j^\beta g^{ij}} dv, \tag{7.1}$$

where dv is the volume element of M with respect to g. In order to derive the Euler-Lagrange equation, we consider a one-parameter family of maps $f_t \in C^\infty(M \times [0, 1], N)$ such that f_t is the endpoint of a segment starting at $f(x)$ determined in length and direction by the vector field $\dot{f}(x)$, and such that the compact support of $\dot{f}(x)$ is contained in the interior of M. Then we have

Y.-J. Chiang, *Developments of Harmonic Maps, Wave Maps and Yang-Mills Fields into Biharmonic Maps, Biwave Maps and Bi-Yang-Mills Fields*, Frontiers in Mathematics, DOI 10.1007/978-3-0348-0534-6_7, © Springer Basel 2013

$$\begin{aligned}\frac{d}{dt}E(f_t)\Big|_{t=0} &= 2\int_M e^{|df_t|^2}(df_t, \nabla_t df_t)\Big|_{t=0} dv = 2\int_M e^{|df|^2}(df, \nabla \dot{f})dv \\ &= 2\left(\int_M div\, w\, dv - \int_M e^{|df|^2}((\tau f, \dot{f}) + ((\nabla|df|^2, df), \dot{f}))\right) dv \\ &= -2\int_M e^{|df|^2}(\tau f + (\nabla|df|^2, df), \dot{f})dv = 0, \quad \forall \dot{f} \end{aligned} \tag{7.2}$$

by the Divergence Theorem, which implies that $\tau f + (\nabla|df|^2, df) = 0$, where $\tau^\alpha(f) = g^{ij} f^\alpha_{i|j} = g^{ij}((f^\alpha_{ij} - \Gamma^k_{ij} f^\alpha_k) + \Gamma'^\alpha_{\beta\gamma} f^\beta_i f^\gamma_j)$ is the tension field, ∇ (we save the notation D for other use in the next section) is the connection on $T^*(M) \otimes f^{-1}TN$ induced by the Levi-Civita connections on M and N, $div\, w = w^j_{|j}$, with $w^j = e^{h_{\alpha\beta} f^\alpha_i f^\beta_j g^{ij}} h_{\alpha\beta} f^\alpha_i \dot{f}^\beta g^{ij}$ a vector field on M.

Definition 7.1.1. A map $f : M \to N$ between two Riemannian manifolds is *exponentially harmonic* if it satisfies

$$\tau f + (\nabla|df|^2, df) = 0, \tag{7.3}$$

i.e., in terms of local coordinates it satisfies

$$\begin{aligned} g^{ij}\Big(\frac{\partial^2 f^\alpha}{\partial x^i \partial x^j} - \Gamma^k_{ij}\frac{\partial f^\alpha}{\partial x^k} + \Gamma'^\alpha_{\beta\gamma}\frac{\partial f^\beta}{\partial x^i}\frac{\partial f^\gamma}{\partial x^j}\Big) + g^{il}g^{jm}h_{\beta\gamma}\frac{\partial f^\alpha}{\partial x^l}\frac{\partial f^\gamma}{\partial x^m}\frac{\partial^2 f^\beta}{\partial x^i \partial x^j} \\ - g^{il}g^{jm}h_{\beta\gamma}\Gamma^k_{ij}\frac{\partial f^\alpha}{\partial x^l}\frac{\partial f^\beta}{\partial x^m}\frac{\partial f^\gamma}{\partial x^k} + g^{ij}g^{lm}h_{\beta\gamma}\Gamma'^\beta_{\mu\nu}\frac{\partial f^\mu}{\partial x^i}\frac{\partial f^\nu}{\partial x^l}\frac{\partial f^\gamma}{\partial x^m}\frac{\partial f^\alpha}{\partial x^j} = 0, \end{aligned} \tag{7.4}$$

where Γ^k_{ij} and $\Gamma'^\alpha_{\beta\gamma}$ are the Christoffel symbols of the Levi-Civita connections on M and N, respectively.

We first observe that, by (7.3), when $|df|^2$ is constant, f is exponentially harmonic if and only if it is harmonic. Some properties of exponentially harmonic maps are different from those of usual harmonic maps. When $dim\, M = m = 2$, if we perform a conformal shift on the metric, $g \mapsto \rho g$, then both the energy $\int_M |df|^2 dv$ and the harmonic map are conformally invariant. However, for an exponentially harmonic map, the energy (7.1) changes completely.

Example 1. If $u : \mathbf{R}^2 \to \mathbf{R}$ is an exponentially harmonic function, (7.4) leads to

$$(1 + u_x^2)u_{xx} + 2u_x u_y u_{xy} + (1 + u_y^2)u_{yy} = 0. \tag{7.5}$$

By the method of separable variables, the solutions are in the form $u(x, y) = F(x) + G(y)$. It follows from (7.5) that $(1 + (F_x)^2)F_{xx} = -(1 + (G_y)^2)G_{yy} = \lambda$. Let $p = F_x$, $q = G_y$ and substitute these into the equation. By straightforward computation, we can get

$$F(x) = \frac{1}{4\lambda}\{[H_+(x;\lambda;c_1)+H_-(x:\lambda;c_1)]^4+2[H_+(x;\lambda;c_1)+H_-(x;\lambda;c_1)]^2\}-k_1,$$

where $H_\pm(x;\lambda;c_1) = \{\frac{3}{2}(c_1+\lambda x) \pm (1+\frac{9}{4}(c_1+\lambda x)^2)^{1/2}\}^{1/3}$. Similarly, we have

$$G(y) = -\frac{1}{4\lambda}\{[H_+(y;-\lambda;c_2)+H_-(y;-\lambda;c_2)]^4+2[H_+(y;-\lambda;c_2) \\ + H_-(y;-\lambda;c_2)]^2\} - k_2.$$

We also have $p = H_+(x;\lambda;c_1) + H_-(x;\lambda;c_1)$, $q = H_+(y;-\lambda;c_2) + H_-(y;-\lambda;c_2)$. Therefore, $u(x,y)$ can be written in parametric form as

$$x = \frac{1}{\lambda}(\frac{p^3}{3}+p-c_1), \quad y = \frac{-1}{\lambda}(\frac{q^3}{3}+q-c_2),$$
$$u(x,y) = \frac{1}{4\lambda}(p^4+2p^2-q^4-2q^2) + \text{constant}.$$

It is easy to check that $u(x,y)$ is not harmonic. □

Assume that $f = f_0$ is exponentially harmonic and that $\xi = \frac{\partial f}{\partial t}$ has compact support contained in the interior of M. The components of $\nabla_t \tau f$ are $f^\alpha_{i|j|t} = \frac{\partial f^\alpha_{i|j}}{\partial t} + \Gamma'^\alpha_{\mu\gamma} f^\mu_{i|j}\xi^\gamma$. Using the curvature formula on $M\times[0,1]\to N$, we have $f^\alpha_{i|j|t} = f^\alpha_{i|t|j} + R'^\alpha_{\beta\gamma\mu} f^\beta_i f^\gamma_j \xi^\mu$. But $f^\alpha_{i|t} = f^\alpha_{t|i} = \xi^\alpha_{|i}$, therefore the trace of $\nabla_t\tau f$ has components $g^{ij}\xi^\alpha_{|i|j} + R'^\alpha_{\beta\gamma\mu} f^\beta_i f^\gamma_j g^{ij}\xi^\mu$. Denote the first term by $(\Delta\xi)^\alpha$. Then we can compute the second variation of the energy from (7.2)

$$\begin{aligned}\frac{1}{2}\frac{d^2}{dt^2}E(f_t)\Big|_{t=0} &= -\int_M \frac{d}{dt}\Big[e^{|df_t|^2}(\tau f_t + (\nabla|df_t|^2, df_t),\xi)\Big]\Big|_{t=0} dv \\ &= -\int_M \Big[e^{|df|^2}(\nabla_t(\tau f + (\nabla|df_t|^2, df_t),\xi)|_{t=0} \\ &\quad + (\tau f + (\nabla|df|^2, df), \nabla_t\xi)) \\ &\quad + e^{|df|^2} 2(\nabla_t df, df)(\tau f + (\nabla|df|^2, df),\xi)\Big]dv. \qquad (7.6)\end{aligned}$$

Since f is exponentially harmonic at $t = 0$, the second and third terms of (7.6) vanish. Substituting the components of $\nabla_t\tau f$, we obtain

$$\begin{aligned}\frac{1}{2}\frac{d^2}{dt^2}E(f_t)\Big|_{t=0} &= -\int_M \Big[e^{|df|^2}(\Delta\xi + R'^\alpha_{\beta\gamma\mu} f^\beta_i f^\gamma_j g^{ij}\xi^\mu, \xi) + (\nabla(\nabla|\dot f|^2, \dot f),\xi)\Big]dv \\ &= \int_M e^{|df|^2}\Big[(\nabla\xi,\nabla\xi) - R'_{\alpha\beta\gamma\mu}\xi^\alpha f^\beta_i f^\gamma_j \xi^\mu + 2(\nabla\xi,\xi)^2\Big]dv, \qquad (7.7)\end{aligned}$$

where we use integration by parts $d(\nabla\xi,\xi) = (\Delta\xi,\xi) + (\nabla\xi,\nabla\xi)$ for the first term, and the fact that $(\nabla(\nabla|\dot{f}|^2,\dot{f}),\xi) = (\nabla((2\nabla\dot{f},\dot{f}),\dot{f}),\xi) = (\nabla((2\nabla\xi,\xi),\xi),\xi) = 2(\Delta\xi,\xi^3) + 4(((\nabla\xi,\nabla\xi),\xi),\xi)$ and integration by parts $d(\nabla\xi,\xi^3) = (\Delta\xi,\xi^3) + 3(\nabla\xi,\xi^2\nabla\xi)$ for the second term. Thus we can rewrite (7.7) as

$$\frac{1}{2}\frac{d^2}{dt^2}E(f_t)\bigg|_{t=0} = \int_M e^{|df|^2}\Big[-(J_f(\xi),\xi) + 2(\nabla\xi,\xi)^2\Big]dv, \tag{7.8}$$

where

$$J_f(\xi) = \Delta\xi + R'(df,df)\xi = g^{ij}\xi^\alpha_{|i|j} + R'^\alpha_{\beta\gamma\mu}f^\beta_i f^\gamma_j \xi^\mu g^{ij}, \tag{7.9}$$

which is a linear equation for ξ. Solutions of $J_f(\xi) = 0$ are called Jacobi fields.

Definition 7.1.2. Let $f : M \to N$ be an exponentially harmonic map. If $\frac{d^2}{dt^2}E(f_t)\Big|_{t=0} \geq 0$, then f is *stable.*

Theorem 7.1.3. *Let $f : M \to N$ be an exponentially harmonic map. If N has non-positive sectional curvature (i.e. $R'_{\alpha\beta\gamma\mu}\lambda^\alpha\eta^\beta\lambda^\gamma\eta^\mu \leq 0$ for arbitrary λ, η), then f is stable.*

Proof. The assertion follows from (7.7). □

Theorem 7.1.4. *Let $f : M \to N$ be an exponentially harmonic map. If $\xi = \dot{f}$ is a Jacobi field, then f is stable.*

Proof. The assertion follows from (7.8).

We can state for exponentially harmonic maps a result similar to Theorem 1 for harmonic maps by Eells and Sampson [129]. □

Proposition 7.1.5 ([188]). *Let $f : M \to N$ be an exponentially harmonic map, where M is compact without boundary, $Ricc^M \geq 0$, and $Riem^N \leq 0$. Then*

(i) *f is totally geodesic.*
(ii) *If $Ricc^M$ is positive at least one point of M, f is constant.*
(iii) *If $Riem^N$ is everywhere negative, f is either constant or maps M onto a closed geodesic of N.*

Proposition 7.1.6 ([188]). *Let $M^m \subset \mathbf{R}^{m+1}$ be a hypersurface which has m principal curvatures with $0 < \lambda_1 \leq \lambda_2 \leq \cdots \leq \lambda_m$, satisfying $\lambda_m < \lambda_1 + \cdots + \lambda_{m-1}$. If $f : N \to M$ is a stable exponentially harmonic map with $|df|^2 < \frac{1}{2\lambda_m^2}\min_{1\leq i\leq m}\{\lambda_i(\sum_{j=1}^m \lambda_j - 2\lambda_i)\}$, then f is constant.*

Proposition 7.1.7 (Liouville [188]). *Let $f : \mathbf{R}^m \to N$ be an exponentially harmonic map. If f has finite energy and $|df|^2 \leq \frac{m}{2} - 1$, then f is constant.*

7.2 Regularity of Exponentially Harmonic Functions

The regularity of exponentially harmonic functions was studied by Eells and Duc [106] and all the following theorems and results were obtained in this paper. It would be also interesting to study the regularity of exponentially harmonic maps.

Let $u : (M, g_{ij}) \to \mathbf{R}$ be a function on a smooth m-dimensional Riemannian manifold. The exponential energy functional of u is defined by

$$E(u) = \int_M e(u)dv = \int_M exp(|du|^2/2)dv, \tag{7.10}$$

where $e(u) = e^{|du|^2/2}$ is the energy density, $|Du(x)|^2 = g^{ij}D_iu(x)D_ju(x)$ and $D_i = \frac{\partial}{\partial x^i}$.

Let $\mathcal{F}(M) = \{u \in W^{1,2}(M) : E(u) < \infty\}$. We say that $u \in \mathcal{F}(M)$ is a *local minimum* if for every $v \in \mathcal{F}(M)$ there is an $\epsilon > 0$ such that

$$E(u) \leq E(u + t(v - u)) \text{ for all } t \in [0, \epsilon]. \tag{7.11}$$

Theorem 7.2.1 ([106]). *Every local E-minimum is in $C^\infty(M)$.*

The Euler-Lagrange operator associated with E is the quasi-linear strictly elliptic operator

$$\Delta u = div(e(u)Du) = g^{ij}\nabla_j(e(u)D_iu) = e(u)Q^{ij}(u)\nabla_j D_i u, \tag{7.12}$$

where $Q^{ij}(u) = u^{ij} + g^{ik}g^{jl}D_kuD_lu$ and ∇_j denotes the covariant derivative, so $\nabla_j D_i u = D_{ij}u - \Gamma^k_{ij}D_ku$. A C^2 function $u : M \to \mathbf{R}$ is *exponentially harmonic* if $\Delta u = 0$.

In the text of Theorem 7.2.3, it is known from [156] that if $\phi \in C^3(M)$, then $u \in C^3$ is a solution of the Dirichlet problem

$$\Delta u = 0 \text{ with } u = \phi \text{ on } \partial M \tag{7.13}$$

iff u is the unique E-minimum in

$$\mathcal{F}(M, \phi) = \{w \in C^3(M) | \, w = \phi \text{ on } \partial M\}.$$

In fact, that problem is equivalent to solving

$$Q^{ij}(u)\nabla_j D_i u = 0 \text{ with } u = \phi \text{ on } \partial M \ in \, C^3(M).$$

Lemma 7.2.2. *For any $\phi \in C^3(M)$ the Dirichlet problem (7.13) has a unique solution $u \in C^0(M) \cap C^3(M - \partial M)$; furthermore, u is the unique E-minimum in $\mathcal{F}(M, \phi)$. Also, for any open relatively compact subset M_1 of M there is $\alpha > 0$ such that α and $||u||_{C^{1,\alpha}(M_1)}$ depend only on M_1 and $E_{M_1(\phi)}$.*

Proof (Proof of Theorem 7.2.1). Let u be a local E-minimum. One can find a sequence $(u_n)_{n\geq 1}$ in $C^3(M)$ which converges to $u \in W^{1,2}(M)$ and for which $\lim_{n\to\infty} E(u_n) = E(u)$. Take a small geodesic disc M_0 in M and let

$$\mathcal{F}(M_0, u_n) = \{w|_{M_0} : w \in \mathcal{F}(M) \text{ and } w = u_n \text{ on } \partial M_0\}.$$

By Lemma 7.2.2, there is a unique E_{M_0}-minimum $w_n \in \mathcal{F}(M_0, u_n)$ such that $w_n \in C^{1,\alpha}(M_1)$ for any relatively compact M_1 in M_0, where α and $||w_n||_{C^{1,\alpha}(M_1)}$ depend only on $dist(M_1, \partial M_0)$ and $E(u_n)$.

Hence, one can find a subsequence of (w_n), still denoted by (w_n), which converges weakly to some $w \in W^{1,p}(M_0)$ for each $p > 1$, and for any relatively compact M_1 in M_0 there is $\beta > 0$ such that $(w_n|_{M_1})$ converges to $w|_{M_1}$ in $C^{1,\beta}(M_1)$. Consequently, $E_{M_0}(w) \leq \liminf E_{M_0}(w_n)$, and w is an E_{M_0}-minimum in $\mathcal{F}(M_0, u)$.

Because $w \in \bigcap\{W^{1,p}(M_0) : p > 1\}$ and $u \in \bigcap\{W^{1,p}(M) : p > 1\}$, one observes that

$$w = u \text{ on } \partial M_0,\ w \in C^{0,\alpha}(\bar{M_0}),\ \ u \in C^{0,\alpha}(M).$$

Therefore, the two functions

$$v_1(x) = \begin{cases} w(x), & \text{if } x \in M_0, \\ u(x), & \text{if } x \in M - M_0, \end{cases}$$

$$v_2(x) = \begin{cases} u(x) + \epsilon(v_1(x) - u(x)), & \text{if } x \in M_0, \\ u(x), & \text{if } x \in M - M_0, \end{cases}$$

are in $\mathcal{F}(M)$, where ϵ is chosen from the Definition (7.11) of u being a local E-minimum.

Obviously, $e(v_1) \in C^{0,\alpha}(\bar{M_0})$ and

$$E(v_1) \leq E(u). \tag{7.14}$$

The strict convexity of the exponential function implies that

$$e(v_2) \leq (1 - \epsilon)e(u) + \epsilon e(v_1)$$

at every point of M; moreover, inequality is strict if $|Du(x)|^2 \neq |Dv_1(x)|^2$. Combining (7.11) and (7.14), we have $e(u) = e(v_1)$ a.e. on M. Therefore, the solution of the Dirichlet problem

$$div(e(v_1)Du) = 0,\ u = \phi \text{ on } \partial M_0 \tag{7.15}$$

is smooth, i.e., our local E-minimum $u \in C^\infty(M)$. □

Theorem 7.2.3 ([106]). *Assume that M is compact and has smooth boundary ∂M. If $\phi \in \mathcal{F}(M)$, then there is a unique E-minimum $u \in \mathcal{F}(M, \phi) = \{w \in \mathcal{F}(M) : w = \phi \in \partial M\}$. Moreover, $u \in C^\infty(M - \partial M)$.*

Proof. Let $\phi \in \mathcal{F}(M)$ and let (u_n) in $C^0(M) \cap C^3(M - \partial M)$ be a minimizing sequence in $\mathcal{F}(M, \phi)$. Therefore, (u_n) is bounded in $W^{1,p}(M - \partial M)$ for every p, and we can assume that (u_n) converges weakly to u there. It follows that u is an E-minimum in $\mathcal{F}(M)$, by Serrin's theorem [328]. The arguments are similar to the proof of Theorem 7.2.1. □

Chapter 8
Exponential Wave Maps

Exponential wave maps are exponentially harmonic maps on Minkowski spaces, and were first investigated by Chiang and Yang [88] in 2007. We provide a few examples of exponential wave maps, and present their properties. We obtain Propositions 8.2.1 and 8.2.2 concerning the stability of exponential wave maps by applying Theorems 7.1.3 and 7.1.4. We prove Theorem 8.2.3 which relates wave maps, exponential wave maps, and the conservation law of second-order symmetric tensors. Afterwards, we verify in Theorem 8.2.5 that if f is an exponential wave map, then the associated energy-momentum tensor is conserved. We then utilize this theorem to prove Proposition 8.2.6 that if f is an exponential wave and pseudo-weakly conformal map, then f is homothetic. We finally discuss applications of exponential wave maps in relativity in two cases – de Sitter spaces and Friedmann-Lemaître spaces, by approximating exponential wave maps using wave maps for the first, and by coupling them with gravitational fields with exponential scalar fields for the latter.

8.1 Definition and Examples

Let $\mathbf{R}^{1+m}$ be an $m+1$ dimensional Minkowski space with the metric $g_{ij} = diag(-1, 1, \cdots, 1)$ and the coordinates $x^0 = t, x^1, x^2, \cdots, x^m$, and $(N, h_{\alpha\beta})$ be an n-dimensional Riemannian manifold. Recall that a wave map is a harmonic map on $\mathbf{R}^{1+m}$ with the energy

$$E(f) = \int_{\mathbf{R}^{1+m}} h_{\alpha\beta}(-f_t^\alpha f_t^\beta + \sum_{i=1}^{m} f_i^\alpha f_i^\beta) dt\, dx. \tag{8.1}$$

Y.-J. Chiang, *Developments of Harmonic Maps, Wave Maps and Yang-Mills Fields into Biharmonic Maps, Biwave Maps and Bi-Yang-Mills Fields*, Frontiers in Mathematics, DOI 10.1007/978-3-0348-0534-6_8, © Springer Basel 2013

The Euler-Lagrange equation describing the critical points of (8.1) is

$$\tau_{\square}^{\alpha}(f) = -\square f^{\alpha} + \Gamma'^{\alpha}_{\beta\gamma}(-f_t^{\beta} f_t^{\gamma} + \sum_{i=1}^{m} f_i^{\beta} f_i^{\gamma}) = 0 \tag{8.2}$$

which is the wave map equation, where $\square = \frac{\partial^2}{\partial t^2} - \frac{\partial^2}{\partial x^{i\,2}}$ is the d'Alembertian. The wave map equation is invariant with respect to the dimensionless scaling $f(t,x) \mapsto f(ct, cx)$, $c \in \mathbf{R}$. However, the energy is scale invariant only in dimension $m = 2$.

An exponential wave map $f : \mathbf{R}^{1+m} \to N$ is an exponentially harmonic map on $\mathbf{R}^{1+m}$ with the exponential energy from (7.1)

$$E(f) = \int_{\mathbf{R}^{1+m}} e^{h_{\alpha\beta}(-\frac{\partial f^{\alpha}}{\partial t}\frac{\partial f^{\beta}}{\partial t} + \sum_{i=1}^{m} \frac{\partial f^{\alpha}}{\partial x^i}\frac{\partial f^{\beta}}{\partial x^i})} dt\, dx. \tag{8.3}$$

The Euler-Lagrange equation describing the critical points of (8.3) from (7.3) is

$$\tau_{\square}(f) + < \nabla(-|\partial_t f|_h^2 + \sum_{i=1}^{m} |\partial_{x^i} f|_h^2), df > = 0, \tag{8.4}$$

i.e., in local coordinates

$$-\frac{\partial^2 f^{\alpha}}{\partial t^2} + \sum_{i=1}^{m} \frac{\partial^2 f^{\alpha}}{\partial x^{i\,2}} + \Gamma'^{\alpha}_{\beta\gamma}\left(-\frac{\partial f^{\beta}}{\partial t}\frac{\partial f^{\gamma}}{\partial t} + \sum_{i=1}^{m} \frac{\partial f^{\beta}}{\partial x^i}\frac{\partial f^{\gamma}}{\partial x^i}\right)$$
$$+ \sum_{i=0}^{m}\sum_{j=0}^{m} g^{ii} g^{jj} h_{\beta\gamma} \frac{\partial f^{\alpha}}{\partial x^i}\frac{\partial f^{\gamma}}{\partial x^j}\frac{\partial^2 f^{\beta}}{\partial x^i \partial x^j} + \sum_{i=0}^{m}\sum_{j=0}^{m} g^{ii} g^{jj} h_{\beta\gamma} \Gamma'^{\beta}_{\mu\nu} \frac{\partial f^{\mu}}{\partial x^i}\frac{\partial f^{\nu}}{\partial x^j}\frac{\partial f^{\gamma}}{\partial x^i}\frac{\partial f^{\alpha}}{\partial x^j} = 0. \tag{8.5}$$

Example 1. If the energy density $e(f) = -f_t^{\alpha} f_t^{\beta} + \sum_{i=1}^{m} f_i^{\alpha} f_i^{\beta}$ is constant, then $\tau_{\square}(f) = 0$ if and only if $\tau_{\square}(f) + \langle \nabla(-|\partial_t f|_h^2 + \sum_{i=1}^{m} |\partial_{x^i} f|_h^2), df\rangle = 0$. Therefore, f is a wave map with constant energy if and only if it is an exponential wave map.

Example 2. If $u : \mathbf{R}^{1+1} \to \mathbf{R}$ is an exponential wave function, (8.5) becomes

$$(1 + u_x^2)u_{xx} - 2u_t u_x u_{tx} - (1 - u_t^2)u_{tt} = 0. \tag{8.6}$$

By the method of separable variables, the solutions are $u(t,x) = F(t) + G(x)$. We have from (8.6) that $(1 + F_t^2)F_{tt} = (1 - G_x^2)G_{xx} = \lambda'$. By a computation similar to that in Example 1 in Sect. 7.1, $u(t,x)$ can be written in the parametric form:

$$t = \frac{1}{\lambda'}(\frac{p^3}{3} + p - c_3), \quad x = \frac{1}{\lambda'}(-\frac{q^3}{3} + q - c_4)$$

$$u(t,x) = \frac{1}{4\lambda'}(p^4 + 2p^2 - q^4 + 2q^2) + \text{constant}.$$

It is easy to check that $u(x,y)$ is not a wave function.

Example 3. Let $M = \mathbf{R}^{1+1}$ and N be a surface of revolution in 3-dimensional Euclidean space with the metric

$$ds^2 = [1 + (dh/dz)^2]dz^2 + h^2(z)d\phi^2,$$

where $r = h(z)$ is the equation of N in cylindrical coordinates. We can generalize the example of a wave map by Gu [163] to an exponential wave map. The first equation of (8.5) becomes

$$\begin{aligned}
&-\frac{\partial^2 z}{\partial t^2} + \frac{\partial^2 z}{\partial x^2} + \frac{h'h''}{1+h'^2}\Big[-(\frac{\partial z}{\partial t})^2 + (\frac{\partial z}{\partial x})^2\Big] - \frac{hh'}{1+h'^2}\Big[-(\frac{\partial \phi}{\partial t})^2 + (\frac{\partial \phi}{\partial x})^2\Big] \\
&+ (1+h'^2)\Big[(\frac{\partial z}{\partial t})^2\frac{\partial^2 z}{\partial t^2} - \frac{\partial z}{\partial t}\frac{\partial z}{\partial x}\frac{\partial^2 z}{\partial t\partial x} - \frac{\partial z}{\partial x}\frac{\partial z}{\partial t}\frac{\partial^2 z}{\partial x\partial t} + \frac{\partial z}{\partial x}\frac{\partial z}{\partial x}\frac{\partial^2 z}{\partial x^2}\Big] \\
&+ h^2\Big[(\frac{\partial z}{\partial t}\frac{\partial \phi}{\partial t}\frac{\partial^2 \phi}{\partial t^2} - \frac{\partial z}{\partial t}\frac{\partial \phi}{\partial x}\frac{\partial^2 \phi}{\partial t\partial x} - \frac{\partial z}{\partial x}\frac{\partial \phi}{\partial t}\frac{\partial^2 \phi}{\partial t\partial x} + \frac{\partial z}{\partial x}\frac{\partial \phi}{\partial x}\frac{\partial^2 \phi}{\partial x^2})\Big] \\
&+ (1+h'^2)\Big[\frac{h'h''}{1+h'^2}((\frac{\partial z}{\partial t})^4 - \frac{\partial z}{\partial t}\frac{\partial z}{\partial x}\frac{\partial z}{\partial t}\frac{\partial z}{\partial x} - \frac{\partial z}{\partial x}\frac{\partial z}{\partial t}\frac{\partial z}{\partial x}\frac{\partial z}{\partial t} + (\frac{\partial z}{\partial x})^4) \\
&- \frac{hh'}{1+h'^2}((\frac{\partial \phi}{\partial t})^2(\frac{\partial z}{\partial t})^2 - \frac{\partial \phi}{\partial t}\frac{\partial \phi}{\partial x}\frac{\partial z}{\partial t}\frac{\partial z}{\partial x} - \frac{\partial \phi}{\partial x}\frac{\partial \phi}{\partial t}\frac{\partial z}{\partial x}\frac{\partial z}{\partial t} + (\frac{\partial \phi}{\partial x})^2(\frac{\partial z}{\partial x})^2)\Big] \\
&+ h^2\Big[\frac{h'}{h}((\frac{\partial z}{\partial t})^2(\frac{\partial \phi}{\partial t})^2 - \frac{\partial z}{\partial t}(\frac{\partial \phi}{\partial x})\frac{\partial \phi}{\partial t}\frac{\partial z}{\partial x} - \frac{\partial z}{\partial x}\frac{\partial \phi}{\partial t}\frac{\partial z}{\partial t}\frac{\partial \phi}{\partial x} + (\frac{\partial z}{\partial x})^2(\frac{\partial \phi}{\partial x})^2) \\
&- \frac{h'}{h}(\frac{\partial \phi}{\partial t})^2(\frac{\partial z}{\partial t})^2 - (\frac{\partial \phi}{\partial t})^2(\frac{\partial z}{\partial x})^2 - (\frac{\partial \phi}{\partial t})^2(\frac{\partial z}{\partial x})^2 + (\frac{\partial \phi}{\partial x})^2(\frac{\partial z}{\partial x})^2\Big] = 0.
\end{aligned} \tag{8.7}$$

Consider the following special initial conditions at $t = 0$:

$$\phi = x, \ \frac{\partial \phi}{\partial t} = 0, \ z = k, \ \frac{\partial z}{\partial t} = \alpha, \tag{8.8}$$

where k, α are constants. The solution is invariant with respect to rotations around the z-axis, and therefore $\phi = x, z = z(t)$. The equation (8.7) for $z(t)$ has the form

$$-\frac{d^2 z}{dt^2} - \frac{h'h''}{1+h'^2}(\frac{dz}{dt})^2 - \frac{hh'}{1+h'^2} + (1+h'^2)(\frac{dz}{dt})^2\frac{d^2 z}{dt^2} + h'h''(\frac{dz}{dt})^4 = 0, \tag{8.9}$$

and the initial conditions are $z(0) = k, (\frac{dz}{dt})(0) = \alpha$. Equation (8.9) admits the first integral

$$
\begin{aligned}
&- (1+h'^2)(\frac{dz}{dt})^2 - h^2 + (1+h'^2)^2\frac{1}{2}(\frac{dz}{dt})^4 \\
&= \Big(\frac{1}{\sqrt{2}}(1+h'^2)(\frac{dz}{dt})^2 - \frac{1+\sqrt{1+2h^2}}{\sqrt{2}}\Big)\Big(\frac{1}{\sqrt{2}}(1+h'^2)(\frac{dz}{dt})^2 - \frac{1-\sqrt{1+2h^2}}{\sqrt{2}}\Big) \\
&= \Big(\frac{1}{\sqrt{2}}(1+h'^2(k))\alpha^2 - \frac{1+\sqrt{1+2h^2(k)}}{\sqrt{2}}\Big)\Big(\frac{1}{\sqrt{2}}(1+h'^2(k))\alpha^2 - \frac{1-\sqrt{1+2h^2(k)}}{\sqrt{2}}\Big).
\end{aligned} \tag{8.10}
$$

The solutions can be represented as

$$
\int_k^z \frac{(1+h'^2)^{1/2}dz}{\sqrt{(1+h'^2(k))\alpha^2 - \sqrt{1+2h^2(k)} + \sqrt{1+2h^2(z)}}} = t, \tag{8.11}
$$

or

$$
\int_k^z \frac{(1+h'^2)^{1/2}dz}{\sqrt{(1+h'^2(k))\alpha^2 + \sqrt{1+2h^2(k)} - \sqrt{1+2h^2(z)}}} = t. \tag{8.12}
$$

If $\alpha^2(1+h'^2(k)) - \sqrt{1+2h^2(k)} + \sqrt{1+2h^2(z)} > 0$ in (8.11) for all z, then all of N can be covered. Otherwise, the surface is covered partially. Similarly for (8.12). The second equation of (8.5) is satisfied under the special initial conditions (8.8).

8.2 Properties

Let $f : \mathbf{R}^{1+m} \to N$ be an exponential wave map. The map f is *stable* if $\frac{d^2}{dt^2}E(f_t)\big|_{t=0} \geq 0$.

Proposition 8.2.1. *If $f : \mathbf{R}^{1+m} \to N$ is an exponential wave map such that $R'_{\alpha\beta\gamma\mu}\,\xi^\alpha(\sum_{i=1}^m f_i^\beta f_i^\gamma - f_t^\beta f_t^\gamma)\xi^\mu \leq 0$, then f is stable.*

Proof. By (7.7), we have

$$
\begin{aligned}
\frac{1}{2}\frac{d^2}{dt^2}E(f_t)\Big|_{t=0} = \int_{\mathbf{R}^{1+m}} & e^{h_{\alpha\beta}(\sum_{i=1}^m f_i^\alpha f_i^\beta - f_t^\alpha f_t^\beta)}\Big((\nabla\xi,\nabla\xi) \\
& - R'_{\alpha\beta\gamma\mu}\xi^\alpha(\sum_{i=1}^m f_i^\beta f_i^\gamma - f_t^\beta f_t^\gamma)\xi^\mu + 2(\nabla\xi,\xi)^2\Big)dt\,dx,
\end{aligned}
$$

and the result follows from the assumption. □

Proposition 8.2.2. *Let $f : \mathbf{R}^{1+m} \to N$ be an exponential wave map. If $\xi = \dot{f}$ is a Jacobi field on the Minkowski space $\mathbf{R}^{1+m}$, then f is stable.*

Proof. By (7.8) we have

$$\frac{1}{2}\frac{d^2}{dt^2}E(f_t)\Big|_{t=0} = \int_{\mathbf{R}^{1+m}} e^{h_{\alpha\beta}(\sum_{i=1}^m f_i^\alpha f_i^\beta - f_t^\alpha f_t^\beta)}\Big(- (J_f(\xi),\xi) + 2(\nabla\xi,\xi)^2\Big)dt\,dx,$$

where $J_f^\alpha(\xi) = -\Box\xi^\alpha + R'^\alpha_{\beta\gamma\mu}(-f_t^\beta f_t^\gamma \xi^\mu + \sum_{i=1}^m f_i^\beta f_i^\gamma \xi^\mu)$. If $J_f(\xi) = 0$, then $\frac{1}{2}\frac{d^2}{dt^2}E(f_t)\big|_{t=0} \geq 0$. □

We next study the relationships among wave maps, exponential wave maps, and the conservation law of second order symmetric tensors.

Theorem 8.2.3 ([88]). *Let $f : \mathbf{R}^{1+m} \to N$ be a non-degenerate map (i.e. $df \neq 0$). Then any two of the following conditions imply the third:*

(1) *f is a wave map.*
(2) *f is an exponential wave map.*
(3) *The second order symmetric tensor $S_f = |df|^2(f^*h - \frac{1}{4}|df|^2 g)$ is conserved, i.e., $div(S_f) = 0$, where $g = (-1, 1, \cdots, 1)$ and $|df|^2 = -|\frac{\partial f}{\partial t}|^2 + \sum_{i=1}^m |\frac{\partial f}{\partial x^i}|^2$.*

Proof. I. (1) and (2) ⇒(3): Let $x^0 = t, x^1, x^2, \cdots, x^m$ be the coordinates in $\mathbf{R}^{1+m}$, and $e_0 = \frac{\partial}{\partial t}, e_1 = (1, 0, \cdots, 0), e_2 = (0, 1, 0, \cdots, 0), \cdots, e_m = (0, 0, \cdots, 1)$. Set

$$S = |df|^2\left(f^*h - \frac{1}{4}|df|^2\begin{pmatrix}-1 & 0\\ 0 & I\end{pmatrix}\right),$$

where I is the $m \times m$ identity matrix. For $X \in T(\mathbf{R}^{1+m}) = \mathbf{R}^{1+m}$, we compute

$$(div\, S_f)(X) = (\nabla_{e_i} S)(e_i, X) = \nabla_{e_i}\left(|df|^2\left(f^*h - \frac{1}{4}|df|^2\begin{pmatrix}-1 & 0\\ 0 & I\end{pmatrix}\right)\right)(e_i, X)$$

$$= \nabla_{e_i}|df|^2\left((f_*e_i, f_*X) - \frac{1}{4}|df|^2\begin{pmatrix}-1 & 0\\ 0 & I\end{pmatrix}\right)(e_i, X) + |df|^2(f_*\nabla_{e_i}e_i, f_*X)$$

$$+ (f_*e_i, f_*\nabla_{e_i}X) - \frac{1}{4}\nabla_{e_i}|df|^2\begin{pmatrix}-1 & 0\\ 0 & I\end{pmatrix}(e_i, X).$$

Since $\tau_\Box(f) = (\nabla df)(e_i, e_i) = (\nabla_{e_i} df)(e_i)$, we have

$$(div\, S_f)(X) = \Big((\nabla|df|^2, df) + |df|^2\tau_\Box(f),\ f_*X\Big) + |df|^2\Big(f_*e_i,\ (\nabla_{e_i}df)X\Big)$$
$$- \frac{1}{2}\nabla_X\left(-|\frac{\partial f}{\partial t}|^2 + |\frac{\partial f}{\partial x^i}|^2\right)\left(-|\frac{\partial f}{\partial t}|^2 + |\frac{\partial f}{\partial x^i}|^2\right)$$

$$= \Big((\nabla |df|^2, df) + |df|^2 \tau_\square(f), f_* X\Big) + \frac{1}{2}(\nabla_X |df|^2)|df|^2$$
$$- \frac{1}{2}\nabla_X\Big(- |\frac{\partial f}{\partial t}|^2 + |\frac{\partial f}{\partial x}|^2\Big)\Big(-|\frac{\partial f}{\partial t}|^2 + |\frac{\partial f}{\partial x^i}|^2\Big)$$
$$= \Big(\nabla(-|\frac{\partial f}{\partial t}|^2 + |\frac{\partial f}{\partial x}|^2), df\Big) + \Big((-|\frac{\partial f}{\partial t}|^2 + |\frac{\partial f}{\partial x}|^2)\tau_\square(f), f_* X\Big), \tag{8.13}$$

where the second and third terms cancel out. Therefore,

$$div\, S_f = \Big(\nabla(-|\frac{\partial f}{\partial t}|^2 + |\frac{\partial f}{\partial x}|^2), df\Big) + \Big((-|\frac{\partial f}{\partial t}|^2 + |\frac{\partial f}{\partial x}|^2)\tau_\square(f), df\Big).$$

Hence, (1) and (2) imply (3).
II. (2) and (3) $\Rightarrow$ (1): If f is an exponential wave map and S is conserved, then

$$\tau_\square(f) + \Big(\nabla(-|\frac{\partial f}{\partial t}|^2 + \frac{\partial f}{\partial x}|^2), df\Big) = 0, \tag{8.14}$$

and

$$\Big((-|\frac{\partial f}{\partial t}|^2 + \frac{\partial f}{\partial x}|^2)\tau_\square(f) + (\nabla(-|\frac{\partial f}{\partial t}|^2 + \frac{\partial f}{\partial x}|^2), df), df\Big) = 0. \tag{8.15}$$

Since f is non-degenerate (i.e. $df \neq 0$), we have

$$(-|\frac{\partial f}{\partial t}|^2 + \frac{\partial f}{\partial x}|^2)\tau_\square(f) + (\nabla(-|\frac{\partial f}{\partial t}|^2 + |\frac{\partial f}{\partial x}|^2), df) = 0,$$

and thus

$$\Big(\nabla(-|\frac{\partial f}{\partial t}|^2 + |\frac{\partial f}{\partial x}|^2), df\Big) = -\Big(- |\frac{\partial f}{\partial t}|^2 + |\frac{\partial f}{\partial x}|^2\Big)\tau_\square(f) = -|df|^2\tau_\square(f).$$

Substituting this into (8.14), we obtain $(1 - |df|^2)\tau_\square(f) = 0$. Suppose that f is not a wave map. Then there exists a point $p \in \mathbf{R}^{1+m}$ such that $\tau_\square(f) \neq 0$ by the continuity of $\tau_\square(f)$. Hence, there exists a neighborhood U of p such that $|df|^2_U = 1$; but then (8.4) implies $\tau_\square(f)|_U = 0$, and we get a contradiction!
III. (1) and (3) $\Rightarrow$ (2) is obvious. □

Definition 8.2.4. $f : \mathbf{R}^{1+m} \to N$ is *pseudo-weakly conformal* if there is a smooth function $\mu : \mathbf{R}^{1+m} \to \mathbf{R}$ ($\mu \neq 0$) such that $f^*h = \mu \begin{pmatrix} -1 & 0 \\ 0 & I \end{pmatrix}$. f is *homothetic* if μ is constant.

If $f : \mathbf{R}^{1+m} \to N$ is *pseudo-weakly conformal*, then we have

$$S_f = \frac{1}{4}(m-1)(5-m)\mu^2 g, \quad g = \begin{pmatrix} -1 & 0 \\ 0 & I \end{pmatrix}. \tag{8.16}$$

Proposition. *(1) $S_f = 0$ if and only if $m = 1$ or 5 and f is pseudo-weakly conformal. (2) If $f : \mathbf{R}^{1+m} \to N$ is pseudo-weakly conformal such that $m \neq 1, 5$ and S_f is conserved, then f is homothetic.*

Proof. By Theorem 8.2.3 (3), assertion $div\, S_f = 0$ and (8.16), we find

$$0 = \frac{1}{2}(m-1)(5-m)\mu\, \mu_{,j}\, g_{ij} = 0 \quad (0 \le i \le m),$$

whence $d\mu = 0$ on $\mathbf{R}^{1+m}$. Therefore, μ is constant. □

The energy-momentum tensor associated with $f : \mathbf{R}^{1+m} \to N$ is defined by $T(f) = e^{|df|^2}(g - 2f^*h)$, where $g = diag(-1, 1, \cdots, 1)$, $|df|^2 = -|\frac{\partial f}{\partial t}|^2 + \sum_{i=1}^m |\frac{\partial f}{\partial x^i}|^2$.

Theorem 8.2.5 ([88]). *If $f : R^{1+m} \to N$ is an exponential wave map, then $T(f)$ is conserved.*

Proof. Let $x^0 = t, x^1, x^2, \cdots, x^m$ be the coordinates in R^{1+m}, and $e_0 = \frac{\partial}{\partial t}$, $e_1 = (1, 0, \cdots, 0), \cdots, e_m = (0, \cdots, 0, 1)$. Set

$$T(f) = e^{|df|^2}\left(\begin{pmatrix} -1 & 0 \\ 0 & I \end{pmatrix} - 2f^*h\right) \tag{8.17}$$

For $X \in \mathbf{R}^{1+m}$ we compute

$$\begin{aligned}
div\, T(f)(X) &= \nabla_{e_i} T(f)(e_i, X) = \nabla_{e_i}\left[e^{|df|^2}\left(\begin{pmatrix} -1 & 0 \\ 0 & I \end{pmatrix} - 2f^*h\right)(e_i, X)\right] \\
&= \nabla_{e_i} e^{|df|^2}\begin{pmatrix} -1 & 0 \\ 0 & I \end{pmatrix}(e_i, X) - 2(f_*e_i, f_*X) - 2e^{|df|^2}(\nabla_{e_i} f^*h)(e_i, X) \\
&= e^{|df|^2}\left[\nabla|df|^2\begin{pmatrix} -1 & 0 \\ 0 & I \end{pmatrix}(e_i, X) - 2\nabla_{e_i}|df|^2(f_*e_i, f_*X) - 2\nabla_{e_i}(f_*e_i, f_*X)\right] \\
&= e^{|df|^2}\Big[2((\nabla\frac{\partial f}{\partial t}, \frac{\partial f}{\partial t}) + (\nabla\frac{\partial f}{\partial x_i}, \frac{\partial f}{\partial x_i})) - 2\nabla_{e_i}|df|^2(f_*e_i, f_*X) - 2(\nabla_{e_i} f_*e_i, f_*X) \\
&\qquad -2(f_*e_i, \nabla_{e_i} f_*X)\Big] \\
&= 2e^{|df|^2}\left[((\nabla_X df)e_i, f_*e_i) - (\nabla|df|^2, df), f_*X) - (f_*e_i, \nabla_{e_i} f_*X) - (\tau_\Box(f), f_*X)\right] \\
&= -2e^{-|\frac{\partial f}{\partial t}|^2 + |\frac{\partial f}{\partial x_i}|^2}\left(\tau_\Box(f) + (\nabla(-|\frac{\partial f}{\partial t}|^2 + |\frac{\partial f}{\partial x_i}|^2), df), f_*X\right),
\end{aligned} \tag{8.18}$$

where the first and third terms cancel out and $\tau_{\Box}(f) = \nabla_{e_i} f_* e_i$. Hence, if f is an exponential wave map, then $T(f)$ is conserved, i.e., $div\, T(f) = 0$. □

Proposition 8.2.6. *If $f : \mathbf{R}^{1+m} \to N$ is an exponential wave and psuedo-weakly conformal map such that $\mu \neq \frac{m-3}{2(m-1)}$ $(m \neq 1)$, then f is homothetic.*

Proof. By (8.17) $T(f) = e^{(m-1)\mu} g(1 - 2\mu), g = \begin{pmatrix} -1 & 0 \\ 0 & I \end{pmatrix}$ due to the pseudo-weakly conformality of f. Since f is exponential wave, then by Theorem 8.2.5 $divT(f) = 0$ we have

$$e^{(m-1)\mu}(m-1)\mu_{,j}\, g_{ij}(1-2\mu) + e^{(m-1)\mu}(-2\mu_{,j} g_{ij})$$
$$= e^{(m-1)\mu}\mu_{,j}\, g_{ij}((m-1)(1-2\mu)-2) = 0 \quad (0 \leq i \leq m).$$

If $\mu \neq \frac{m-3}{2(m-1)}$ $(m \neq 1)$, then $\mu_{,j} g_{ij} = 0\, (0 \leq i \leq m)$, which implies $d\mu = 0$, and hence μ is constant. □

The proof of the Liouville-type theorem for an exponentially harmonic in Proposition 7.1.7 depends on the assumption that f has finite energy, i.e., $\int_{\mathbf{R}^m} |df|^2 dv < \infty$, which implies $|df|^2 = 0$, and therefore, f is constant. If we apply it to an exponential wave map $f : \mathbf{R}^{m,1} \to N$ under the assumption : $\int_{\mathbf{R}^{1+m}} (-|f_t|^2 + \sum_{i=1}^m |f_{x_i}|^2) dt\, dx < \infty$, which implies $\sum_{i=1}^m |f_{x_i}|^2 = |f_t|^2$. Then f is not necessarily constant.

8.3 Applications

We discuss applications of exponential wave maps in relativity theory in two cases – de Sitter spaces and Friedmann-Lemaître spaces, by approximating exponential wave maps using wave maps, and by coupling them with gravitational fields with exponential scalar fields. The results presented are based on Chiang and Yang [88].

Let $f : \mathbf{R}^{1+m} \to (N, h_{\alpha\beta})$ be a C^∞ map between two Riemannian manifolds. If we want to relate our context with physics, we need to modify the exponential energy (8.3) as follows:

$$\begin{aligned} E'_\lambda(f) &= \int_{\mathbf{R}^{1+m}} e^{\lambda h_{\alpha\beta}(-\frac{\partial f^\alpha}{\partial t}\frac{\partial f^\beta}{\partial t} + \frac{\partial f^\alpha}{\partial x^i}\frac{\partial f^\beta}{\partial x^i})} dt\, dx \\ &\approx \lambda \int_{\mathbf{R}^{1+m}} \Big[h_{\alpha\beta}(-f_t^\alpha f_t^\beta + f_i^\alpha f_i^\beta) + \frac{\lambda}{2}(h_{\alpha\beta}(-f_t^\alpha f_t^\beta + f_i^\alpha f_i^\beta))^2 \\ &+ \frac{\lambda^2}{6}(h_{\alpha\beta}(-f_t^\alpha f_t^\beta + f_i^\alpha f_i^\beta))^3 + \cdots \Big] dt\, dx. \end{aligned} \tag{8.19}$$

When λ is sufficiently small, the Euler-Lagrange equations for $E'_\lambda(f)$ lead to equations which approximate those of usual wave maps. The equations derived from E'_λ can be obtained from (8.4) via the transformation $f \mapsto \sqrt{\lambda} f$ $(\lambda > 0)$.

General relativistic solutions can be locally embedded in Ricci-flat 5-dimensional spaces. This is important in establishing local generality for the recent work by Wesson [293], in which $(1+4)$-dimensional vacuum field equations give rise to $(1+3)$-dimensional equations with sources. We first describe the mathematical structure of the Wesson's schemes by the following two postulates.

Postulate 8.3.1. *The fundamental space in which an ordinary 4-dimensional space-time is locally and isometrically embedded is a 5-dimensional manifold M_5. The line element of this space is given by $d\tilde{s}^2 = g_{ab}dx^a dx^b$ and can be put, at least locally, in the form $d\tilde{s}^2 = g_{ij}dx^i dx^j + \epsilon\phi^2 d\psi^2$, where $\{a,b\}$ and $\{i,j\}$ run from 0 to 4, and 0 to 3 respectively, $x^a = (x^i, \psi)$ are coordinates, $g_{ij} = g_{ij}(x^i), \phi = \phi(x^a), \epsilon^2 = 1$.*

Postulate 8.3.2. *The fundamental 5-dimensional space satisfies the vacuum field equations ${}^{(5)}\tilde{R}_{ab} = 0$.*

Theorem 8.3.3 (Campbell). *Any analytic n-dimensional Riemannian space can be locally embedded in a (n+1)-dimensional Ricci-flat space.*

Let us discuss some applications of exponential wave maps in relativity.

Case 1: Let S_4 be the 4-dimensional de Sitter space-time with the metric

$$ds^2 = dt^2 - e^{2\sqrt{\Lambda/3}t}(dx^2 + dy^2 + dz^2), \tag{8.20}$$

where Λ is the cosmological constant. We consider an exponential wave map $f : S_4 \to \mathbf{R}$ approximated by a usual wave map which is an extremal of the functional E'_λ in (8.19). It satisfies a modified version of the wave map equation (8.2) via $f \mapsto \sqrt{\lambda} f (\lambda > 0)$, which reads

$$\ddot{f}(1 + \lambda \dot{f}^2) + 6\sqrt{\Lambda/3}\dot{f} = 0 \ (\dot{f} = f_t), \tag{8.21}$$

if the map is restricted to $f = f(t)$. This yields $ln(\dot{f}) + \frac{\lambda}{2}\dot{f}^2 = -6\sqrt{\Lambda/3}t + c_1$, and thus, $\dot{f}e^{\frac{\lambda}{2}\dot{f}^2} = c_2 e^{-6\sqrt{\Lambda/3}t}$.

(i) When $t \to \infty$, $f(t) \to$ const.
(ii) When $t \to 0$ and λ is small, $f(t) \approx c_2 t + c_3$, which is regular at $t = 0$.

By Postulates 8.3.1, 8.3.2 and Theorem 8.3.3, S_4 can be embedded in a 5-dimensional Ricci-flat space S_5 (cf. [293] p. 333) with the metric

$$d\tilde{s}^2 = \Lambda(\psi^2/3)dt^2 - \Lambda(\psi^2/3)e^{2\sqrt{\Lambda/3}t}(dx^2 + dy^2 + dz^2) - d\psi^2, \tag{8.22}$$

which induces the metric (8.20) on the hypersurface $\psi = \psi_0 = \pm\sqrt{3/\Lambda}$, $\psi^2 = 3/\Lambda$.

Case 2: (1) Let M_4 be a Friedmann-Lemaître space with the metric

$$ds^2 = dt^2 - a^2(t)\left(\frac{dr^2}{1-kr^2} + r^2 d\Omega^2\right), \; d\Omega^2 = d\theta^2 + \sin^2\theta d\gamma^2. \tag{8.23}$$

It is known that an exponentially harmonic map $f : M_4 \to \mathbf{R}$ (globally, $k = 0$) is not regular at $t = 0$. Kanfon, Fuzfa and Lambert [212] considered such an exponentially harmonic map $f : M_4 \to \mathbf{R}$ on the Friedmann-Lemaître space without matter coupled with an exponentially scalar field which can make f regular at $t = 0$.

(2) (a) By Postulates 8.3.1, 8.3.2 and Theorem 8.3.3, the Friedmann-Lemaître space M_4 can be locally embedded in a 5-dimensional space M_5 with the metric

$$^{(5)}d\tilde{s}^2 = dt^2 - a^2(t)\left(\frac{dr^2}{1-kr^2} + r^2 d\Omega^2\right) + \epsilon\phi^2 d\psi^2, \; \epsilon^2 = 1. \tag{8.24}$$

In particular, if $k = 0$, $a^2(t) = t$, $\phi^2(t) = 1/t$, the space M_5 has the metric [308]

$$d\tilde{s}^2 = dt^2 - t(dx^2 + dy^2 + dz^2) + \frac{\epsilon}{t} d\psi^2.$$

We assume that M_5 is equipped with the metric ($a(t)$ is a function of t and $\phi(t) = 1/a(t)$)

$$d\tilde{s}^2 = dt^2 - a^2(t)(dx^2 + dy^2 + dz^2) + \frac{\epsilon}{a^2(t)} d\psi^2. \tag{8.25}$$

Take $\epsilon = -1$ (space-like). We consider an exponential wave map $f : M_5 \to \mathbf{R}$ (locally) approximated by a usual wave map which is an extremal of the functional E'_λ in (8.19). It satisfies the following modified version of wave map equation via the transformation $f \mapsto \sqrt{\lambda} f (\lambda > 0)$:

$$\ddot{f}(1 + \lambda \dot{f}^2) + 4\frac{\dot{a}}{a}\dot{f} = 0$$

if the map is restricted to $f = f(t)$. This gives $a^4(t) = \frac{c_4}{|\dot{f}|} e^{-\frac{\lambda}{2}\dot{f}^2}$; for instance, take $a(t) = a_0 (t/t_0)^{1/2}$, and $t_0 = \frac{1}{2H_0}$, where H_0 is the present Hubble constant ($\frac{\dot{a}}{a}|_{t=t_0} = H_0$). Then we have

$$|\dot{f}| e^{\frac{\lambda}{2}\dot{f}^2} = \frac{1}{dt^2}, \; d = \frac{a_0^4}{c_4 t_0^2}.$$

(i) When $t \to \infty$, $f(t) \to$ const.

(ii) When $t \to 0$, and λ is small, $f(t) \approx f_0 \pm \frac{1}{dt}$, which is not regular at $t = 0$.

(b) If we consider $f : M_5 \to \mathbf{R}$ coupled with an exponentially scalar field using the metric (8.24):

$$S(f) = -\frac{1}{2\kappa}\int \sqrt{-g}d^4xdy\{(\tilde{R} - exp(\frac{\lambda}{2}\partial_a f\partial^b f) - \Lambda) + \mathcal{L}_{mat}\}, \qquad (8.26)$$

where $y = \psi$ represents the fifth new coordinate and the integration restricts to the hypersurface Σ_4 defined by $\psi = \psi_0 = constant$, κ is a coupling constant, Λ is a modified cosmological constant: $\Lambda = 2\kappa(2\Lambda_0 - 1)$, with Λ_0 is the usual cosmological constant, and $\mathcal{L}_{mat}$ is the Lagrangian density for matter. By Sect. 3 of [293], (8.26) reduces to

$$S(f) = -\frac{1}{2\kappa}\int \sqrt{-g}d^4x\left\{(R\phi - exp(\frac{\lambda}{2}\partial_i f\partial^j f) - \Lambda) + \mathcal{L}_{mat}\right\}, \quad \phi = 1/a(t). \qquad (8.27)$$

Taking the variation of $S(f)$ leads to Einstein's equations:

$$(R_{ij} - \frac{1}{2}Rg_{ij})\phi = \frac{1}{2}g_{ij}\left\{(-e^{\frac{\lambda}{2}\partial_i f\partial^j f} - \Lambda) + \lambda\partial_i f\partial_j f e^{\frac{\lambda}{2}\partial_i f\partial^j f}\right\} - R\delta\phi + \kappa T_{ij}^{(mat)} \qquad (8.28)$$

where $T_{ij}^{(mat)}$ is the energy-momentum tensor for matter. Let us assume that $f = f(t)$. Then the field equations can be written as

$$\frac{1}{a}3(\frac{\dot{a}}{a})^2 + \frac{1}{a}(\frac{3k}{a^2}) = \rho - \frac{1}{2}e^{\frac{\lambda}{2}\dot{f}^2}(1 - \lambda\dot{f}^2) - \frac{\Lambda}{2} - 6(\frac{\ddot{a}}{a} + \frac{\dot{a}^2}{a^2})(-a^{-2})\dot{a}, \qquad (8.29)$$

$$\frac{1}{a}((\frac{\dot{a}}{a})^2 + 2\frac{\ddot{a}}{a}) + \frac{1}{a}(\frac{k}{a^2}) = -p - \frac{1}{2}e^{\frac{\lambda}{2}\dot{f}^2} - \frac{\Lambda}{2}, \qquad (8.30)$$

$$\ddot{f}(1 + \lambda\dot{f}^2) + 4\frac{\dot{a}}{a}\dot{f} = 0, \qquad (8.31)$$

where ρ is the mass-energy density of matter, and p is the pressure of the fluid.

In particular, if $k = 0$ and one considers the Friedmann-Lemaître space without matter, the above field equations become

$$\frac{1}{a}3(\frac{\dot{a}}{a})^2 = -\frac{1}{2}e^{\frac{\lambda}{2}\dot{f}^2}(1 - \lambda\dot{f}^2) - \frac{\Lambda}{2} - 6(\frac{\ddot{a}}{a} + \frac{\dot{a}^2}{a^2})(-a^{-2})\dot{a}, \qquad (8.32)$$

$$\frac{1}{a}((\frac{\dot{a}}{a})^2 + 2\frac{\ddot{a}}{a}) = -\frac{1}{2}e^{\frac{\lambda}{2}\dot{f}^2} - \frac{\Lambda}{2}, \qquad (8.33)$$

Let $y = \dot{f}$ and let $H = \frac{\dot{a}}{a}$ be the Hubble constant. We have $\dot{H} + H^2 = \frac{\ddot{a}}{a}$. Then we can rewrite (8.32), (8.33) and (8.31) as

$$3H^2 = (-\frac{1}{2}e^{\frac{\lambda}{2}\dot{f}^2}(1-\lambda\dot{f}^2) - \frac{\Lambda}{2})a + 6(\dot{H} + H^2)H + 6H^3, \tag{8.34}$$

$$a = \frac{H^2 + 2(\dot{H} + H^2)}{-\frac{1}{2}e^{\frac{\lambda}{2}\dot{f}^2} - \frac{\Lambda}{2}}, \tag{8.35}$$

$$H = \frac{-\dot{y}(1+\lambda y^2)}{4y}. \tag{8.36}$$

Substituting (8.35) and (8.36) into (8.34) we get, since λ is very small,

$$\frac{1}{2}\frac{\ddot{y}}{y} + \frac{3}{8}\frac{\ddot{y}\dot{y}}{y^2} + \frac{9}{16}(\frac{\dot{y}}{y})^3 - \frac{1}{2}(\frac{\dot{y}}{y}) \approx 0. \tag{8.37}$$

Let $z = \frac{\dot{y}}{y}$, and we have $\dot{z} + z^2 = \frac{\ddot{y}}{y}$. We can rewrite (8.37) as

$$\frac{1}{2}\dot{z} + \frac{3}{8}\dot{z}z + \frac{15}{16}z^3 \approx 0,$$

or

$$\frac{dz}{dt} = \frac{-15z^3}{8+6z}.$$

By integrating we get $(t+c)z^2 - \frac{6}{15}z - \frac{4}{15} = 0$. Substituting $z = \dot{y} = \frac{dy}{dt}$ and solving for $\frac{dy}{dt}$, we have

$$\frac{dy}{dt} = \frac{16 \pm \sqrt{256 + 240(t+c)}}{30(t+c)}y.$$

This gives solutions

$$y = c_5 e^{\frac{16\pm\sqrt{256+240(t+c)}}{30(t+c)}}, \quad a(t) = c_6 e^{Ht}.$$

When $t = 0$, $f'(0) = c_5 e^{\frac{16\pm\sqrt{256+240c}}{30c}}$ exists if $c \neq 0$. Hence, $f(t)$ is differentiable and regular at $t = 0$. It is interesting to note that the coupling of f with the gravitational field can make f regular at $t = 0$, which is not the case in the un-coupled situation.

Chapter 9
Exponential Yang-Mills Connections

We introduce exponential Yang-Mills connections and their relationships with Yang-Mills connections, which were first studied by Matsuura and Urakawa [260] in 1995. We first compute the first variation of the exponential Yang-Mills functional, and discuss the minimizer. We then narrate the existence of exponential Yang-Mills connections. We finally compute the second variation of the exponential Yang-Mills functional. All the theorems and results are based on [39, 117, 187, 260].

9.1 First Variation and Minimizer

9.1.1 First Variation

Let (M, g) be an m-dimensional compact Riemannian manifold, G a compact Lie group and E a G-vector bundle over M. For $D \in \mathcal{C}(E)$, let R^D be its curvature tensor and define the *exponential Yang-Mills functional* by $\mathcal{YM}_e(D) = \int_M exp(\frac{1}{2}||R^D||^2)dv$. Consider a smooth family of G-connections D^t, $-\epsilon < t < \epsilon$, such that $D^0 = D$. Write $D^t = D + A^t$, where $A^t \in \Omega^1(\mathcal{G}_E)$ for $|t| < \epsilon$ and $A^0 = 0$. We have that $R^{D^t} = R^D + d^D A^t + \frac{1}{2}[A^t \wedge A^t]$, where for $\phi, \psi \in \Omega^1(\mathcal{G}_E)$,

$$[\phi \wedge \psi](X, Y) = [\phi(X), \psi(Y)] - [\phi(Y), \psi(X)]$$

for vector fields X and Y on M. We compute the first variation of the exponential Yang-Mills functional:

Y.-J. Chiang, *Developments of Harmonic Maps, Wave Maps and Yang-Mills Fields into Biharmonic Maps, Biwave Maps and Bi-Yang-Mills Fields*, Frontiers in Mathematics, DOI 10.1007/978-3-0348-0534-6_9,

$$\begin{aligned}\frac{d}{dt}\Big|_{t=0}\mathcal{YM}_e(D^t) &= \int_M exp(\frac{1}{2}||R^D||^2)\frac{d}{dt}\Big|_{t=0}(\frac{1}{2}||R^{D^t}||^2)dv \\ &= \int_M exp(\frac{1}{2}||R^D||^2) < d^D\alpha,\, R^D > dv \\ &= \int_M < \alpha,\, \delta^D(exp(\frac{1}{2}||R^D||^2)R^D) > dv,\end{aligned}$$

where $\alpha = \frac{d}{dt}\big|_{t=0}D^t \in \Omega^1(\mathcal{G}_E)$, and δ^D was defined in Sect. 3.1.

Theorem 9.1.1. *The first variation of the exponential Yang-Mills functional is given by*

$$\frac{d}{dt}\Big|_{t=0}\mathcal{YM}_e(D^t) = \int_M < \alpha,\, \delta^D(exp(\frac{1}{2}||R^D||^2)R^D) > dv,$$

where $\alpha = \frac{d}{dt}\big|_{t=0}D^t$. *Hence, D is an exponential Yang-Mills field if and only if*

$$\delta^D(exp(\frac{1}{2}||R^D||^2)R^D) = 0. \tag{9.1}$$

In particular, if $||R^D||$ is constant and D is a smooth connection, then D is an exponential Yang-Mills connection if and only if it is a Yang-Mills connection.

9.1.2 Minimizer

Fixing D^0 as a C^∞ G-connection of E (i.e., $D^0 \in \mathcal{C}(E)$), define the L^p space of G-connections of E by $L^p(E) = \{D = D^0 + A : A \in L^p(T^*M \otimes \mathcal{G}_E)\}$, $1 < p < \infty$, where $L^p(T^*M \otimes \mathcal{G}_E)$ is the completion of $\Omega^1(\mathcal{G}_E)$ with respect to the norm $||A||_p = (\int_M ||A|^p dv)^{1/p}$. Let $L^p_1(E)$ denote the Sobolev space of the L^p_1 G-connections, where L^p_1 means the Sobolev functions with first derivative which are p-integrable. Alternatively,

$$L^p_1(E) = \{D = D^0 + A : A \in L^p_1(T^*M \otimes \mathcal{G}_E)\},$$

where $L^p_1(T^*M \otimes \mathcal{G}_E)$ is the completion of $\Omega^1(\mathcal{G}_E)$ with respect to the norm

$$||A||_{1,p} = \Big(\int_M ||DA||^p dv\Big)^{1/p} + \Big(\int_M ||A||^p dv\Big)^{1/p}.$$

Furthermore, we define the space of α-Hölder continuous G-connections of E by

$$\mathcal{C}^\alpha(E) = \{D = D^0 + A :\ A \in \mathcal{C}^\alpha(T^*M \otimes \mathcal{G}_E)\},\ 0 < \alpha < 1,$$

where $C^\alpha(T^*M \otimes \mathcal{G}_E)$ is the completion of $\Omega^1(\mathcal{G}_E)$ with respect to the norm

$$||A||_\alpha = \inf_{x \neq y \in M} \inf_{\sigma} \inf_{X \in T_x M, Y \in T_y M} \frac{||T_\sigma^{-1}(A(X)) - A(Y)||}{r(x,y)^\alpha}.$$

Here $r(x, y)$ is the Riemannian distance in (M, g) between two points x and y, σ runs through the smooth curves $[0, 1] \to M$ such that $\sigma(0) = x$, $\sigma(1) = y$, and $T_\sigma : End(E_x) \to End(E_y)$ is the parallel transport along σ with respect to the connection induced by D^0. By the Sobolev imbedding theorem, the embedding $L_1^p(E) \hookrightarrow C^\alpha(E)$ is a compact operator for $0 < \alpha < 1 - dim(M)/p$. We then define the space

$$\mathcal{W}(E) = \bigcap_{p \geq 1} L_1^p(E) \bigcap \{D : \mathcal{YM}_e(D) < \infty\}.$$

Theorem 9.1.2. *Suppose that D is a minimizer in $\mathcal{W}(E)$ of the Yang-Mills functional $\mathcal{YM}$ and the norm of the curvature $||R^D||$ is almost everywhere constant. Then D is also a minimizer of the exponential Yang-Mills functional $\mathcal{YM}_e$ and for any minimizer D' of the exponential Yang-Mills functional $\mathcal{YM}_e$ in $\mathcal{W}(E)$, the norm $||R^{D'}||$ is almost everywhere constant.* (See the proof in [260].)

Thanks to the convexity of the function $exp(\frac{1}{2}x^2)$, the exponential Yang-Mills functional is lower semi-continuous, and then by a direct method we can show the existence of a minimizer for it.

Theorem 9.1.3. *The exponential Yang-Mills functional admits a minimizing connection D which is C^α-Hölder continuous for all $0 < \alpha < 1$.*

Proof. We need the following lemma. □

Lemma. *Let $\{D_i\}_{i=1}^\infty$ be a sequence of connections in $L_1^p(E)$ which converges weakly to a connection D in $L_1^p(E)$. Then we have*

$$\mathcal{YM}_e(D) \leq \liminf_{i \to \infty} \mathcal{YM}_e(D_i).$$

Now let $\{D_i = D_0 + A_i\}_{i=1}^\infty$ be a minimizing sequence of the exponential Yang-Mills functional $\mathcal{YM}_e$ in $\mathcal{W}(E)$, where $\{D_i\}_{i=1}^\infty$ is bounded in $L_1^p(E)$ and $\{||R^{D_i}||^2\}_{i=1}^\infty$ is bounded in $L^p(E)$ for all $1 < p < \infty$. By the definition,

$$\mathcal{YM}_e(D) = \int_M \sum_{j=0}^\infty \frac{1}{j!}(\frac{1}{2}||R^D||^2)^j \, dv.$$

For each $0 < \alpha < 1$ select p such that $0 < \alpha < 1 - \frac{dim(M)}{p}$. Applying the compact Sobolev embedding $L_1^p(E) \hookrightarrow C^\alpha(E)$ and a diagonal process, there exist a subsequence of $\{D_i\}_{i=1}^\infty$ (still denoted the same), and a connection D such that

$\{D_i\}_{i=1}^{\infty}$ converges weakly to D in $L_1^p(E)$, and converge strongly to D in $L^p(E)$ and $\mathcal{C}^{\alpha}(E)$. Using the above lemma for $\{D_i\}_{i=1}^{\infty}$, we have

$$\mathcal{YM}_e(D) \leq \liminf_{i\to\infty} \mathcal{YM}_e(D_i).$$

Thus D achieves a minimum of $\mathcal{YM}_e$ and lies in $\mathcal{YM}_e$ and in $\mathcal{C}^{\alpha}(E)$. □

9.2 Existence of Exponential Yang-Mills Connections

In order to show the existence of exponential Yang-Mills connections, we need to show the existence of Yang-Mills connections first.

Theorem 9.2.1 (Katagiri [214]). *Let (M, g) be an m-dimensional manifold, G a compact Lie group and E a G-vector bundle over M. Suppose that $m \geq 5$. Then there exist a C^{∞} Riemannian $\tilde{g}$ on M conformal to g and a C^{∞} G-connection D on E such that D is a Yang-Mills connection with respect to $\tilde{g}$.*

Note that if $m = 4$, the Yang-Mills functional $\mathcal{YM}$ is invariant under the conformal change from g to $\tilde{g} = \rho g$ for a positive smooth function ρ of M. If $m = 2$ or 3, Yang-Mills connections exist for any G-vector bundle over any Riemannian manifold (M, g) (see [305]).

Proof. For a positive C^{∞} function ρ, let $\tilde{g} = \rho g$ be a new Riemannian metric on M. Then we have

$$\int_M ||R^D||_{\tilde{g}}^2 \, dv_{\tilde{g}} = \int_M \rho^{(m-4)/2} ||R^D||_g^2 \, dv_g.$$

In terms of the Euler-Lagrange equation, we get

$$\delta_{\tilde{g}}^D \, R^D = 0 \text{ if and only if } \delta_g^D \, (\rho^{(m-4)/2} ||R^D||_g^2) = 0,$$

where $\delta_{\tilde{g}}^D$, $\delta_{\tilde{g}}^D$ are the formal adjoints of d^D with respect to g and $\tilde{g}$, respectively. Furthermore, the functional $H_p(D) = \frac{1}{2}\int_M (1+||R^D||_g^2)^{p/2} dv_g$ satisfies the Palais-Smale condition and attains a minimum if $2p > dim(M)$ [1]. In terms of Euler-Lagrange equation, we have

$$\delta_g^D((1 + ||R^D||_g^2)^{(p-2)/2} R^D) = 0. \tag{9.2}$$

For $A \in \Omega^1(\mathcal{G}_E)$, we compute

$$\begin{aligned} \frac{d}{dt}\Big|_{t=0} H_p(D + tA) &= \frac{d}{dt}\Big|_{t=0} \int_M (1 + ||R^{D+tA}||_g^2)^{p/2} dv_g \\ &= \frac{p}{2} \int_M (1 + ||R^D||_g^2)^{(p-2)/2} < d^D A, R^D > dv_g. \end{aligned}$$

Therefore, (9.2) has a solution D for $2p > dim(M)$. For this solution, defining

$$\rho = (1 + ||R^D||_g^2)^{(p-2)/(m-4)}$$

and $\tilde{g} = \rho g$, we have $\delta_{\tilde{g}}^D R^D = 0$, so that D is a Yang-Mills connection with respect to $\tilde{g}$. □

The following three theorems were obtained by Matsuura and Urakawa [260]. Theorem 9.2.2 follows from Theorems 9.2.1 and 9.2.3.

Theorem 9.2.2. *Let (M, g) be an m-dimensional manifold, G a compact Lie group and E a G-vector bundle over M. Suppose that $m \geq 5$. Then there exist a C^∞ Riemannian metric $\tilde{g}$ on M conformal to g and a C^∞ G-connection D on E such that D is an exponential Yang-Mills connection with respect to $\tilde{g}$.*

Theorem 9.2.3. *Let (M, g) be an m-dimensional manifold, G a compact Lie group and E a G-vector bundle over M. Suppose that $m \geq 5$. Then there exist a C^∞ Riemannian metric $\tilde{g}$ on M conformal to g and a C^∞ G-connection D on E such that D is an exponential Yang-Mills connection with respect to $\tilde{g}$.*

In order to prove the theorem, we require the following two lemmas:

Lemma 1. *The function $\rho \mapsto log \frac{log \rho}{\rho^2}$ is a strictly increasing function on the interval $[1, \sqrt{e})$. The inverse function $\rho = \phi(y)$ exists on the interval $[0, 1/2e)$ and is smooth.*

Lemma 2. *Under the assumptions of Theorem 9.2.2, suppose that $m \geq 5$ and D is a Yang-Mills connection. Then for any ϵ there exists a C^∞ Riemannian metric $\tilde{g}$ on M which is homothetic to g, such that D is a Yang-Mills connection with respect to $\tilde{g}$ and $||R^D||_{\tilde{g}}^2 < \epsilon$.* (Note that $\tilde{g} = Cg$ and $||R^D||_{\tilde{g}}^2 = C^{-2}||R^D||_g^2$, where C is a constant. We have $||R^D||_g < \epsilon$ if C is sufficiently large, since M is compact.)

Proof (Proof of Theorem 9.2.3). By Lemma 2, we may assume that a Yang-Mills connection D satisfies $||R^D||^2 < \epsilon < (m-4)/2e$. For a positive smooth function ρ on M, define $\tilde{g} = \rho^{-1}g$. Thus $\delta_{\tilde{g}}^D R^D = 0$ if and only if $\delta_g^D (\rho^{(m-4)/2} R^D) = 0$. Since $||R^D||_{\tilde{g}}^2 < (m-4)/2e$, we can define the function ρ on M by $\rho = \phi(||R^D||_g^2/(m-4))$ and then $\rho > 0$, by Lemma 1. Therefore, we obtain

$$\begin{aligned}\rho^{(m-4)/2} &= (exp(\rho^2||R^D||_g^2/(m-4)))^{(m-4)/2} \\ &= exp(\rho^2||R^D||_g^2/2) = exp(||R^D||_{\tilde{g}}^2/2).\end{aligned}$$

It follows that $\delta_{\tilde{g}}^D(exp(\frac{1}{2}||R^D||_{\tilde{g}}^2)R^D) = 0$, which implies that D is an exponential Yang-Mills connection with respect to $\tilde{g}$. □

Theorem 9.2.4. *Let (M, g) be a 4-dimensional compact Riemannian manifold, G be a compact Lie group and E be a G-vector bundle over M. Then there exist a C^0 (continuous) Riemannian metric $\tilde{g}$ on M conformal to g and a C^∞ G-connection D*

on E such that D is an exponential Yang-Mills connection with respect to $\tilde{g}$ in the weak sense.

Note that a C^∞ G-connection D is called an exponential Yang-Mills connection *in the weak sense* if it satisfies

$$\int_M < d^D A,\ exp(\frac{1}{2}||R^D||^2_{\tilde{g}})R^D >_{\tilde{g}} dv_{\tilde{g}} = 0,$$

for all $A \in \Omega^1(\mathcal{G}_E)$, where $\tilde{g}$ is a C^0 Riemannian metric.

Proof. For an m-dimensional Riemannian manifold (M, g) and any positive C^∞ function ρ on M, set $\tilde{g} = \rho g$. The corresponding exponential Yang-Mills functional is

$$\mathcal{YM}_{e,\tilde{g}}(D) = \int_M exp(\frac{1}{2}||R^D||^2_{\tilde{g}})dv_{\tilde{g}} = \int_M \rho^{m/2} exp(\frac{1}{2}\rho^{-2}||R^D||^2_g)dv_g.$$

In terms of the Euler-Lagrange equation, for any $A \in \Omega^1(\mathcal{G}_E)$ we have

$$\frac{d}{dt}\Big|_{t=0}\mathcal{YM}_{e,\tilde{g}}(D+tA) = \int_M \rho^{(m-4)/2} exp(\frac{1}{2}\rho^{-2}||R^D||^2_g) < d^D A,\ R^D >_g dv_g.$$

It follows that $\delta^D_{\tilde{g}}(exp(\frac{1}{2}||R^D||^2_{\tilde{g}})R^D) = 0$ if and only if $\delta^D_g(\rho^{(m-4)/2} exp(\frac{1}{2}\rho^{-2}||R^D||^2_g)R^D) = 0$. When $m = 4$, the Euler-Lagrange equation with respect to $\tilde{g}$ is

$$\delta^D_g(exp(\frac{1}{2}\rho^{-2}||R^D||^2_g)R^D) = 0.$$

Note that a C^∞ solution of (9.2),

$$\delta^D_g(exp(\frac{1}{2}\rho^{-2}||R^D||^2_g)R^D) = 0$$

exists for $p > 2$ when $dim(M) = 4$. For the solution D, we define a C^0 function ρ on M by

$$\rho = \begin{cases} \sqrt{\frac{||R^D||^2_g}{log((1+||R^D||^2_g)^{p-2})}}, & \text{if } ||R^D|| \neq 0, \\ \sqrt{\frac{1}{p-2}}, & \text{if } ||R^D|| = 0, \end{cases}$$

and set $\tilde{g} = \rho g$. It follows that for any $A \in \Omega^1(\mathcal{G}_E)$,

$$\int_M < d^D A,\ exp(\frac{1}{2}||R^D||^2_{\tilde{g}})R^D >_{\tilde{g}} dv_{\tilde{g}} = 0.$$

This completes the proof. □

9.3 Second Variation

Let us compute the second variation of the exponential Yang-Mills functional

$$\mathcal{YM}_e(D) = \int_M exp(\frac{1}{2}||R^D||^2)dv.$$

Following the notations from last section, let (M, g) be an m-dimensional compact Riemannian manifolds, G a compact Lie group and E a G-vector bundle over M. Suppose that D^t, $|t| < \epsilon$, is a smooth family of G-connections on E such that $D = D^0$ is an exponential Yang-Mills connection. Set $D^t = D + A^t$, where $A^t \in \Omega^1(\mathcal{G}_E)$ for all t and $A^0 = 0$. The infinitestimal variation of the connection associated to D^t at $t = 0$ is $\alpha = \frac{d}{dt}\big|_{t=0} D^t \in \Omega^1(\mathcal{G}_E)$. As in Sect. 3.1.2, we define an endomorphism $\mathcal{R}^D$ of $\Omega^1(\mathcal{G}_E)$ by

$$\mathcal{R}^D(\phi)(X) = \sum_{i=1}^{m}[R^D(e_i, X), \phi(e_i)]$$

for $\phi \in \Omega^1(\mathcal{G}_E)$, where $\{e_i\}_{i=1}^m$ is a local orthonormal frame of a point in (M, g).

Theorem 9.3.1. *Let (M, g), G, E, D and D^t be as above. Then the second variation of the exponential Yang-Mills functional is*

$$\begin{aligned} \frac{d^2}{dt^2}\Big|_{t=0}\mathcal{YM}_e(D^t) &= \int_M exp(\frac{1}{2}||R^D||^2)\Big[< d^D\alpha,\, R^D >^2 + < d^D\alpha,\, d^D\alpha > + < \alpha, \mathcal{R}^D(\alpha) >\Big]dv \\ &= \int_M < \mathcal{Q}^D(\alpha), \alpha > dv, \end{aligned}$$

where $\alpha = \frac{d}{dt}\big|_{t=0} D^t$, *and* $\mathcal{Q}^D$ *is a differential operator acting on* $\Omega^1(\mathcal{G}_E)$ *defined by*

$$\begin{aligned} \mathcal{Q}^D(\alpha) &= \delta^D(exp(\frac{1}{2}||R^D||^2) < d^D\alpha,\, R^D > R^D) \\ &\quad + \delta^D(exp(\frac{1}{2}||R^D||^2)d^D\alpha) + exp(\frac{1}{2}||R^D||^2)\mathcal{R}^D(\alpha). \end{aligned}$$

Proof. We first have

$$\frac{1}{2}\frac{d^2}{dt^2}\Big|_{t=0}||R^{D^t}||^2 = < d^D\alpha,\, d^D\alpha > + < d^D F + [\alpha \wedge \alpha],\, R^D >,$$

where $F = \frac{d^2}{dt^2}\big|_{t=0} D^t$. Thus we arrive at

$$\frac{d^2}{dt^2}\Big|_{t=0} \mathcal{YM}_e(D^t) = \frac{d}{dt}\Big|_{t=0} \int_M \frac{1}{2} exp(\frac{1}{2}||R^D||^2) \frac{d}{dt} ||R^{D^t}||^2 dv$$
$$= \frac{1}{4}\int_M exp(\frac{1}{2}||R^D||^2)\Big[(\frac{d}{dt}\Big|_{t=0}||R^{D^t}||^2)^2 + 2\frac{d^2}{dt^2}\Big|_{t=0}||R^{D^t}||^2\Big]dv$$
$$= \int_M exp(\frac{1}{2}||R^D||^2)\Big[< d^D\alpha,\, R^D >^2 + < d^D F + [\alpha \wedge \alpha], R^D > + < d^D\alpha,\, d^D\alpha > \Big]dv.$$

Moreover, because D is an exponential Yang-Mills connection, we have

$$\int_M exp(\frac{1}{2}||R^D||^2) < d^D F,\, R^D > dv = \int_M < F,\, \delta^D (exp(\frac{1}{2}||R^D||^2) R^D) > dv = 0.$$

It follows from (6.7) in [39] that $< [\alpha \wedge \alpha],\, R^D > = < \alpha,\, \mathcal{R}^D(\alpha) >$. □

The index, nullity and stability of an exponential Yang-Mills connection D can be defined similarly to those of a Yang-Mills connection. However, they are very difficult to analyze since the operator $\mathcal{Q}^D$ is much more complicated than in the case of a Yang-Mills connection.

Corollary 9.3.2. *Suppose that D is an exponential Yang-Mills connection such that* $||R^D||$ *is constant. Then the stability of D as a Yang-Mills connection implies the stability of D as an exponential Yang-Mills connection.*

Bibliography

1. R. Ababou, P. Baird, J. Brossard, Polynômes semi-conformes et morphismes harmoniques. Math. Z. **231**(3), 589–604 (1999)
2. J.F. Adams, On the groups J(X). Topology **3**, 137–171 (1965); Topology **5**, 21–71 (1966)
3. J.F. Adams, Maps from a surface to the projective plane. Bull. Lond. Math. Soc. **14**(6), 533–534 (1982)
4. A.D. Aleksandrov, Uniqueness theorems for surfaces in the large. I. Am. Math. Soc. Transl. (2) **21**, 412–416 (1982)
5. W.K. Allard, An integrality theorem and a regularity theorem for surfaces whose first variation with respect to a parametric elliptic integrand is controlled. Proc. Symp. Pure Math. **44**, 1–28 (1986)
6. F. Almgren, Q-valued functions minimizing Dirichlet's integrals and the regularity of area-minimizing rectifiable currents up to codimension two. Bull. Am. Math. Soc. **8**(2), 327–328 (1983)
7. H.W. Alt, Verzweigungspunkte von H-Flächen II. Math. Ann. **201**, 33–55 (1973)
8. M. Ara, Geometry of F-harmonic maps. Kodai Math. J. **22**(2), 243–263 (1999)
9. M. Ara, Stability of F-harmonic maps into pinched manifolds. Hiroshima Math. J. **31**(1), 171–181 (2001)
10. N. Aronszajn, A unique continuation theorem for solutions of elliptic differential equations or inequalities. J. Math. Pures Appl. **36**, 235–249 (1957)
11. K. Arslan, R. Ezentas, C. Murathan, T. Sasahara, Biharmonic submanifolds in 3-dimensional (k, μ)-manifolds. Int. J. Math. Math. Sci. **22**, 3575–3586 (2005)
12. K. Arslan, R. Ezentas, C. Murathan, T. Sasahara, Biharmonic anti-invariant submanifolds in Sasakian space forms. Beitr. Algebra Geom. **48**(1), 191–207 (2007)
13. M.F. Atiyah, R. Bott, The Yang-Mills equations over Riemann surfaces. Philos. Trans. R. Soc. Lond. A **308**(1505), 523–615 (1982)
14. M.F. Atiyah, R.S. Ward, Instantons and algebraic geometry. Commun. Math. Phys. **55**(2), 117–124 (1977)
15. M.F. Atiyah, N.J. Hitchin, V.G. Drinfel'd, Yu.I. Manin, Construction of instantons. Phys. Lett. A **65**(3), 185–187 (1978)
16. M.F. Atiyah, N.J. Hitchin, I.M. Singer, Self-duality in four-dimensional Riemannian geometry. Proc. R. Soc. Lond. A **362**(1711), 425–461 (1978)
17. P. Aviles, The Dirichlet problem and Fatou's theorem for harmonic maps on regular domains, in *Recent Developments in Geometry* (Los Angeles, 1987). Contemporary Mathematics, vol. 101 (American Mathematical Society, Providence, 1989), pp. 79–96
18. P. Baird, J. Eells, *A Conservation Law for Harmonic Maps*. Lecture Notes in Mathematics, vol. 894 (Springer, Berlin, 1981), pp. 1–25

Y.-J. Chiang, *Developments of Harmonic Maps, Wave Maps and Yang-Mills Fields into Biharmonic Maps, Biwave Maps and Bi-Yang-Mills Fields*, Frontiers in Mathematics, DOI 10.1007/978-3-0348-0534-6, © Springer Basel 2013

19. P. Baird, D. Kamissoko, On constructing biharmonic maps and metrics. Ann. Glob. Anal. Geom. **23**(1), 65–75 (2003)
20. P. Baird, J.C. Wood, Bernstein theorems for harmonic morphisms from $\mathbf{R}^3$ and S^3. Math. Ann. **280**(4), 579–603 (1988)
21. P. Baird, J.C. Wood, Hermitian structures and harmonic morphisms in higher-dimensional Euclidean spaces. Int. J. Math. **6**(2), 161–192 (1995)
22. P. Baird, J.C. Wood, Weierstrass representations for harmonic morphisms on Euclidean spaces and spheres. Math. Scand. **81**(2), 283–300 (1997)
23. P. Baird, J.C. Wood, *Harmonic Morphisms Between Riemannian Manifolds*. London Mathematical Society Monographs (N.S.), vol. 29 (Oxford University Press, Oxford, 2003)
24. A. Balmuç, Biharmonic properties and conformal changes. An. Stiint. Univ. Al. I. Cuza Iaçi Mat. (N.S.) **50**(2), 361–272 (2004)
25. A. Balmuç, S. Montaldo, C. Oniciuc, Biharmonic maps between warped product manifolds. J. Geom. Phys. **57**(2), 449–466 (2007)
26. A. Balmusç, On the biharmonic curves of the Euclidean and Berger 3-dimensional spheres. Sci. Ann. Univ. Agric. Sci. Vet. Med. **47**(1), 87–96 (2004)
27. S. Bando, Y.-T. Siu, Stable sheaves and Einstein-Hermitian metrics, in *Geometry and Analysis on Complex Manifolds*, ed. by T. Mabuchi, J. Noguchi, T. Ochiai (World Science Publication, River Edge, 1994)
28. V. Benci, J.-M. Coron, The Dirichlet problem for harmonic maps from the disk into the Euclidean n-sphere. Ann. Inst. H. Poincaré Anal. Non Lineairé **2**(2), 119–141 (1985)
29. J. Berndt, Real hypersurfaces in quaternionic space space forms. J. Reine Angew. Math. **419**, 9–26 (1991)
30. J. Berndt, F. Tricerri, L. Vanhecke, *Generalized Heisenberg Groups and Damek-Ricci Harmonic Spaces*. Lecture Notes in Mathematics (Springer, Berlin, 1598)
31. M. Bertola, D. Gouthier, Lie triple systems and warped products. Rend. Mat. Appl. (7) **21**, 275–293 (2001)
32. F. Bethuel, On the singular set of stationary harmonic maps. Manuscr. Math. **78**(4), 417–443 (1993)
33. F. Bethuel, X. Zheng, Sur la densité des fonctions régulières entre deux variétés dans des espaces de Sobolev. C. R. Acad. Sci. Paris A **303**, 447–449 (1986)
34. P. Bizoń, T. Chmaj, Z. Tabor, Formation of singularities for equivariant $(2 + 1)$-dimensional wave maps into the 2-sphere. Nonlinearity **14**(5), 1041–1053 (2001)
35. B. Bojarski, T. Ivaniec, p-harmonic equation and quasilinear mappings, in *Partial Differential Equations* (Warsaw, 1984). Banach Center Publications, vol. 19 (PWN, Warsaw, 1987)
36. H.-J. Borchers, W.D. Garber, Local theory of solutions for the O(2k+1) σ-model. Commun. Math. Phys. **72**(1), 77–102 (1980)
37. J.-P. Bourguignon, Formules des Weitzenböck en dimension 4, in *Riemannian Geometry in Dimension 4*. Textes Mathématiques, vol. 3 (CEDIC, Paris, 1981), pp. 308–333
38. J.-P. Bourguignon, H.B. Lawson, Stability and gap phenomena for Yang-Mills fields. Proc. Natl. Acad. Sci. USA **76**(4), 1550–1553 (1979)
39. J.-P. Bourguignon, H.B. Lawson, Stability and isolation phenomena for Yang-Mills fields. Commun. Math. Phys. **79**(2), 189–230 (1981)
40. H. Brezis, J.-M. Coron, Large solutions for harmonic maps in two dimensions. Commun. Math. Phys. **13**(2), 203–215 (1983)
41. H. Brezis, L. Nirenberg, Degree theory and BMO. I. Compact manifolds without boundaryies. Sel. Math. (N.S.) **1**(2), 197–263 (1995)
42. H. Brezis, J.-M. Coron, E.H. Lieb, Estimation d'énergie pour les applications de $\mathbf{R}^3$ a valeurs dans S^2. C. R. Acad. Sci. Paris **303**(5), 207–210 (1986)
43. R.L. Bryant, Submanifolds and special structures on the octonions. J. Differ. Geom. **17**, 185–232 (1982)
44. R.L. Bryant, Minimal surfaces of constant curvature in S^n. Trans. Am. Math. Soc. **290**(1), 259–271 (1985)

45. R.L. Bryant, Harmonic morphisms with fibers of dimension one. Commun. Anal. Geom. **8**(2), 219–265 (2000)
46. D. Burns, Harmonic maps from $\mathbf{C}P^1$ to $\mathbf{C}P^n$. in *Proceedings of the Tulane Conference*. Lecture Notes in Mathematics, vol. 949 (Springer, Berlin, 1982), pp. 48–56
47. F.E. Burstall, Harmonic maps of finite energy from non-compact manifolds. J. Lond. Math. **30**(2), 361–370 (1984)
48. F.E. Burstall, Non-linear functional analysis and harmonic maps, Ph.D. thesis, University of Warwick, 1984
49. F.E. Burstall, A twistor description of harmonic maps of a 2-sphere into a Grassmannian. Math. Ann. **274**(1), 61–74 (1986)
50. F.E. Burstall, J. Rawnsley, Sphères harmoniques dans les groupes de Lie compacts et courbes holomorphes dans les espaces homogènes. C. R. Acad. Sci. Paris A **302**(20), 709–721 (1986)
51. F.E. Burstall, S. Salamon, Tournaments, flags and harmonic maps. Math. Ann. **277**(2), 249–265 (1987)
52. F.E. Burstall, J.C. Wood, The construction of harmonic maps into complex Grassmannians. J. Differ. Geom. **23**(3), 255–297 (1986)
53. F.E. Burstall, D. Ferus, F. Pedit, U. Pinkall, Harmonic tori in symmetric spaces and commuting Hamiltonian systems on loop algebras. Ann. Math. (II) **138**(1), 173–212 (1993)
54. R. Caddeo, S. Montaldo, C. Oniciuc, Biharmonic submanifolds of S^3. Int. J. Math. **12**(8), 867–876 (2001)
55. R. Caddeo, S. Montaldo, P. Piu, On biharmonic maps. Contemp. Math. Am. Math. Soc. **288**(3), 286–290 (2001)
56. R. Caddeo, S. Montaldo, C. Oniciuc, Biharmonic submanifolds in spheres. Isr. J. Math. **130**, 109–123 (2002)
57. R. Caddeo, C. Oniciuc, P. Piu, Explicit formulas for non-geodesic biharmonic curves of the Heisenberg group. Rend. Sem. Mat. Univ. Politec. Torino **62**(3), 265–278 (2004)
58. E. Calabi, Minimal immersions of surfaces in Euclidean spheres. J. Differ. Geom. **1**, 111–125 (1967)
59. E. Calabi, An intrinsic characterization of harmonic one-forms, in *Global Analysis*, ed. by K. Kodaira, S. Iyanaga, D.C. Spencer (University Press, Tokyo, 1969), pp. 101–117
60. D.M.J. Calderbank, Self-dual Einstein metrics and conformal submersions. Edinburgh University (2000). arXiv:mathDG0001041
61. J. Carlson, D. Toledo, Harmonic mappings of Kähler manifolds to locally symmetric spaces. Inst. Hautes Étues Sci. Publ. (69), 173–201 (1989)
62. S.-Y.A. Chang, L. Wang, P.C. Yang, Regularity of harmonic maps. Commun. Pure Appl. Math. **LII 52**(9), 1099–1111 (1999)
63. S.-Y.A. Chang, L. Wang, P.C. Yang, Regularity of biharmonic maps. Commun. Pure Appl. Math. **LII 52**(9), 1113–1137 (1999)
64. J. Cheeger, Singular Riemannian spaces. J. Differ. Geom. **18**(4), 575–657 (1983)
65. J. Cheeger, M. Goresky, R. MacPherson, L^2-cohomology and intersection homology of algebraic varieties, in *Seminars on Differential Geometry*, ed. by S.T. Yau. Annals of Mathematics Studies, vol. 102 (Princeton University Press, Princeton, 1982), pp. 303–340
66. B.Y. Chen, Some open problems and conjectures on submanifolds of finite type. Soochow J. Math. **17**(2), 169–188 (1991)
67. J.H. Chen, Compact 2-harmonic hypersurfaces in $S^{n+1}(1)$. Acta Math. Sin. **36**(3), 341–347 (1993)
68. B.Y. Chen, A report on submanifolds of finite type. Soochow J. Math. **22**(2), 117–337 (1996)
69. B.Y. Chen, S. Ishikawa, Biharmonic pseudo-Riemannian submanifolds in pseudo-Euclidean spaces. Kyushu J. Math. **52**(1), 167–185 (1988)
70. S.S. Chern, E. Spanier, A theorem on orientable surfaces in four-dimensional space. Comment. Math. Helv. **25**, 205–209 (1951)
71. S.S. Chern, J.G. Wolfson, Harmonic maps of S^2 into a complex Grassmannian manifold. Proc. Natl. Acad. Sci. USA **82**(8), 2217–2219 (1985)

72. S.S. Chern, J.G. Wolfson, Harmonic maps of the two-spheres into a complex Grassmannian manifold II. Ann. Math. (2) **125**(2), 301–335 (1987)
73. Y.J. Chiang, Harmonic maps of V-manifolds. Ann. Glob. Anal. Geom. **8**(3), 315–344 (1990)
74. Y.J. Chiang, Spectral geometry of V-manifolds and its applications to harmonic maps. Proc. Symp. Pure Math. Am. Math. Soc. **54**(Part 1), 93–99 (1993)
75. Y.J. Chiang, Biwave maps into manifolds. Int. J. Math. Math. Sci. **2009**, Article ID 104274, 1–14 (2009)
76. Y.J. Chiang, Some properties of biwave maps. J. Geom. Phys. **62**(4), 839–850 (2012)
77. Y.J. Chiang, f-biharmonic maps between Riemannian manifolds. J. Geom. Symm. Phys. **27**, 45–58 (2012)
78. Y.J. Chiang, Harmonic and biharmonic maps of Riemann surfaces. Glob. J. Pure Appl. Math. **9**(2), 109–124 (2013)
79. Y.J. Chiang, A. Ratto, Harmonic maps on spaces with conical singularities. Bull. Soc. Math. Fr. **120**(2), 251–262 (1992)
80. Y.J. Chiang, H.A. Sun, 2-harmonic totally real submanifolds in a complex projective space. Bull. Inst. Math. Acad. Sin. **27**(2), 99–107 (1999)
81. Y.J. Chiang, H.A. Sun, 2-harmonic maps between V-manifolds. J. Math. (Wuhan, China) **20**(2), 139–144 (2000)
82. Y.J. Chiang, H.A. Sun, Biharmonic maps on V-Manifolds. Int. J. Math. Math. Sci. **27**(8), 477–484 (2001)
83. Y.J. Chiang, R. Wolak, Transversally biharmonic maps between foliated Riemannian manifolds. Int. J. Math. **19**(8), 981–996 (2008)
84. Y.J. Chiang, R. Wolak, Transversal biwave maps. Arch. Math. (Brno) **46**(3), 211–226 (2010)
85. Y.J. Chiang, R. Wolak, Transversally f-harmonic and transversally f-biharmonic maps between foliated manifolds. JP J. Geom. Topol. **13**(1), 93–117 (2013)
86. Y.J. Chiang, R. Wolak, Transversally F-harmonic maps between foliated manifolds (preprint)
87. Y.J. Chiang, R. Wolak, Transversal wave maps and transversal exponential wave maps (preprint)
88. Y.J. Chiang, Y.H. Yang, Exponential wave maps. J. Geom. Phys. **57**(12), 2521–2532 (2007)
89. J.T. Cho, J. Inoguchi, J.E. Lee, Biharmonic curves in 3-dimensional Sasakian space forms. Ann. Mat. Pura. Appl. (4) **186**(1), 685–700 (2007)
90. D. Christodoulou, S. Tahvildar-Zadeh, On the regularity of spherically symmetric wave maps. Commun. Pure Appl. Math. **46** (1993), no. 7, 1041–1091.
91. C. Constantinescu, A. Cornea, *Potential Theory on Harmonic Spaces*. Grundlehren der Mathematischen Wissenschaften, vol. 158 (Springer, New York, 1972)
92. L.A. Cordero, R. Wolak, Examples of foliations with foliated geometric structures. Pac. J. Math. **142**(2), 265–276 (1990)
93. J.-M. Coron, F. Hélein, Harmonic diffeomorphisms, minimizing harmonic maps and rotational symmetry. Compos. Math. **69**(2), 175–228 (1989)
94. N. Course, f-harmonic maps. Ph.D. thesis, University of Warwick, 2004
95. N. Course, f-harmonic maps which map the boundary of the domain to one point in the target. N. Y. J. Math. **13**, 423–435 (2007) (electronic)
96. Y. Dai, M. Shoji, H. Urakawa, Harmonic maps into Lie groups and homogeneous spaces. Differ. Geom. Appl. **7**(2), 143–160 (1997)
97. I. Dimitric, Submanifolds of E^m with harmonic mean curvature vector. Bull. Inst. Math. Acad. Sin. **20**(1), 53–65 (1992)
98. M. Din, W.J. Zakrzewski, General classical solution in the $\mathbf{C}P^{n-1}$ model. Nucl. Phys. B. **174**(2–3), 397–406 (1980)
99. S.K. Donaldson, Anti-self-dual Yang-Mills connections on complex algebraic surfaces and stable vector bundles. Proc. Lond. Math. Soc. (3) **50**(1), 1–26 (1985)
100. S.K. Donaldson, Twisted harmonic maps and the self-duality equations. Proc. Lond. Math. Soc. (3) **55**(1), 127–131 (1987)
101. S.K. Donaldson, Mathematical uses of gauge theory, in *The Encyclopedia of Mathematical Physics*, ed. by J.-P. Francoise, G. Naber, Tsou Sheung Tsun (Elsevier, 2006)

102. S.K. Donaldson, P.B. Kronheimer, *The Geometry of Four-Manifolds* (Oxford University Press, New York, 1990)
103. S.K. Donaldson, R.P. Thomas, Gauge theory in higher dimensions, in *The Geometric Universe*, ed. by S.A. Huggett et al. (Oxford University Press, Oxford, 1998), pp. 31–47
104. J. Dorfmeister, F. Pedit, H. Wu, Weierstrass type representation of harmonic maps into symmetric spaces. Commun. Anal. Geom. **6**(4), 633–668 (1998)
105. V.A. Drinfeld, Yu.I. Manin, A description of instantons. Commun. Math. Phys. **63**(2), 177–192 (1978)
106. D.M. Duc, J. Eells, Regularity of exponentially harmonic functions. Int. J. Math. **2**(1), 395–398 (1991)
107. F. Duheille, On the range of $\mathbf{R}^2$ or $\mathbf{R}^3$-valued harmonic morphisms. Ann. Probab. **26**(1), 308–315 (1998)
108. N. Dunford, J.T. Schwartz, *Linear operators I, II* (Wiley-Interscience, London/New York, 1958/1963)
109. T. Duyckaerts, C. Kenig, F. Merle, Universality of blow-up profile for small type II blow-up solutions of energy-critical wave equations: the non-radial case. J. Eur. Math. Soc. **13**(3), 533–539 (2011)
110. C.J. Earle, J. Eells, Deformations of Riemann Surfaces, Lecture Notes in Mathematics, vol. 103 (Springer, Berlin, 1969), pp. 122–149
111. P. Eberlein, When is a geodesic flow of Anosov type? II. J. Differ. Geom. **8**, 565–577 (1973)
112. A.L. Edmonds, Deformations of maps to branched coverings in dimension two. Ann. Math. (2) **110**(1), 113–125 (1979)
113. J. Eells, On equivariant harmonic maps, in *Proceedings of the 1981 Shanghai-Hefei Symposium on Differential Geometry and Differential Equations* (Science Press, Beijing, 1984), pp. 55–73
114. J. Eells, Regularity of certain harmonic maps, in *Global Riemannian Geometry* (Durham, 1983). Horwood Series in Mathematics and Its Applications (Horwood, Chichester, 1984), pp. 13–147
115. J. Eells, Gauss maps of surfaces, in *Perspective in Mathematics*, ed. by W. Jäger, J. Moser, R. Remmert (Birkhäuser, Basel, 1984), pp. 111–129
116. J. Eells, Minimal branched immersions into three-manifolds, in *Proceedings of the University of Maryland (1983–1984)* Lecture Notes in Mathematics, vol. 1167 (Springer, Berlin, 1985), pp. 81–94
117. J. Eells, M.J. Ferreira, On representing homotopy classes by harmonic maps. Bull. Lond. Math. Soc. **23**(2), 160–162 (1991)
118. J. Eells, B. Fuglede, *Harmonic Maps Between Riemannian Polyhedra*. Cambridge Tracts in Mathematics, vol. 42 (Cambridge University Press, Cambridge, 2001)
119. J. Eells, L. Lemaire, A report on harmonic maps. Bull. Lond. Math. Soc. **10**(1), 1–68 (1978)
120. J. Eells, L. Lemaire, On the construction of harmonic and holomorphic maps between surfaces. Math. Ann. **252**(1), 27–52 (1980)
121. J. Eells, L. Lemaire, Deformations of metrics and associated harmonic maps, in *Patodi Memorial Vol., Geometry and Analysis* (Indian Academy of Science, Baugalore, 1981), pp. 33–45
122. J. Eells, L. Lemaire, *Selected Topics in Harmonic Maps*. CBMS Regional Conference Series in Mathematics, vol. 150 (American Mathematical Society, Providence, 1983)
123. J. Eells, L. Lemaire, Examples of harmonic maps from disks to hemispheres. Math. Z. **5**(4), 517–519 (1984)
124. J. Eells, L. Lemaire, Another report on harmonic maps. Bull. Lond. Math. Soc. **20**(5), 385–524 (1988)
125. J. Eells, L. Lemaire, Some properties of exponentially harmonic maps, in *Partial Differential Equations, Part 1, 2* (Warsaw, 1990). Banach Center Publications, vol. 27 (Polish Academy of Sciences, Warsaw, 1992), pp. 129–136
126. J. Eells, J.C. Polking, Removable singularities of harmonic maps. Indiana Univ. Math. J. **33**(6), 859–871 (1984)

127. J. Eells, A. Ratto, Harmonic maps between spheres and ellipsoids. Int. J. Math. **1**(1), 1–27 (1990)
128. J. Eells, A. Ratto, *Harmonic Maps and Minimal Immersions with Symmetries*. Methods of Ordinary Differential Equations Applied to Elliptic Variational Problems. Annals of Mathematics Studies, vol. 130 (Princeton University Press, Princeton, 1993)
129. J. Eells, J.H. Sampson, Harmonic maps of Riemannian manifolds. Am. J. Math. **86**, 109–160 (1964)
130. J. Eells, J.H. Sampson, Énergie et déformations en géométrie différentielle. Ann. Inst. Fourier (Grenoble) **14**, 61–69 (1965)
131. J. Eells, J.H. Sampson, Variational theory in fibre bundles, in *1966 Proceedings of the U.S.-Japan Seminar in Differential Geometry*, Kyoto, 1965, pp. 22–33
132. J. Eells, A. Verjovsky, Harmonic and Riemannian foliations. Bol. Soc. Mat. Mex. (3) **4**(1), 1–12 (1998)
133. J. Eells, J.C. Wood, Restrictions on harmonic maps of spheres. Topology **15**(3), 263–266 (1976)
134. J. Eells, J.C. Wood, Maps of minimum energy. J. Lond. Math. **23**(2), 303–310 (1981)
135. J. Eells, J.C. Wood, The existence and construction of certain harmonic maps, in *Symposia Mathematica, Vol. XXVI* (Rome, 1980) (Academic, London, 1982), pp. 123–138
136. J. Eells, J.C. Wood, Harmonic maps from surfaces to complex projective spaces. Adv. Math. **49**(3), 217–263 (1983)
137. L.P. Eisenhart, *Riemannian Geometry* (Princeton University Press, Princeton, 1926)
138. A. El Kacimi Alaoui, E.G. Gómez, Applications Harmoniques Feuilletées. Ill. J. Math. **40**(1), 115–122 (1996)
139. L.C. Evans, Partial regularity for stationary harmonic maps into spheres. Arch. Ration. Mech. Anal. **116**(2), 101–113 (1991)
140. L.C. Evans, *Partial Differential Equations*. Graduate Studies in Mathematics, vol. 19 (American Mathematical Society, Providence, 1998)
141. H. Federer, *Geometric Measure Theory* (Springer, New York, 1969)
142. H. Federer, The singular sets of area minimizing rectifiable currents with codimension one and of area minimizing flat chains modulo two with arbitrary codimension. Bull. Am. Math. Soc. **79**, 761–771 (1970)
143. D. Fetcu, Biharmonic curves in the generalized Heisenberg group. Beitr. Algebra Geom. **46**(2), 513–521 (2005)
144. D. Fetcu, Biharmonic curves in Cartan-Vranceanu $(2n + 1)$-dimensional spaces. Bull. Acad. Stiinte Repub. Mold. Mat. **53**(1), 59–65 (2007)
145. A.P. Fordy, J.C. Wood (eds.), *Harmonic Maps and Integrable Systems*. Aspects of Mathematics, E. vol. 23 (Friedr. Vieweg & Sohn, Braunschweig, 1994)
146. J. Frehse, A discontinuous solution of a mildly nonlinear elliptic system. Math. Z. **134**, 229–230 (1973)
147. A. Friedman, *Partial Differential Equations of Parabolic Type* (Prentice-Hall, Englewood Cliffs, 1964)
148. B. Fugledge, Harmonic morphisms between Riemannian manifolds. Ann. Inst. Fourier (Grenoble) **28**(2), vi, 107–144 (1978)
149. B. Fugledge, A criterion of nonvanishing differential of a smooth map. Bull. Lond. Math. Soc. **14**(2), 98–102 (1982)
150. F.B. Fuller, Harmonic mappings. Proc. Natl. Acad. Sci. USA **40**, 987–991 (1954)
151. A. Futaki, Non-existence of minimizing harmonic maps from 2-spheres. Proc. Jpn. Acad. A **56**(6), 291–293 (1980)
152. S. Gallot, D. Hulin, J. Lafontaine, *Riemannian Geometry* Universitext, 3rd edn. (Springer, 2004)
153. M. Giaquinta, E. Giusti, On the regularity of the minima of variational integrals. Acta Math. **148**, 31–46 (1982)
154. M. Giaquinta, E. Giusti, The singular set of the minima of certain quadratic functional. Ann. Scuola Norm. Super. Pisa (4) **11**(1), 45–55 (1984)

155. G.W. Gibbons, S.W. Hawking, Gravitational and multi-instantons. Phys. Lett. B. **78**(4), 430–432 (1978)
156. D. Gilbarg, N.S. Trudinger, *Elliptic Partial Differential Equations of Second Order*, 2nd edn. (Springer, Berlin, 1983)
157. V. Glasa, R. Stora, Regular solutions of the $\mathbf{C}P^n$ models and further generalization. CERN (preprint, 1980)
158. W.B. Gordon, Convex functions and harmonic maps. Proc. Am. Math. Soc. **33**, 433–437 (1972)
159. V. Grigoryan, Stability of geodesic wave maps. Ph.D. thesis, University of Massachusetts, Amherst, 2008
160. M. Grillakis, A priori estimates and regularity of nonlinear waves, in *Proceedings of the International Congress of Mathematicians*, Zürich, vols. 1, 2 (1994), pp. 1187–1194
161. M. Grillakis, Classical solutions for the equivariant wave map in $1+2$ dimensions, preprint
162. M. Gromov, Pseudo-holomorphic curves in symplectic manifolds. Invent. Math. **82**(2), 307–347 (1985)
163. C.H. Gu, On the Cauchy problem for harmonic maps defined on two-dimensional Minkowski space. Commun. Pure Appl. Math. **33**(6), 727–737 (1980)
164. S. Gudmundsson, M. Svensson, On the existence of harmonic morphisms from certain symmetric spaces. J. Geom. Phys. **57**(2), 353–366 (2007)
165. S. Gudmundsson, J.C. Wood, Multi-valued harmonic morphisms. Math. Scand. **73**(1), 127–155 (1993)
166. M. Guest, *Harmonic Maps, Loop Groups, and Integrable Systems*. London Mathematical Society Student Texts, vol. 38 (Cambridge University Press, Cambridge, 1997)
167. R. Gulliver, Regularity of minimizing surfaces of prescribed mean curvature. Ann. Math. (2) **97**, 275–305 (1973)
168. R. Gulliver, Index and total curvature of complete minimal surfaces, in *Proceedings of Symposia in Pure Mathematics*, vol. 44 (American Mathematical. Society, Providence, 1986), pp. 207–211
169. A. Haefliger, Pseudogroups of local isometries, differential geometry, in *Proceedings of the Vth International Colloquium on Differential Geometry*, Santiago de Compostela, 1984, ed. by L.A. Cordero (Pitman, Boston, 1985)
170. R. Hamilton, *Harmonic Maps of Manifolds with Boundary*. Lecture Notes in Mathematics, vol. 471 (Springer, Berlin/New York, 1975)
171. R. Hardt, F.H. Lin, Mappings minimizing the L^p-norm of the gradient. Commun. Pure Appl. Math. **40**(5), 555–588 (1987)
172. R. Hardt, D. Kinderlehrer, F.H. Lin, Energy bounds for minimizing maps (preprint, 1987)
173. P. Hartman, On homotopic harmonic maps. Can. J. Math. **19**, 673–687 (1967)
174. R. Harvey, H.B. Lawson, Calibrated geometries. Acta Math. **148**, 47–157 (1982)
175. R. Harvey, B. Shiffman, A characterization of holomorphic chains. Ann. Math. (2) **99**, 553–587 (1974)
176. T. Hasanis, T. Vlachos, Hypersurfaces in E^4 with harmonic mean curvature vector field. Math. Nachr. **72**, 145–169 (1995)
177. F. Hélein, Regularite des applications failblement harmoniques entreune sur face et une varitee Riemanniane. C. R. Acad. Sci. Paris I **312**, 591–596 (1991)
178. F. Hélein, *Constant Mean Curvature Surfaces, Harmonic Maps and Integrable Systems*. Lectures in Mathematics, ETH Zürich (Birkhäuser, Basel, 2001)
179. F. Hélein, *Harmonic maps, Conservation Laws and Moving Frames*. Cambridge Tracts in Mathematics, vol. 150 (Cambridge University Press, Cambridge, 2002)
180. R. Hermann, A sufficient condition that a mapping of Riemannian manifolds can be a fiber bundle. Proc. Am. Math. Soc. **11**, 236–242 (1960)
181. D. Hilbert, Die Grundlagen der Physik. Math. Ann. **92**(1–2), 1–32 (1924)
182. S. Hildebrandt, Liouville theorems for harmonic mappings and an approach to Bernstein theorems, in *Seminar on Differential Geometry*. Annals of Mathematics Studies, vol. 102 (Princeton University Press, Princeton, 1982), pp. 107–131

183. S. Hildebrandt, H. Kaul, K.-O. Widman, An existence theorem for harmonic mappings of Riemannian manifolds. Acta Math. **138**(1–2), 1–16 (1977)
184. S. Hildebrandt, J. Jost, K.-O. Widman, Harmonic mappings and minimal submanifolds. Invent. Math. **62**(2), 269–298 (1980)
185. N. Hitchin, The self-duality equations on a Riemann surface. Proc. Lond. Math. Soc. (3) **55**(1), 59–126 (1987)
186. N. Hitchin, Harmonic maps from a 2-torus to the 3-sphere. J. Differ. Geom. **31**(3), 627–710 (1990)
187. M.C. Hong, On the conformal equivalence of harmonic maps and exponential harmonic maps. Bull. Lond. Math. Soc. **24**(5), 488–492 (1992)
188. J.-Q. Hong, Y. Yang, Some results on exponentially harmonic maps. Chin. Ann. Math. A **14**(6), 686–691 (1993)
189. H. Hopf, Über Flächen mit einer Relation zwischen den Hauptkrümmungen. Math. Nachr. **4**, 232–249 (1950–1951)
190. W.Y. Hsiang, H.B. Lawson Jr., Minimal submanifolds of low cohomology. J. Differ. Geom. **5**, 1–38 (1971)
191. T. Ichiyama, J.-I. Inoguchi, H. Urakawa, Biharmonic maps and bi-Yang-Mills fields. Note Mat. **28**(suppl. 1), 233–275 (2009)
192. T. Ichiyama, J.-I. Inoguchi, H. Urakawa, Classification and isolation phenomena of biharmonic maps and bi-Yang-Mills fields. arXiv:0912.4806v1 [math.DG] 24 Dec 2009
193. J.-I. Inoguchi, Submanifolds with harmonic mean curvature in contact 3-manifolds. Colloq. Math. **101**(2), 163–179 (2004)
194. T. Ishihara, A mapping of Riemannian manifolds which preserves harmonic functions. J. Math. Kyoto Univ. **19**(2), 215–229 (1979)
195. C.G.J. Jacobi, *Ü*ber eine L*ö*sung der partiellen Differentialgleichung $\frac{\partial^2 V}{\partial x^2} + \frac{\partial^2 V}{\partial y^2} + \frac{\partial^2 V}{\partial z^2} = 0$. J. Reine Angew. Math. **36**, 113–134 (1848)
196. G.Y. Jiang, 2-harmonic maps and their first and second variational formulas. Chin. Ann. Math. A **7**(4), 389–402 (1986)
197. G.Y. Jiang, 2-harmonic isometric immersions between Riemannian manifolds. Chin. Ann. Math. A **7**(2), 130–144 (1986)
198. G.Y. Jiang, The conservation law of 2-harmonic maps between Riemannian manifolds. Acta Math. Sin. **30**(2), 220–225 (1987)
199. F. John, L. Nirenberg, On functions of bounded mean oscillation. Commun. Pure Appl. Math. **14**, 415–426 (1961)
200. P.E. Jones, K.P. Tod, Minitwistor spaces and Einstein-Weyl spaces. Class. Quantum Gravity **2**(4), 565–577 (1985)
201. J. Jost, Ein Existenzbeweis f*ü*r harmonische Abbildungen, die ein Dirichlet problem l*ö*sen, mittels Methode des W*ä*rmeflusses. Manuscr. Math. **34**(1), 17–25 (1981)
202. J. Jost, The Dirichlet problem for harmonic maps from a surfaces with boundary onto a 2-sphere with non-constant boundary values. J. Differ. Geom. **19**(2), 393–401 (1984)
203. J. Jost, *Harmonic Mappings Between Surfaces*. Lecture Notes in Mathematics, vol. 1062 (Springer, Berlin, 1984)
204. J. Jost, A note on harmonic maps between surfaces. Ann. Inst. H. Poincaré Anal. Non Lineairé **2**(6), 397–405 (1985)
205. J. Jost, *Two-Dimensional Geometric Variational Problems*. Pure and Applied Mathematics (Wiley-Interscience, Chichester, 1991)
206. J. Jost, X. Li-Jost, *Calculus of Variations*. Cambridge Studies in Advanced Mathematics, vol. 64 (Cambridge University Press, Cambridge, 1998)
207. J. Jost, R. Schoen, On the existence of harmonic diffeomorphisms. Invent. Math. **66**(2), 353–359 (1982)
208. J. Jost, S.T. Yau, Harmonic mappings and Kähler manifolds. Math. Ann. **262**(2), 145–166 (1983)

209. J. Jost, S.T. Yau, A strong rigidity theorem for a certain class of compact complex analytic surfaces. Math. Ann. **271**(1), 143–152 (1985)
210. J. Jost, S.T. Yau, The strong rigidity of locally symmetric complex manifolds of rank one and finite volume. Math. Ann. **275**(2), 291–304 (1986)
211. M. Kalka, Deformation of submanifolds of strongly negatively curved manifolds. Math. Ann. **251**(3), 243–248 (1980)
212. A.D. Kanfon, A. Füzfa, D. Lambert, Some examples of exponentially harmonic maps. J. Phys. A (35), 7629–7639 (2002)
213. N. Kapouleas, Constant mean curvature surfaces in Euclidean three-space. Bull. Am. Math. Soc. **17**(2), 318–320 (1987)
214. U. Katagiri, On the existence of Yang-Mills connections by conformal changes in higher dimensions. J. Math. Soc. Jpn. **46**(1), 139–145 (1994)
215. H. Karcher, U. Pinkall, I. Sterling, New minimal surfaces in S^3. J. Differ. Geom. **28**(2), 169–185 (1988)
216. M. Keel, T. Tao, Endpoint Strichartz estimates. Am. J. Math. **120**(5), 955–980 (1998)
217. M. Keel, T. Tao, Local and global well-posedness of wave maps on R^{1+1} for rough data. Int. Math. Res. Not. (21), 1117–1156 (1998)
218. C. Kenig, F. Merle, Global well-posedness, scattering and blow-up for the energy-critical, focusing non-linear wave equations. Acta Math. **201**(2), 147–212 (2008)
219. C. Kenig, F. Merle, Scattering for $\dot{H}^{1/2}$ bounded solutions to the cubic, defocusing NLS in 3 dimensions. Trans. Am. Math. Soc. **362**(4), 1937–1962 (2010)
220. K. Kenmotsu, Weirstrass formula for surfaces of prescribed mean curvature. Math. Ann. **245**(2), 89–99 (1979)
221. J.R. King, The currents defined by analytic varieties. Acta Math. **127**(3–4), 185–220 (1971)
222. S. Klainerman, M. Machedon, Space-time estimates for null forms and the local existence theorem. Commun. Pure Appl. Math. **46**(9), 1221–1268 (1993)
223. S. Klainerman, M. Machedon, Smoothing estimates for null forms and applications. Duke Math. J. **81**(1), 99–133 (1995)
224. S. Klainerman, M. Machedon, On the optimal local regularity for gauge fields theories. Differ. Integral Equ. **10**(7), 1019–1030 (1997)
225. S. Klainerman, M. Machedon, On the algebraic properties of the $H^{n/2,1/2}$ spaces. Int. Math. Res. Not. (15), 765–774 (1998)
226. S. Klainerman, I. Rodnianski, On the global regularity of wave maps in the critical Sobolev norm. Int. Math. Res. Not. (13), 655–677 (2001)
227. S. Klainerman, S. Selberg, Remark on the optimal regularity for equations of wave maps. Commun. Partial Differ. Equ. **22**(5–6), 901–918 (1997)
228. S. Klainerman, D. Tataru, On the optimal local regularity for Yang-Mills equations in $\mathbf{R}^{4+1}$. J. Am. Math. Soc. **12**(1), 93–116 (1999)
229. S. Kobayashi, K. Nomizu, *Foundations of Differential Geometry*, vols. I, II (Wiley, New York, 1963/1969)
230. J.J. Konderak, R. Wolak, Transversally harmonic maps between manifolds with Riemannian foliations. Q. J. Math. **54**(3), 335–354 (2003)
231. J.J. Konderak, R. Wolak, Some remarks on transversally harmonic maps. Glasg. J. Math. **50**(1), 1–16 (2008)
232. J.L. Koszul, B. Malgrange, Sur certaines structures fibreés complexes. Arch. Math. (Basel) **9**, 102–109 (1958)
233. J. Krieger, Global regularity of wave maps from $\mathbf{R}^{3+1}$ to surfaces. Commun. Math. Phys. **238**(1), 333–366 (2003)
234. J. Krieger, Global regularity of wave maps from $\mathbf{R}^{2+1}$ to H^2. Small energy. Commun. Math. Phys. **250**(3), 507–580 (2004)
235. J. Krieger, *Stability of Spherically Symmetric Wave Maps*, vol. 181, no. 853 (Am. Math. Soc., Providence, 2006)
236. J. Krieger, W. Schlag, D. Tataru, Renormalization and blow up for charge one equivariant critical wave maps. Invent. Math. **171**(3), 543–615 (2008)

237. Y. Ku, Interior and boundary regularity of intrinsic biharmonic maps to spheres. Pac. J. Math. **234**(1), 46–67 (2008)
238. O.A. Ladyzenskaya, N.N. Ural'tseva, *Linear and Quasilinear Elliptic Equations* (Academic, New York/London, 1968)
239. T. Lamm, Heat flow for extrinsic biharmonic maps with small initial energy. Ann. Glob. Anal. Geom. **26**(4), 369–384 (2004)
240. D. Laugwitz, *Differential and Riemannian Geometry* (Academic, New York/London, 1965)
241. H.B. Lawsōn, Complete minimal surfaces in S^3. Ann. Math. (2) **92**, 335–374 (1970)
242. P.D. Lax, Integrals of non-linear equations of evolution and solitary waves. Commun. Pure. Appl. Math. **21**, 467–490 (1968)
243. C. LeBrun, Complete Ricci-flat Kähler metrics on $\mathbf{C}^n$ need not be flat, in *Several Complex Variables and Complex Geometry* (Santa Cruz, 1989). Proceedings of Symposia in Pure Mathematics, vol. 52, Part 2 (American Mathematical Society, Providence, 1991), pp. 297–304
244. J. Lelong-Ferrand, Construction de modules continuité dans de cas limite de Soboleff et applications à la géométrie différentielle. Arch. Ration. Mech. Anal. **52**, 297–311 (1973)
245. L. Lemaire, Applications harmoniques de variétés produits. Comment. Math. Helv. **52**(1), 11–24 (1977)
246. L. Lemaire, Applications harmoniques de surfaces riemanniennes. J. Differ. Geom. **13**(1), 51–78 (1978)
247. L. Lemaire, Harmonic nonholomorphic maps from a surface to a sphere. Proc. Am. Math. Soc. **71**(2), 299–304 (1978)
248. L. Lemaire, Existence des applications harmoniques et courbure des variétés, in *Bourbaki Seminar Vol. 1979/80*. Lecture Notes in Mathematics, vol. 842 (Springer, Berlin, 1981), pp. 174–195
249. L. Lemaire, Boundary value problems for harmonic and minimal maps of surfaces into manifolds. Ann. Scuola Norm. Super. Pisa (4) **9**(1), 91–103 (1982)
250. P.F. Leung, On the stability of harmonic maps, in *Harmonic Maps* (New Orleans). Lecture Notes in Mathematics, vol. 949 (Springer, Berlin, 1982), pp. 122–129
251. A.M. Li, J.M. Li, An inequality for matrices and its applications in differential geometry. Adv. Math. (China) **20**(3), 375–376 (1991)
252. A. Lichnerowicz, Applications harmoniques et variétiés Kählieriennes, in *1968/1969 Symposia Mathematica, Vol. III*, INDAM, Rome (Academic, London, 1970), pp. 341–402
253. F.H. Lin, Gradient estimates and blow-up analysis for stationary harmonic maps. Ann. Math. (2) **149**(3), 785–829 (1999)
254. E. Loubau, Y.-L. Ou, Biharmonic maps and morphisms from conformal mappings. Tohoku Math. J. (2) **62**(1), 55–73 (2010)
255. E. Loubeau, C. Oniciuc, On the biharmonic and harmonic indices of the Hopf map. Trans. Am. Math. Soc. **359**(11), 5239–5256 (2007)
256. E. Loubeau, Y.-L. Ou, The characterization of biharmoinic morphisms, in *Differential Geometry and Its Applications*, (Opava, 2001). Math. Publ. **3** (2001), pp. 31–41
257. E. Loubeau, R. Pantilie, Harmonic morphisms between Weyl spaces and twistorial maps. Commun. Anal. Geom. **14**(5), 847–881 (2006)
258. E. Loubeau, R. Pantilie, Harmonic morphisms between Weyl spaces and twistorial maps II. Ann. Inst. Fourier (Grenoble) **60**(2), 433–453 (2010)
259. E. Loubeau, S. Montaldo, C. Oniciuc, The stress-energy tensor for biharmonic maps. Math. Z. **259**(3), 503–524 (2008)
260. F. Matsuura, H. Urakawa, On exponential Yang-Mills connections. J. Geom. Phys. **17**(1), 73–89 (1995)
261. W.H. Meeks III, A survey of the geometric results in the classical theory of minimal surfaces. Bol. Soc. Bras. Mat. **12**(1), 29–86 (1981)
262. M. Meier, Removable singularities of harmonic maps and an application to minimal submanifolds. Indiana Univ. Math. J. **35**(4), 705–726 (1986)

263. M.J. Micallef, Stable minimal surfaces in Euclidean space. J. Differ. Geom. **19**(1), 57–84 (1984)
264. M.J. Micallef, Stable minimal surfaces in flat tori, in *Complex Differential Geometry and Nonlinear Differential Equations* (Brunswick, Maine, 1984). Contemporary Mathematics, vol. 49 (American Mathematical Society, Providence, 1986), pp. 73–78
265. I. Min-Oo, An L_2-isolation theorem for Yang-Mills fields. Compos. Math. **47**, 153–163 (1982)
266. N. Mok, The holomorphic or antiholomorphic character of harmonic maps into irreducible compact quotients of polydiscs. Math. Ann. **272**(2), 197–216 (1985)
267. N. Mok, The uniformization theorem for compact Kähler manifolds of non-negative holomorphic bisectional curvature. J. Differ. Geom. **27**(2), 179–214 (1988)
268. P. Molino, *Riemannian Foliations* (Birkhäuser, Boston, 1988)
269. S. Montaldo, C. Oniciuc, A short survey on biharmonic maps between Riemannian manifolds. Rev. Union Mat. Argent. **47**(2), 1–22 (2006)
270. C.B. Morrey, The problem of Plateau on a Riemannian manifold. Ann. Math. (2) **49**, 807–851 (1948)
271. C.B. Morrey, The analytic embedding of abstract real-analytic manifolds. Ann. Math. (2) **68**, 159–201 (1958)
272. C.B. Morrey, *Multiple Integrals in the Calculus of Variations*. Grundlehren der mathematischen Wissenschaften (Springer, New York, 1966)
273. G.D. Mostow, *Strong Rigidity of Locally Symmetric Spaces*. Annals of Mathematics Studies, vol. 78 (Princeton University Press, Princeton, 1973)
274. G.D. Mostow, Y.-T. Siu, A compact Kähler surface of negative curvature not covered by the ball. Ann. Math. (2) **112**, 321–360 (1980)
275. A. Nahmod, A. Stefanov, K. Uhlenbeck, On the well-posedness of the wave map problem in high dimensions. Commun. Anal. Geom. **11**(1), 49–83 (2003)
276. H. Nakajima, Compactness of the moduli space of Yang-Mills connections in higher dimensions. J. Math. Soc. Jpn. **40**(3), 383–392 (1988)
277. N. Nakauchi, H. Urakawa, Removable singularities and bubbling of biharmonic maps. arXiv:0912.4086[math.DG]17 Jan 2011
278. M. Obata, The Gauss map of immersions of Riemannian manifolds in spaces of constant curvature. J. Differ. Geom. **2**, 217–223 (1968)
279. B. O'Neill, *Semi-Riemannian Geometry with Applications to Relativity* (Academic, New York/London, 1983)
280. C. Oniciuc, Biharmonic maps between Riemannian manifolds. An. Stiint. Univ. Al. I. Cuza Iaçi Mat. (N.S.) **48**(2), 237–248 (2002)
281. C. Oniciuc, On the second variation formula for biharmonic maps to a sphere. Publ. Math. Debr. **61**(3–4), 613–622 (2002)
282. V. Oproiu, Some classes of natural almost Hermitian structures on the tangent bundles. Publ. Math. Debr. **62**(3–4), 561–576 (2003)
283. R. Osserman, Minimal surfaces in the large. Comment. Math. Helv. **35**, 65–76 (1961)
284. Y.-L. Ou, Quadratic harmonic morphisms and O-systems. Ann. Inst. Fourier (Grenoble) **47**(2), 687–713 (1997)
285. Y.-L. Ou, p-harmonic morphisms, biharmonic morphisms and nonharmonic biharmonic maps. J. Geom. Phys. **56**(3), 358–374 (2006)
286. Y.-L. Ou, On conformal biharmonic immersions. Ann. Glob. Anal. Geom. **36**(2), 133–142 (2009)
287. Y-L. Ou, Conformally biharmonic immersions into 3-dimensional manifolds (preprint)
288. Y.-L. Ou, Some constructions of biharmonic maps and Chen's conjecture on biharmonic hypersurfaces. J. Geom. Phys. **62**, 751–762 (2012)
289. Y.-L. Ou, S. Lu, Biharmonic maps in two dimensions. Ann. Mat. doi:10.1007/s10231-011-0215-0
290. Y.-L. Ou, L. Tang, The generalized Chen's conjecture on biharmonic submanifolds is false. arXiv:1006.1838v2 [math.DG] 1 Jan 2011

291. Y.-L. Ou, J.C. Wood, On the classification of quadrtic harmonic morphisms between Euclidean spaces. Algebras Groups Geom. **13**(1), 41–53 (1996)
292. S. Ouakkas, R. Nasri, M. Djaa, On the f-harmonic and f-biharmonic maps. J. P. J. Geom. Topol. **10**(1), 11–27 (2010)
293. J.M. Overduin, P.S. Wesson, Kaluza-Klein gravity. Phys. Rep. **283**(5–6), 303–378 (1997)
294. R. Pantilie, Harmonic morphisms with 1-dimensional fibres on 4-dimensional Einstein manifolds. Commun. Anal. Geom. **10**(4), 779–814 (2002)
295. R. Pantilie, Harmonic morphisms between Weyl spaces, in *Modern Trends in Geometry and Topology* (Cluj University Press, Cluj-Napoca, 2006), pp. 321–332
296. R. Pantilie, J.C. Wood, A new construction of Einstein self-dual manifolds. Asian J. Math. **6**(2), 337–348 (2002)
297. R. Pantilie, J.C. Wood, Twistorial harmonic morphisms with one-dimensional fibres on self-dual four-manifolds. Q. J. Math. **57**(1), 105–132 (2003)
298. T.H. Parker, A Morse theory for equivariant Yang-Mills fields. Duke Math. J. **66**(2), 337–356 (1992)
299. U. Pinkall, I. Sterling, On the classification of constant mean curvature tori. Ann. Math. (2) **130**(2), 407–451 (1989)
300. I. Pluzhnikov, Some properties of harmonic mappings in the case of spheres and Lie groups. Sov. Math. Dokl. **27**, 246–248 (1983)
301. W. Pogorzelski, Propriétés des intégrales de l'équation parabolique normale. Ann. Polin. Math. **4**, 61–92 (1957)
302. D. Preiss, Geometry of measure in $\mathbf{R}^n$: distribution, rectifiability and density. Ann. Math. (2) **125**(3), 537–643 (1987)
303. A. Pressley, G. Segal, *Loop Groups*. Oxford Mathematical Monograph (Oxford Science Publications, 1988), pp. 1–328
304. P. Price, A monotonicity formula for Yang-Mills fields. Manuscr. Math. **43**(2–3), 131–166 (1983)
305. J. Råde, On the Yang-Mills heat equation in two and three dimensions. J. Reine Angew. Math. **431**, 123–163 (1992)
306. J.H. Rawnsley, Noether's theorem for harmonic maps, in *Differential Geometric Methods in Mathematical Physics*. Mathematical Physics Studies, vol. 6 (Riedel, Dordrecht, 1984), pp. 197–202
307. I. Rodnianski, J. Sterbenz, On the formation of singularities in the critical O(3) σ-model. Ann. Math. (2) **172**(1), 187–242 (2010)
308. C. Romero, R. Tavakol, R. Zalaletdinov, The embedding of general relativity in five dimensions. Gen. Relativ. Gravit. **28**(3), 365–376 (1996)
309. A. Ruh, J. Vilms, The tension field of the Gauss map. Trans. Am. Math. Soc. **149**, 569–573 (1970)
310. J. Sacks, K. Uhlenbeck, The existence of minimal immersions of 2-spheres. Ann. Math. (2) **113**(1), 1–24 (1981)
311. J. Sacks, K. Uhlenbeck, Minimal immersions of closed Riemann surfaces. Trans. Am. Math. Soc. **271**(2), 639–652 (1982)
312. J.H. Sampson, *Cours de Topologie Algébrique* (Département de Mathématique, Strasbourg, 1969)
313. J.H. Sampson, Some properties and applications of harmonic mappings. Ann. Sci. Ecole Norm. Sup. (4) **11**(2), 211–228 (1978)
314. J.H. Sampson, On harmonic mappings, in *Symposia Mathematica, Vol. XXVI* (Rome, 1980) (Academic, London/New York, 1982), pp. 197–210
315. J.H. Sampson, Harmonic maps in Kähler geometry, in *Harmonic Mapings and Minimal Immersions* (Montecatini, 1984). Lecture Notes in Mathematics, vol. 1161 (Springer, Berlin, 1985), pp. 193–205
316. J.H. Sampson, Applications of harmonic maps to Kähler geometry, in *Complex Differential Geometry and Nonlinear Differential Equations* (Brunswick, Maine, 1984). Contemporary Mathematics, vol. 49 (American Mathematical Society, Providence, 1986), pp. 125–134

317. T. Sasahara, Legendre surfaces in Sasakian space forms whose mean curvature vectors are eigenvectors. Publ. Math. Debr. **67**(3–4), 285–303 (2005)
318. T. Sasahara, Stability of biharmonic Legendre submanifolds in Sasakian space forms. Can. Math. Bull. **51**(3), 448–459 (2008)
319. R. Schoen, Analytic aspects of the harmonic map problem, in *Seminars on Nonlinear Partial Differential Equations* (Berkeley, 1983). Mathematical Sciences Research Institute Publications, vol. 2 (Springer, New York, 1984), pp. 321–358
320. R. Schoen, K. Uhlenbeck, A regularity theory for harmonic maps. J. Differ. Geom. **17**(2), 307–335 (1982)
321. R. Schoen, K. Uhlenbeck, Boundary regularity and the Dirichlet problem for harmonic maps. J. Differ. Geom. **18**(2), 253–268 (1983)
322. R. Schoen, K. Uhlenbeck, Regularity of minimizing harmonic maps into sphere. Invent. Math. **78**(1), 89–100 (1984)
323. R. Schoen, S.T. Yau, Harmonic maps and the topology of stable hypersurfaces and manifolds with non-negative Ricci curvature. Comment. Math. Helv. **51**(3), 333–341 (1976)
324. R. Schoen, S.T. Yau, Existence of incompressible minimal surfaces and the topology of three-dimensional manifolds with nonnegative scalar curvature. Ann. Math. (2) **110**(1), 127–142 (1979)
325. H.C.J. Sealey, Harmonic diffeomorphisms of surfaces, in *Harmonic Maps* Proceedings, Tulane. Lecture Notes in Mathematics, vol. 949 (Springer-Verlag, Berlin-New York, 1982), pp. 140–145
326. N. Seiberg, E. Witten, Electric-magnetic duality, monopole condensation and confinement in $N = 2$ super-symmetric Yang-Mills theory. Nucl. Phys. B **426**(1), 19–52 (1994)
327. N. Seiberg, E. Witten, Monopoles, duality and chiral symmetry breaking in $N = 2$ super-symmetric QCD. Nucl. Phys. B **431**(3), 484–550 (1994)
328. J. Serrin, The problem of Dirichlet for quasi-linear elliptic differential equations with many independent variables. Philos. Trans. R. Soc. Lond. A **264**, 413–496 (1969)
329. J. Shatah, Geometric wave equations, in *Recent Advances in Partial Differential Eequations* (El Escorial, 1992). Ram Research in Applied Mathematics, vol. 30 (Masson, Paris, 1994), pp. 99–114
330. J. Shatah, Regularity results for semilinear and geometric wave equations, in *Mathematics of Gravitation* (Warsaw, 1996). Banach Center Publications, vol. 41, Part I (Polish Academy of Sciences, Warsaw, 1997), pp. 69–90
331. J. Shatah, M. Struwe, Regularity results for nonlinear wave equations. Ann. Math. (2) **138**(3), 503–518 (1993)
332. J. Shatah, M. Struwe, *Geometric Wave Equations*. Courant Lecture Notes in Mathematics, vol. 2 (New York University/Courant Institue of Mathematical Sciences, New York; American Mathematical Society, Providence, 1998)
333. J. Shatah, M. Struwe, The Cauchy problem for wave maps. Int. Math. Res. Not. (11), 555–571 (2002)
334. J. Shatah, S. Tahvildar-Zadeh, On the Cauchy problem for equivariant wave maps. Commun. Pure Appl. Math. **47**(5), 719–753 (1994)
335. J. Shatah, S. Tahvildar-Zadeh, On the stability of stationary wave maps. Commun. Math. Phys. **185**(1), 231–256 (1997)
336. Y.B. Shen, Totally real n-dimensional minimal submanifold in complex n-dimensional projective space. Adv. Math. (Beijing) **13**(1), 65–70 (1984)
337. K. Shibata, On the existence of a harmonic mapping. Osaka Math. J. **15**, 173–211 (1963)
338. T. Sideris, Global existence of harmonic maps in Minkowski space. Commun. Pure Appl. Math. **42**(1), 1–13 (1989)
339. L. Simon, *Lectures on Geometric Measure Theory*. Proceedings of the Centre for Mathematical Analysis, vol. 3 (Australian National University Press, Canberra, 1983)
340. L. Simon, *Theorems on Regularity and Singularity of Energy Minimizing Maps*. Lectures in Mathematics ETH Zurich (Birkhäuser, Basel, 1996)

341. I.M. Singer, Infinitesimally homogeneous spaces. Commun. Pure. Appl. Math. **13**, 685–697 (1960)
342. Y.-T. Siu, Some remarks on the complex analyticity of harmonic maps. Southeast Asian Bull. Math. **3**(2), 240–253 (1979)
343. Y.-T. Siu, The complex-analyticity of harmonic maps and the strong rigidity of compact Kähler manifolds. Ann. Math. (2) **112**(1), 73–111 (1980)
344. Y.-T. Siu, Curvature characterization of the hyperquadrics. Duke Math. J. **47**(3), 641–654 (1980)
345. Y.-T. Siu, Strong rigidity of compact quotients of exceptional bounded symmetric domain. Duke Math. J. **48**(4), 857–871 (1981)
346. Y.-T. Siu, Complex-analyticity of harmonic maps, vanishing and Lefschetz theorems. J. Differ. Geom. **17**(1), 55–138 (1982)
347. Y.-T. Siu, S.T. Yau, Complete Kähler manifolds with non-positive curvature of faster than quadratic decay. Ann. Math. (2) **105**(2), 255–264 (1977)
348. Y.-T. Siu, S.T. Yau, Compact Kähler manifolds of positive bisectional curvature. Invent. Math. **59**(2), 189–204 (1980)
349. R. Smith, Harmonic maps of spheres. Am. J. Math. **97**, 364–385 (1975)
350. R. Smith, The second variation formula for harmonic maps. Proc. Am. Math. Soc. **47**, 229–236 (1975)
351. B. Smyth, Stationary minimal surfaces. Invent. Math. **76**(3), 411–420 (1984)
352. M. Struwe, *Regularity Results for Harmonic Maps of Minkowski Space, Mathematics* (Orsay, 1990). NATO Advanced Science Institutes Series C: Mathematical and Physical Sciences, vol. 332 (Kluwer, Dordrecht, 1991), pp. 357–369
353. M. Struwe, Geometric evolution problems, in *Nonlinear Partial Differential Equations in Differential Geometry* (Park City, 1992). IAS/Park City Mathematics Series, vol. 2 (American Mathematical Society, Providence, 1996), pp. 257–339
354. M. Struwe, Wave maps, in *Nonlinear Partial Differential Equations in Geometry and Physics* (Knoxville, 1995). Progress in Nonlinear Differential Equations and Their Applications, vol. 29 (Birkhäuser, Basel, 1997), pp. 113–153
355. M. Struwe, Radially symmetric wave maps from $(1+2)$-dimensional Minkowski space to the sphere. Math. Z. **242**(3), 407–414 (2002)
356. M. Struwe, Equivariant wave maps in two space dimension. Commun. Pure. Appl. Math. **56**(7), 815–823 (2003)
357. R. Takagi, On homogeneous real hypersurfaces in a complex projective space. Osaka J. Math. **10**, 495–506 (1973)
358. R. Takagi, Real hypersurfaces in a complex projective space with constant principal curvatures. J. Math. Soc. Jpn. **27**, 43–53 (1975)
359. R. Takagi, Real hypersurfaces in a complex projective space with constant principal curvatures II. J. Math. Soc. Jpn. **27**(4), 507–516 (1975)
360. T. Tao, Global regularity of wave maps. I. Small critical Sobolev norm in high dimension. Int. Math. Res. Not. (6), 299–328 (2001)
361. T. Tao, Global regularity of wave maps. II. Small energy in two dimension. Commun. Math. Phys. **224**(2), 443–544 (2001)
362. T. Tao, Endpoint bilinear restriction theorems for the cone and some sharp null form estimates. Math. Z. **238**(2), 215–268 (2002)
363. D. Tataru, Local and global results for wave maps I. Commun. Partial Differ. Equ. **23**(9–10), 1781–1793 (1998)
364. D. Tataru, On the equation $\Box u = |\nabla u|^2$ in 5+1 dimensions. Math. Res. Lett. **6**(5–6), 469–485 (1999)
365. D. Tataru, On global existence and scattering for the wave maps equation. Am. J. Math. **123**(1), 37–77 (2001)
366. D. Tataru, The wave maps equations. Bull. Am. Math. Soc. (N. S.) **41**(2), 185–204 (2004)
367. D. Tataru, Rough solutions for the wave maps equation. Am. J. Math. **127**(2), 293–377 (2005)

368. C.H. Taubes, Self-dual Yang-Mills connections over non-self-dual 4-manifolds. J. Differ. Geom. **17**(1), 139–170 (1982)
369. C.H. Taubes, A framework for Morse theory for the Yang-Mills functional. Invent. Math. **94**(2), 327–402 (1988)
370. C.H. Taubes, SW=>Gr: from the Seiberg-Witten equations to pseudo-holomorphic curves. J. Am. Math. **9**(3), 845–918 (1996)
371. C.H. Taubes, in *Seiberg Witten and Gromov Invariants for Symplectic 4-Mainfolds*, ed. by R. Wentworth. First International Press Lecture Series, vol. 2 (International Press, Somerville, 2000)
372. C.H. Taubes, The Seiberg-Witten equations and the Weinstein conjecture. Geom. Topol. **11**, 2117–2202 (2007)
373. C.H. Taubes, Embedded contact homology and Seiberg-Witten Floer cohomology I. Geom. Topol. **14**(5), 2497–2581 (2010)
374. G. Tian, Gauge theory and calibrated geometry, I. Ann. Math. **151**(1), 193–268 (2000)
375. P. Tolksdorf, A parametric variational principle for minimal surfaces of varying topological type. J. Reine Angew Math. **354**, 16–49 (1984)
376. Ph. Tondeur, *Geometry of Foliation* (Birkhäuser, Basel, 1997)
377. G. Tóth, On rigidity of harmonic mappings into spheres. J. Lond. Math. Soc. (2) **26**(3), 475–486 (1982)
378. G. Tóth, Construction des applications harmoniques non riegides d'un tore dans la sphère. Ann. Glob. Anal. Geom. **1**(2), 105–118 (1982)
379. G. Tóth, On classification of orthogonal multiplications *à* la do Carmo-Wallach. Geom. Dedicata **22**(2), 251–254 (1987)
380. K. Uhlenbeck, Regularity for a class of non-linear elliptic systems. Acta Math. **138**(3–4), 219–240 (1977)
381. K. Uhlenbeck, Morse theory by perturbation methods in hyperbolic manifolds. Trans. Math. Soc. **267**, 569–583 (1981)
382. K. Uhlenbeck, Removable singularities in Yang-Mills fields. Commun. Math. Phys. **83**(1), 11–29 (1982)
383. K. Uhlenbeck, Connections with L^p-bounds on curvature. Commun. Math. Phys. **83**(1), 31–42 (1982)
384. K. Uhlenbeck, Minimal spheres and other conformal variational problems, in *Seminars on Minimal Submanifolds*, ed. by E. Bombieri. Annals of Mathematics Studies, vol. 103 (Princeton University Press, Princeton, 1983), pp. 169–248
385. K. Uhlenbeck, Closed minimal hypersurfaces in hyperbolic manifolds, in *Seminars on Minimal Submanifolds*, ed. by E. Bombieri. Annals of Mathematics Studies, vol. 103 (Princeton University Press, Prineton, 1983), pp. 147–168
386. H. Urakawa, Stability of harmonic maps and eigenvalues of the Laplacian. Trans. Am. Math. Soc. **301**(2), 557–589 (1987)
387. K. Uhlenbeck, Harmonic maps into Lie groups. J. Differ. Geom. **30**(1), 1–50 (1989)
388. K. Uhlenbeck, S.-T. Yau, On the existence of Hermitian Yang-Mills connections in stable bundles. Commun. Pure Appl. Math. **39**(5 Suppl.) S 257–S 293 (1986)
389. H. Urakawa, *Calculus of Variations and Harmonic Maps*. Translations of Mathematical Monographs, vol. 132 (American Mathematical Society, Providence, 1993)
390. H. Urakawa, A discrete analogue of the harmonic morphism and Green kernel comparison theorems. Glasg. Math. J. **42**(3), 319–334 (2000)
391. H. Urakawa, A discrete analogue of the harmonic morphism, in *Harmonic Morphisms, Harmonic Maps and Related Topics* (Brest, 1997). Research Notes in Mathematics, vol. 413 (Chapman and Hall/CRC, Boca Raton, 2000), pp. 97–108
392. H. Urakawa, Biharmonic maps into compact Lie groups and the integrable systems. arXiv:0910.0692v2 [math.DG] 5 31 Jan 2012, 1–27
393. G. Valli, Some remarks on geodesics in Gauge groups and harmonic maps. J. Geom. Phys. **4**(3), 335–359 (1987)

394. G. Valli, On the energy spectrum of harmonic 2-spheres in unitary groups. Topology **27**(2), 129–136 (1988)
395. M. Ville, Harmonic morphisms from Einstein 4-manifolds to Riemann surfaces. Int. J. Math. **14**(3), 327–337 (2003)
396. W. von Wahl, The continuity or stability method for nonlinear elliptic and parabolic equations and systems. Rend. Semin. Mat. Fis. Milano **62**, 157–183 (1992)
397. N. Wallach, Minimal immersion of symmetric spaces into spheres, in *Symmetric Spaces*. Pure and Applied Mathematics, vol. 8 (Marcel Dekker, New York, 1972), pp. 1–40
398. C. Wang, Biharmonic maps from $\mathbf{R}^4$ in to a Riemannian manifold. Math. Z. **247**(1), 65–87 (2004)
399. C. Wang, Stationary biharmonic maps from $\mathbf{R}^m$ into a Riemannian manifold. Commun. Pure Appl. Math. **57**(4), 419–444 (2004)
400. Z.-L. Wang, Y.-L. Ou, Biharmonic Riemannian submanifolds from 3-manifolds. Math. Z. **269**, 917–925 (2011)
401. K. Wehrheim, *Uhlenbeck Compactness*. EMS Series of Lectures in Mathematics (European Mathematical Society, Zürich, 2004)
402. B. White, Existence of least area mappings of N-dimensional domains. Ann. Math. (2) **118**(1), 179–185 (1983)
403. B. White, Mappings that minimize area in their homotopy classes. J. Differ. Geom. **20**(2), 433–446 (1984)
404. B. White, Homotopy class in Sobolev spaces and energy minimizing maps. Bull. Am. Math. Soc. (N. S.) **13**(2), 166–168 (1985)
405. H.C. Wente, Counterexample to a conjecture of H. Hopf. Pac. J. Math. **121**(1), 193–243 (1986)
406. H.C. Wente, The capillary problem for an infinite trough. Calc. Var. Partial Differ. Equ. **3**(2), 155–192 (1995)
407. E. Witten, Monopoles and 4-manifolds. Math. Res. Lett. **1**(6), 769–796 (1994)
408. R. Wolak, Foliated and associated geometric structures on foliated manifolds. Ann. Fac. Sci. Toulouse Math. (5) **10**(3), 337–360 (1989)
409. T. Wolff, A sharp bilinear cone restriction estimate. Ann. Math (2) **153**(3), 661–698 (2001)
410. J.G. Wolfson, On minimal surfaces in Kähler manifolds of constant holomorphic sectional curvature. Trans. Am. Math. Soc. **290**(2), 627–646 (1985)
411. J.G. Wolfson, Harmonic maps of the two-sphere into the complete hyperquadric. J. Differ. Geom. **24**(2), 141–152 (1986)
412. J.G. Wolfson, Harmonic sequences and harmonic maps of surfaces into complex Grassmann manifolds. J. Differ. Geom. **27**(1), 161–178 (1988)
413. J.C. Wood, Holomorphicity of certain harmonic maps from a surface to complex projective n-space. J. Lond. Math. Soc. (2) **20**(1), 137–142 (1979)
414. C.M. Wood, The Gauss section of a Riemannian immersion. J. Lond. Math. (2) **33**(1), 157–168 (1986)
415. J.C. Wood, Harmonic morphisms, foliations and Gauss maps, in *Complex Differential Geometry and Nonlinear Differential Equations* (Brunswick, Maine, 1984). Contemporary Mathematics, vol. 49, (American Mathematical Society, Providence, 1986), pp. 145–184
416. C.M. Wood, Harmonic sections and Yang-Mills fields. Proc. Lond. Math. Soc. (3) **54**(3), 544–558 (1987)
417. J.C. Wood, Twistor constructions for harmonic maps, in *Differential Geometry and Differential Equations*. Lecture Notes in Mathematics, vol. 1255 (Springer, Berlin, 1987)
418. J.C. Wood, The explicit construction and parametrization of all harmonic maps from the two-sphere to a complex Grassmannian. J. Reine Angew. Math. **386**(1), 1–31 (1988)
419. J.C. Wood, Explicit construction and parametrization of harmonic two-spheres in the unitary group. Proc. Lond. Math. Soc. (3) **58**(3), 608–624 (1990)
420. J.C. Wood, Harmonic morphisms and Hermitian structures on Einstein 4-manifolds. Int. J. Math. **3**(3), 415–439 (1992)

421. J.C. Wood, On the construction of harmonic morphisms from Euclidean spaces, in *Harmonic Morphisms, Harmonic Maps and Related Topics* (Brest, 1997). Research Notes in Mathematics, vol. 413 (Chapman and Hall/CRC, Boca Raton, 2000), pp. 47–60
422. H.H. Wu, The Bochner technique in differential geometry. Math. Rep. **3**(2), 289–538 (1988)
423. Y.L. Xin, Some results on stable harmonic maps. Duke Math. **47**(3), 609–613 (1980)
424. Y.L. Xin, Non-existence and existence for harmonic maps in Riemannian manifolds, in *Proceedings of the Shanghai Symposium on Differential Geometry and Differential Equations*, 1984 (Science Press, Beijing, 1985), pp. 529–538
425. Y.L. Xin, *Geometry of Harmonic Maps*. Progress in Nonlinear Differential Equations and Their Applications, vol. 23 (Birkhäuser, Boston, 1996)
426. Y.L. Xin, X.P. Chen, The hypersurfaces in the Euclidean sphere with relative affine Gauss maps. Acta Math. Sin. **28**(1), 131–139 (1985)
427. K. Yosida, *Functional Analysis* 4th edn. (Springer, New York, 1974)

Index

Y.-J. Chiang, *Developments of Harmonic Maps, Wave Maps and Yang-Mills Fields into Biharmonic Maps, Biwave Maps and Bi-Yang-Mills Fields*, Frontiers in Mathematics, DOI 10.1007/978-3-0348-0534-6,

MIX
Papier aus verantwortungsvollen Quellen
Paper from responsible sources
FSC® C105338

If you have any concerns about our products,
you can contact us on
ProductSafety@springernature.com

In case Publisher is established outside the EU,
the EU authorized representative is:
Springer Nature Customer Service Center GmbH
Europaplatz 3, 69115 Heidelberg, Germany

Printed by Libri Plureos GmbH
in Hamburg, Germany